BRIEF CALCULUS
A Graphing Calculator Approach

BRIEF CALCULUS
A Graphing Calculator Approach

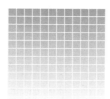

Ruric Wheeler
Samford University

Karla Neal
Louisiana State University

Roseanne Hofmann
Montgomery County Community College

JOHN WILEY & SONS, INC.
New York   Chichester   Brisbane   Toronto   Singapore

ACQUISITIONS EDITOR   **Ruth Baruth**
DEVELOPMENT EDITOR   **Madalyn Stone**
MARKETING MANAGER   **Carl White**
SENIOR PRODUCTION EDITOR   **Cathy Ronda**
PRODUCTION ASSISTANT   **Raymond Alvarez**
DESIGN   **A Good Thing, Inc.**
MANUFACTURING MANAGER   **Mark Cirillo**
INDEXER   **Julie S. Palmeter**
ILLUSTRATION EDITOR   **Sigmund Malinowski**
ELECTRONIC ILLUSTRATION   **Radiant Illustration & Design**

This book was set in Times Roman by Progressive Information Technologies and
printed and bound by R. R. Donnelley/Crawfordville.
The cover was printed by Leigh Press

Library of Congress Cataloging in Publication Data:
Wheeler, Ruric E., 1923–
    Brief calculus : a graphing calculator approach / Ruric Wheeler,
Karla Neal, Roseanne Hofmann.
        p.   cm.
    Includes bibliographical references (p.   –   )
    ISBN 0-471-05721-5 (cloth : alk. paper)
    1. Calculus—Data processing.   2. Graphic calculators.   I. Neal,
Karla.   II. Hofmann, Roseanne Salvatore.   III. Title.
QA303.5.D37W47   1996
515—dc20   95-41658
                                                                       CIP

Printed in the United States of America

10   9   8   7   6   5   4   3   2   1

PREFACE

Our primary goal in writing this book has been to create a book that is easy to read and comprehend by the average student, and yet also contains quality mathematics for students majoring in business management, social sciences, and life sciences. To accomplish this seemingly impossible goal, we have made extensive use of a calculator approach.

Most students will use throughout their lives the knowledge they gain to solve problems by using a calculator. We are assuming that every student using this book will have access to an inexpensive graphing calculator, for both class activities and homework.

The graphing calculator is revolutionizing the teaching and learning of mathematics. The time that used to be spent on pencil-and-paper computations can now be spent learning concepts and applications. Many people regard the graphing calculator as a tool, and it is true that it saves a substantial amount of time in performing matrix operations, drawing graphs, and other such operations. However, the graphing calculator is much more than a tool.

USE OF GRAPHING CALCULATOR

The graphing calculator can assist a teacher in using an investigative, exploratory approach to the teaching of mathematics. In this book, we use the graphing calculator in four ways.

1. **To investigate or explore new ideas, which we then validate by algebraic procedures.**
2. **To develop or define concepts algebraically, which we then support with numerical or graphical procedures on the calculator.**
3. **To mix calculating and algebraic procedures in solving problems.**
4. **To work a problem numerically or graphically.**

The teacher can choose to emphasize or not to emphasize any of these four uses of the graphing calculator.

We have attempted to introduce material independent of the specific calculator to be used. However, introductions to the use of the TI-82, the Sharp EL-9200, and the Casio 7700 GB are found in Appendix A. Whenever it is necessary to list calculator steps, these steps will be given for the TI-82.

DISTINGUISHING FEATURES

In addition to emphasizing a calculator approach, how else does this book differ from the typical book on this subject? The answer is summarized in five ways.

Problem Solving The ability to analyze problems and translate them into mathematical language is an important and, for the average student, a difficult skill. This book begins with a section on problem-solving and encourages the student to use problem-solving procedures throughout. Thought processes and algorithmic procedures are introduced to improve problem solving.

Computer Use Although no specific reference is given to computer use, this book can be used with as much emphasis on the computer as desired. In most places in this book the words "graphing calculator" can be replaced with the word "computer," and the given statement will still be valid; that is, the computer can be used to carry out the same functions as the graphing calculator.

Ample Review The beginning of the book contains ample review of the basic topics of algebra and the use of graphing calculators for students not adequately prepared.

Future Usability Students often ask, "Why should I study mathematics?" In this book, we attempt to answer this question in two ways. First, every section of this book contains numerous applications, classified as "Business and Economics" or "Social and Life Sciences." Then, CPA, CMA, and actuarial exam questions are scattered throughout the exercise sets to indicate that knowledge of the material is important in professional exams. These are denoted by the word "EXAM."

STUDENT-BASED APPROACH

First and foremost, this book is student oriented. A distinct effort is made to base each new concept on the student's prior experience or prior knowledge from the textbook. The book has a more intuitive than formal approach. To further help the student, the following features and teaching aids are included.

Gradual Development This book begins with very easy material, giving the student the opportunity to develop mathematical maturity, problem solving procedures, and calculator skills. Once the student has developed this maturity, the material gradually increases in difficulty. This increase is so gradual that most students never realize it is happening. At the end of the course, students will be solving problems as complicated as those found in any textbook for this subject.

Practice Problems Practice problems are scattered throughout each section as a check for understanding. That is, immediately after a topic is discussed, a simple problem is given that uses the material of the discussion. These practice problems allow students to evaluate for themselves whether they have understood the content before studying new material.

End-of-Chapter Tests At the end of each chapter is a short test on the chapter material to assist students in evaluating their comprehension. All the answers to these problems are given in the back of the book.

Notes Throughout the text, you will find the term ''NOTE,'' directed to the student and used to draw attention to an unusual idea or subject.

Exercises The exercise sets contain more than 2500 problems that are arranged within each exercise set in order of difficulty from easy to medium to challenging. The problems are usually arranged in matching pairs with the answers to odd-numbered problems in the back of the book.

SUPPLEMENTS

Teacher's Manual to Accompany *Brief Calculus: A Graphing Calculator Approach*
0471-13665-4

Testbank to Accompany *Brief Calculus: A Graphing Calculator Approach*
0471-13666-2

Computerized Testbank IBM 3.5 to Accompany *Brief Calculus: A Graphing Calculator Approach*
0471-13667-0

Computerized Testbank MAC 3.5 to Accompany *Brief Calculus: A Graphing Calculator Approach*
0471-13669-7

Student Solutions Manual to Accompany *Brief Calculus: A Graphing Calculator Approach*
0471-13649-2

ACKNOWLEDGMENTS

We would like to thank the following individuals who made valuable contributions to this book:

Debbie Garrison, Valencia Community College East
Stefan Hui, San Diego State University
Gary S. Itzkowitz, Rowan College of New Jersey
Rose Marie Kinik, Sacred Heart University
Anne Landry, Dutchess Community College

David Lesley, San Diego State University
Kenneth A. Retzer, Abilene Christian University
Burla J. Sims, University of Arkansas-Little Rock

The production of this book has been a team effort, involving valuable contributions from many people, including Wiley staff members Ruth Baruth, Mathematics Editor, Madalyn Stone, Senior Developmental Editor, Cathy Ronda, Senior Production Editor, Sigmund Malinowski, Illustration Editor.

We are indeed indebted to our secretaries and student assistants, especially Jill Bailey.

We encourage all teachers to send us suggestions as they use this book. In many ways we, too, are exploring the use of the graphing calculator to enhance our methods of teaching.

RURIC WHEELER
KARLA NEAL
ROSEANNE HOFMANN

CONTENTS

LEARNING GUIDE FOR STUDENTS

Before you start reading, take a moment to look over the next few pages, which provide an overview of some of the book's built-in learning devices. Becoming familiar with these unique features can make your march through the material a lot easier.

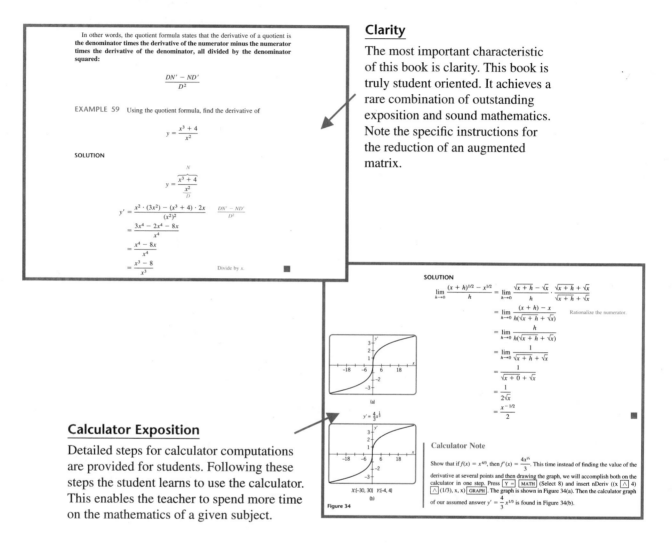

In other words, the quotient formula states that the derivative of a quotient is the denominator times the derivative of the numerator minus the numerator times the derivative of the denominator, all divided by the denominator squared:

$$\frac{DN' - ND'}{D^2}$$

EXAMPLE 59   Using the quotient formula, find the derivative of

$$y = \frac{x^3 + 4}{x^2}$$

SOLUTION

$$y = \frac{\overset{N}{\overbrace{x^3 + 4}}}{\underset{D}{\underbrace{x^2}}}$$

$$y' = \frac{x^2 \cdot (3x^2) - (x^3 + 4) \cdot 2x}{(x^2)^2} \qquad \frac{DN' - ND'}{D^2}$$

$$= \frac{3x^4 - 2x^4 - 8x}{x^4}$$

$$= \frac{x^4 - 8x}{x^4}$$

$$= \frac{x^3 - 8}{x^3} \qquad \text{Divide by } x. \qquad \blacksquare$$

Clarity

The most important characteristic of this book is clarity. This book is truly student oriented. It achieves a rare combination of outstanding exposition and sound mathematics. Note the specific instructions for the reduction of an augmented matrix.

Calculator Exposition

Detailed steps for calculator computations are provided for students. Following these steps the student learns to use the calculator. This enables the teacher to spend more time on the mathematics of a given subject.

SOLUTION

$$\lim_{h \to 0} \frac{(x + h)^{1/2} - x^{1/2}}{h} = \lim_{h \to 0} \frac{\sqrt{x + h} - \sqrt{x}}{h} \cdot \frac{\sqrt{x + h} + \sqrt{x}}{\sqrt{x + h} + \sqrt{x}}$$

$$= \lim_{h \to 0} \frac{(x + h) - x}{h(\sqrt{x + h} + \sqrt{x})} \qquad \text{Rationalize the numerator.}$$

$$= \lim_{h \to 0} \frac{h}{h(\sqrt{x + h} + \sqrt{x})}$$

$$= \lim_{h \to 0} \frac{1}{\sqrt{x + h} + \sqrt{x}}$$

$$= \frac{1}{\sqrt{x + 0} + \sqrt{x}}$$

$$= \frac{1}{2\sqrt{x}}$$

$$= \frac{x^{-1/2}}{2} \qquad \blacksquare$$

$$y' = \frac{4}{3}x^{\frac{1}{3}}$$

X:[-30, 30] Y:[-4, 4]

(a)

(b)

Figure 34

Calculator Note

Show that if $f(x) = x^{4/3}$, then $f'(x) = \frac{4x^{1/3}}{3}$. This time instead of finding the value of the derivative at several points and then drawing the graph, we will accomplish both on the calculator in one step. Press $\boxed{Y =}$ $\boxed{\text{MATH}}$ (Select 8) and insert nDeriv ((x $\boxed{\wedge}$ 4) $\boxed{\wedge}$ (1/3), x, x) $\boxed{\text{GRAPH}}$. The graph is shown in Figure 34(a). Then the calculator graph of our assumed answer $y' = \frac{4}{3}x^{1/3}$ is found in Figure 34(b).

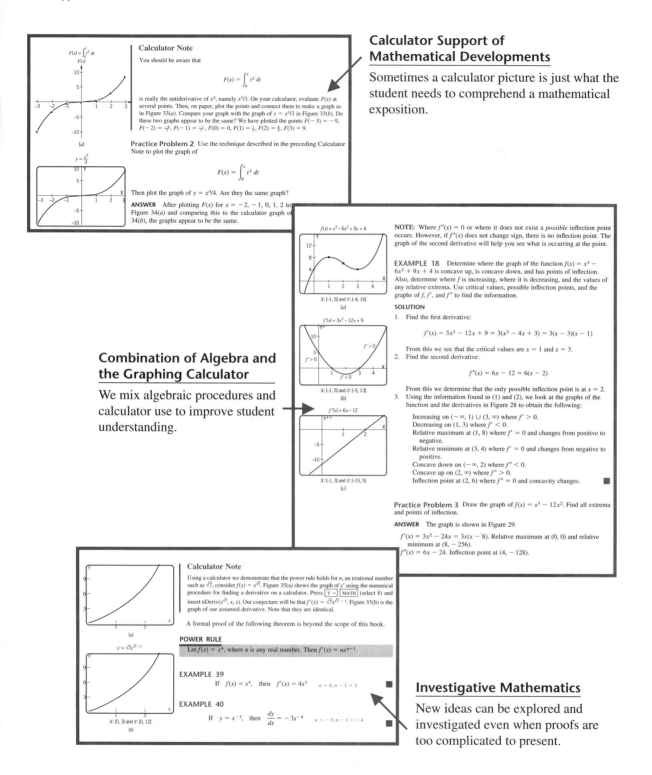

Calculator Note

You should be aware that

$$F(x) = \int_0^x t^2\, dt$$

is really the antiderivative of x^2, namely $x^3/3$. On your calculator, evaluate $F(x)$ at several points. Then, on paper, plot the points and connect them to make a graph as in Figure 33(a). Compare your graph with the graph of $y = x^3/3$ in Figure 33(b). Do these two graphs appear to be the same? We have plotted the points $F(-3) = -9$, $F(-2) = -\frac{8}{3}$, $F(-1) = -\frac{1}{3}$, $F(0) = 0$, $F(1) = \frac{1}{3}$, $F(2) = \frac{8}{3}$, $F(3) = 9$.

Practice Problem 2 Use the technique described in the preceding Calculator Note to plot the graph of

$$F(x) = \int_0^x t^3\, dt$$

Then plot the graph of $y = x^4/4$. Are they the same graph?

ANSWER After plotting $F(x)$ for $x = -2, -1, 0, 1, 2$ to ... Figure 34(a) and comparing this to the calculator graph of ... 34(b), the graphs appear to be the same.

Calculator Support of Mathematical Developments

Sometimes a calculator picture is just what the student needs to comprehend a mathematical exposition.

NOTE: Where $f''(x) = 0$ or where it does not exist a *possible* inflection point occurs. However, if $f''(x)$ does not change sign, there is no inflection point. The graph of the second derivative will help you see what is occurring at the point.

EXAMPLE 18 Determine where the graph of the function $f(x) = x^3 - 6x^2 + 9x + 4$ is concave up, is concave down, and has points of inflection. Also, determine where f is increasing, where it is decreasing, and the values of any relative extrema. Use critical values, possible inflection points, and the graphs of f, f', and f'' to find the information.

SOLUTION

1. Find the first derivative:

$$f'(x) = 3x^2 - 12x + 9 = 3(x^2 - 4x + 3) = 3(x - 3)(x - 1)$$

From this we see that the critical values are $x = 1$ and $x = 3$.

2. Find the second derivative:

$$f''(x) = 6x - 12 = 6(x - 2)$$

From this we determine that the only possible inflection point is at $x = 2$.

3. Using the information found in (1) and (2), we look at the graphs of the function and the derivatives in Figure 28 to obtain the following:

Increasing on $(-\infty, 1) \cup (3, \infty)$ where $f' > 0$.
Decreasing on $(1, 3)$ where $f' < 0$.
Relative maximum at $(1, 8)$ where $f' = 0$ and changes from positive to negative.
Relative minimum at $(3, 4)$ where $f' = 0$ and changes from negative to positive.
Concave down on $(-\infty, 2)$ where $f'' < 0$.
Concave up on $(2, \infty)$ where $f'' > 0$.
Inflection point at $(2, 6)$ where $f'' = 0$ and concavity changes. ∎

Practice Problem 3 Draw the graph of $f(x) = x^3 - 12x^2$. Find all extrema and points of inflection.

ANSWER The graph is shown in Figure 29.

$f'(x) = 3x^2 - 24x = 3x(x - 8)$. Relative maximum at $(0, 0)$ and relative minimum at $(8, -256)$.
$f''(x) = 6x - 24$. Inflection point at $(4, -128)$.

Combination of Algebra and the Graphing Calculator

We mix algebraic procedures and calculator use to improve student understanding.

Calculator Note

Using a calculator we demonstrate that the power rule holds for n, an irrational number such as $\sqrt{7}$, consider $f(x) = x^{\sqrt{7}}$. Figure 35(a) shows the graph of y' using the numerical procedure for finding a derivative on a calculator. Press [Y =] [MATH] (select 8) and insert nDeriv($x^{\sqrt{7}}$, x, x). Our conjecture will be that $f'(x) = \sqrt{7}x^{\sqrt{7}-1}$. Figure 35(b) is the graph of our assumed derivative. Note that they are identical.

A formal proof of the following theorem is beyond the scope of this book.

POWER RULE

Let $f(x) = x^n$, where n is any real number. Then $f'(x) = nx^{n-1}$.

EXAMPLE 39
If $f(x) = x^4$, then $f'(x) = 4x^3$ $n = 4; n - 1 = 3$ ∎

EXAMPLE 40
If $y = x^{-3}$, then $\dfrac{dy}{dx} = -3x^{-4}$ $n = -3; n - 1 = -4$ ∎

Investigative Mathematics

New ideas can be explored and investigated even when proofs are too complicated to present.

Future Usability

Students are interested in the mathematics they will use in the future. The exercise sets contain problems from professional examinations, denoted by *Exam.* Application problems, many of them taken from advanced textbooks in business, economics, and the social and life sciences, are abundant in this book.

40. ***Exam.*** Find $\lim\limits_{x \to 3} \dfrac{x^2 - x - 6}{x^2 - 9}$.

   (a) 0 (b) $\dfrac{5}{6}$ (c) 1 (d) $\dfrac{5}{3}$

   (e) Increases without bound

41. Using a calculator, locate points of discontinuity of

   $$f(x) = \frac{|x^2 + 3x + 2|}{x - 2}$$

42. Locate points of discontinuity of

   $$g(x) = \frac{x^2 - 4x + 3}{x^2 - 1}$$

Applications (Business and Economics)

43. ***Cost Functions.*** A cost function $C(x)$ has the following graph for $0 \le x \le 8$.
   (a) Where is the cost function discontinuous?
   (b) Find $\lim\limits_{} C(x)$. (c) Find $C(2)$.
   (d) Find $\lim\limits_{x \to 4} C(x)$. (e) Find $C(4)$.
   (f) Find $\lim\limits_{x \to 7} C(x)$. (g) Find $C(7)$.

44. ***Revenue Functions.*** A revenue function $R(x)$, giving revenue in terms of the number of items sold, x, is found to be

   $$R(x) = 10x - \frac{x^2}{100}, \qquad 0 \le x \le 1000$$

   (a) Find $R(10)$. (b) Find $R(100)$.
   (c) Find $R(500)$. (d) Draw the graph of $R(x)$.
   (e) Is $R(x)$ continuous on $0 < x < 1000$?

45. ***Demand Functions.*** The demand function for an item is given by

   $$p(x) = \frac{80}{x - 16}, \qquad 16 < x \le 96$$

   where p is the price per unit and x is the number of items. Draw the graph of the demand function and discuss the continuity of $p(x)$.

Applications (Social and Life Sciences)

46. ***Agronomy.*** The height of a plant is given by

   $$h(t) = 3\sqrt{t}$$

   ᵣe h is measured in inches and t in days of ᵥth. Graph this function and find any points of ᴏntinuity.

47. ***Laboratory Experiment.*** A laboratory uses mice in its experiments. The number N of mice available is a function of time t. The following chart was drawn for 6 days.
   (a) Where is this function discontinuous?
   (b) Find $N(1)$. (c) Find $\lim\limits_{t \to 1} N(t)$.
   (d) Find $\lim\limits_{t \to 2^+} N(t)$.

48. ***Learning Functions.*** A person learns $L(x)$ items in x hours according to the function

   $$L(x) = 40\sqrt{x}, \qquad 0 \le x \le 16$$

   (a) Find $L(4)$. (b) Find $L(9)$.
   (c) Find $L(16)$. (d) Draw the graph of $L(x)$.
   (e) Discuss the continuity of $L(x)$.

49. ***Learning Functions.*** Suppose that a learning function is given as

   $$L(x) = 30x^{2/3}, \qquad 0 \le x \le 8$$

   (a) Draw the graph of $L(x)$.
   (b) Discuss the continuity of $L(x)$.

50. ***Voting.*** The registered voting population of a city, $P(t)$, in thousands, is given for the next 8 years as

   $$P(t) = 40 + 9t^2 - t^3, \qquad 0 \le t \le 8$$

   (a) Find $P(2)$. (b) Find $P(4)$.
   (c) Find $P(6)$. (d) Draw the graph of $P(t)$.
   (e) Discuss the continuity of $P(t)$.

Review Exercises

Find the following limits if they exist.

51. $\lim\limits_{x \to 2} \dfrac{x}{x + 1}$ 52. $\lim\limits_{x \to -1} \dfrac{x + 1}{x}$

53. $\lim\limits_{x \to 0} \dfrac{5}{x^2}$ 54. $\lim\limits_{x \to 1/3} \left(6 - \dfrac{2}{x}\right)$

55. $\lim\limits_{x \to 2} \dfrac{4}{x - 2}$ 56. $\lim\limits_{x \to -1} \dfrac{2}{x + 1}$

Steps in Optimization

1. Use any or all of the techniques and strategies for solving problems you have learned in previous chapters.
2. Work toward a functional relationship between variables to be optimized and one other variable. (If two or more independent variables arise, it will be necessary in this chapter to eliminate all but one.)
3. Find critical values and locate absolute maxima and minima.

For the problems in this section, you should check your work by drawing the graph of the function being optimized. The first two examples should give some understanding of optimization before we consider the more important applications of this section.

The coefficient of inequality can be used to compare income distributions for various countries, or to check to see if there is a change in the income distribution in a given country. For example, a few years ago the coefficient of inequality for the United States was 0.26 while that for Sweden was 0.18 and that for Brazil was 0.34. From these values we can see that among these three countries income was more equitably distributed in Sweden than in the United States and was the least equitably distributed in Brazil.

According to the *1993 World Almanac*, the gap between rich and poor did not widen in 1991, but the long-term trend pointed to an increasing inequality of income. The richest 20% of all households got 46.5% of all household income in 1991, up from 43.5% in 1971 and 44.4% in 1981. The poorest 20% got 3.8% of all income in 1991, contrasted with 4.1% in 1971 and 1981. (Note that most government assistance to the poor is not included and does affect the income inequity.)

Problem Solving Strategies

Polya's suggestions for solving problems, introduced in Chapter 0, are helpful throughout the book. The concept of estimating is a recommended procedure for all calculator computations.

Applications

An unusually large number of applications of the mathematics of a section to real-world problems illustrates the importance of a section.

Calculator Notes

Found throughout the book are suggestions or instructions on using a calculator for a particular problem.

Calculator Note

Again, we need to take time to emphasize the usefulness of the calculator for problems such as Example 52. The manager of the cola plant has her cost function, which she places in her graphing calculator as $y = 600 + 1000x - 2x^2$. Under the CALC menu she can obtain the cost for any production and the marginal cost for any level of production as she moves the cursor along the graph. Also, she can use TABLE with $Y_1 = 600 + 1000X - 2X^2$ and $Y_2 = 1000 - 4X$ to obtain a table similar to the following.

Production (number of cases of cola)	5	25	50	100
Approximate Cost	$5550	$24,350	$45,600	$80,600
Approximate Marginal Cost	$980	$900	$800	$600

Then, if the manager wishes, she can change parts of the cost function and note changes in production costs and marginal costs.

Color Utilization

Color is used for instructional purposes throughout the book.

Special Helps

Special helps (set apart by a second color) are given to assist students through critical stages of a problem. These aids provide additional explanation of steps in the solution of a problem.

EXAMPLE 31   If $x^2y = x^3 + 4y$, find $\dfrac{dy}{dx}$.

SOLUTION   We first take the derivative with respect to x of each term to obtain

$$\frac{d}{dx}(x^2y) = \frac{d}{dx}(x^3) + \frac{d}{dx}(4y)$$

Now, as mentioned above, x^2y is a product and we must use the product formula.

$$x^2\frac{d}{dx}(y) + 2xy = 3x^2 + 4\frac{d}{dx}(y) \qquad \text{Derivative of a product}$$

$$x^2\frac{dy}{dx} - 4\frac{dy}{dx} = 3x^2 - 2xy$$

$$\frac{dy}{dx}(x^2 - 4) = 3x^2 - 2xy \qquad \text{Factor.}$$

$$\frac{dy}{dx} = \frac{3x^2 - 2xy}{x^2 - 4} \qquad \text{Divide by } x^2 - 4 \neq 0. \quad \blacksquare$$

graph as $y = \ln 6 + \ln x$ in part (b); that is, $\ln 6x = \ln 6 + \ln x$. The graph of $y = \ln (x/6)$ in part (c) is the same as the graph of $y = \ln x - \ln 6$ in part (d); that is, $\ln (x/6) = \ln x - \ln 6$. In part (e) the graph of $y = \ln x^3$ is the same as the graph of $y = 3 \ln x$ in part (f); that is, $\ln x^3 = 3 \ln x$.

Visual Aids

Students will find an unusually large number of diagrams to assist in understanding an explanation.

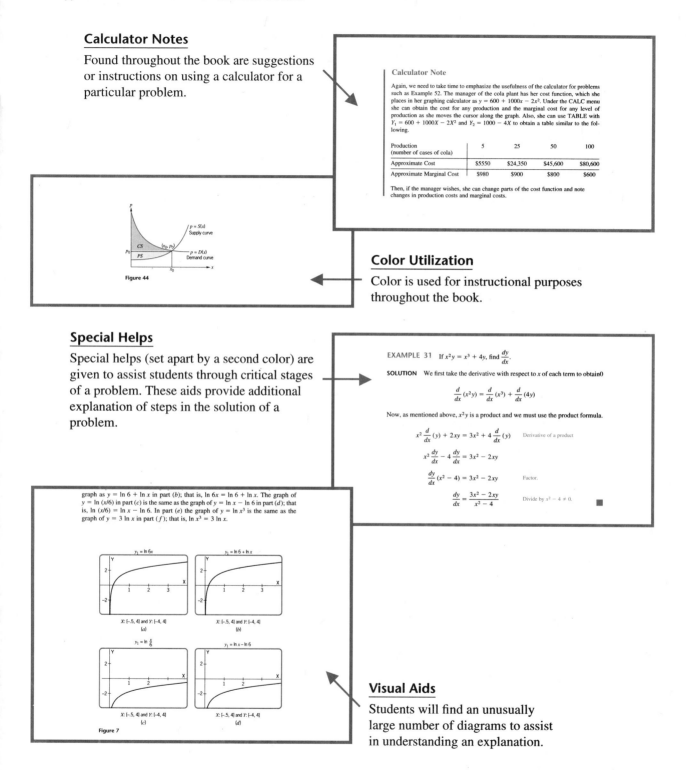

Figure 44

Figure 7

Overviews

Each section begins with an overview, which enables the authors to make the material more personal and less formal in approach. The overviews summarize the goals of a section and relate the section to previous studied material or real-world experiences.

10.1 FINDING DERIVATIVES OF EXPONENTIAL FUNCTIONS

OVERVIEW   This is an important section because exponential functions are used extensively in applications. Growth and decay models are based on exponentials. Exponential functions are used by economists to study the growth rate of the money supply. They are needed by businesspeople to study the rate of change in sales, by biologists to study the rate of growth of organisms, and by social scientists to study the rate of population growth. In this section we

- Find the derivative of exponential functions
- Use the chain rule with exponential functions
- Study graphing techniques
- Introduce applications of exponential functions

Summary

At the end of each section is a summary of key ideas or formulas in the section.

SUMMARY

The derivative of a function f, at a number x in its domain, is defined to be

$$f'(x) = \lim_{h \to 0} \frac{f(x + h) - f(x)}{h}$$

Since the difference quotient

$$\frac{[f(x + h) - f(x)]}{h}$$

is the slope of the line joining $(x, f(x))$ and $((x + h), f(x + h))$, the derivative may be interpreted as the slope of a tangent line to the curve. Economists often refer to the derivative of a function as the marginal value of that function.

We can now complete the problem.

$$\int_0^4 x^2\sqrt{1 + 2x}\, dx = \frac{x^2(1 + 2x)^{3/2}}{3}\bigg|_0^4 - \frac{2x(1 + 2x)^{5/2}}{15}\bigg|_0^4$$
$$+ \int_0^4 \frac{(1 + 2x)^{5/2}}{15} \cdot 2\, dx$$
$$= \left[\frac{x^2(1 + 2x)^{3/2}}{3} - \frac{2x(1 + 2x)^{5/2}}{15} + \frac{2(1 + 2x)^{7/2}}{105}\right]\bigg|_0^4$$
$$= \left[\frac{16(9)^{3/2}}{3} - \frac{8(9)^{5/2}}{15} + \frac{2(9)^{7/2}}{105}\right] - \left[\frac{2}{105}\right]$$
$$= 144 - \frac{648}{5} + \frac{1458}{35} - \frac{2}{105}$$
$$= \frac{5884}{105} \approx 56.04$$

☑ Calculator Check: fnInt($x \wedge 2 \sqrt{} (1 + 2x)$, x, 0, 4) $\approx$ 56.04

Draw the graph of $y = x^2\sqrt{1 + 2x}$ with your graphing calculator and shade the area being found (see Figure 2). ■

$y = x^2\sqrt{1 + 2x}$

50
40
30
20
10
Y
1 2 3 4 X
X: [0, 4] and Y: [0, 50]

Using the Calculator as a Tool

The most common use of the graphical calculator is as a tool. Throughout much of the book a graphical picture of the problem is most helpful to students.

Common Errors

Throughout the book a word of warning is given to the student about common errors made on the material being studied.

COMMON ERROR   Many times students will differentiate like this:

$$y(x) = (3x^2 + 4x)^5, \quad \text{so} \quad y'(x) = 5(3x^2 + 4x)^4$$

What is wrong with this? The answer is $5(3x^2 + 4x)^4(6x + 4)$.

Functions and Their Graphs

We begin this chapter with a study of linear equations and discuss how to determine an equation of a line. Then we study one of the most important concepts in elementary mathematics, the **function**. After functions are defined, we consider several special functions. One objective of this book is to assist you in understanding certain characteristics of the graphs of functions and in interpreting those graphs. This ability is important to the economist and the businessperson as well as someone in the social or life sciences. It is essential that relationships among quantities in a business be expressed efficiently and precisely. Functions enable us to do this. A profit function can be used to analyze what change in profit occurs if production levels are changed. Furthermore, a function can tell a doctor what the level of concentration of a drug will be at a certain time after being administered.

In this chapter we study special cases of the polynomial function in one variable, that is,

$$f(x) = a_n x^n + a_{n-1} x^{n-1} + \cdots + a_1 x + a_0$$

where $a_n \neq 0$ and n is a positive integer. Later in this chapter, we study functions such as

$$f(x) = 3^x$$

You will learn to recognize the graphs of these functions and investigate applications in various fields of study.

1.1 SLOPES AND LINEAR EQUATIONS

OVERVIEW   The ability to develop an equation that accurately describes a real-life situation is an important skill. This section is important in your accumulation of strategies to solve application problems. If a company can make 20 air conditioners for $15,000 and 10 air conditioners for $7950, the points (20, 15,000) and (10, 7950) can be plotted in a plane. If we draw a line between these two points, then other points on that segment can be used to represent the number of units and the cost. Since we are dealing with air conditioners, only integer values are useful in this problem, and the line gives an overall view of the relationship between number of units and cost. If we want to know the cost of producing 15 units, we need to find the equation of the line. In this section we

- Develop equations of lines from data representing real-life situations
- Define the slope of a line
- Use slope to graph a line and to find an equation of the line
- Learn to find an equation of a line given two points on the line

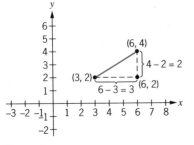

Figure 1

Any two points in a plane can be considered as endpoints of some line segment. Let (3, 2) and (6, 4) be endpoints of the line segment in Figure 1. If we now construct a line parallel to the x-axis through (3, 2) and a line parallel to the y-axis through (6, 4), the lines meet at the point (6, 2). The change in x as we move from (3, 2) to (6, 2) is $6 - 3 = 3$, and the change in y as we move from (6, 2) to (6, 4) is $4 - 2 = 2$. The ratio of the change in y to the change in x is $\frac{2}{3}$ (see Figure 1).

Practice Problem 1 For $P_1(1, 2)$ and $P_2(4, 4)$, find $y_2 - y_1$ (the change in y), $x_2 - x_1$ (the change in x), and the ratio of the change in y to the change in x.

ANSWER $y_2 - y_1 = 2$; $x_2 - x_1 = 3$, and the ratio of the change is $\frac{2}{3}$.

The preceding concepts are important in discussing the inclination of a line, which is measured by comparing the **rise** (the change in y: $y_2 - y_1$) to the **run** (the

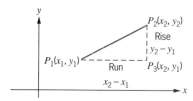

Figure 2

change in x: $x_2 - x_1$), as shown in Figure 2. This inclination, called the **slope**, is a useful characteristic of a line, telling us both the direction and the relative steepness of the line. In business, for example, this slope could represent the rate of change in profit.

DEFINITION: SLOPE OF A LINE SEGMENT

The ratio of the rise to the run of a line segment is called the **slope** of the line segment and is designated by the letter m. The slope of the line segment from $P_1(x_1, y_1)$ to $P_2(x_2, y_2)$ is

$$m = \frac{y_2 - y_1}{x_2 - x_1} \qquad (x_1 \neq x_2)$$

Since the ratio of the rise to the run on a line is always constant, the slope of a line segment is always the same no matter which two points on the line are selected to compute the slope. The slope of a line is defined as the slope of any of its line segments.

If P_1 is to the left of P_2, $x_2 - x_1$ will necessarily be positive, and the slope will be positive or negative as $y_2 - y_1$ is positive or negative. Consequently, positive slope indicates that a line rises from left to right; negative slope indicates that a line falls from left to right. For example, a slope of 2 means that y increases by 2 when x increases by 1. A slope of $-\frac{2}{3}$ means that y decreases by 2 when x increases by 3.

There is no restriction on which point is labeled P_1 and which is labeled P_2, since

$$\frac{y_2 - y_1}{x_2 - x_1} = \frac{-(y_1 - y_2)}{-(x_1 - x_2)} = \frac{y_1 - y_2}{x_1 - x_2}$$

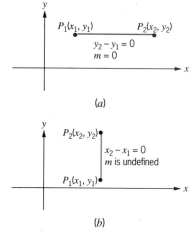

(a)

(b)

Figure 3

and the order in which the points are considered is immaterial in determining the slope. However,

$$\frac{y_2 - y_1}{x_2 - x_1} \neq \frac{y_2 - y_1}{x_1 - x_2}$$

Slopes of Horizontal and Vertical Lines   Two points on the line $y = 3$ are $(1, 3)$ and $(2, 3)$. The slope is

$$m = \frac{3 - 3}{2 - 1} = 0$$

The slope of any horizontal line is 0 because the value of y does not change as x changes [see Figure 3(a)].

Two points on the line $x = 2$ are $(2, 1)$ and $(2, 4)$, but when we try to calculate the slope we find that the slope is undefined:

$$m = \frac{4 - 1}{2 - 2} = \frac{3}{0} \quad \text{(undefined)}$$

A vertical line has no slope; it is undefined because the value of x does not change and the denominator is always 0 [see Figure 3(b)].

EXAMPLE 1   Find the slope of the line shown in Figure 4.

SOLUTION

$$m = \frac{y_2 - y_1}{x_2 - x_1} = \frac{5 - 2}{2 - 1} = \frac{3}{1} = 3 \quad \text{or} \quad m = \frac{2 - 5}{1 - 2} = \frac{-3}{-1} = 3 \quad \blacksquare$$

Practice Problem 2   Find the slope of the line through points $(-1, 2)$ and $(3, -2)$.

ANSWER   The slope is -1.

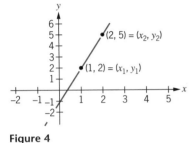

Figure 4

Calculator Note

Under the DRAW DRAW menu (obtained from 2nd DRAW), select 2:Line (to draw a line segment from $(-1, 2)$ to $(3, -2)$. Line $(-1, 2, 3, -2)$ gives the segment in Figure 5.

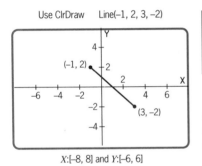

X:[–8, 8] and Y:[–6, 6]

Figure 5

The relationship between two quantities in a business can often be expressed as a linear function in the form $ax + dy + c = 0$. For example, if y is the total cost to produce x units and the relationship is linear, the function may be linear and in this form. Since the slope of a line is important in analyzing a graph, let's look at slopes in relation to linear equations of the form $ax + dy + c = 0$, where a and d are not both zero. The graphs of equations such as $y = x + 1$, $x + y = 4$, and $2x - 3y = 6$ are straight lines. In fact, the graph of any linear equation with not more than two unknowns is a straight line. (If there are more than two unknowns, the graph is not a line.) Any line may be described by an equation of the form $ax + dy + c = 0$. Therefore, a given graph is a line if and only if it has an equation that can be written in the form $ax + dy + c = 0$, where a and d are not both zero.

A linear equation may be solved for y, if $d \neq 0$, to obtain

$$y = \frac{-a}{d} x + \frac{-c}{d}$$

By letting $m = -a/d$ and $b = -c/d$, the expression becomes

$$y = mx + b$$

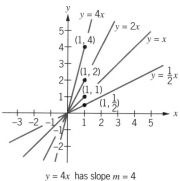

$y = 4x$ has slope $m = 4$
$y = 2x$ has slope $m = 2$
$y = x$ has slope $m = 1$
$y = \frac{1}{2}x$ has slope $m = \frac{1}{2}$

Figure 6

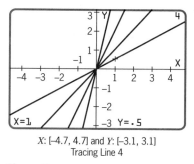

X: [-4.7, 4.7] and Y: [-3.1, 3.1]
Tracing Line 4

Figure 7

Consider the graph of $y = mx$ ($b = 0$ for this example) when m has values $\frac{1}{2}$, 1, 2, and 4 (see Figure 6). Notice the change in the lines as the value of m increases.

Practice Problem 3 Using a calculator, draw the lines $y = 4x$, $y = 2x$, $y = x$, and $y = \frac{1}{2}x$.

ANSWER We can enter each line as Y_1, Y_2, Y_3, Y_4. By pressing the Trace key, we can scroll through each of the lines. Notice the small number in the upper right-hand corner of the screen (see Figure 7). That number indicates which line we are tracing.

By studying the graphs in Figures 6 and 7, we can make some observations. Since (0, 0) and (1, $\frac{1}{2}$) are two points on the line $y = \frac{1}{2}x$, the slope can be evaluated by $(\frac{1}{2} - 0)/(1 - 0) = \frac{1}{2}$. We can also see from studying the two points that the value of y increases by $\frac{1}{2}x$. You may have already realized that in the form $y = mx + b$, the value of m is the slope of the line.

If (x_1, y_1) is a fixed point on a given nonvertical, straight line, and (x, y) is any point on the line, then the slope from (x_1, y_1) to (x, y) is

$$m = \frac{y - y_1}{x - x_1} \quad \text{or} \quad y - y_1 = m(x - x_1)$$

Since the coordinates x and y are variables denoting any point on the line, the equation $y - y_1 = m(x - x_1)$ represents the relationship between x and y. Therefore, the equation of the line with slope m passing through the fixed point (x_1, y_1) is $y - y_1 = m(x - x_1)$. A linear equation written in this form is said to be in **point-slope form**. If the slope and one point on the line are known, then the equation of the line can be obtained.

DEFINITION: POINT-SLOPE FORM

If a line has slope m and passes through the point (x_1, y_1), then the equation of the line in **point-slope form** is given by $y - y_1 = m(x - x_1)$.

EXAMPLE 2 Find an equation of the line through (2, 1) with a slope of 3.

SOLUTION

$$y - y_1 = m(x - x_1)$$
$$y - 1 = 3(x - 2) \qquad x_1 = 2, y_1 = 1, m = 3$$
$$y = 1 + 3x - 6$$
$$y = 3x - 5$$

As a special case, the fixed point can be chosen as the point where the line crosses the y-axis (the y-intercept). The coordinates of this point are usually written as (0, b). Then the equation of the line becomes $y = mx + b$, which is the equation that we discussed earlier. The b in this equation is the value of y when $x = 0$, or the y-intercept.

DEFINITION: SLOPE-INTERCEPT FORM

If a line has slope m and y-intercept b, then the equation of the line in **slope-intercept form** is given by $y = mx + b$.

EXAMPLE 3 If the slope of a line is 3 and the y-intercept is 2, what is an equation of the line?

SOLUTION Since $m = 3$ and $b = 2$, an equation is $y = mx + b$ or $y = 3x + 2$.

Practice Problem 4 Find an equation of the line that crosses the y-axis at $(0, -5)$ and has a slope of 2.

ANSWER An equation is $y = 2x - 5$.

Suppose that we are given two points that are on a line. How can we find an equation of the line? Since we can find the slope with two points, we can then select one of the points and the slope to put into the point-slope form to find the equation.

EXAMPLE 4 Find an equation of the line that contains the points $(2, 3)$ and $(-1, 4)$.

SOLUTION We can use the two points in either order to find the slope:

$$m = \frac{4 - 3}{-1 - 2} = -\frac{1}{3}$$

Now select either one of the points (but only one) and put that point and the slope into the point-slope form. Then simplify the resulting equation. Let's use the point $(2, 3)$. Note that we could also use the point $(-1, 4)$, which would yield the same simplified equation.

$$y - 3 = -\frac{1}{3}(x - 2) \qquad x_1 = 2, y_1 = 3, m = -1/3$$

$$y = -\frac{1}{3}x + \frac{11}{3}$$

The equation $y = -\frac{1}{3}x + \frac{11}{3}$ is the equation in slope-intercept form. Substitute the point $(-1, 4)$ to verify that the resulting equations are the same.

Practice Problem 5 Find an equation of the line through the points $(-1, 2)$ and $(3, -2)$.

ANSWER $y = -x + 1$

The graph of a linear equation can be a horizontal line, which we illustrate using a calculator.

Practice Problem 6 Draw the line $y = 3$ using a calculator.

ANSWER   We can enter the function $Y_1 = 3$ and use an appropriate viewing rectangle to get a graph such as that in Figure 8.

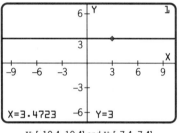

X: [–10.4, 10.4] and Y: [–7.4, 7.4]

Figure 8

The equation of a vertical line is of the form $x = h$, where h is a constant and the slope is undefined. The equation of the line shown in Figure 9 is $x = 3$. By selecting two points on the line, say $(3, 5)$ and $(3, 2)$, we can see that the slope is undefined. That is,

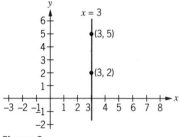

Figure 9

$$m = \frac{5 - 2}{3 - 3} = \frac{3}{0}$$

which does not exist.

Practice Problem 7 Find an equation of the vertical line through $(1, -2)$.

ANSWER $x = 1$

Calculator Note

A vertical line such as $x = 1$ can be drawn on a calculator using the DRAW DRAW menu (obtained from $\boxed{\text{2nd}}\ \boxed{\text{DRAW}}$). Three different graphs are shown in Figure 10. For (a) select 4: Vertical and insert 1 then $\boxed{\text{ENTER}}$. For (b) select 2: Line (and insert the y values of the endpoints of a line segment (say $y = -2$ to $y = 2$). Use Line $(1, -2, 1, 2)$. For (c) follow the same procedure as in (b) except use Ymin and Ymax for the endpoints , or Line (1, Ymin, 1, Ymax).

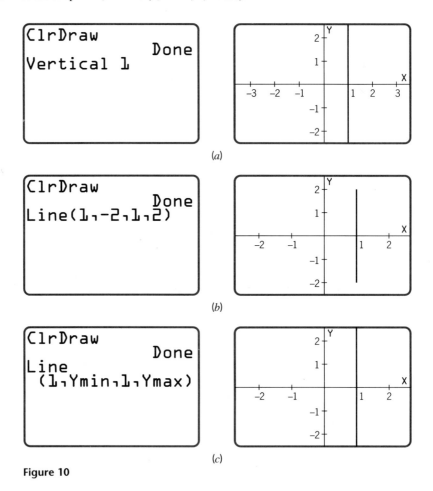

(a)

(b)

(c)

Figure 10

Parallel and Perpendicular Lines

It can be proved that if two nonvertical lines are **parallel**, then they have the same slope. Conversely, two lines with the same slope and different y-intercepts are parallel. Similarly, if two nonvertical lines are **perpendicular**, then the slope of one line is the negative reciprocal of the other; that is,

$$m_2 = \frac{-1}{m_1} \quad \text{or} \quad m_1 \cdot m_2 = -1$$

Conversely, if the slope of one line is the negative reciprocal of another, then the two lines are perpendicular.

EXAMPLE 5   Find an equation of the line passing through (2, 1) and perpendicular to the line given by $x + 2y = 4$.

SOLUTION   Putting $x + 2y = 4$ into slope-intercept form, we have $y = -\frac{1}{2}x + 2$. Therefore, the slope of this line is $-\frac{1}{2}$ and any line perpendicular to it will have a slope of 2, since $-\frac{1}{2} \cdot 2 = -1$. We are given the point (2, 1) and now we have the slope of 2. Putting this information into point-slope form, we have

$$y - 1 = 2(x - 2) \quad \text{or} \quad y = 2x - 3 \qquad \blacksquare$$

Practice Problem 8 Find the y-intercept of the line containing the points (2, 1) and (4, −2).

ANSWER   The y-intercept is 4.

Calculator Note

The equation of a line can be verified by putting the equation in Y_1, graphing the line, and then using the Trace command to check that the given points are on the line. Or, the Value command under the CALC menu can also be used.

SUMMARY

1. *Point-slope form:* If the slope is m and the line passes through (x_1, y_1), then
$$y - y_1 = m(x - x_1)$$

2. *Slope-intercept form:* If the slope is m and the y-intercept is $(0, b)$, then
$$y = mx + b$$

3. *General form:* $ax + dy + c = 0$.
4. *Horizontal line:* If the y-intercept is $(0, b)$ and the line has a slope of 0, then
$$y = b$$

5. *Vertical line:* If the x-intercept is $(h, 0)$ and the slope of the line is undefined, then
$$x = h$$

Exercise Set 1.1

Compute the slope or indicate that the slope is undefined for the line through each pair of points.

1. (3, 6), (4, 1)
2. (0, 1), (2, 3)
3. (−3, −5), (4, 2)
4. (7, −1), (−3, 1)
5. (0, 4), (4, 0)
6. (−1, −7), (−6, −5)
7. (4, 3), (4, −1)
8. (7, −1), (7, 4)
9. (3, 1), (7, 1)
10. (−1, 2), (7, 2)
11. Find an equation for and graph the line that has the following conditions. Verify using a graphing calculator.
    (a) Slope of 4; passes through the point (2, 3)

(b) Slope of -2; passes through the point $(4, -1)$
(c) Slope of $\frac{1}{2}$; passes through the point $(-1, 1)$
(d) Slope of $-7/2$; passes through the point $(3, 4)$
(e) Slope is undefined; passes through the point $(2, -4)$
(f) Slope is 0; passes through the point $(-3, -5)$

Find the slope and the y-intercept in each of the following linear equations. Graph each equation using a calculator.

12. $y = 3x + 2$ 13. $y + 2x - 1 = 0$

14. $y = 3x - 1$ 15. $2y = 10x$

16. $y = \dfrac{x - 4}{2}$ 17. $x = \dfrac{y - 1}{3}$

18. $4x + 3y - 7 = 0$ 19. $3x - 2y = 5$

20. Classify the following statements as either true or false.
   (a) The slope of the y-axis is 0.
   (b) The line segment joining (a, b) and (c, b) is horizontal.
   (c) A line with a negative slope rises to the right.
   (d) A line that is almost vertical has a slope close to 0.

Find an equation of each line through the given point with the given slope. Verify using a graphing calculator.

21. $(1, 3), \quad m = \dfrac{1}{2}$ 22. $(0, 2), \quad m = 1$

23. $(-1, -2), \quad m = -\dfrac{1}{3}$

24. $(-3, 1), \quad m = 0$

Find an equation of the line through each of the following pairs of points. Verify using a graphing calculator.

25. $(1, 1), (2, 5)$ 26. $(-1, 1), (2, 5)$

27. $(1, 3), (1, -2)$

28. Find an equation of each line with the following characteristics and verify using a graphing calculator.
   (a) The line contains the two points $(1, -3)$ and $(4, 5)$.

(b) The line has a slope of -3 and goes through the point $(7, 1)$.
(c) The line has a slope of 1 and goes through the point $(-7, 1)$.
(d) The line contains the two points $(0, 1)$ and $(4, 3)$.
(e) The line has a y-intercept of 4 and a slope of 5.
(f) The line has a y-intercept of 6 and a slope of -3.

29. Find an equation of the horizontal line through $(-4, -6)$.

30. Find an equation of the vertical line through $(-5, 4)$.

31. Write the equation of the x-axis.

32. Write the equation of the y-axis.

33. Suppose that the equation of a line is written in the form

$$\frac{x}{a} + \frac{y}{b} = 1$$

What is the x-intercept? The y-intercept?

34. Use the intercept form of the equation of a line (see Exercise 33) to find equations for lines with the following intercepts.
   (a) $x = 2$ and $y = -3$ (b) $x = 3$ and $y = 5$

35. Find the y-intercept of the line that passes through the point $(3, -2)$ with a slope of 2.

Applications (Business and Economics)

36. An electric company charges a $6-per-month customer charge plus $0.07186 per kilowatt-hour used during the month. Write an equation that relates the monthly bill, in dollars, to the number of kilowatt-hours used. What would be the charge for 1500 kilowatt-hours?

37. Every Monday, a newsstand sells x copies of a weekly sports magazine for $2.50. The owner of the newsstand buys the magazines for $1.70 a copy, plus a delivery fee of $50.
   (a) Write an equation that relates the profit, in dollars, to the number of copies sold; graph this equation.
   (b) How many copies must be sold to make a profit?
   (c) What will the profit be if 200 copies are sold?

1.2   FUNCTIONS

OVERVIEW   We study functional relationships in just about every chapter of this book: functional relationships between supply and demand, between cost and production, between the value of a piece of equipment and the age of the equipment, between IQ and accomplishment, and in many other applications. In this section, we introduce the mathematical concept of a *function*. We begin our discussion by considering a function as a rule associating one set with another. Numerous examples help to explain the meaning of a function. In this section we

- Define a function as a set of ordered pairs with certain characteristics
- Specify a function by an equation
- Introduce function notation

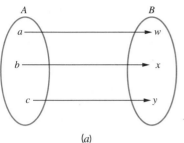

To introduce the concept of a function, we start with a correspondence between the elements in two sets. In Figures 11(a), 11(b), and 11(c), we have set up a correspondence between the elements of set A and the elements of set B. Each correspondence is a relation. A **relation** is a set of ordered pairs. In Figure 11(a), we have the relation defined by the set of ordered pairs $\{(a, w), (b, x), (c, y)\}$. The relations in Figures 11(b) and 11(c) can be similarly listed.

A function is a special type of relation; it is a rule that sets up a correspondence between a set A and a set B so that for every element of A there is a unique element of B. This way of thinking of a function can be demonstrated by what we call input–rule–output. Set A can be considered as the input, called the *domain*, and then from a rule, set B, called the *range*, can be obtained as the output.

Figure 11

EXAMPLE 6   As shown in Figure 12(a), we insert an input (the domain), operate with the rule, and obtain an output (the range). Suppose that we establish the rule to be ''add 4 to the input.'' This rule can be expressed as $x + 4$ when the input is x. For example, when the input is 3, the rule operates to give

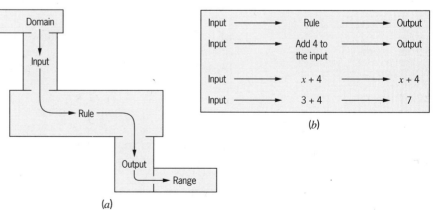

Figure 12

3 + 4 and the output is 7 [see Figure 12(*b*)]. This creates the ordered pair (3, 7). An infinite number of ordered pairs can be created because *x* can be any real number and we can add 4 to any real number.   ■

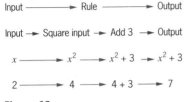

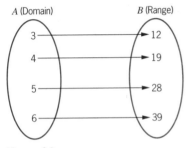

Figure 13

EXAMPLE 7   As a more complex example, consider the rule described in Figure 13. Note that this rule also sets up a correspondence. By examining the correspondence closely, we see that each element of the input is paired with exactly one element of output, creating a set of ordered pairs.   ■

Figure 14 shows some examples of the ordered pairs that are created by the function in Example 7.

From the preceding discussion we can say that a rule describes a function if it produces a correspondence between one set of elements (input set), called the domain, and a second set of elements (output set), called the range, in such a way that to each element in the domain there corresponds one and only one element in the range. Looking back at the rule in Figure 12 (*b*), we can see that the set of ordered pairs

$$\{\ .\ .\ .\ ,\ (\tfrac{1}{2},\ 4\tfrac{1}{2}),\ (1,\ 5),\ (2,\ 6),\ (3,\ 7),\ .\ .\ .\}$$

Figure 14

can be created. Also, from the rule in Figure 13, we can obtain the set

$$\{\ .\ .\ .\ ,\ (\tfrac{1}{2},\ 3\tfrac{1}{4}),\ (1,\ 4),\ (2,\ 7),\ (3,\ 12),\ .\ .\ .\ .\}$$

What is characteristic of both sets of ordered pairs? In each set, no two ordered pairs have the same first element. This is a defining characteristic of a function.

DEFINITION: FUNCTION

A **function** is a set of ordered pairs with the property that no two ordered pairs have the same first element. The set of first elements constitutes the **domain** and the set of second elements constitutes the **range**.

EXAMPLE 8   Consider the following sets of ordered pairs (*x*, *y*), where *x* is an element of the domain {2, 3, 4} and *y* is an element of the range {3, 4, 5, 6}. Which of these relations are functions?

(a) $S = \{(2, 3), (3, 4), (4, 5)\}$
(b) $T = \{(3, 3), (3, 4)\}$
(c) $U = \{(2, 3), (3, 4), (4, 5), (2, 6)\}$
(d) $V = \{(2, 5), (3, 5), (4, 5)\}$

SOLUTION

Using the preceding definition, we see that *S* and *V* are functions, since the first element is not repeated. The sets *T* and *U* are relations but not functions be-

TABLE 1

x	1	3	4	6
y	4	6	12	18

TABLE 2

x	1	0	1	2
y	−2	3	4	5

cause in T the element 3 is paired with both 3 and 4 and in U the 2 is paired with both 3 and 6. ■

NOTE: It is acceptable for the second element in the list of ordered pairs to be repeated. In fact, there are functions that contain only one element in the range.

Sometimes a set of ordered pairs is given in a table. In the two given tables, notice that Table 1 represents a function because for each x there corresponds only one y, and Table 2 does not represent a function because the two ordered pairs $(1, -2)$ and $(1, 4)$ have the same first element and different second elements.

Practice Problem 1 Does the relation $R = \{(2, 1), (3, 2), (4, 5), (4, 6), (5, 9)\}$ represent a function?

ANSWER The relation R does not represent a function because $(4, 5)$ and $(4, 6)$ have the same first element.

In Figure 12(b), the rule could be written as $y = x + 4$ and in Figure 13 as $y = x^2 + 3$. That is, the rule can be expressed in equation form. The equation representing a function assigns to each x in the domain a unique value y in the range. Often, the domain and range are restricted to a subset of the real numbers by either the equation itself or the nature of the function. The variable x in these equations is called the **independent variable** and the variable y the **dependent variable**. We can say that if an equation in two variables specifies exactly one value of the dependent variable for each value of the independent variable, then the equation represents a function.

If the domain of a function is not stated, we assume that it is the largest set of real numbers for which the rule or equation gives a real-valued function.

Calculator Note

When using a calculator to determine the domain and range of a function, it is essential that you graph enough of the function to observe all relevant behavior. You must have an algebraic understanding of the function to use the calculator properly. Using a calculator and your algebraic knowledge together can greatly facilitate your understanding of the function and its behavior.

EXAMPLE 9 Draw the graph of each function on a calculator (see page A-1 in Appendix A) and use the graph to determine the domain and range of each function.

(a) $y = x^2 + 2x + 1$

(b) $y = \dfrac{2}{x + 3}$

(c) $y = \sqrt{x + 2}$

(d) $y = \sqrt{9 - x^2}$

SOLUTION

(a) Since $y = x^2 + 2x + 1$ is a polynomial, it is defined for all values of x and the domain is the set of real numbers. The domain is $D = (-\infty, \infty)$ and the range is $R = [0, \infty)$. The graph is shown in Figure 15(a).

(b) When $x = -3$, $x + 3 = 0$ and $2/(x + 3)$ is undefined. However, for any other real number $2/(x + 3)$ is defined. Therefore, the domain is the set of all real numbers except -3. So $D = (-\infty, -3), (-3, \infty)$ and the range is $R = (-\infty, 0), (0, \infty)$. The fraction cannot equal zero because the numerator is a constant, so zero is not included in the range [see Figure 15(b)].

(c) For $y = \sqrt{x + 2}$ to be a real number, $x + 2$ must be nonnegative. Furthermore, $x + 2 \geq 0$ implies that $x \geq -2$. Therefore, the domain is $D = [-2, \infty)$ and the range is $R = [0, \infty)$, since y is always nonnegative. The graph is shown in Figure 15(c).

(d) Since $9 - x^2 \geq 0$ only over the interval $[-3, 3]$, we see in Figure 15(d) that the domain of the function is $D = [-3, 3]$ and the range is $R = [0, 3]$. ∎

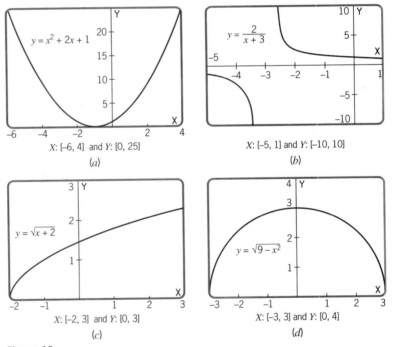

X: [-6, 4] and Y: [0, 25]

(a)

X: [-5, 1] and Y: [-10, 10]

(b)

X: [-2, 3] and Y: [0, 3]

(c)

X: [-3, 3] and Y: [0, 4]

(d)

Figure 15

Practice Problem 2 Find the domain and range of $y = \sqrt{x + 1}$.

ANSWER The domain is $D = [-1, \infty)$ and the range is $R = [0, \infty)$.

EXAMPLE 10 Is the relation $y^2 = 2 + x^2$ a function?

SOLUTION Since $y^2 = 2 + x^2$ implies that $y = \pm\sqrt{2 + x^2}$, the assignment of a real value to x will result in two different values of y (one positive and one negative), and hence the relation is not a function. By restricting y to $\sqrt{2 + x^2}$

or $-\sqrt{2 + x^2}$, we have a function. To graph this relation, you would have to graph $y_1 = \sqrt{2 + x^2}$ and $y_2 = -\sqrt{2 + x^2}$ or $y_1 = \sqrt{2 + x^2}$ and $y_2 = -y_1$.   ■

Graphically, the fact that a function associates each element in its domain with one and only one element in the range implies that no two of the ordered pairs in a function correspond to points on the same vertical line. That is, if a vertical line cuts the graph at more than one point, then the graph does not represent a function. Figures 16(a) and 16(c) show the graphs of functions. Figure 16(b) shows a graph that is not a function, since a vertical line could be drawn to intersect the figure at more than one point.

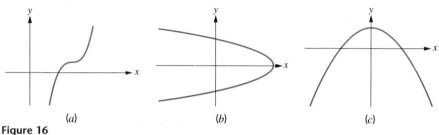

(a)   (b)   (c)

Figure 16

Vertical Line Test

The **vertical line test** can be applied as follows: If there is no vertical line that intersects the graph of an equation in more than one point, then the equation represents a function; if any vertical line passes through two or more points of the graph, the equation does not specify a function.

EXAMPLE 11   Which of the graphs in Figure 17 represent functions?

SOLUTION   The graphs in Figures 17(a) and 17(c) represent functions. The circle in Figure 17(b) is not a function.   ■

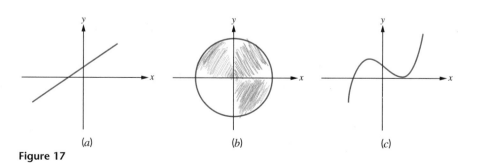

(a)   (b)   (c)

Figure 17

Practice Problem 3   Use a calculator to sketch the graphs of $x^2 - y = 3$ and $y^2 - 2x = 1$. (**Hint:** Use two equations for the second graph.) Do the graphs represent functions?

ANSWER   The equation $x^2 - y = 3$ defines a function but $y^2 - 2x = 1$ does not define a function (see Figure 18).

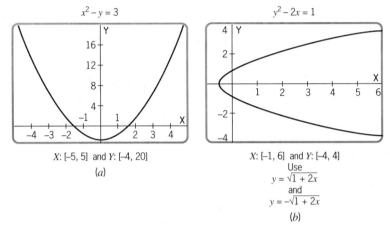

X: [–5, 5] and Y: [–4, 20]

(a)

X: [–1, 6] and Y: [–4, 4]
Use
$y = \sqrt{1 + 2x}$
and
$y = -\sqrt{1 + 2x}$

(b)

Figure 18

Practice Problem 4

Use a calculator to find the domain and range of each function.
(a) $y = \sqrt{x^2 + 2x - 3}$
(b) $y = |x^2 - 9|$

ANSWER

(a) Since a calculator graphs only real values, we can see in Figure 19(a) that the domain is $D = (-\infty, -3], [1, \infty)$ and the range is $R = [0, \infty)$.
(b) As shown in Figure 19(b), the domain is $D = (-\infty, \infty)$ and the range is $R = [0, \infty)$.

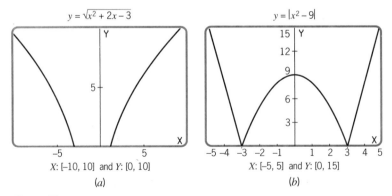

X: [–10, 10] and Y: [0, 10]

(a)

X: [–5, 5] and Y: [0, 15]

(b)

Figure 19

Function Notation

Often, we write a function using function notation. That is, we write $y = f(x)$, which is read "y equals f of x." It is important to remember that f is the name of the function and that $f(x)$ does not indicate multiplication. Any letter can be used to name a function; we often give a function a name relating to its nature. For example, a cost function could be $C(x)$. This notation indicates that with each x there is associated a unique C.

Suppose that we write $y = f(x) = x + 7$. When $x = 2$, we substitute 2 for x into the function to obtain

$$y = f(2) = 2 + 7 = 9$$

Likewise, $f(4) = 11$, and $f(6) = 13$.

EXAMPLE 12   For $y = f(x) = -x^2 + 2x + 1$ find each of the following:

(a) $f(1)$ (b) $f(0)$ (c) $f(-2)$ (d) $f(x + 1)$

SOLUTION

(a) $f(1) = -(1)^2 + 2(1) + 1 = 2$
(b) $f(0) = -(0)^2 + 2(0) + 1 = 1$
(c) $f(-2) = -(-2)^2 + 2(-2) + 1 = -4 - 4 + 1 = -7$
(d) $f(x + 1) = -(x + 1)^2 + 2(x + 1) + 1 = -(x^2 + 2x + 1) + 2x + 2 + 1$
    $= -x^2 - 2x - 1 + 2x + 3 = -x^2 + 2$ ■

Practice Problem 5 For $f(x) = 3x^2 - 2x + 1$, find each of the following:

(a) $f(-3)$ (b) $f(x + 1)$

ANSWER

(a) $f(-3) = 34$ (b) $f(x + 1) = 3x^2 + 4x + 2$

Calculator Note

Some calculators create tables that allow you to see the values of a variable and the function in a table. If your calculator has this feature, study several functions using this feature. See page A-7 of Appendix A.

COMMON ERROR   When evaluating a function such as $f(x) = -x^2 + 3x - 2$, students often include the negative sign as part of the squared term. Why is this not correct?

Correct	Incorrect
$f(4) = -4^2 + 3 \cdot 4 - 2 = -6$	$f(4) = (-4)^2 + 3 \cdot 4 - 2 = 26$

There is an important expression that we will use many times in calculus when we study the way functions look in their graphs and the way that they change. It is called the difference quotient and is given as

$$\frac{f(x + h) - f(x)}{h} \qquad (h \neq 0)$$

Finding the difference quotient for a function is important. It will be done often in calculus as important formulas are developed. Let's look at an example of how a difference quotient is found and simplified.

EXAMPLE 13   For the function $f(x) = x^2 - 2x$, find the difference quotient.

SOLUTION

$$\frac{f(x + h) - f(x)}{h} = \frac{\overbrace{[(x + h)^2 - 2(x + h)]}^{f(x + h)} - \overbrace{[x^2 - 2x]}^{f(x)}}{h}$$

$$= \frac{x^2 + 2xh + h^2 - 2x - 2h - x^2 + 2x}{h}$$

$$= \frac{2xh + h^2 - 2h}{h}$$

$$= \frac{h(2x + h - 2)}{h}$$

$$= 2x + h - 2 \qquad (h \neq 0)$$

Practice Problem 6   Find the difference quotient for $f(x) = 2x - 3$.

ANSWER   2

SUMMARY

Make certain that you understand the definition and concept of a function. It is an important concept and will be used throughout this book.

1. *Function:* A relation that creates a set of ordered pairs, each consisting of an independent variable and a dependent variable and for which each independent variable has a unique value of the dependent variable associated with it.
2. *Notation:* In the equation $y = f(x)$, f is the *name* of the function and its *value* at x is $f(x)$.
3. *Domain:* All values of the independent variable for which the function is defined over the real numbers.
4. *Range:* The set of all values of the dependent variables.

Exercise Set 1.2

1. Which of the following relations are functions?
    (a) $\{(1, 3), (3, 3), (5, 3)\}$
    (b) $\{(1, 3), (3, 3), (5, 7)\}$
    (c) $\{(1, 3), (3, 5), (5, 1)\}$
    (d) $\{(1, 1), (3, 3), (5, 5)\}$
    (e) $\{(3, 4), (5, 10), (6, 4), (7, 1)\}$
    (f) $\{(1, 5), (1, 6), (2, 5), (3, 10)\}$
    (g) $\{(3, 7), (7, 3), (8, 3)\}$

(h) $\{(4, 6), (5, 6)\}$
(i) $\{(5, 3), (5, 4)\}$
(j) $\{(5, 5), (6, 6)\}$

2. Which of the following tables define functions? (The variable x is the independent variable.)

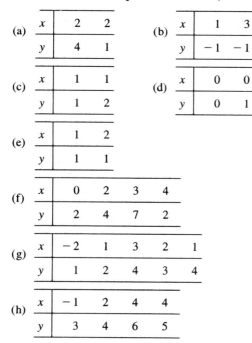

(a)
x	2	2
y	4	1

(b)
x	1	3
y	-1	-1

(c)
x	1	1
y	1	2

(d)
x	0	0
y	0	1

(e)
x	1	2
y	1	1

(f)
x	0	2	3	4
y	2	4	7	2

(g)
x	-2	1	3	2	1
y	1	2	4	3	4

(h)
x	-1	2	4	4
y	3	4	6	5

3. State whether or not each correspondence represents a function.

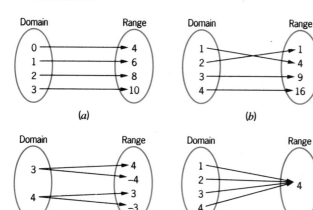

(a)

(b)

(c)

(d)

4. If $y = f(x) = x^3 - 2x$, find $f(-2)$, $f(1)$, and $f(0) - f(3)$.

5. If $y = f(x) = (x + 2)(x - 1)$, find $f(3)$, $f(2)$, and $f(-1) - f(0)/2$.

6. Which of the following graphs represent functions?

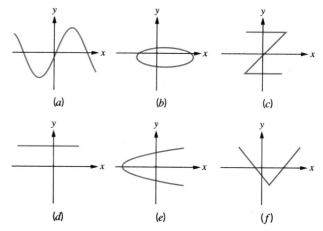

(a)

(b)

(c)

(d)

(e)

(f)

7. If $y = f(x) = x^3 - 2x$, find $f(-1)/f(1)$, $f(2) \cdot f(4)$, $f(w)$, $f(2z)$, $f(t + 2)$, and $f(3x - 1)$.

8. If $y = f(x) = (x + 2)(x - 1)$, find $f(-2)$, $f(3)$, $f(z)$, $f(3w)$, $f(t + 3)$, $f(2w - 1)$, and $f(4 + h)$.

Determine the domain of each function using a graphing calculator.

9. $y = \sqrt{x}$

10. $y = \dfrac{x^2 + 2}{x^2 - 1}$

11. $y = \dfrac{x}{x^2 - 4x - 5}$

12. $y = \dfrac{3}{x - 2}$

13. $y = \sqrt{3 - x^2}$

14. $y = \dfrac{1}{x}$

15. $y = \dfrac{3}{x(x - 3)}$

16. $y = \sqrt{9x^2 - 16}$

17. $y = \sqrt{x(x + 4)}$

18. $y = \sqrt{x^2 - 5x + 4}$

19. $y = \sqrt{\dfrac{x - 1}{x + 1}}$

20. $y = \sqrt{\dfrac{|x - 3|}{x - 2}}$

21. $y = \sqrt{(x - 1)(x + 2)(x - 3)}$

22. $y = \dfrac{x + 7}{x - 1}$

23. $y = x^2 + 6x + 4$

24. $y = \sqrt{x + 1}$

25. Find the expression

$$\frac{f(2 + h) - f(2)}{h}$$

for each function.

(a) $f(x) = 3x - 1$ (b) $f(x) = 2x + 4$

(c) $f(x) = 4x^2$ (d) $f(x) = x^2 - 3$

(e) $f(x) = \sqrt{x}$ (f) $f(x) = 1/x$

26. Find the difference quotient

$$\frac{f(x + h) - f(x)}{h}$$

for each of the functions in Exercise 25.

In Exercises 27–30, match each graph to the correct function listed below. Try to do this without using a graphing calculator. Then verify your answer using a graphing calculator.

(a) $y = |3x + 2|$ (b) $y = x^2 - 4x + 3$

(c) $y = \sqrt{2x + 1}$ (d) $y = \dfrac{x}{x + 1}$

27.

28.

29.

30.

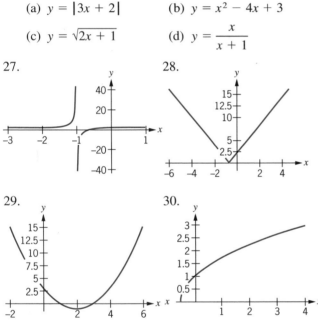

Applications (Business and Economics)

31. **Revenue.**   A parking lot charges by the hour as follows:

$$f(x) = 2 + 1.5(x - 1) \quad \text{for } x \geq 1$$

Find the charge for 1 hour, 2 hours, 10 hours, and 24 hours. Try to state the parking rate in words.

32. **Revenue Function.**   A travel agency books a flight to Europe for a group of college students. The fare in a 200-passenger airplane will be $400 per student plus $2.00 per student for each vacant seat.

(a) Write the total revenue $R(x)$ as a function of empty seats, x.

(b) What is the domain of this function?

(c) Calculate $R(x)$ for 5, 10, 20, 40, and 100 empty seats.

33. **Demand Function.**   Suppose that the demand function (in price per unit) for a certain item is given by

$$p = D(x) = \frac{50 - 5x}{4}$$

when x units are demanded by the consumer at price $p = D(x)$.

(a) What is $D(x)$ when $x = 0$?

(b) What is $D(x)$ when $x = 4$?

(c) What happens to $D(x)$ when $x = 10$?

(d) Sketch the graph of $D(x)$.

34. **Supply Function.**   Suppose that the supply function for the item in Exercise 33 is given by

$$p = S(x) = \frac{5x}{6}$$

where $p = S(x)$ is the price per unit of an item at which the seller is willing to supply x units.

(a) What is $S(x)$ when $x = 3$? When $x = 9$?

(b) Sketch the graph of $S(x)$ on the same axis system that you used in Exercise 33.

(c) Estimate the point of equilibrium from the intersection of the graphs.

(d) Estimate the equilibrium price.

(e) When is supply greater than demand?

35. **Demand and Supply Functions.**   The supply function $S(x)$ and the demand function $D(x)$ for a certain commodity in terms of units available, x, at price p, are

$$p = S(x) = \frac{400 - 5x}{2} \quad \text{and} \quad p = D(x) = \frac{5x}{2}$$

(a) Graph $S(x)$ and $D(x)$ on the same axes.

(b) Find the equilibrium point.

(c) Find the equilibrium price.

(d) When is the supply less than the demand?

Applications (Social and Life Sciences)

36. **Bacteria Count.** The number of bacteria in a culture x hours after an antibacterial treatment has been administered is given by

$$N(x) = 1000 - 150x, \qquad 0 \le x \le 6$$

Find $N(0)$, $N(2)$, $N(4)$, and $N(6)$.

37. **Data.** Write a function that shows the relationship between the gallons of gasoline that your car uses and the miles you drive. What effect would getting more miles per gallon have on the function?

Review Exercises

38. Write an equation of the line meeting the following criteria.
    (a) Has slope -3 and passes through the point $(-1, 2)$
    (b) Has no slope and passes through the point $(1, 1)$
    (c) Passes through the points $(1, -2)$ and $(3, 1)$
    (d) Passes through the point $(2, -4)$ and is parallel to the line having the equation $2x - 3y + 4 = 0$
    (e) Has x-intercept 4 and y-intercept -2

1.3 QUADRATIC FUNCTIONS

OVERVIEW The equation $y = f(x)$, where $f(x)$ is a polynomial of degree n in the variable x, is called a polynomial function. A special case of the polynomial function when $n = 1$ (first-degree equation) has a graph that is always a line. In this section, we study another special case of a polynomial function by letting $n = 2$. This second-degree equation is called a quadratic function. In this section we

- Study the characteristics of graphs of quadratic functions
- Find the vertex and the axis of symmetry of graphs of quadratic functions
- Graph quadratic functions

All functions that can be written in the form

$$y = f(x) = ax^2 + bx + c \qquad (a \ne 0)$$

are called **quadratic functions**. The graphs of such functions are curves called **parabolas**. For example, the path a ball takes when thrown or hit in the air is parabolic in shape. Parabolic shapes are also found in such places as satellite dishes or spotlights. All parabolas are symmetric about a line called the **axis of symmetry**. In addition, all parabolas have a **vertex**, which is the point at which the parabola and the axis of symmetry intersect. This is shown in Figure 20.

We begin our discussion by examining the graph of the simplest quadratic function, $y = x^2$.

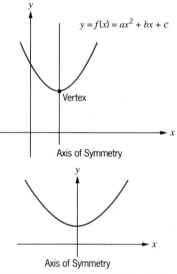

Note the axis of symmetry will be the y-axis for some quadratic functions.

Figure 20

EXAMPLE 14 Draw the graph of $y = x^2$ and then compare it to the graphs of $y = 3x^2$ and $y = \frac{1}{3}x^2$. Describe how changing the value of a (coefficient of x^2) alters the graph.

SOLUTION Using a calculator, we can easily graph $y = x^2$ and $y = 3x^2$ on the same screen. We can then add the graph of $y = \frac{1}{3}x^2$ to the screen (see Figure 21). By examining these graphs, we see that as the coefficient a is changed, the apparent "width" of the parabola is changed. If $|a| > 1$, the parabola is not as wide as when $0 < |a| < 1$. The domain of all the functions is the real

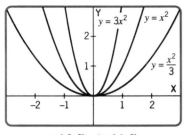

X: [−3, 3] and Y: [−1, 3]

Figure 21

numbers, so in actuality, they all have the same width. What we are referring to here is the appearance of the graph. Instead of "width," we could say that a affects the "rate of change" of the function.

In Figure 21, notice that all of the parabolas pass through the point (0, 0) and are symmetric about the y-axis. Thus, (0, 0) is the vertex and the y-axis is the axis of symmetry for these parabolas. [When the y-axis is the axis of symmetry, the vertex will be the point (0, c).]

EXAMPLE 15   Use a graphing calculator to graph $y = x^2$, and $y = -x^2$; then graph $y = 2x^2$ and $y = -2x^2$. What effect does the change in a (coefficient of x^2) have on the graph?

SOLUTION   We say that the relationship between $y = x^2$ and $y = -x^2$ and between $y = 2x^2$ and $y = -2x^2$ is a reflection across the x-axis. We can see in Figure 22 that when $a < 0$ in $y = -x^2$ and $y = -2x^2$, the parabola is turned downward. The fact that $a = -2$ means that the parabola turns down and that it is not as wide as $y = -x^2$. Again we see that the vertex is (0, 0) and the axis of symmetry is the y-axis for these parabolas.

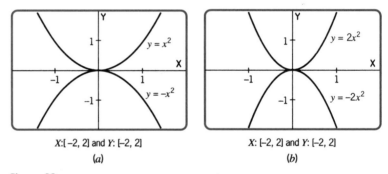

X:[−2, 2] and Y: [−2, 2]	X: [−2, 2] and Y: [−2, 2]
(a)	(b)

Figure 22

We can summarize the characteristics of quadratic functions discussed in Examples 14 and 15 as follows.

CHARACTERISTICS OF $y = ax^2$

1.   If $|a| > 1$, the graph of $y = ax^2$ is not as wide for a fixed x as the graph of $y = x^2$.
2.   If $0 < |a| < 1$, the graph of $y = ax^2$ is wider than that of $y = x^2$.
3.   If $a > 0$, the graph of the function has an upward direction.
4.   If $a < 0$, the graph of the function has a downward direction.
5.   The vertex of $y = ax^2$ is (0, 0).
6.   The axis of symmetry of $y = ax^2$ is the y-axis.

EXAMPLE 16   Draw the graphs of $y = x^2 + 1$ and $y = x^2 - 3$. Discuss the relation of these graphs to that of $y = x^2$.

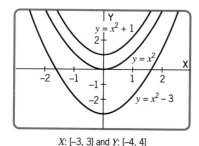

X: [-3, 3] and Y: [-4, 4]

Figure 23

SOLUTION   In Figure 23, we can see that the graphs have been shifted vertically. A **vertical shift** or **translation** occurs when a constant is added to $y = ax^2$, giving $y = ax^2 + c$. If $c > 0$, the shift is upward. If $c < 0$, the shift is downward. The vertex of $y = x^2 + 1$ is (0, 1) and the vertex of $y = x^2 - 3$ is (0, −3). The y-axis is the axis of symmetry for both parabolas. ∎

Vertical Shifts

1.   If $c > 0$, to obtain the graph of $y = x^2 + c$, move the graph of $y = x^2$ **up** c units.
2. $y = x^2 - c$, move the graph of $y = x^2$ **down** c units.

Calculator Note

Experiment with a graphing calculator by entering different values for both a and c in $y = ax^2 + c$. Leave the graph of $y = x^2$ on the viewing rectangle to compare what is happening with each change in a and c.

Up to this point, we have looked only at quadratic functions with the vertex on the y-axis. However, all quadratic functions do not have graphs that are symmetric about the y-axis, as illustrated in the next example.

EXAMPLE 17   Draw the graphs of $y = (x - 3)^2$ and $y = -(x + 2)^2$ and compare these to the graph of $y = x^2$. What is the vertex and axis of symmetry for each function?

SOLUTION   These graphs are examples of **horizontal shifts** or **translations**. The graph of the function $y = (x - 3)^2$ is the same as the graph $y = x^2$ shifted 3 units to the right. Its vertex is (3, 0) and the line $x = 3$ is the axis of symmetry [see Figure 24(a)]. The graph of the function $y = -(x + 2)^2$ is the same as the graph $y = x^2$ shifted 2 units to the left and reflected across the x-axis. Its vertex is (− 2, 0) and the line $x = -2$ is the axis of symmetry [see Figure 24(b)]. ∎

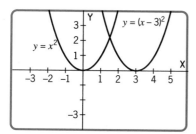

X: [-4, 6] and Y: [-4, 4]

(a)

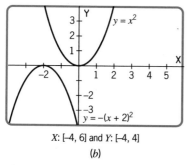

X: [-4, 6] and Y: [-4, 4]

(b)

Figure 24

Horizontal Shifts

1.   For $c > 0$, to obtain the graph of $y = (x - c)^2$, move the graph of $y = x^2$ c units to the **right**.
2. $y = (x + c)^2$, move the graph of $y = x^2$ c units to the **left**.

Practice Problem 1   Draw the graph of $y = (x - 2)^2$, $y = (x + 1)^2$. How do these graphs compare to the graphs of $y = x^2 - 2$ and $y = x^2 + 1$? What is the vertex and axis of symmetry for each parabola?

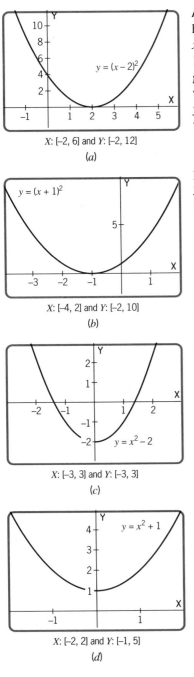

X: [-2, 6] and Y: [-2, 12]

(a)

X: [-4, 2] and Y: [-2, 10]

(b)

X: [-3, 3] and Y: [-3, 3]

(c)

X: [-2, 2] and Y: [-1, 5]

(d)

Figure 25

ANSWER The graph of $y = (x - 2)^2$ in Figure 25(a) shifts the parabola $y = x^2$ horizontally to the right 2 units. The vertex is (2, 0) and the axis of symmetry is $x = 2$. The graph of $y = (x + 1)^2$ Figure 25(b) is a horizontal shift of $y = x^2$ 1 unit to the left. The vertex is (- 1, 0) and the axis of symmetry is $x = - 1$. The graph of $y = x^2 - 2$ in Figure 25(c) is a vertical shift of $y = x^2$ two units downward. It has vertex (0, - 2) and the axis of symmetry is the y-axis. The graph of $y = x^2 + 1$ in Figure 25(d) is a vertical shift of $y = x^2$ one unit upward. It has vertex (0, 1) and the axis of symmetry is the y-axis.

When graphing and analyzing a parabola, the intercepts and the vertex are the points we use most often. The x-intercept is obtained by setting $y = 0$ and the y-intercept is found by setting $x = 0$. We discuss the algebraic techniques for finding the x-intercepts in Section 1.4.

The vertex is not always obvious when looking at a parabola. There are several methods of finding the vertex. Look at the graphs shown in Figures 24 and 25. Can you see any correlation between the equation and the vertex? Notice that when there is only a vertical or horizontal shift, we can determine the vertex very quickly. Thus $y = x^2 + 1$ has its vertex at (0, 1), whereas $y = (x + 1)^2$ has its vertex at (- 1, 0). Before we give a specific method for finding the vertex, let's examine some more graphs to see if we can find a pattern. Let's look at some graphs that have both a horizontal and vertical shift and are in what is called the **standard form of a quadratic**:

$$y = f(x) = a(x - h)^2 + k$$

EXAMPLE 18 Draw the graphs of $y = (x - 2)^2 - 3$ and $y = (x + 2)^2 + 3$. Can you determine the vertex of each parabola? How do the points of the vertex correspond to the equation of the function when compared to the standard form of a quadratic?

SOLUTION We see in Figure 26(a) that the vertex of $y = (x - 2)^2 - 3$ is (2, - 3) and that in Figure 26(b) the vertex of $y = (x + 2)^2 + 3$ is (- 2, 3). These points correspond to the point (h, k) from the standard form given above.

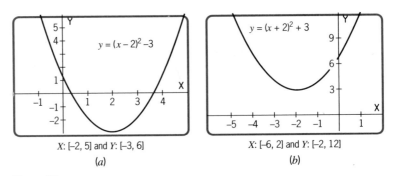

X: [-2, 5] and Y: [-3, 6]

(a)

X: [-6, 2] and Y: [-2, 12]

(b)

Figure 26

Practice Problem 2 Experiment using a calculator with equations in the form $y = f(x) = a(x - h)^2 + k$. What do the changes in a, h, and k do to the graph of $y = x^2$?

ANSWER   Note that a determines the direction and width of the parabola, and changes in (h, k) alter the vertex of the parabola.

We see that the form used in Example 18 and Practice Problem 2 is extremely useful because it helps us to quickly identify the vertex and the direction of the parabola.

DEFINITION: STANDARD FORM OF A QUADRATIC FUNCTION

A quadratic function $y = f(x) = ax^2 + bx + c$, is in **standard form** when it is written as

$$y = f(x) = a(x - h)^2 + k$$

and has as its vertex the point (h, k). Its axis of symmetry is the line $x = h$. If $a > 0$, the parabola turns upward and has a minimum value of k. If $a < 0$, the parabola turns downward and has a maximum value of k.

COMMON ERROR

Students often confuse the y-intercept $f(0) = c$ with the value of k in the standard form. If $h \neq 0$, then the y-intercept is not k.

It is important to note that not all quadratic functions in x have x-intercepts, but they always have a y-intercept. Since the domain of the function is all real numbers, x can take on the value 0 and there will be a y-intercept.

EXAMPLE 19   Find the axis of symmetry, the coordinates of the vertex, the maximum or minimum value of the function, and the y-intercept for the function

$$y = f(x) = x^2 - 4x + 3$$

SOLUTION   The first step is to write the function in standard form by obtaining the expression $(x - h)^2$. We do this by completing the square. (Later, we will learn another way to find the vertex.) Remember, to complete the square, find $(-4/2)^2$. When this is added and subtracted from the function, the net change is zero, but we are able to factor the perfect square trinomial that results.

$$y = [x^2 - 4x + (-2)^2] - (-2)^2 + 3 \qquad \frac{b}{2} = \frac{-4}{2} = -2$$
$$= (x - 2)^2 - 1$$

By comparing this to the standard form, we determine that the vertex is at $(2, -1)$. Since $a > 0$, there is a minimum value, which occurs at $x = 2$ and is $f(2) = -1$. The y-intercept is $f(0) = 3$. The graph is shown in Figure 27. ■

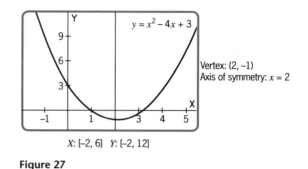

X: [−2, 6]   Y: [−2, 12]

Figure 27

Practice Problem 3 For $y = x^2 - 2x - 3$, find the equation for the axis of symmetry, the vertex, and the y-intercept. Does the function have a maximum value or a minimum value? If so, what is the value?

ANSWER   The equation $x = 1$ is the axis of symmetry. The vertex is $(1, -4)$ and the y-intercept is -3. Since $a > 0$, the minimum value of the function is -4.

As stated earlier, we discuss the algebraic techniques for finding the x-intercepts in Section 1.4. For now, we can use a calculator to locate the values (or approximate values).

Practice Problem 4 Using the equation in Practice Problem 3, $y = x^2 - 2x - 3$, draw the graph and estimate the x-intercepts.

ANSWER   The graph is shown in Figure 28. The x-intercepts are -1 and 3.

We can describe the coordinates of the vertex and the axis of symmetry for a quadratic function as follows.

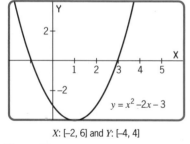

X: [−2, 6] and Y: [−4, 4]

Figure 28

Vertex and Symmetry

For the graph of the quadratic function $y = f(x) = ax^2 + bx + c$:

1.   The vertex is $\left(\dfrac{-b}{2a}, f\left(\dfrac{-b}{2a} \right) \right) = \left(\dfrac{-b}{2a}, c - \dfrac{b^2}{4a} \right)$.

2.   The axis of symmetry is $x = \dfrac{-b}{2a}$.

We verify these characteristics as follows:

$$f(x) = ax^2 + bx + c$$

$$= a\left(x^2 + \frac{b}{a}x\right) + c \qquad \text{Factor out } a.$$

$$= a\left(x^2 + \frac{b}{a}x + \frac{b^2}{4a^2}\right) + c - \frac{b^2}{4a} \qquad \text{Complete the square.}$$

$$= a\left(x + \frac{b}{2a}\right)^2 + \left(c - \frac{b^2}{4a}\right)$$

Comparing this result to $y = a(x - h)^2 + k$, we can see that the axis of symmetry is $x = -b/2a$. The x-coordinate of the vertex is also $-b/2a$ and the y-coordinate is

$$f\left(\frac{-b}{2a}\right) = c - \frac{b^2}{4a}$$

EXAMPLE 20   For $f(x) = 2x^2 - 12x + 19$, find the vertex, y-intercept, and the axis of symmetry and determine whether there are any x-intercepts. Write the function in standard form. What is the maximum or minimum value of f?

SOLUTION   Let's use the technique that we just developed to find the vertex. For $f(x) = 2x^2 - 12x + 19$, we have $a = 2$, $b = -12$, and $c = 19$. Since the x-coordinate of the vertex is $-b/2a$, we have

$$x = \frac{-(-12)}{2(2)} = \frac{12}{4} = 3$$

Since the x-coordinate of the vertex is 3, the y-coordinate is

$$f(3) = 2(3)^2 - 12(3) + 19 = 1$$

Therefore, the vertex is $(3, 1)$ and we can write f in standard form as

$$f(x) = 2x^2 - 12x + 19 = 2(x - 3)^2 + 1$$

The minimum value of f is 1, the axis of symmetry is $x = 3$, and the y-intercept is 19. We can see in Figure 29 that there are no x-intercepts.

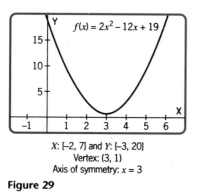

X: [-2, 7] and Y: [-3, 20]
Vertex: (3, 1)
Axis of symmetry: $x = 3$

Figure 29

Practice Problem 5 For $f(x) = -3x^2 + 6x - 5$, find the axis of symmetry, the vertex, y-intercept; put in standard form; and approximate any x-intercepts. What is the maximum or minimum value of f?

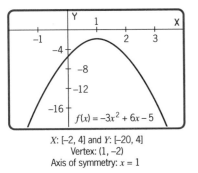

X: [−2, 4] and Y: [−20, 4]
Vertex: (1, −2)
Axis of symmetry: $x = 1$

Figure 30

ANSWER As shown in Figure 30, the axis of symmetry is $x = 1$; the vertex is $(1, -2)$; and we write $f(x) = -3(x - 1)^2 - 2$. The maximum value of f is -2. The y-intercept is -5 and there are no x-intercepts.

SUMMARY

1. The graph of a quadratic equation in standard form, $f(x) = a(x - h)^2 + k$, has vertex (h, k) and axis of symmetry $x = h$.
2. If the quadratic is not written in standard form, the vertex of $f(x) = ax^2 + bx + c$ is

$$\left(\frac{-b}{2a}, f\left(\frac{-b}{2a} \right) \right)$$

with the axis of symmetry $x = -b/2a$. From this, we see that $h = -b/2a$ and $k = f(h)$.
3. The direction and width of a parabola is determined by the value of a. If $a > 0$, the parabola opens upward. If $a < 0$, the parabola opens downward.
4. The maximum value or minimum value of the function is k.

Exercise Set 1.3

Use a calculator to draw the three given functions in the same viewing rectangle and discuss why they are different.

1. $y = 4x^2$, $y = 4(x + 1)^2$, $y = 4(x - 1)^2$
2. $y = 2x^2$, $y = 2(x - 1)^2$, $y = 2(x + 1)^2$
3. $y = -x^2$, $y = -(x - 3)^2$, $y = -(x + 3)^2$
4. $y = -x^2$, $y = -x^2 + 1$, $y = -x^2 - 1$
5. $y = x^2 - 2$, $y = (x - 3)^2$, $y = (x + 3)^2 + 2$
6. $y = x^2 + 1$, $y = 2x^2 + 1$, $y = -2x^2 + 1$

Describe how the graph of each function can be obtained from the graph of $f(x) = x^2$ in terms of horizontal and vertical shifts, magnification, and reflection.

7. $f(x) = 2x^2$
8. $f(x) = -\frac{1}{3}x^2$
9. $f(x) = (x + 1)^2$
10. $f(x) = (x - 3)^2$
11. $f(x) = x^2 + 3$
12. $f(x) = -x^2 - 2$
13. $f(x) = 2(x - 1)^2 + 3$
14. $f(x) = -3(x + 2)^2 - 1$

For each function in Exercises 15–24:

(a) *Find the line of symmetry.*
(b) *Find the vertex.*
(c) *Find the y-intercept.*
(d) *Draw the graph.*
(e) *Approximate any x-intercepts.*
(f) *Determine whether the vertex is a maximum point or minimum point.*

15. $f(x) = x^2 + x - 6$
16. $f(x) = 2x^2 - 9x - 5$
17. $f(x) = -x^2 - 2x + 24$
18. $f(x) = -2x^2 + 7x - 3$
19. $f(x) = 3x^2 + 9x - 12$
20. $f(x) = 2x^2 + 6x - 30$
21. $f(x) = x^2 + 6x$
22. $f(x) = -2x^2 + 4x$
23. $f(x) = 3x^2 - 4$
24. $f(x) = -4x^2 + 2$

Review Exercises

Determine the domain and range of each function using graphs drawn with a calculator.

25. $f(x) = \sqrt{9x^2 - 16}$
26. $y = \dfrac{x + 3}{x - 1}$
27. $f(x) = \dfrac{|4 + x|}{x}$
28. $y = \sqrt{\dfrac{3 - x}{x + 2}}$

1.4   QUADRATIC EQUATIONS AND APPLICATIONS

OVERVIEW   A quadratic function can have zero, one, or two x-intercepts. The number of x-intercepts can be determined graphically; however, to find the exact value of these intercepts, algebraic techniques are used. These intercepts are called **zeros of the function**. Models involving quadratic functions are used in business, economics, and social and life sciences. In this section we

- Find the zeros of a quadratic function (solve quadratic equations)
- Locate a point of equilibrium for supply and demand functions
- Locate break-even points for cost and revenue functions

We use two procedures in this section for solving quadratic equations: factoring and the quadratic formula. If a quadratic equation $ax^2 + bx + c = 0 \ (a \neq 0)$ can be factored, the solutions can be attained easily. If the product of two numbers is zero, then one or both of the numbers must be zero. If, for example, $(x + 3)(x + 4) = 0$, then either $x + 3 = 0$ or $x + 4 = 0$.

COMMON ERROR   When solving an equation such as $(x + 3)(x + 7) = 10$, you cannot set each factor equal to 10 and solve. The equation must be written so that one side is zero.

EXAMPLE 21   Solve $2x^2 = 32$ by factoring.

SOLUTION

$$2x^2 - 32 = 0 \qquad \text{Rewrite the equation so one side is 0.}$$

$$2(x^2 - 16) = 0 \qquad \text{Factor out the common factor of 2.}$$

$$2(x - 4)(x + 4) = 0 \qquad \text{Factor as the difference of two squares.}$$

This equation is only true if $x - 4 = 0$ or $x + 4 = 0$. Hence

$$x = 4 \quad \text{or} \quad x = -4$$

The factor 2 is not considered because only factors that can attain a value of 0 will yield the solutions.

Check: $2(-4)^2 = 2(16) = 32$ and $2(4)^2 = 2(16) = 32$

Thus, the solution set is $\{-4, 4\}$ (see Figure 31).

EXAMPLE 22   Find the solution set of $x^2 + x - 12 = 0$.

SOLUTION   The equation $x^2 + x - 12 = 0$ is equivalent to $(x + 4)(x - 3) = 0$. This equation is only true if $x + 4 = 0$ or $x - 3 = 0$. If $x + 4 = 0$, then $x = -4$ and if $x - 3 = 0$, then $x = 3$.

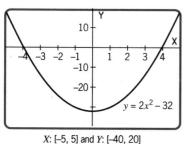

X: [-5, 5] and Y: [-40, 20]
x-intercepts: $x = -4$ and $x = 4$

Figure 31

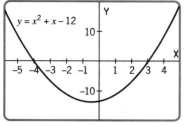

$y = x^2 + x - 12$

X: [-6, 5] and Y: [-20, 20]
x-intercepts: $x = -4$ and $x = 3$

Figure 32

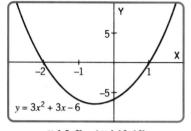

$y = 3x^2 + 3x - 6$

X: [-3, 2] and Y: [-10, 10]
x-intercepts: $x = -2$ and $x = 1$

Figure 33

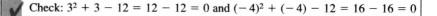 Check: $3^2 + 3 - 12 = 12 - 12 = 0$ and $(-4)^2 + (-4) - 12 = 16 - 16 = 0$

Thus, the solution set is $\{-4, 3\}$ (see Figure 32). ■

Practice Problem 1 Solve $3x^2 + 3x = 6$ and draw the graph using a calculator to reinforce your answer.

ANSWER The solution set is $\{-2, 1\}$ (see Figure 33).

There are quadratics that do not factor easily and some that do not factor at all over the rational numbers. Also, some quadratic equations have no real solutions. In order to work effectively with these quadratics, let's return to the work we did in Section 1.3—completing the square of the quadratic $f(x) = ax^2 + bx + c$ to find the vertex. From this equation, we obtained

$$f(x) = a\left(x + \frac{b}{2a}\right)^2 + c - \frac{b^2}{4a}$$

which can also be written as

$$f(x) = a\left(x + \frac{b}{2a}\right)^2 - \left(\frac{b^2 - 4ac}{4a}\right)$$

Now let $f(x) = 0$ (for the equation $ax^2 + bx + c = 0$) and solve for x.

$$\left(x + \frac{b}{2a}\right)^2 = \frac{b^2 - 4ac}{4a^2} \qquad \text{Divide by } a \neq 0.$$

$$x + \frac{b}{2a} = \pm\sqrt{\frac{b^2 - 4ac}{4a^2}} \qquad \text{Take the square root.}$$

$$= \pm\frac{\sqrt{b^2 - 4ac}}{2a} \qquad \text{Simplify the denominator.}$$

$$x = \frac{-b \pm \sqrt{b^2 - 4ac}}{2a}$$

This result is called the **quadratic formula** and can be used to solve any quadratic equation.

THE QUADRATIC FORMULA

If a, b, and c are real numbers and $a \neq 0$, then the solutions of $ax^2 + bx + c = 0$, if they exist, are

$$x = \frac{-b + \sqrt{b^2 - 4ac}}{2a} \quad \text{or} \quad \frac{-b - \sqrt{b^2 - 4ac}}{2a}$$

You can program the quadratic formula into your graphing calculator and use it to get a decimal approximation of the roots, but you will need to use the quadratic formula without a calculator or program your calculator to give you the exact expression if a problem requires that the answer be in exact form (i.e., a rational number or a radical expression).

Calculator Note

You must be aware of the capabilities of your graphing calculator. As mentioned previously, some graphing calculators are always in complex mode and give complex answers as an ordered pair.

EXAMPLE 23 Solve $2x^2 + 2x = 12$.

SOLUTION In order to use the quadratic formula, we need to put the equation in the form $ax^2 + bx + c = 0$. Thus, we have $2x^2 + 2x - 12 = 0$, where $a = 2$, $b = 2$, and $c = -12$. Using the quadratic formula, the two solutions are

$$x = \frac{-2 + \sqrt{(2)^2 - 4(2)(-12)}}{2(2)} = \frac{-2 + \sqrt{4 + 96}}{4}$$

$$= \frac{-2 + \sqrt{100}}{4} = \frac{-2 + 10}{4} = 2$$

$$x = \frac{-2 - \sqrt{(2)^2 - 4(2)(-12)}}{2(2)} = \frac{-2 - \sqrt{4 + 96}}{4}$$

$$= \frac{-2 - \sqrt{100}}{4} = \frac{-2 - 10}{4} = -3$$

 Check: $2(2)^2 + 2(2) = 8 + 4 = 12$ and $2(-3)^2 + 2(-3) = 18 - 6 = 12$

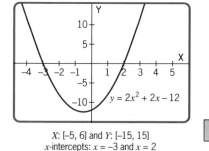

X: [−5, 6] and Y: [−15, 15]
x-intercepts: $x = -3$ and $x = 2$

Figure 34

The solution set is $\{-3, 2\}$ (see Figure 34).

EXAMPLE 24 Solve $x^2 - 6x + 25 = 0$.

SOLUTION We have $a = 1$, $b = -6$, and $c = 25$, so

$$x = \frac{6 \pm \sqrt{36 - 100}}{2} = \frac{6 \pm \sqrt{-64}}{2}$$

Since $\sqrt{-64}$ is not a real number, this equation has no real solution. The graph is shown in Figure 35. Notice that there are no x-intercepts; hence, there are no real solutions to the equation.

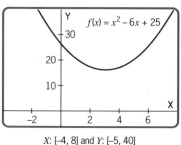

X: [−4, 8] and Y: [−5, 40]
No x-intercept: no real solution

Figure 35

Practice Problem 2 Solve $3x^2 - 0.1x - 12 = 0$ using a graphing calculator. Draw the graph and locate the x-intercepts using the CALC menu and selecting root.

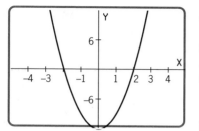

X: [-5, 5] and Y: [-12, 12]
The roots are approximately
-1.983402777 and 2.01673611

Figure 36

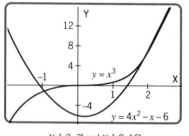

X: [-1, 5] and Y: [-1, 5]

Figure 37

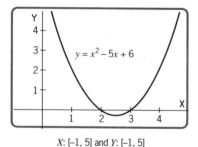

X: [-2, 3] and Y: [-8, 16]

(a)

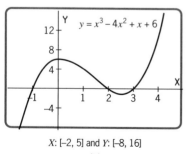

X: [-2, 5] and Y: [-8, 16]

(b)

Figure 38

ANSWER   Using the range $X: [-5, 5]$ and $Y: [-12, 12]$, we locate roots close to -2 and 2 (see Figure 36). You can change the viewing rectangle to get a better approximation of the roots. Try to locate the roots in this manner (see page A-3 of Appendix A). After this, you can have your graphing calculator state the intercepts. The solution set of approximate answers is $\{-1.983402777, 2.01673611\}$.

The material on quadratic equations can be easily extended to inequalities or higher degree equations by using a graphing calculator. For example, we can use a graphing calculator to solve inequalities and to solve cubics and higher-degree equations and inequalities.

EXAMPLE 25   Use a graphing calculator to solve the inequality $x^2 - 5x + 6 < 0$.

SOLUTION   Draw the graph and locate the x-intercepts. Then determine the intervals over which the graph is below the x-axis. We can see from Figure 37 that the solution interval is the interval $(2, 3)$. ■

Practice Problem 3   Solve the inequality $-2x^2 + 3x + 4 \leq 0$.

ANSWER $(-\infty, -0.85], [2.35, \infty)$

EXAMPLE 26   Solve the equation $x^3 = 4x^2 - x - 6$.

SOLUTION   If we graph the two equations $y_1 = x^3$ and $y_2 = 4x^2 - x - 6$, we can locate the points of intersection. However, as shown in Figure 38(a), the points of intersection are not clear and we must take precautions to find all the points of intersection. This problem can be approached in another way. We know that if $y_1 = y_2$, then $y_1 - y_2 = 0$; so we graph the equation $y = y_1 - y_2 = x^3 - 4x^2 + x + 6$ and then solve the equation for the zeros. From Figure 38(b) we see that the solutions are $x = -1$, $x = 2$, and $x = 3$. ■

Using algebra to find the point of intersection of the graphs of two quadratic functions, or of a linear function and a quadratic function, we can eliminate y and obtain a quadratic in x.

EXAMPLE 27   Find the intersection of $y = -x + 1$ and $y = x^2 - 4x + 3$.

SOLUTION   Set the two equations equal and simplify to a quadratic:

$$x^2 - 4x + 3 = -x + 1$$

$$x^2 - 3x + 2 = 0$$

$$(x - 2)(x - 1) = 0$$

Thus $x = 2$ and $x = 1$. Substituting into $y = -x + 1$, we get the points of intersection $(1, 0)$ and $(2, -1)$ (see Figure 39). ■

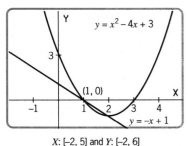

X: [–2, 5] and Y: [–2, 6]

Figure 39

Supply and Demand Curves

In economics, a supply curve represents the price p that is necessary for manufacturers to produce x number of an item and is written as $p = S(x)$. A demand curve represents the price necessary for consumers to purchase x items produced and is written as $p = D(x)$. Since all values of x and p are nonnegative in the first quadrant, the parts of parabolas that are in the first quadrant can often be used for supply and demand curves. The point at which these curves are equal is called the equilibrium point or point of equilibrium, that is, where supply equals demand.

EXAMPLE 28   If the supply function for a commodity is given by $p = S(x) = 2x^2 + 30x$ and the demand function by $p = D(x) = -15x + 5000$, where x is the number of units of the commodity and p is the price, find the equilibrium point.

SOLUTION   At the point of equilibrium, we have

$$D(x) = S(x)$$

Supply = Demand at equilibrium point

$$-15x + 5000 = 2x^2 + 30x$$

Substitute for S and D.

$$2x^2 + 45x - 5000 = 0$$

$$(x - 40)(2x + 125) = 0$$

Factor and solve.

Eliminate $x = -125/2$, since x is the number of units of a commodity and must be positive.

$$x = 40 \quad \text{and} \quad x = -\frac{125}{2}$$

$$D(x) = -15(40) + 5000 = 4400$$

Find the equilibrium point.

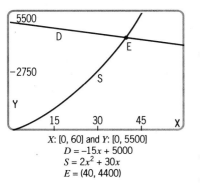

X: [0, 60] and Y: [0, 5500]
$D = -15x + 5000$
$S = 2x^2 + 30x$
$E = (40, 4400)$

Figure 40

The equilibrium quantity is 40 units and the equilibrium price is \$4400, since $S(40) = D(40) = 4400$. The graphs of both the supply function and the demand function are shown in Figure 40.

Break-Even Points and Profit Maximization

Some costs tend to decrease sharply after a certain level of production. A quadratic function is often used to represent the total cost of producing a product. For example,

$$C(x) = 400 + 30x - 0.4x^2$$

might represent the cost of producing x units of a product. A revenue function is found by the following product: Number of units sold · Price per unit. Often, revenue functions are quadratic functions. When both are sketched on a coordinate system, the intersections of the two graphs should represent the **break-even** points—the points where the functions are equal. Algebraically, by setting $C(x) = R(x)$, we obtain a quadratic equation that we solve either by factoring or by the quadratic formula. Furthermore, when $R(x) > C(x)$, a profit is made, but when $R(x) < C(x)$, there is a loss.

EXAMPLE 29   The cost of producing x small portable vacuum cleaners is given by

$$C(x) = 400 + 8x + 0.1x^2$$

when the price per unit is

$$p = 32 - 0.1x$$

Find the break-even point, draw the graphs of both cost and revenue, and show areas of profit and loss.

SOLUTION   Since Revenue = Units sold · Price per unit, the revenue equation is calculated as $R(x) = x(32 - 0.1x) = 32x - 0.1x^2$. To find the break-even point, set

$$C(x) = R(x)$$

$$400 + 8x + 0.1x^2 = 32x - 0.1x^2$$

$$0.2x^2 - 24x + 400 = 0$$

$$x^2 - 120x + 2000 = 0 \qquad \text{Divide by 0.2.}$$

$$(x - 20)(x - 100) = 0 \qquad \text{Factor.}$$

$$x = 20 \quad \text{and} \quad x = 100 \qquad \begin{array}{l}\text{Therefore, the break-}\\ \text{even points occur at}\\ x = 100 \text{ and } x = 20.\end{array}$$

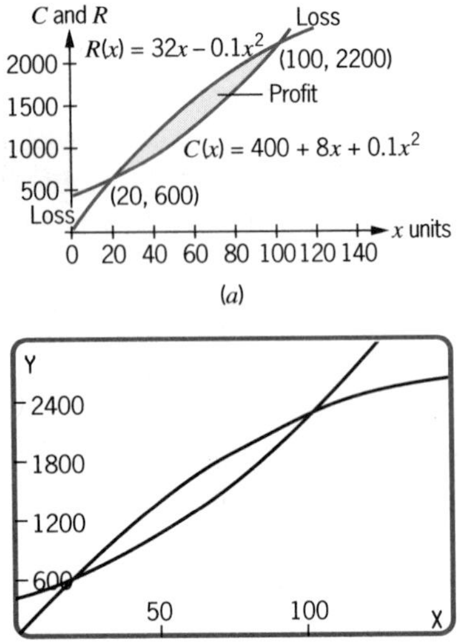

Thus, at $x = 100$ we have

$$C = R = 32(100) - 0.1(100)^2$$
$$= \$2200$$

and at $x = 20$ we have

$$C = R = 32(20) - 0.1(20)^2$$
$$= \$600$$

(a)

X: [0, 150] and Y: [0, 3000]

(b)

Figure 41

Figure 41(a) shows the areas of profit and loss, and Figure 41(b) shows how the graphs look on a graphing calculator. ■

SUMMARY

1. The solutions of $y = ax^2 + bx + c = 0$ $(a \neq 0)$ are

$$x = \frac{-b \pm \sqrt{b^2 - 4ac}}{2a}$$

2. Revenue = Number of units sold · Price per unit
3. A break-even point occurs when Revenue = Cost.
4. Profit occurs when revenue is greater than cost.
5. The equilibrium point occurs when supply and demand are equal.

Exercise Set 1.4

Find only real solutions for the following exercises. If possible, solve by factoring. Verify by drawing a graph on a calculator.

1. $x^2 - 16 = 0$
2. $3x^2 - 27 = 0$
3. $2x^2 - 14 = 0$
4. $3x^2 - 15 = 0$
5. $x^2 - 5x = 0$
6. $3x^2 - 7x = 0$
7. $3x^2 - 75 = 0$
8. $4x^2 - 36 = 0$
9. $x^2 - 2x = 3$
10. $x^2 - 2x = 15$
11. $2x^2 - 6x = 56$
12. $x^2 + 2x = 24$
13. $x^2 + x = 2$
14. $x^2 - 5x - 10 = 4$
15. $x^2 = 6 - x$
16. $x^2 - 2x = 3$

Use the quadratic formula to solve each equation.

17. $x^2 + 8x - 9 = 0$
18. $x^2 - 4x = 4$
19. $x^2 + 11x = -30$
20. $2x^2 = 5x + 2$

In Exercises 21–23, determine the point(s) of intersection of the graphs of the functions. Verify your answers using a calculator.

21. $f(x) = x^2 - 4x + 7$; $g(x) = x + 1$
22. $f(x) = -2x^2 + 5x + 3$; $g(x) = -3x^2 + x$
23. $f(x) = 6x^2 + 2x - 2$; $g(x) = 3x^2 + x$

Solve each of the following quadratic inequalities using a calculator to draw the graph.

24. $(x - 2)(x + 1) > 0$
25. $x^2 \geq 4$
26. $r^2 \leq 9$
27. $b^2 + 5b + 6 > 0$
28. $w^2 + 6 < 5w$
29. $6k^2 - k \leq 2$
30. $2y^2 \geq 5y + 3$

Find the zeros (x-intercepts) and discuss the sign of y between the zeros. Use a calculator.

31. $y = x(x + 2)(x - 2)$
32. $y = x^2(x - 4)$
33. $y = x(x + 1)^2$
34. $y = x(x + 1)(x^2 - 4)$
35. $y = x^2(x^2 - 9)$
36. $y = 3x^4 + x^3 - 2x^2$
37. $y = x^4 + x^3 - 2x^2$

Applications (Business and Economics)

38. **Maximum Profit.** Find the maximum profit for the following profit functions.
    (a) $P(x) = 4 + 6x - 3x^2$
    (b) $P(x) = 80x - 8x^2$

39. **Maximum Revenue.** Find the maximum revenue for the revenue function $R(x) = 386x - x^2$, where x is the number of units sold.

40. **Equilibrium Point.** Sketch the graph of the demand function $p = D(x) = (x - 6)^2$ for $x \geq 6$ and the supply function $p = S(x) = x^2 + x + 10$ on the same coordinate system and locate the equilibrium point. What is the price at this point?

41. **Equilibrium Point.** The demand function for a commodity is $p = D(x) = (x - 4)^2$ and the supply function is $p = S(x) = 2x$, where $x \geq 4$. Sketch the graphs of these two functions on a coordinate system and locate the equilibrium point. What is the price at this point?

42. **Profit and Loss Regions.** Suppose that a company has a fixed cost of $150 per day and a variable cost of $x^2 + x$. Further suppose that the revenue function is $R(x) = xp$ and the price per unit is given by $p = 57 - x$. Sketch the cost and revenue functions and locate the regions of profit and loss.

43. **Maximum Profit.** In Exercise 42, where does the maximum profit occur? What is the maximum profit? (**Hint**: $P = R - C$.)

Applications (Social and Life Sciences)

44. **Blood Velocity.** In the formula

$$V_r = V_m \left(1 - \frac{r^2}{R^2} \right)$$

V_r is the velocity of a blood corpuscle r units from the center of the artery, R is the constant radius of the artery, and V_m is the constant maximum velocity of the corpuscle. If $V_r = \frac{1}{2}V_m$, solve for r in terms of R and then draw the graph.

45. **Pollution.** An environmental study is used to approximate the annual growth of carbon monoxide in the air (parts per million) as a quadratic function. The approximating function is $M(t) = 0.01t^2 + 0.05t + 1.6$, where M is parts per million and t is measured in minutes starting at 7:00 A.M. ($0 \le t \le 180$). Show this relationship with a graph.

46. **Population Growth.** A small city is expecting a decline in growth for a few years and then a rapid growth. This trend has been approximated by $P(t) = 2t^2 - 100t + 10{,}000$ ($t \ge 0$), where $P(t)$ is the population and t is time in years from now. Show this growth trend with a graph.

47. **Laffer Curve.** At the endpoints of the Laffer curve shown in the figure, there is no revenue (R) for given tax rates (r). Suppose that a Laffer curve is approximated by

$$R = \frac{r(50r - 5000)}{r - 110} \quad \text{for } 0 \le r \le 100$$

What tax rates provide no revenue? What rate provides maximum revenue?

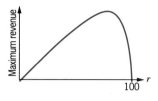

Review Exercises

48. Which of the following specify functions?
    (a) $y = 2x + 3$ (b) $3x^2 + 4y^2 = 25$
    (c)   (d)

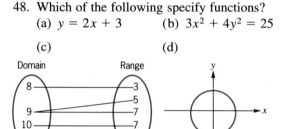

49. Find an equation of the line passing through the points $(-2, 1)$ and $(7, -4)$.

50. Determine the vertex and sketch the graph for $y = -2(x + 1)^2 - 3$.

1.5 EXPONENTIAL FUNCTIONS

OVERVIEW   We have been studying functions such as $f(x) = x^2$, where the base is the variable and the exponent is a constant; we now study functions such as $f(x) = 2^x$, where the base is constant and the exponent is a variable. These are called exponential functions. Although this type of function may be unfamiliar to you, such functions are used by economists to study the growth of money supply, by businesspeople to study declines in sales, by biologists to study the growth or decay of organisms in a laboratory culture, by social scientists to study population growth, and by archeologists to study the carbon dating of fossils. In this section we consider

- Exponential functions with any base $a > 0$ $(a \neq 1)$
- Exponential functions with base e
- Graphs of exponential functions
- Applications involving exponential functions

We know how to evaluate the expression 2^x when x assumes integral values. For example,

$$2^0 = 1 \qquad 2^1 = 2 \qquad 2^2 = 4$$

$$2^{-1} = \frac{1}{2} \qquad 2^{-2} = \frac{1}{2^2} = \frac{1}{4} \qquad 2^{-3} = \frac{1}{2^3} = \frac{1}{8}$$

However, in this section we allow x to be any real number. We call the function $f(x) = 2^x$ an exponential function with base 2.

DEFINITION: EXPONENTIAL FUNCTIONS

Let a be any positive number, where $a \neq 1$. The function

$$f(x) = a^x$$

is called the **exponential function f with base a**. The domain of the independent variable x is any real number.

There are two important things to note here: the domain of f is the set of real numbers and a must be positive, $a > 0$. If a were negative, then x could not be any real number. The function would not have meaning when $x = \frac{1}{2}$ and $a < 0$. For example, if $a = -6$, then $(-6)^{1/2} = \sqrt{-6}$ has no meaning over the set of real numbers. The base a cannot be 1 because $f(x) = 1^x = 1$ is a constant function and not an exponential function.

Exponential functions can be evaluated easily on a calculator. We have already evaluated expressions such as $16^{3/4} = 8$ and $16^2 = 256$. Now we look at irrational exponents, such as $a^{\sqrt{3}}$ or $4^{\pi + 1}$. We will not go into a technical definition for terms like $a^{\sqrt{3}}$, but we will state that we can get as close as we like with closer and closer approximations for $\sqrt{3}$.

Calculator Note

Use $\boxed{\wedge}$ when evaluating values of an exponential function. However, be sure to use some care with irrational and negative exponents. By using parentheses around the exponent, you should be able to get the correct answers consistently.

EXAMPLE 30   Calculate each of the following. Round to three decimal places.

(a) $3^{\sqrt{2}}$ (b) 5^{π} (c) $7 \cdot 8^{2/5}$ (d) $4^{-3/7}$

SOLUTION

(a) $3^{\sqrt{2}} \approx 4.729$ (b) $5^{\pi} \approx 156.993$

(c) $7 \cdot 8^{2/5} \approx 16.082$ (d) $4^{-3/7} \approx 0.552$ ■

In the last section, we looked at the graphs of quadratic functions. Now let's turn to graphs of exponential functions.

EXAMPLE 31   On a calculator draw the graphs of $y = 2^x$ and $y = 3^x$.

SOLUTION   The graphs are shown in Figure 42. Table 3 shows some values for both functions. Notice that as x increases, the functions also increase. Use the Trace key to move the cursor along the graph of each function and locate the values shown in Table 3. Note that you can create this table on a calculator. We can see that the y-intercept for both functions is 1, since $a^0 = 1$ and $a \neq 0$.   ■

TABLE 3

x	-2	-1	0	1	2
$y = 2^x$	$\frac{1}{4}$	$\frac{1}{2}$	1	2	4
$y = 3^x$	$\frac{1}{9}$	$\frac{1}{3}$	1	3	9

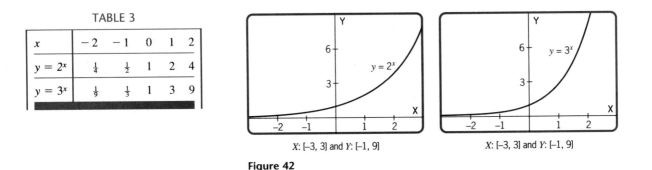

X: [−3, 3] and Y: [−1, 9]   X: [−3, 3] and Y: [−1, 9]

Figure 42

EXAMPLE 32   Using a calculator, draw the graphs of $y = 2^{-x}$ and $y = 3^{-x}$.

SOLUTION   The graphs are shown in Figure 43, and Table 4 shows some values for both functions. Notice that as the value of x increases, the value of the function decreases. Also, note that the y-intercept is 1, as was the case in Example 31. Use the Trace key to move the cursor along the graph of each function and locate the values shown in Table 4. Create this table with a calculator.   ■

TABLE 4

x	-2	-1	0	1	2
$y = 2^{-x}$	4	2	1	$\frac{1}{2}$	$\frac{1}{4}$
$y = 3^{-x}$	9	3	1	$\frac{1}{3}$	$\frac{1}{9}$

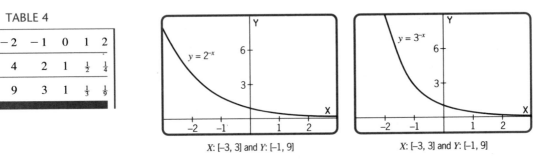

X: [−3, 3] and Y: [−1, 9]   X: [−3, 3] and Y: [−1, 9]

Figure 43

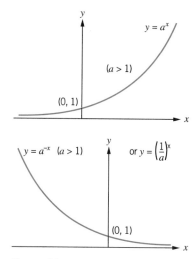

Figure 44

Practice Problem 1 Using a calculator, draw the graphs of several functions of the form $y = a^x$: for example, $y = 4^x$, $y = 4^{-x}$, $y = (\frac{1}{4})^x$, and $y = (\frac{1}{4})^{-x}$. Remember that $a > 0$. What do you observe about these graphs? Try some other numbers as bases.

ANSWER Note that when $0 < a < 1$, the function decreases as x increases. When $a > 1$, the function increases as x increases. What happens when $a = 1$, $a = 0$, $a < 1$? Consider these cases.

The graphs in Figures 42 and 43 are typical exponential graphs. Now let's look at the properties of exponential functions (see Figure 44).

PROPERTIES OF EXPONENTIAL FUNCTIONS

Each of the following is true for the exponential function

$$y = f(x) = a^x \qquad (a > 0, a \neq 1, x \in \mathcal{R})$$

(a) The domain of the function is all real numbers.
(b) The graph lies above the x-axis for all x; thus, the range of the function is $(0, \infty)$.
(c) If $x = 0$, then $a^x = 1$ or $f(0) = 1$, so the y-intercept is at $(0, 1)$.
(d) There are no x-intercepts; the graph does not cross the x-axis.
(e) If $a > 1$, the function increases as x increases.
(f) If $0 < a < 1$, the function decreases as x increases.
(g) By the rules of exponents we can write $f(x) = a^{-x} = (1/a)^x$.

EXAMPLE 33 Draw the graphs of $y = 2^x$ and $y = 2^{-x}$ on the same coordinate axes.

SOLUTION The calculator graphs of these functions are shown in Figure 45. From the rules of exponents, we know that $2^{-x} = (\frac{1}{2})^x$. We can also see in Figure 45 that the graph of $y = 2^{-x}$ is the reflection of the graph of $y = 2^x$ across the y-axis. (The y-axis is the axis of symmetry for the two graphs.) ∎

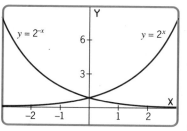

X: [−3, 3] and Y: [−1.6, 9]

Figure 45

Obviously, the graphs of exponential functions can be moved around just as those of quadratic functions. What functions produce such changes? For example, how do the graphs of $f(x) = 2^{x+1}$ and $f(x) = 2^x + 1$ compare to the graph of $f(x) = 2^x$? Let's look at this in the following example.

EXAMPLE 34 Draw the calculator graphs of $f(x) = 2^x$, $f(x) = 2^{x+1}$, and $f(x) = 2^x + 1$. What conclusions can you draw from these graphs?

SOLUTION In Figure 46(a) we can see that the graph of $f(x) = 2^{x+1}$ is a horizontal shift to the left one unit, just as we saw in the last section that $f(x) =$

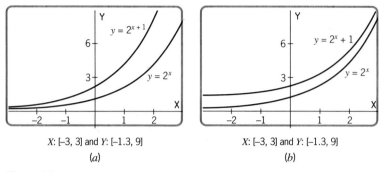

X: [-3, 3] and Y: [-1.3, 9]   X: [-3, 3] and Y: [-1.3, 9]
(a)   (b)

Figure 46

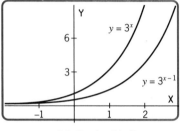

X: [-2, 3] and Y: [-1, 9]
(a)

$(x + 1)^2$ shifts the graph of $f(x) = x^2$ one unit to the left. Likewise, in Figure 46(b) we can see that $f(x) = 2^x + 1$ is a vertical shift upward 1 unit.

Practice Problem 2 Compare the graphs of $f(x) = 3^{x-1}, f(x) = 3^x - 2$, and $f(x) = -3^x$ with the graph of $f(x) = 3^x$. Check by drawing the calculator graphs.

ANSWER

$f(x) = 3^{x-1}$ is a horizontal shift 1 unit to the right in Figure 47(a).
$f(x) = 3^x - 2$ is a vertical shift 2 units down in Figure 47(b).
$f(x) = -3^x$ is a reflection across the x-axis in Figure 47(c).

Doubling Concept

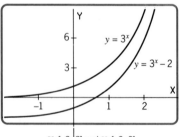

X: [-2, 3] and Y: [-3, 9]
(b)

Certain bacteria split at regular intervals. That is, where there is one, after a given amount of time there are two [see Table 5(a)].

TABLE 5

(a)

Time	0	1	2	3	4 . . .
Bacteria	1	2	4	8	16 . . .

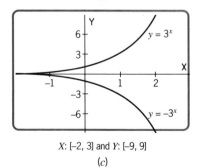

X: [-2, 3] and Y: [-9, 9]
(c)

Figure 47

(b) After $t = t_0$

Time	0	d	$2d$	$3d$
Bacteria	$N(t_0)$	$2N(t_0)$	$4N(t_0)$	$8N(t_0)$

In general, after a given time $t = d$, the number of bacteria doubles. Then, if $N(t_0)$ is the number at time t_0, the number at time t is equal to

$N(t) = N(t_0)2^{t/d}$ t is the time after t_0; d is the time after t_0 for the population to double.

Note that when $t = d$, $N(t) = 2N(t_0)$; that is, the number has doubled [see Table 5(b)]. This concept applies to the growth of populations involving bacteria, animals, insects, or people. This same concept can be applied to the increase in value of a purchase.

EXAMPLE 35   An abstract painting was purchased in 1930 for $600. Its value has doubled every 10 years. What was its value in 1990?

SOLUTION

$$V(t) = V(t_0) \cdot 2^{t/d}$$
$$= 600 \cdot 2^{t/10} \qquad \text{$V(t_0) = 600$ and $d = 10$ years}$$

$$V(60) = 600 \cdot 2^{60/10} \qquad \text{From 1930 to 1990 is 60 years.}$$
$$= \$38{,}400$$

The value of the painting in 1990 is $38,400.   ■

A good example of exponential growth is compound interest. The formula for compound interest is

$$A = P(1 + i)^n$$

where P is the principal, i is the interest rate per compounding period, and n is the number of compounding periods.

EXAMPLE 36   If $1000 is invested at 7% interest compounded annually for 3 years, how much will be in the account?

SOLUTION   The amount will be

$$A = 1000(1 + 0.07)^3 = \$1225.04$$ ■

The Natural Exponent

A logical question to ask is, What happens to the amount A if the interest is compounded more and more frequently? Later in the book we will show that as n increases without bound, the value of $[1 + (1/n)]^n$ approaches an irrational number that is denoted by e. The function $f(x) = e^x$ is called the **natural exponential function**. This number occurs naturally in many areas of study. We use the approximation

$$e \approx 2.71828$$

We often see the natural exponential function in the form

$$f(x) = ce^{bx}$$

This function is one of the most useful exponential functions in mathematics, economics, and applications in the fields of life science and social science.

Calculator Note

Graphing calculators have an e^x key. It is most likely a 2nd function. Look above the $\boxed{\text{ln}}$ key and see if e^x is written there. Try to evaluate e^2 to see if you get ≈ 7.38906.

Practice Problem 3 Use a calculator to evaluate e^{-3}, $e^{2/3}$, and $e^{-5/6}$. Round to five decimal places.

ANSWER
$$e^{-3} \approx 0.049787$$
$$e^{2/3} \approx 1.94773$$
$$e^{-5/6} \approx 0.43460$$

The graph of $f(x) = e^x$ is shown in Figure 48(a). In Figure 48(b), the graph of $f(x) = e^x$ is shown with the graphs of $f(x) = 2^x$ and $f(x) = 3^x$.

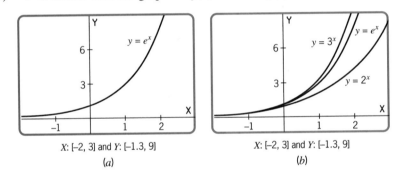

X: [–2, 3] and Y: [–1.3, 9]　　　　X: [–2, 3] and Y: [–1.3, 9]

(a)　　　　　　　　　　(b)

Figure 48

Practice Problem 4 With a calculator, draw the graphs of several functions of the form $y = ce^{bx}$. Use such functions as $y = -2e^{3x}$ or $y = e^{x-1}$. What can you determine from the graphs?

ANSWER Regardless of the functions that are graphed, notice that the graph of $y = ce^{bx}$ behaves the same as the graphs of the exponential functions that we graphed at the beginning of this section.

Practice Problem 5 Use a calculator to locate the value of x for $y = 1$ on the graph of $y = e^{x^2 - 2x - 3}$ over the interval $[-1.5, .5]$.

ANSWER The graph is shown in Figure 49.

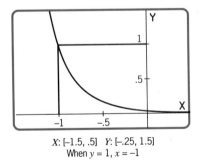

X: [–1.5, .5]　Y: [–.25, 1.5]
When $y = 1$, $x = -1$

Figure 49

The natural exponential function for money compounded continuously is as follows.

DEFINITION: COMPOUNDED CONTINUOUSLY

The compound amount of an investment at the end of t years is given by

$$A = Pe^{rt}$$

where r is the annual rate (expressed as a decimal) **compounded continuously**, and P is the principal or money invested.

EXAMPLE 37   A person wishes to place $10,000 in a savings account for $2\frac{1}{2}$ years. A savings and loan association advertises that the interest rate is 7% compounded continuously. To what amount will the $10,000 accumulate?

SOLUTION   Since $A = Pe^{rt}$, we have that $P = 10,000$, $r = 0.07$, and $t = 2\frac{1}{2} = 2.5$. The $10,000 will accumulate according to the equation

$$A = 10,000e^{0.07 \cdot 2.5} = \$11,912.46$$

Note that the answer is rounded to the nearest cent. ■

Many problems that describe the growth of bacteria or the growth of a city's population, the decrease of sales activity, or the decay of radioactive materials can be formulated to use e as the base of the exponential function. We illustrate this concept with an example involving decay.

EXAMPLE 38   Suppose that the amount in grams of a radioactive material present at a time t is given by

$$A(t) = 100e^{-0.002t}$$

where t is measured in days and 100 grams is the original amount. Find the amount of radioactive material at the end of 80 days.

SOLUTION   The amount of radioactive material remaining after 80 days is

$$A(80) = 100e^{-0.002 \cdot 80} = 100e^{-0.16} = 85.2144 \text{ grams}$$ ■

Practice Problem 6   If you deposit $10,000 today in a bank that pays interest at a rate of 8% compounded continuously, how much would you have at the end of 6 years?

ANSWER $16,160.74

Equations with Exponential Functions

Equations often contain exponential functions. Solving some exponential equations, such as $2^x = 5^{2-x}$, requires the use of topics that we will cover later;

however, if the bases are the same, the equations can be solved using only the properties that we have covered in this chapter. For a^x and a^y, where $a > 0$ and $a \neq 1$, if $a^x = a^y$, then $x = y$ (since a^x and a^y are one-to-one functions). Likewise, if $x = y$, then $a^x = a^y$.

EXAMPLE 39 Solve $\dfrac{1}{27} = 9^x$.

SOLUTION

$$\frac{1}{27} = 9^x$$

$$\frac{1}{3^3} = (3^2)^x \qquad \text{27 and 9 are powers of 3.}$$

$$3^{-3} = 3^{2x} \qquad \text{Write as } a^x = a^y.$$

$$-3 = 2x \qquad \text{Exponents are equal.}$$

$$x = -\frac{3}{2}$$

■

SUMMARY

Exponential functions have wide applications in business to describe growth of money as well as decrease in sales. In science, exponential functions are used to describe the growth of bacteria and populations and the decay of radioactive materials.

1. An exponential function is given as $f(x) = a^x$, where $a > 0$ and $a \neq 1$.
2. The amount of an investment at the end of t years and rate r, if interest is compounded continuously, is given by $A = Pe^{rt}$.

Exercise Set 1.5

Evaluate each of the following using a calculator.

1. 3^{-6}
2. $5^{3/4}$
3. $1000(2.01)^{-3}$
4. $20^{\sqrt{3}}$
5. $e^{-1/3}$
6. $5e^{2/5}$
7. $-e^{1.10}$
8. $6^{-\pi+1}$
9. $7^{(1/2)+\sqrt{2}}$

Evaluate each function at $t = 0$, $t = 0.5$, and $t = -2$.

10. $y = 2^{-t}$
11. $y = \left(\dfrac{1}{2}\right)^t$
12. $y = 2^t - 2$
13. $y = 10^t - 1$

14. $y = 4(3)^{-2t}$
15. $y = 3(2)^{-t}$
16. $y = 3e^{-2t}$
17. $y = e^{-t} + e^t$

Draw the graphs of each exponential function.

18. $y = 4^x$
19. $y = 3^{-x}$
20. $y = 4e^x$
21. $y = 2^x + 1$
22. $y = 2^{x+1} - 2$
23. $y = -e^x$
24. $y = 3^{x-2} + 1$
25. $y = -2^{-x}$
26. $y = 4^{x+1} - 2$

Graph each exponential function. What is the range of each function?

27. $y = 4e^{-0.05x}$
28. $y = 3.2^{2x}$
29. $y = 5 - 3e^{2x}$
30. $y = 1 - e^{-x}$

31. $y = 6e^{0.02x}$ 32. $y = 3e^{-0.13x}$

33. Use a calculator to graph $y = 2^x$ and $y = 4^x$ on the same viewing screen. From this determine where $2^x < 4^x$.

34. Use a calculator to graph $y = 2^{-x}$ and $y = 4^{-x}$ on the same viewing screen. From this determine where $2^{-x} < 4^{-x}$.

Graph each of the following functions. Determine maximum or minimum values that may occur.

35. $f(x) = x^2 e^{-x}$ 36. $f(x) = x3^{2+x}$

37. $f(x) = e^{x^2-1}$

Solve each equation.

38. $5^{x^2} = 25^4$ 39. $3^{2x+1} = 27^{-x}$

40. $e^{x^2+1} = e^{2x}$ 41. $16^{2x} = 64^{x-3}$

Applications (Business and Economics)

42. **Depreciation.** A machine is depreciated by a given amount each year. The value of the machine (usually designated by S) at the end of n years, when the machine depreciates r percent per year and C is the original cost is

$$S = C\left(1 - \frac{r}{100}\right)^n$$

Find the value at the end of 8 years of a machine costing $100,000 at an annual rate of depreciation of 12%.

43. **Sales Functions.** The sales for a product are represented by the following exponential function, where t represents the number of years the product has been on the market:

$$S(t) = 600 - 200e^{-t}$$

Graphically represent sales as a function of time t. From the graph describe how sales change with time.

44. **Money Doubling.** An antique automobile was purchased in 1970 for $4000. Its value has doubled every 8 years. What is its value in 1990?

45. **Demand Function.** The demand function for a certain product is given as

$$p = D(x) = 4500\left(1 - \frac{3}{3 + e^{-0.003x}}\right)$$

(a) What is the price p for a demand of $x = 100$ and $x = 400$?

(b) Draw the graph using a graphing calculator. What are the domain and range of the function?

46. **Inflation.** If the annual rate of inflation averages 3% over the next 5 years, then the approximate cost C of goods or services during any year in that interval is given by

$$C(t) = P(1.03)^t$$

where t is the time in years and P is the present cost. If the price of a set of tires is presently $269, estimate the price in 5 years.

47. **Depreciation.** After t years, the value of a car that originally cost $28,000 is given by

$$V(t) = 28,000(0.75)^t$$

Use a graph of the function to estimate the value of the car (a) 2 years from the date of purchase and (b) 5 years from the date of purchase. When will the car be worth less than half of its original cost?

Applications (Social and Life Sciences)

48. **Bacteria Culture.** A certain bacteria splits every 10 hours. If a culture starts with 10,000 bacteria, what is the size of the culture after 24 hours?

49. **Bacteria Culture.** A bacteria culture starts with 10,000 bacteria. After 3 hours the estimated count is 160,000. Find the period necessary for doubling the number of bacteria by estimating with a graph on your calculator. (**Hint**: Use X: [0, 1], Y: [0, 170,000].)

50. **Learning Curves.** The graph of an equation in the form

$$y = k - ke^{-ct}$$

is called a learning curve. Learning increases rapidly with time and then levels out and tends to an upper limit. Suppose that $c = 0.1098$ and $y = 200$ when $t = 10$ minutes.

(a) Find k.

(b) Find y when $t = 5$ minutes.

(c) Find y when $t = 20$ minutes.

(d) Draw the graph of the function. At approximately what point does learning level out?

51. **Learning Curves.** A special learning curve is found to be represented by

$$R = 100(1 - e^{-ct})$$

where R is the number of responses and t is the time involved. Suppose that $c = 0.0695$.
(a) What is R when $t = 0$? Explain.
(b) Find R when $t = 1$.
(c) Draw the graph of the function. Approximately where does the curve level out?

52. **Bacteria Culture.** A certain type of bacteria increases according to the function

$$P(t) = 100e^{0.3124t}$$

where t is the time in hours. Find (a) $P(0)$, (b) $P(10)$, and (c) $P(24)$.

53. **Population Growth.** In 1993, the population of a small town is found to be growing according to the function

$$P(t) = 3200e^{0.0157t}$$

where t is the time in years and $t = 0$ corresponds to the year 1993. Use the function to approximate the population in (a) 1975 and (b) 2000.

Review Exercises

54. Solve each quadratic equation.
    (a) $9x^2 - 5 = 0$ (b) $7x^2 - 14 = 0$
    (c) $(y - 4)^2 = 9$ (d) $(k - 2)^2 = 16$

55. Solve each equation by factoring.
    (a) $x^2 - 3x - 4 = 0$ (b) $2x^2 - 5x + 3 = 6$

56. Solve each equation.
    (a) $3x^2 + x = 2$ (b) $4x^2 - 5x - 7 = 0$

Sketch the graph of each equation using a graphing calculator. Locate the vertex and all intercepts.

57. $y = -3x^2 + 4x$ 58. $y = x^2 - 7x + 10$
59. $y = 2x^2 - 4x + 3$

Chapter Review

Review the properties of slopes of lines, the solution of quadratic equations by factoring, the properties of exponential functions, and the quadratic formula

$$x = \frac{-b \pm \sqrt{b^2 - 4ac}}{2a}$$

Functions	Quadratic functions and graphs
Rule	Line of symmetry
Ordered pair	Upward or downward
Table	Point of maximum or minimum
Notation	Intercepts

Chapter Test

Find an equation of the line with the given conditions.

1. The line passes through the points $(2, -4)$ and $(7, 1)$.
2. The line passes through the point $(3, 1)$ and is perpendicular to the line having the equation $y = 2x + 3$.

Solve each equation algebraically and then verify your answers using a calculator.

3. $2x^2 - 14 = 0$ 4. $(x + 3)^2 - 5 = 0$ 5. $6x^2 + x = 1$

Solve each equation using a graphing calculator.

6. $1.26 = 2^x$ 7. $3x^2 - 11x = 10$ 8. $x^3 - 2x + 1 = 0$

9. If $f(x) = x^2 + 2x$, find
    (a) $f(2) - f(3)$
    (b) $\dfrac{f(x + h) - f(x)}{h}$

Find the domain of each function using a calculator if necessary.

10. $y = \dfrac{2x}{x^2 - 4x - 5}$ 11. $y = \sqrt{25 - x^2}$

12. $y = x^3 - 2x + 3$ 13. $y = 3^{x + 1}$

14. For $y = x^2 - 2x - 3$:
    (a) Find the axis of symmetry.
    (b) Does the curve turn upward or downward?
    (c) Find the point of maximum or minimum value.
    (d) Find the x- and y-intercepts.
    (e) Put the function in standard form for graphing.
    (f) Draw the graph by hand and then verify using a calculator.

15. Draw the graph of $y = 3^x + 1$ by hand and then verify using a calculator.

Determine whether each relation is a function.

16. $\{(2, 1), (3, 1), (4, 2), (5, 3)\}$ 17. $x^2 + y^2 = 36$

18. The profit function for a firm making widgets is $P(x) = 88x - x^2 - 1200$. Find the number of units at which maximum profit is achieved. Draw the graph using a calculator to verify your answer.

19. The demand equation for a certain product is given by $p = D(x) = 600 - 0.6e^{0.003x}$. Find the price p for $x = 100$.

CHAPTER

2

Differential Calculus

What is calculus? Simply put, calculus is the branch of mathematics that is used to study change. Since we live in a changing world, the study of calculus is very important. There are changes in inventory, wages, population, temperature, and so on. Quite often, we are not just interested in the fact that something is changing but also with how fast it is changing. For example, if your wages increase by $100, it is important to know whether that increase is per year, per month, or per week.

It may be surprising to learn that only a knowledge of algebra is needed for this study. One of our major objectives is to present calculus concepts with a minimum of prerequisite study. Isaac Newton (1642–1727) and Gottfried Leibniz (1646–1716) are given credit as being the creators of calculus. Two general problems characterize the study of elementary calculus: the tangent line problem and the area under a curve problem. How do we find the slope of the tangent line to the graph of a function $f(x)$ at the point P in Figure 1(a)? We consider this question in the study of differential calculus, which involves a multitude of ideas, all having to do with the rate of change of a function. How do we find the area of the region R bounded by the graph of $f(x)$ and the x-axis for $a \leq x \leq b$ in Figure 1(b)? We answer this question in the study of integral calculus with related topics in later chapters.

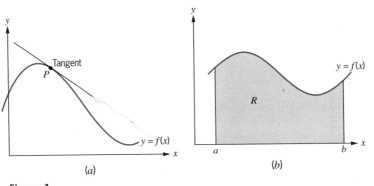

(a)　　　　(b)

Figure 1

2.1 LIMITS

OVERVIEW　In this section we introduce a concept that opens the way for us to consider each of the two major goals of calculus. Whether we begin our study with differential calculus or with integral calculus, the notion of a limit must be considered. That is, we must investigate what happens to the value of a function as the variable gets closer and closer to some number. This concept helps distinguish between average rates of change and instantaneous rates of change. For example, in traveling from New York to Chicago you might compute your average speed. However, a glance at your speedometer gives an instantaneous speed. This instantaneous speed can be obtained as a **derivative**, which is defined in terms of a limit. We will find that derivatives (giving us instantaneous changes) are useful in economics and business management as well as in the social and life sciences. In this section we

- Introduce an intuitive concept of a limit
- Consider limits from the right and from the left
- Discuss the properties of limits

We introduce the concept of a limit by considering the graph of $f(x) = x + 1$ as x gets closer and closer to 3, but not equal to 3 (Figure 2). Recall that to determine $f(x)$ at a value x, proceed vertically from the x value on the x-axis to the graph of f,

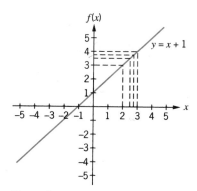

Figure 2

and then proceed horizontally to the corresponding value, $f(x)$, on the y-axis, as indicated by the dashed line for $x = 2$. In Figure 2, $f(2) = 3$.

What happens to $f(x)$ when x is chosen closer and closer to 3 for values of x just smaller than 3? If you let $x = 2.5$, then $f(x) = 3.5$; if $x = 2.7$, then $f(x) = 3.7$; and if $x = 2.9$, then $f(x) = 3.9$ (see Figure 2). In fact, if $x = 2.9999$, then $f(x) = 3.9999$. We see that $f(x)$ approaches 4 as x approaches 3 through values of x less than 3. In a similar manner we can demonstrate that $f(x)$ approaches 4 as x approaches 3 through values of x greater than 3.

EXAMPLE 1   Consider the function $f(x) = 3x + 4$ and a sequence of x values 1, 1.9, 1.99, 1.999, . . . , which are approaching the x-coordinate 2 from the left. That is, the x values are all less than 2 but are getting closer and closer to 2, as shown in the table. (The symbol $\rightarrow$ is read "approaches.")

x	1.000	1.900	1.990	1.999	$\cdots$	$\rightarrow$	2
$f(x)$	7.000	9.700	9.970	9.997	$\cdots$	$\rightarrow$	10

The sequence of corresponding $f(x)$ values 7, 9.7, 9.97, 9.997, . . . seems to be approaching (getting closer and closer to) 10. We say that the limit of $f(x)$ as x approaches 2 from the left (from the "less than 2" side) is 10, and we write this symbolically as

$$\lim_{x \to 2^-} f(x) = \lim_{x \to 2^-} (3x + 4) = 10$$

With your calculator, draw the graph of $y = f(x) = 3x + 4$, for $0 \le x \le 4$ (see Figure 3). For any value of x just less than 2 (no matter how close to 2), y is just less than 10. That is, as $x \to 2^-$, $y \to 10$.   ■

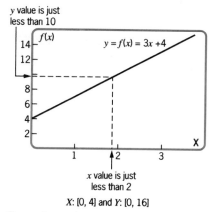

X: [0, 4] and Y: [0, 16]

Figure 3

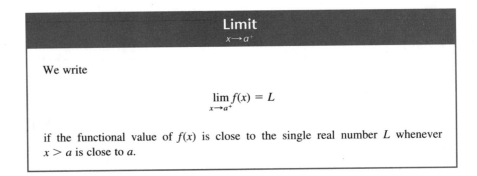

Limit
$x \to a^-$

We write

$$\lim_{x \to a^-} f(x) = L$$

if the functional value of $f(x)$ is close to the single real number L whenever $x < a$ is close to a.

EXAMPLE 2   Use the $\boxed{\text{TRACE}}$ feature of your calculator to discover what happens as you let x approach 2 from the left for the function $f(x) = 3x + 4$. Use the Zoom feature for x to approach the 2 rapidly. Now let x approach 2 from the right and see what happens to the y coordinates of $f(x)$.

SOLUTION   Note that numbers will vary with different calculators and with different original ranges.

x	1.9574468	1.9972838	1.9995377	1.9999947	→	2
$f(x)$	9.8723404	9.9812695	9.9997986	9.9999283	→	10
x	2 ←	2.0003945	2.004045	2.0059054	2.0425532	
$f(x)$	10 ←	10.001222	10.002215	10.022893	10.127666	

As x approaches 2 from the right, the sequence of corresponding $f(x)$ values is approaching 10. We say that the limit of $f(x)$ as x approaches 2 from the right (from the ''greater side of 2'') is 10, and we write this as

$$\lim_{x \to 2^+} f(x) = \lim_{x \to 2^+} (3x + 4) = 10$$

Limit
$x \to a^+$

We write

$$\lim_{x \to a^+} f(x) = L$$

if the functional value of $f(x)$ is close to the single real number L whenever $x > a$ is close to a.

In the simple example above, the corresponding y-coordinates of f, as $x \to 2$, approach 10, which is the same value you would get if you merely evaluated $f(x)$

at $x = 2$, namely $f(2) = 3(2) + 4 = 10$. But it is not always this simple. Let's consider a case where the function does not exist at the x value that is being approached.

EXAMPLE 3   Discover what value the function $g(x) = (x^2 - 4)/(x - 2)$ approaches as x approaches 2 (a) from the left, and (b) from the right.

SOLUTION

(a)   From the table we can see that

$$\lim_{x \to 2^-} g(x) = \lim_{x \to 2^-} \frac{x^2 - 4}{x - 2} = 4$$

x	1.9	1.9900	1.9990	1.9999	1.99999 $\cdots \to 2$
$g(x)$	3.9	3.9900	3.9990	3.9999	3.99999 $\cdots \to 4$

(b)   From the following table we can see that

$$\lim_{x \to 2^+} g(x) = \lim_{x \to 2^+} \frac{x^2 - 4}{x - 2} = 4$$

x	$2 \leftarrow \cdots$ 2.0001	2.0010	2.010	2.1
$g(x)$	$4 \leftarrow \cdots$ 4.0001	4.0010	4.010	4.1

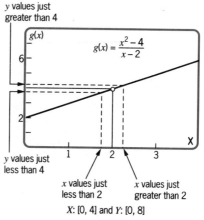

Figure 4

y values just greater than 4

$g(x)$

$$g(x) = \frac{x^2 - 4}{x - 2}$$

y values just less than 4

x values just less than 2

x values just greater than 2

X: [0, 4] and Y: [0, 8]

Calculator Note

With your calculator, draw the graph of $g(x) = (x^2 - 4)/(x - 2)$ for $0 \leq x \leq 4$ and $0 \leq y \leq 8$ in Figure 4. Use the Trace key to locate the value of y at $x = 2$. On your calculator, the value of y is left blank. This means that there is no value of y for $x = 2$. In fact, there is a hole in the graph at $x = 2$ because $(2^2 - 4)/(2 - 2)$ is not a number [the denominator is zero and $g(2)$ is undefined]. So the point (2, 4) is not a point on the graph, indicated by an empty circle. However, this does not change the fact that 4 is the value that the y-coordinate is approaching. As x approaches 2 from the left and right, the limit from the left and the limit from the right are both 4, as we see in Figure 4.

NOTE: In Figure 4, as the values of x approach 2, the values of $y = (x^2 - 4)/(x - 2)$ approach 4. The word *approach* means that we can make all y values be within a certain distance of 4 (and this distance can be made as small as we wish) by keeping the x values within a certain distance of 2.

Practice Problem 1   On a calculator, draw the graph of $y = (x^2 - 9)/(x + 3)$ and use the Zoom feature to discover the values of

$$\lim_{x \to -3^-} \frac{x^2 - 9}{x + 3} \quad \text{and} \quad \lim_{x \to -3^+} \frac{x^2 - 9}{x + 3}$$

ANSWER -6 and -6

Calculator Note

We will form tables many times to explore the possibility of a limit. Practice using the table capabilities of your calculator by forming the table in Example 3. See page 7 of Appendix A.

Limits do not always equal the same value from the left and from the right. When the limits from the left and from the right are equal, we define a limit as follows.

DEFINITION: LIMIT

When both the left limit, $\lim\limits_{x \to a^-} f(x)$, and the right limit, $\lim\limits_{x \to a^+} f(x)$, exist and are equal to some real number L, we say the limit of $f(x)$ as x approaches a is equal to L, or

$$\lim_{x \to a} f(x) = L$$

If the limit from the left and from the right are not equal or do not exist, we say the limit of $f(x)$ as x approaches a does not exist.

Using this definition, in Examples 1 and 2, $\lim\limits_{x \to 2} (3x + 4) = 10$, and in Example 3,

$$\lim_{x \to 2} \frac{x^2 - 4}{x - 2} = 4$$

As we will discover, limits do not always exist or equal the same value from the left and from the right.

EXAMPLE 4   Find $\lim\limits_{x \to 1} \dfrac{1}{x-1}$.

SOLUTION

x	0.9	0.99	0.999	0.9999	$\to$	1^{-}
$f(x)$	-10	-100	-1000	$-10{,}000$	$\to$	$-$ (large)

x	1.1	1.01	1.001	1.0001	$\to$	1^{+}
$f(x)$	10	100	1000	$10{,}000$	$\to$	(large)

Intuitively, as x is close to 1 but less than 1, $f(x)$ becomes numerically large but negative; likewise, as x is close to 1 but greater than 1, $f(x)$ gets numerically large but positive. Thus,

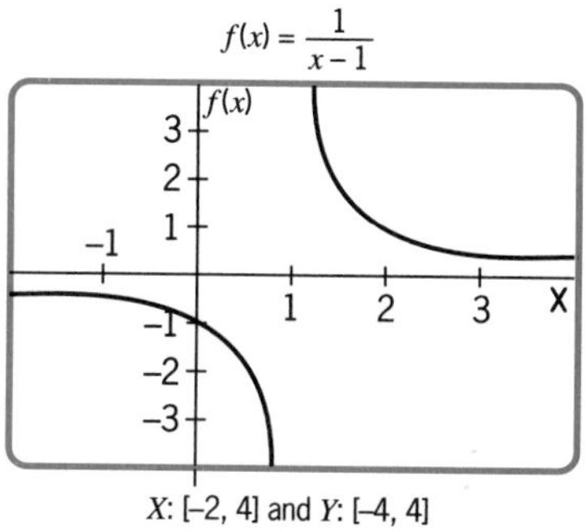

$$\lim_{x \to 1^{-}} \frac{1}{x-1} \quad \text{and} \quad \lim_{x \to 1^{+}} \frac{1}{x-1}$$

do not exist. We say that $f(x)$ is **unbounded** at $x = 1$. The calculator graph of this function is shown in Figure 5 for $-2 \le x < 1, 1 < x \le 4$, and $-4 \le y \le 4$. ■

$f(x) = \dfrac{1}{x-1}$

X: [–2, 4] and Y: [–4, 4]

Figure 5

Calculator Note

Graph $f(x) = 1/(x - 1)$ again. This time use a very small horizontal viewing window, such as [0.9, 1.1], and a very large vertical window, [−1000, 1000]. Do you observe what we discussed in Example 4?

Practice Problem 2 With the Zoom feature of your graphing calculator, investigate what happens when $x \to 1$ on the graph of the function $y = 1/(x - 1)$ from the left and from the right.

ANSWER   The graph is unbounded as $x \to 1$ from both the left and the right. ■

From the table in Example 4, the graph in Figure 5, and Practice Problem 2, we see that

$$\lim_{x \to 1^{-}} \frac{1}{x-1} \quad \text{and} \quad \lim_{x \to 1^{+}} \frac{1}{x-1}$$

do not exist.

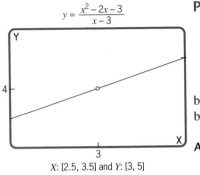

$$y = \frac{x^2 - 2x - 3}{x - 3}$$

X: [2.5, 3.5] and Y: [3, 5]

Figure 6

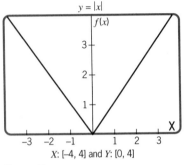

$y = |x|$

X: [-4, 4] and Y: [0, 4]

Figure 7

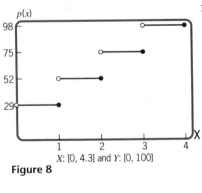

$p(x)$

X: [0, 4.3] and Y: [0, 100]

Figure 8

Practice Problem 3 Use a calculator to investigate

$$\lim_{x \to 3} \frac{x^2 - 2x - 3}{x - 3}$$

by decreasing the size of the interval about $x = 3$ (or on some calculators let the box containing the limit point decrease in size).

ANSWER   The graph is shown in Figure 6.

$$\lim_{x \to 3} \frac{x^2 - 2x - 3}{x - 3} = 4$$

Sometimes graphs are in sections, with a different formula for each section. Such a graph represents what is called a **piecewise function**. The graph of $f(x) = |x|$ is the graph of a piecewise function because its two sections are $f(x) = -x$ when $x < 0$ and $f(x) = x$ when $x \geq 0$. On a calculator, with $-4 \leq x \leq 4$ and $0 \leq y \leq 4$, verify the graph in Figure 7. For this function $\lim_{x \to 0^-} |x| = 0$ and $\lim_{x \to 0^+} |x| = 0$, so $\lim_{x \to 0} |x| = 0$.

EXAMPLE 5   At one time postage for a first-class letter was $0.29 for the first ounce, or fraction thereof, and $0.23 more for each additional ounce, or fraction thereof. The piecewise function (using whole numbers for cents) is

$$p(x) = \begin{cases} 29, & \text{if } 0 < x \leq 1 \\ 52, & \text{if } 1 < x \leq 2 \\ 75, & \text{if } 2 < x \leq 3 \\ 98, & \text{if } 3 < x \leq 4 \\ & \text{etc.} \end{cases}$$

As shown in Figure 8, the graph of this function can be obtained by using the greatest integer function on your calculator. First select the dot mode. Press MODE and select "Dot"; then press 2nd QUIT. To graph press Y = 29 + 23 × MATH ▶ (select 4) X GRAPH to obtain Figure 7. Note that the dots and circles will not appear on your calculator screen; use the trace key to see the values change. In Figure 8, notice that $\lim_{x \to 2^-} p(x) = 52$ but $\lim_{x \to 2^+} p(x) = 75$. That is, they are not equal. However, $\lim_{x \to 1.5^-} p(x) = 52$ and $\lim_{x \to 1.5^+} p(x) = 52$, so $\lim_{x \to 1.5} p(x) = 52$. ∎

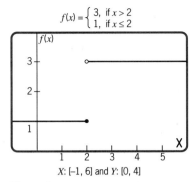

$$f(x) = \begin{cases} 3, & \text{if } x > 2 \\ 1, & \text{if } x \le 2 \end{cases}$$

X: [-1, 6] and Y: [0, 4]

Figure 9

EXAMPLE 6   Suppose that $f(x) = 1$ when $x \le 2$ and $f(x) = 3$ when $x > 2$. The graph of this function is given in Figure 9 for $-1 \le x \le 6$ and $0 \le y \le 4$. Find each of the following.

(a) $\lim\limits_{x \to 0^-} f(x)$ (b) $\lim\limits_{x \to 0^+} f(x)$ (c) $\lim\limits_{x \to 0} f(x)$

(d) $\lim\limits_{x \to 2^-} f(x)$ (e) $\lim\limits_{x \to 2^+} f(x)$ (f) $\lim\limits_{x \to 2} f(x)$

(g) $\lim\limits_{x \to 4^-} f(x)$ (h) $\lim\limits_{x \to 4^+} f(x)$ (i) $\lim\limits_{x \to 4} f(x)$

SOLUTION

Since $f(x)$ is a constant 1 for all x close to 0, from Figure 9 it is clear that

(a, b, c) $\lim\limits_{x \to 0^-} f(x) = 1$, $\lim\limits_{x \to 0^+} f(x) = 1$,   and $\lim\limits_{x \to 0} f(x) = 1$

(d)   From Figure 9, we see that as x approaches 2 from the left, $f(x)$ approaches 1; that is,

$$\lim\limits_{x \to 2^-} f(x) = 1 \qquad f(x) = 1 \text{ to the left of } x = 2, \text{ no matter how close to } x = 2.$$

(e)   As x approaches 2 from the right, $f(x)$ approaches 3; that is,

$$\lim\limits_{x \to 2^+} f(x) = 3 \qquad f(x) = 3 \text{ to the right of } x = 2.$$

(f)   Since $f(x)$ does not approach the same number as x approaches 2 from the left as it does from the right, $f(x)$ does not approach a unique limit as $x \to 2$.

(g, h, i)   From Figure 9, we see that as x approaches 4 from either direction, $f(x)$ approaches 3; that is,

$$\lim\limits_{x \to 4^-} f(x) = 3 \qquad \text{As } x \text{ gets close to 4, } f(x) \text{ is a constant 3.}$$

$$\lim\limits_{x \to 4^+} f(x) = 3$$

$$\lim\limits_{x \to 4} f(x) = 3$$

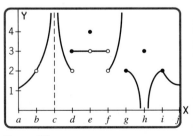

Figure 10

EXAMPLE 7   Find the limits at x for $x = b, c, d, e, f, g, h,$ and i for the function $g(x)$ in Figure 10, or state why the limit does not exist. Again, the black dots indicate the values of the function for the given values of x.

SOLUTION   At x values b, e, and i, both one-sided limits exist and agree, so the following limits exist:

$$\lim\limits_{x \to b} g(x) = 2, \quad \lim\limits_{x \to e} g(x) = 3, \quad \text{and} \quad \lim\limits_{x \to i} g(x) = 2$$

At x values c, g, and h, at least one of the one-sided limits does not exist, so the limit does not exist at c, g, or h. At x values d and f, both one-sided limits exist, but they are unequal, so the limit does not exist.   ■

In general, we can say that a function $f(x)$ has a finite limit L as x approaches the value a from either direction, which is written as

$$\lim_{x \to a} f(x) = L$$

if we can cause $f(x)$ to be as close to L as desired by restricting x to a sufficiently small interval surrounding a, but not necessarily including a. L is the y value we are *expecting* while on our approach to a, even though L may not turn out to be $f(a)$ when we arrive at a.

The following properties of limits are listed without proofs. It may be easier to remember the properties as stated in words; consequently, we give both the statement of the property and the symbolic form.

PROPERTIES OF LIMITS

Assume that a and c are constants, n is a positive integer, and

$$\lim_{x \to a} f(x) = L \quad \text{and} \quad \lim_{x \to a} g(x) = M$$

1. The limit of a constant function is that constant.

$$\lim_{x \to a} c = c$$

2. The limit of x^n is a^n

$$\lim_{x \to a} x^n = a^n$$

3. The limit of a constant times a function is the constant times the limit of the function.

$$\lim_{x \to a} cf(x) = c \lim_{x \to a} f(x) = cL$$

4. The limit of a sum (or difference) is equal to the sum (or difference) of the limits.

$$\lim_{x \to a} [f(x) \pm g(x)] = \lim_{x \to a} f(x) \pm \lim_{x \to a} g(x) = L \pm M$$

5. The limit of a product of two functions is the product of the limits of the functions provided that they both exist.

$$\lim_{x \to a} [f(x) \cdot g(x)] = \lim_{x \to a} f(x) \cdot \lim_{x \to a} g(x) = LM$$

PROPERTIES OF LIMITS *(Continued)*

6. The limit of a quotient of two functions is the quotient of the limits of the functions whenever both limits exist and the limit of the denominator is not zero.

$$\lim_{x \to a} \frac{f(x)}{g(x)} = \frac{\lim_{x \to a} f(x)}{\lim_{x \to a} g(x)} = \frac{L}{M}, \quad \text{if } M \neq 0$$

7. The limit of the nth power is the nth power of the limit.

$$\lim_{x \to a} [f(x)]^n = L^n, \quad \text{where } n \text{ is a positive integer}$$

8. [The domain of $f(x)$ is restricted so that $\sqrt[n]{f(x)}$ is always real.] The limit of a root is the root of the limit.

$$\lim_{x \to a} \sqrt[n]{f(x)} = \sqrt[n]{\lim_{x \to a} f(x)} = \sqrt[n]{L}$$

9. Let $P(x)$ be a polynomial function. Then $\lim_{x \to a} P(x) = P(a)$.

For practice in using these properties, Examples 8 through 15 illustrate separate properties rather than property 9, which could also be applied. Each example can be verified by using a calculator and an appropriate window.

EXAMPLE 8

$$\lim_{x \to 2} 7 = 7 \qquad \text{Property 1}$$

EXAMPLE 9

$$\lim_{x \to 3} 9x^2 = 9 \lim_{x \to 3} x^2 = 9 \cdot 3^2 = 81 \qquad \text{Properties 2, 3}$$

EXAMPLE 10

$$\lim_{x \to 2} (8x^3 - 3x) = \lim_{x \to 2} 8x^3 - \lim_{x \to 2} 3x \qquad \text{Property 4}$$
$$= 8 \lim_{x \to 2} x^3 - 3 \lim_{x \to 2} x \qquad \text{Property 3}$$
$$= 8 \cdot 2^3 - 3 \cdot 2 \qquad \text{Property 2}$$
$$= 64 - 6 = 58$$

EXAMPLE 11

$$\lim_{x \to 1} (3x + 5)^2 = \left[\lim_{x \to 1} (3x + 5) \right]^2 = 8^2 = 64 \qquad \text{Property 7}$$

EXAMPLE 12

$$\lim_{x \to 4} [(3x^2)(2x^3)] = \left(\lim_{x \to 4} 3x^2 \right) \cdot \left(\lim_{x \to 4} 2x^3 \right) \qquad \text{Property 5}$$

$$= \left(3 \lim_{x \to 4} x^2 \right) \cdot \left(2 \lim_{x \to 4} x^3 \right) \qquad \text{Property 3}$$

$$= (3 \cdot 4^2) \cdot (2 \cdot 4^3) \qquad \text{Property 2}$$

$$= 6144$$

EXAMPLE 13

$$\lim_{x \to 3} \frac{5x^3 + 5}{3x} = \frac{\lim\limits_{x \to 3}(5x^3 + 5)}{\lim\limits_{x \to 3} 3x} \qquad \text{Property 6}$$

$$= \frac{5 \lim\limits_{x \to 3} (x^3 + 1)}{3 \lim\limits_{x \to 3} x} \qquad \text{Property 3}$$

$$= \frac{5 \cdot 28}{3 \cdot 3} \qquad \text{Property 2}$$

$$= \frac{140}{9}$$

EXAMPLE 14

$$\lim_{x \to 2} \sqrt{(3x + 5)} = \sqrt{\lim_{x \to 2} (3x + 5)} \qquad \text{Property 8}$$

$$= \sqrt{\lim_{x \to 2} 3x + \lim_{x \to 2} 5} \qquad \text{Property 4}$$

$$= \sqrt{3 \lim_{x \to 2} x + \lim_{x \to 2} 5} \qquad \text{Property 3}$$

$$= \sqrt{3 \cdot 2 + 5} \qquad \text{Property 1, 2}$$

$$= \sqrt{11}$$

EXAMPLE 15

$$\lim_{x \to 2} (3x^3 - 4x^2 + x - 2) = 3(2)^3 - 4(2)^2 + 2 - 2 = 8 \qquad \text{Property 9}$$

Practice Problem 4 Find

$$\lim_{x \to 2} \frac{x^2 - 2x + 5}{2x - 3}$$

ANSWER

$$\frac{\lim_{x \to 2} (x^2 - 2x + 5)}{\lim_{x \to 2} (2x - 3)} = \frac{2^2 - 2 \cdot 2 + 5}{2 \cdot 2 - 3} = 5$$

SUMMARY

We introduced the following important concepts in this section:

1. The limit of function $f(x)$ as x approaches a is denoted by $\lim_{x \to a} f(x)$.

2. The limit of $f(x)$ as x approaches a from the left is denoted by $\lim_{x \to a^-} f(x)$ and as x approaches a from the right by $\lim_{x \to a^+} f(x)$.

3. If $\lim_{x \to a^-} f(x) = L$ and $\lim_{x \to a^+} f(x) = L$, then $\lim_{x \to a} f(x)$ exists and equals L.

Exercise Set 2.1

Use your calculator to complete the tables and obtain the limits intuitively.

1. $f(x) = 2x - 3$; find $\lim_{x \to 2} f(x)$.

x	1.900	1.990	1.999	$\to 2 \leftarrow$	2.001	2.010	2.100
$f(x)$	____	____	____	$\to ? \leftarrow$	____	____	____

2. $g(x) = \dfrac{3}{x^2}$; find $\lim_{x \to 0} g(x)$.

x	-0.100	-0.010	-0.001	$\to 0 \leftarrow$	0.0010	0.0100	0.1000
$g(x)$	____	____	____	$\to ? \leftarrow$	____	____	____

3. $h(x) = \dfrac{x^2 - 16}{x - 4}$; find $\lim_{x \to 4} h(x)$.

x	3.900	3.990	3.999	$\to 4 \leftarrow$	4.001	4.010	4.100
$h(x)$	____	____	____	$\to ? \leftarrow$	____	____	____

4. $r(x) = \dfrac{x}{x - 1}$; find $\lim_{x \to 1} r(x)$.

x	0.9000	0.9900	0.9990	$\to 1 \leftarrow$	1.001	1.010	1.100
$r(x)$	____	____	____	$\to ? \leftarrow$	____	____	____

Use the Zoom feature of your graphing calculator to find the limits of the following functions intuitively.

5. $\lim\limits_{x\to 1} \dfrac{3x^2 + 4x}{x + 2}$

6. $\lim\limits_{x\to 2} \dfrac{x^2 - 4}{x - 2}$

Find the following limits.

7. $\lim\limits_{x\to 4} -8$

8. $\lim\limits_{x\to 3} 0$

9. $\lim\limits_{x\to 3} 2x$

10. $\lim\limits_{x\to 2} 5x$

11. $\lim\limits_{x\to 2} \dfrac{3x}{2}$

12. $\lim\limits_{x\to -1} -\dfrac{2x}{3}$

In Exercises 13–22, use the following graph.

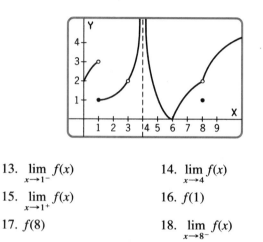

13. $\lim\limits_{x\to 1^-} f(x)$

14. $\lim\limits_{x\to 4} f(x)$

15. $\lim\limits_{x\to 1^+} f(x)$

16. $f(1)$

17. $f(8)$

18. $\lim\limits_{x\to 8^-} f(x)$

19. $f(3)$

20. $\lim\limits_{x\to 8^+} f(x)$

21. $\lim\limits_{x\to 3} f(x)$

22. $\lim\limits_{x\to 8} f(x)$

Find the following limits.

23. $\lim\limits_{x\to 3} (2x^2 - 4)$

24. $\lim\limits_{x\to 5} (3x + 5)$

25. $\lim\limits_{x\to 2} (3x^2 - 5)(x + 4)$

26. $\lim\limits_{x\to 2} (2x + 7)(3x^2 - 1)$

27. $\lim\limits_{x\to -3} \dfrac{7x}{2x + 3}$

28. $\lim\limits_{x\to 0} \dfrac{9x^2 - x + 1}{x^2 + 2x + 5}$

29. $\lim\limits_{x\to 1} \dfrac{5x}{2 + x^2}$

30. $\lim\limits_{x\to 2} (x + 1)^2(2x - 1)$

31. $\lim\limits_{x\to 2} x^2(x^2 + 1)^3$

32. Determine whether $\lim\limits_{x\to 2} f(x)$ exists for the following functions. If the limit exists, find it.

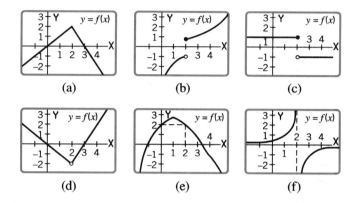

(a)   (b)   (c)

(d)   (e)   (f)

33. Determine whether $\lim\limits_{x\to 3} f(x)$ exists for the following functions. If the limit exists, find it.

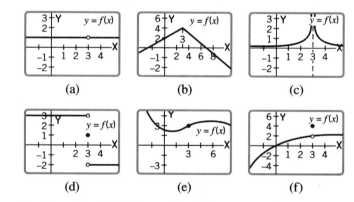

(a)   (b)   (c)

(d)   (e)   (f)

For Exercises 34–39, use the following graphs.

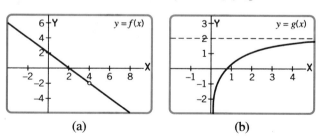

(a)   (b)

34. $\lim\limits_{x \to 0} g(x)$

35. $\lim\limits_{x \to 0} f(x)$

36. $\lim\limits_{x \to 2} f(x)$

37. $\lim\limits_{x \to 4} f(x)$

38. $\lim\limits_{x \to -1} f(x)$

39. $\lim\limits_{x \to 2} g(x)$

In Exercises 40–43, find the limit, if it exists, for each function as $x \to 2$. (Note that parts of the graphs of the functions are found in Exercise 32.)

40. $y = \begin{cases} x, & \text{for } x \leq 2 \\ 6 - 2x, & \text{for } x > 2 \end{cases}$

41. $y = \begin{cases} -2/x, & \text{for } x < 2 \\ (x - 1)^2, & \text{for } x > 2 \end{cases}$

42. $y = \begin{cases} 1, & \text{for } x \leq 2 \\ -1, & \text{for } x > 2 \end{cases}$

43. $y = \begin{cases} -x, & \text{for } x < 2 \\ 2x - 6, & \text{for } x > 2 \end{cases}$

Use the Zoom feature on your graphing calculator to find the following limits if they exist.

44. $\lim\limits_{x \to -2} \dfrac{x^3 + 8}{x + 2}$

45. $\lim\limits_{x \to 3} \dfrac{x^3 - 27}{x - 3}$

46. $\lim\limits_{x \to 2} \dfrac{x^3 - x^2 - 4x + 4}{x - 2}$

47. $\lim\limits_{x \to -3} \dfrac{x^2 - 9}{x + 3}$

48. Graph

$$f(x) = \frac{(x + 3)^2 - 9}{x}$$

Magnify the graph around $x = 0$. What do you discover? Validate your observation algebraically.

49. Graph

$$y = \frac{x^3 - 8}{x - 2}$$

Magnify the graph near $x = 2$. What do you ob-serve? How can you reach the same conclusion algebraically?

50. Graph

$$y = \frac{x^3 + 8}{x - 1}$$

Use a very small horizontal viewing window and a very large vertical window near $x = 1$. What do you observe? Can you verify your conjecture algebraically?

51. Graph

$$y = \frac{x^3 + 1}{x - 2}$$

Follow the instructions for Exercise 50 near $x = 2$.

Applications (Business and Economics)

52. **Telephone Rates.** The cost of a direct call between Atlanta and Nashville is $0.64 for the first 3 minutes and $0.23 for each additional minute or fraction thereof. Sketch this function, where x is the number of minutes, and determine the following.
   (a) $\lim\limits_{x \to 3} f(x)$ (b) $\lim\limits_{x \to 5.5} f(x)$

53. **Cost function.** A cost function C is given as

$$C(x) = \begin{cases} 5, & \text{for } 0 \leq x < 4 \\ 6, & \text{for } 4 \leq x < 6 \\ x, & \text{for } 6 \leq x < 10 \end{cases}$$

where x is the number of units. Find the following limits if they exist.
   (a) $\lim\limits_{x \to 1} C(x)$ (b) $\lim\limits_{x \to 4} C(x)$
   (c) $\lim\limits_{x \to 6} C(x)$ (d) $\lim\limits_{x \to 8} C(x)$

54. **Demand Function.** A price for a product is given as a function of x, the number of units available:

$$p(x) = \frac{900}{x^2} + \frac{800}{x}$$

Find each of the following.

(a) $\lim\limits_{x \to 1} p(x)$ (b) $\lim\limits_{x \to 0^+} p(x)$

(c) $\lim\limits_{x \to 0} p(x)$

Applications (Social and Life Sciences)

55. **Growth function.** A growth function is given as

$$n(t) = \frac{80,000t}{100 + t}$$

in terms of time t in minutes. Find the value of each of the following.

(a) $\lim\limits_{t \to 100} n(t)$

(b) $\lim\limits_{t \to 0} n(t)$

(c) $\lim\limits_{t \to -100} n(t)$

2.2 LIMITS AND CONTINUITY

OVERVIEW   A continuous function over an open interval is a function whose graph does not have any breaks in it. Another way of stating this concept is to say that a function is continuous on an interval if we can draw the graph of the function on the complete interval without taking our pen off the paper. These intuitive concepts are explained and defined in terms of limits in this section. In this section we

- Study additional techniques for finding limits
- Define what is meant by a function being continuous at a point
- Consider examples of discontinuous functions
- List properties of continuous functions

In our study of $\lim\limits_{x \to a} f(x) = L$, we were not concerned with the value of the function at $x = a$. We were interested in what happens to the function values as x approaches a. We now shift our interest to the value of the function at $x = a$. If the function is defined at $x = a$, if $\lim\limits_{x \to a} f(x)$ exists, and if $\lim\limits_{x \to a} f(x) = f(a)$, then the function is **continuous** at $x = a$.

DEFINITION: CONTINUOUS FUNCTION

A function $f(x)$ is **continuous** at $x = a$ if and only if

1. $f(a)$ is defined.
2. $\lim\limits_{x \to a} f(x)$ exists.
3. $\lim\limits_{x \to a} f(x) = f(a)$.

If any one of these conditions is not satisfied, the function $f(x)$ is discontinuous at $x = a$. The point $x = a$ is then called a **point of discontinuity**.

DEFINITION: CONTINUOUS ON AN OPEN INTERVAL

A function is **continuous on an open interval** if it is continuous at each point of the interval.

That is, a function is continuous on an open interval if there are no points of discontinuity in the interval.

EXAMPLE 16   State why the following functions are discontinuous at point P.

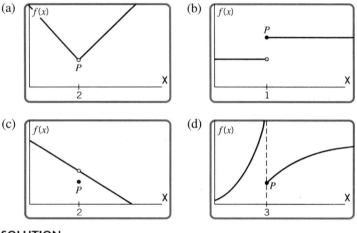

(a)

SOLUTION

(a) $f(2)$ is not defined.

(b) $\lim_{x \to 1} f(x)$ does not exist because $\lim_{x \to 1^-} f(x) \neq \lim_{x \to 1^+} f(x)$.

(c) $\lim_{x \to 2} f(x) \neq f(2)$.

(d) $\lim_{x \to 3} f(x)$ does not exist because $\lim_{x \to 3^-} f(x) \neq \lim_{x \to 3^+} f(x)$.   ■

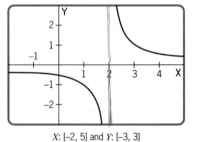

X: [–2, 5] and Y: [–3, 3]

Figure 11

EXAMPLE 17   Graph $f(x) = 1/(x - 2)$. Is $f(x)$ continuous at $x = 2$?

SOLUTION   A calculator graph of $f(x) = 1/(x - 2)$ is given in Figure 11. Since $f(x)$ is not defined at $x = 2$, condition 1 of the definition is not satisfied. Thus, $f(x)$ is discontinuous at $x = 2$. It is continuous at all other values of x; hence, $f(x)$ is continuous on the interval $(-\infty, 2)$ and on the interval $(2, \infty)$.   ■

Practice Problem 1   With a graphing calculator, draw the graph of

$$y = \frac{1}{(x - 1)^2}$$

on the intervals $-6 \leq x \leq 6$ and $0 \leq y \leq 6$. Is the function continuous at $x = 1$?

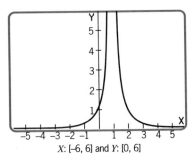

X: [-6, 6] and Y: [0, 6]

Figure 12

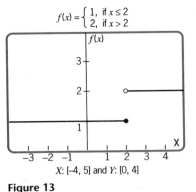

X: [-4, 5] and Y: [0, 4]

Figure 13

ANSWER   The graph is shown in Figure 12. The function is not continuous because a point of discontinuity exists at $x = 1$.

EXAMPLE 18   Graph

$$f(x) = \begin{cases} 1, & \text{if } x \le 2 \\ 2, & \text{if } x > 2 \end{cases}$$

SOLUTION   The graph is shown in Figure 13. Note that $f(2) = 1$ but $\lim\limits_{x \to 2} f(x)$ does not exist. Since condition 2 of the definition is not satisfied, $f(x)$ is not continuous at $x = 2$. It is continuous at all other values of x; consequently, $f(x)$ is continuous on the open interval $(-\infty, 2)$ and on the open interval $(2, \infty)$.   ■

Practice Problem 2 Discuss the continuity of f at $x = c$ for each of the following.

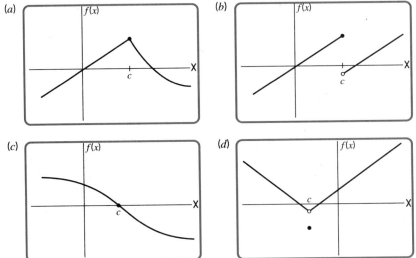

ANSWER

(a) Continuous at $x = c$.
(b) Discontinuous at $x = c$ because $\lim\limits_{x \to c} f(x)$ does not exist.

(c) Continuous at $x = c$.
(d) Discontinuous at $x = c$ because $\lim\limits_{x \to c} f(x) \ne f(c)$.

Practice Problem 3 Locate all points of discontinuity of

$$f(x) = \begin{cases} 2 + x, & \text{if } x \le 1 \\ 6 - x, & \text{if } x > 1 \end{cases}$$

ANSWER A point of discontinuity is at $x = 1$, since $\lim\limits_{x \to 1} f(x)$ does not exist.

$$\lim_{x \to 1^-} f(x) = \lim_{x \to 1^-} (2 + x) = 3 \quad \text{but} \quad \lim_{x \to 1^+} f(x) = \lim_{x \to 1^+} (6 - x) = 5$$

Practice Problem 4 Using a graphing calculator find all points of discontinuity of

$$f(x) = \frac{|x|}{x - 2} \qquad \text{for } -6 \le x \le 6$$

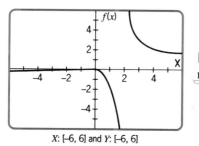

X: [–6, 6] and Y: [–6, 6]

Figure 14

ANSWER As shown in Figure 14, the function is discontinuous at $x = 2$.

For the next two examples on continuity, we need to discuss an additional technique for finding limits. Property 6 on limits in Section 8.1 does not always give an answer, as we see in the following example.

EXAMPLE 19 Find $\lim\limits_{x \to 3} \dfrac{x^2 - 9}{x - 3}$.

SOLUTION If we apply property 6, we have

$$\lim_{x \to 3} \frac{x^2 - 9}{x - 3} = \frac{\lim\limits_{x \to 3} (x^2 - 9)}{\lim\limits_{x \to 3} (x - 3)} = \frac{0}{0}$$

The form 0/0 is called an **indeterminate form**. When this occurs you might perform an algebraic simplification to determine whether the limit exists.

$$\lim_{x \to 3} \frac{x^2 - 9}{x - 3} = \lim_{x \to 3} \frac{(x + 3)(x - 3)}{x - 3}$$

If $x \ne 3$, then $(x - 3)/(x - 3) = 1$. Therefore

$$\lim_{x \to 3} \frac{x - 3}{x - 3} = 1$$

and the limit of a product is the product of the limits. Thus,

$$\lim_{x \to 3} \frac{x^2 - 9}{x - 3} = \lim_{x \to 3} \left[\frac{x - 3}{x - 3} \cdot \frac{x + 3}{1} \right]$$

$$= \lim_{x \to 3} \left[\frac{x - 3}{x - 3} \right] \cdot \lim_{x \to 3} (x + 3)$$

$$= 1 \cdot \lim_{x \to 3} (x + 3) = 6$$

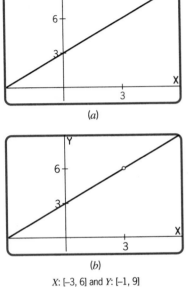

(a)

(b)

X: [–3, 6] and Y: [–1, 9]

Figure 15

The graph of $y = x + 3$ is shown in Figure 15(a), and the graph of $y = \dfrac{x^2 - 9}{x - 3}$ is shown in Figure 15(b). Actually, they are the same except that the point (3, 6) is not on the graph of $y = (x^2 - 9)/(x - 3)$. ■

Calculator Note

Your calculator may show a hole in the graph at (3, 6), but if it does not, you can use the TRACE key and note that no value for y is given at $x = 3$.

Practice Problem 5 Find $\displaystyle\lim_{x \to -2} \frac{x^2 - 4}{x + 2}$.

ANSWER $\displaystyle\lim_{x \to -2} \frac{(x + 2)(x - 2)}{x + 2} = \lim_{x \to -2} (x - 2) = -4$

EXAMPLE 20 Find

$$\lim_{x \to 3} \frac{\dfrac{1}{x} - \dfrac{1}{3}}{x - 3}$$

SOLUTION In this example, we again use algebra to write an equivalent expression for one that results in an indeterminate form. First we clear the numerator of fractions by multiplying both numerator and denominator by $3x$ ($3x$ is the lcd).

$$\lim_{x \to 3} \frac{\dfrac{3x}{x} - \dfrac{3x}{3}}{3x(x - 3)} = \lim_{x \to 3} \frac{3 - x}{3x(x - 3)}$$

$$= \lim_{x \to 3} \frac{-(x - 3)}{3x(x - 3)}$$

$$= \lim_{x \to 3} \frac{-1}{3x}$$

$$= \frac{-1}{3(3)} = \frac{-1}{9}$$ ■

Practice Problem 6 Find

$$\lim_{x \to -4} \frac{\dfrac{1}{x} + \dfrac{1}{4}}{x + 4}$$

ANSWER

$$\lim_{x \to -4} \frac{\dfrac{4x}{x} + \dfrac{4x}{4}}{4x(x + 4)} = \lim_{x \to -4} \frac{4 + x}{4x(x + 4)} = \lim_{x \to -4} \frac{1}{4x} = \frac{-1}{16}$$

EXAMPLE 21   Find $\displaystyle\lim_{x \to 4} \frac{\sqrt{x} - 2}{x - 4}$.

SOLUTION   We rationalize the numerator by multiplying the numerator and denominator by $\sqrt{x} + 2$.

$$\lim_{x \to 4} \frac{\sqrt{x} - 2}{x - 4} = \lim_{x \to 4} \frac{(\sqrt{x} - 2)(\sqrt{x} + 2)}{(x - 4)(\sqrt{x} + 2)}$$

$$= \lim_{x \to 4} \frac{x - 4}{(x - 4)(\sqrt{x} + 2)}$$

$$= \lim_{x \to 4} \frac{x - 4}{x - 4} \cdot \lim_{x \to 4} \frac{1}{\sqrt{x} + 2} \qquad x \neq 4$$

$$= (1) \cdot \frac{1}{\sqrt{4} + 2} = \frac{1}{4}$$

This problem could also be worked by factoring $x - 4 = (\sqrt{x} + 2)(\sqrt{x} - 2)$. ■

Practice Problem 7 Find $\displaystyle\lim_{x \to 9} \frac{x - 9}{\sqrt{x} - 3}$. (**Hint**: Rationalize the denominator.)

ANSWER $\displaystyle\lim_{x \to 9} \frac{x - 9}{x - 9} \cdot \lim_{x \to 9} (\sqrt{x} + 3) = 1 \cdot (6) = 6$

The technique of using algebraic manipulation to produce equivalent expressions for finding limits will be useful as we continue our discussion of continuity.

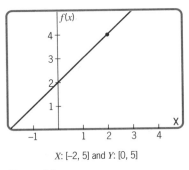

X: [−2, 5] and Y: [0, 5]

Figure 16

EXAMPLE 22   Graph

$$f(x) = \begin{cases} \dfrac{x^2 - 4}{x - 2}, & \text{if } x \neq 2 \\ 4, & \text{if } x = 2 \end{cases}$$

Is this function continuous at $x = 2$?

SOLUTION   The calculator graph of the function is given in Figure 16. Note that if $x \neq 2$, then

$$\lim_{x \to 2} \frac{x^2 - 4}{x - 2} = \lim_{x \to 2} \frac{(x + 2)(x - 2)}{x - 2} = \lim_{x \to 2} (x + 2) = 4$$

Since $f(2) = 4$ and $\lim_{x \to 2} f(x)$ exists and $\lim_{x \to 2} f(x) = f(2)$, this function is continuous at $x = 2$. It is also continuous at all other values of x. ∎

EXAMPLE 23   The cost of a telephone call between two cities on a pay phone is $2.70 for the first 3-minute period (or fraction thereof) and $0.60 for each additional minute (or fraction thereof). Let $C(x)$ be the cost of a telephone call that lasts x minutes.

$$C(x) = \begin{cases} \$2.70, & \text{if } 0 < x \leq 3 \\ \$3.30, & \text{if } 3 < x \leq 4 \\ \$3.90, & \text{if } 4 < x \leq 5 \\ \$4.50, & \text{if } 5 < x \leq 6 \\ \$5.10, & \text{if } 6 < x \leq 7 \end{cases}$$

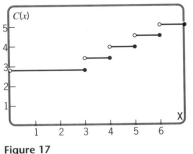

Figure 17

Determine whether this function is continuous at each of the following.

(a) $x = 4\frac{1}{2}$ minutes   (b) $x = 3$ minutes
(c) $x = 5$ minutes   (d) $x = 6.3$ minutes

SOLUTION   The graph is shown in Figure 17. Note the breaks at $x = 3$, $x = 4$, $x = 5$, and $x = 6$.

(a)   Since $\lim_{x \to 4\frac{1}{2}} C(x) = C(4\frac{1}{2})$, $C(x)$ is continuous at $x = 4\frac{1}{2}$.

(b)   Since $\lim_{x \to 3} C(x)$ does not exist, $C(x)$ is not continuous at $x = 3$.

(c)   Since $\lim_{x \to 5} C(x)$ does not exist, $C(x)$ is not continuous at $x = 5$.

(d)   Since $\lim_{x \to 6.3} C(x) = C(6.3)$, $C(x)$ is continuous at $x = 6.3$.

We observe that $C(x)$ is continuous on the open interval $0 < x < 3$, on the open interval $3 < x < 4$, on the open interval $4 < x < 5$, on the open interval $5 < x < 6$, and on the open interval $6 < x < 7$. ∎

Some of the properties or theorems of continuous functions are listed below.

PROPERTIES OF CONTINUOUS FUNCTIONS

(a) *Constant function:* $f(x) = c$, where c is any constant, is continuous for all real values of x.
(b) $f(x) = x^n$ is continuous for all real values of x.

If we assume that $f(x)$ and $g(x)$ are continuous functions, then

(c) $f(x) \pm g(x)$ is continuous.
(d) $f(x) \cdot g(x)$ is continuous.
(e) $f(x)/g(x)$ is continuous, except where $g(x) = 0$.
(f) $[f(x)]^n$, where n is a positive integer, is continuous.
(g) $\sqrt{f(x)}$ is continuous when x is restricted to those values that make $f(x) \geq 0$.

Practice Problem 8 Assume that $f(x) = 2x^2 + 4$ and $g(x) = x + 3$ are both continuous functions. Graph the following functions and determine whether they are continuous or discontinuous.

(a) $y = f + g = (2x^2 + 4) + (x + 3)$ (b) $y = f \cdot g = (2x^2 + 4)(x + 3)$

(c) $y = \dfrac{g}{f} = \dfrac{x + 3}{2x^2 + 4}$ (d) $y = f^3 = (2x^2 + 4)^3$

(e) $y = \sqrt{f} = \sqrt{2x^2 + 4}$

Note the graphs that have no breaks (i.e., graphs that represent continuous functions).

ANSWER   (a), (b), (c), (d), and (e) are continuous.

Two commonly used functions, $f(x) = c$, where c is any constant, and $f(x) = x^n$, where n is any positive integer, are continuous over $(-\infty, \infty)$. From the preceding properties, we can see that all polynomials are continuous functions for all real numbers; all rational functions are continuous except where the denominators are zero; for n, an odd positive integer greater than 1, $\sqrt[n]{f(x)}$ is continuous wherever $f(x)$ is continuous, and for n, an even positive integer, $\sqrt[n]{f(x)}$ is continuous whenever $f(x)$ is continuous and nonnegative.

SUMMARY

For a function f on an open interval, f is defined to be continuous at $x = a$ if and only if $\lim_{x \to a} f(x) = f(a)$. This implies that

1. $\lim_{x \to a} f(x)$ exists.
2. $f(a)$ is defined.
3. $\lim_{x \to a} f(x)$ must equal $f(a)$.

In order to show that a function is continuous at $x = a$, we need to verify that the three conditions stated above are satisfied.

Exercise Set 2.2

Determine whether the following functions are continuous at the given points.

1. $f(x) = 5$;   at $x = -1, x = 0, x = 2$
2. $f(x) = -3$;   at $x = 1, x = 2$
3. $f(x) = 3x$;   at $x = -2, x = 0, x = 1$
4. $f(x) = -2x$;   at $x = -1, x = 3$
5. $f(x) = 2x^2$;   at $x = -1, x = 0, x = 2$
6. $f(x) = -2x^2$;   at $x = -2, x = 3$
7. $f(x) = \dfrac{1}{x}$;   at $x = -1, x = 0, x = 2$

8. $f(x) = \dfrac{2}{x}$;   at $x = 0, x = 1$

9. $f(x) = \dfrac{1}{x + 2}$;   at $x = -2, x = 0, x = 1$

10. $f(x) = \dfrac{x}{x + 2}$;   at $x = -2, x = 1$

11. $f(x) = \dfrac{x - 1}{x - 2}$;   at $x = -1, x = 0, x = 2$

12. $f(x) = \dfrac{1 - 2x}{x - 2}$;   at $x = 1, x = 2$

13. $f(x) = \dfrac{x^2 - 9}{x - 3}$;   at $x = -3, x = 0, x = 3$

14. $f(x) = \dfrac{x^2 - 9}{x + 3}$;   at $x = -3, x = 3$

15. $f(x) = \dfrac{1}{x(x - 1)}$;   at $x = -1, x = 0, x = 1$

16. $f(x) = \dfrac{1}{2x(x - 1)}$;   at $x = 0, x = 1$

Are the following functions continuous or discontinuous at $x = 2$? Explain your answer.

17. (a)

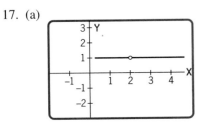

(b)

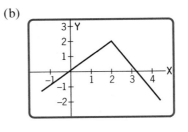

18. (a)

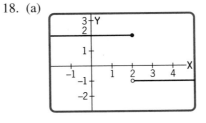

(b)

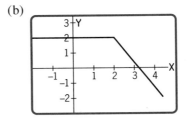

Find the following limits if they exist.

19. $\lim\limits_{x \to 3} \dfrac{x^2 - 9}{x - 3}$

20. $\lim\limits_{x \to 3} \dfrac{x^2 - 2x - 3}{x - 3}$

21. $\lim\limits_{x \to 0} \dfrac{4x^2 - 3x}{x}$

22. $\lim\limits_{x \to 0} \dfrac{(1/x) - 1}{1/x}$

23. $\lim\limits_{x \to 4} \dfrac{x - 4}{\sqrt{x} - 2}$

24. $\lim\limits_{x \to 16} \dfrac{x - 16}{\sqrt{x} - 4}$

25. $\lim\limits_{x \to 0} \dfrac{(2/x) + 1}{3/x}$

Find all points where the graphed functions are discontinuous.

26.

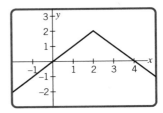

27.

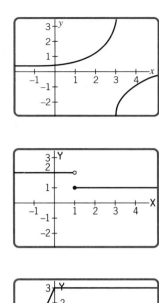

28.

29.

36. $f(x) = \begin{cases} x^2, & \text{if } -1 \leq x < 1 \\ 3x, & \text{if } x \geq 1 \end{cases}$

37. $f(x) = \dfrac{x-1}{x^2-4}$

38. Where are the following functions continuous and discontinuous?

(a) $y = \begin{cases} 2, & \text{if } x < 2 \\ 2, & \text{if } x > 2 \end{cases}$

(b) $y = \begin{cases} x, & \text{if } x \leq 2 \\ 6-2x, & \text{if } x > 2 \end{cases}$

39. Where are the following functions continuous and discontinuous?

(a) $y = \begin{cases} 2, & \text{if } x \leq 2 \\ -1, & \text{if } x > 2 \end{cases}$

(b) $y = \begin{cases} 2, & \text{if } x < 2 \\ 6-2x, & \text{if } x \geq 2 \end{cases}$

40. **Exam.**   Find $\lim\limits_{x \to 3} \dfrac{x^2-x-6}{x^2-9}$.

(a) 0   (b) $\dfrac{5}{6}$ (c) 1   (d) $\dfrac{5}{3}$

(e) Increases without bound

41. Using a calculator, locate points of discontinuity of

$$f(x) = \frac{|x^2+3x+2|}{x-2}$$

42. Locate points of discontinuity of

$$g(x) = \frac{x^2-4x+3}{x^2-1}$$

Applications (Business and Economics)

43. **Cost Functions.**   A cost function $C(x)$ has the following graph for $0 \leq x \leq 8$.

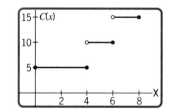

(a) Where is the cost function discontinuous?

(b) Find $\lim\limits_{x \to 2} C(x)$.   (c) Find $C(2)$.

30. Discuss whether the function is continuous at each of the eight points.

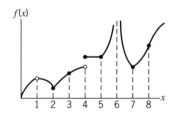

Use your graphing calculator to find all points of discontinuity for each function.

31. $g(x) = \dfrac{1}{3x}$

32. $f(x) = \dfrac{1}{x-3}$

33. $g(x) = \dfrac{x}{x+2}$

34. $f(x) = \dfrac{1}{x(x-2)}$

35. $g(x) = \dfrac{x+1}{x^2-4}$

(d) Find $\lim\limits_{x \to 4} C(x)$.   (e) Find $C(4)$.

(f) Find $\lim\limits_{x \to 7} C(x)$.   (g) Find $C(7)$.

44. ***Revenue Functions.***   A revenue function $R(x)$, giving revenue in terms of the number of items sold, x, is found to be

$$R(x) = 10x - \frac{x^2}{100}, \qquad 0 \le x \le 1000$$

(a) Find $R(10)$.   (b) Find $R(100)$.
(c) Find $R(500)$.   (d) Draw the graph of $R(x)$.
(e) Is $R(x)$ continuous on $0 < x < 1000$?

45. ***Demand Functions.***   The demand function for an item is given by

$$p(x) = \frac{80}{x - 16}, \qquad 16 < x \le 96$$

where p is the price per unit and x is the number of items. Draw the graph of the demand function and discuss the continuity of $p(x)$.

Applications (Social and Life Sciences)

46. ***Agronomy.***   The height of a plant is given by

$$h(t) = 3\sqrt{t}$$

where h is measured in inches and t in days of growth. Graph this function and find any points of discontinuity.

47. ***Laboratory Experiment.***   A laboratory uses mice in its experiments. The number N of mice available is a function of time t. The following chart was drawn for 6 days.

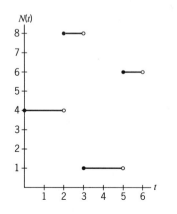

(a) Where is this function discontinuous?
(b) Find $N(1)$.   (c) Find $\lim\limits_{t \to 2^-} N(t)$.

(d) Find $\lim\limits_{t \to 2^+} N(t)$.

48. ***Learning Functions.***   A person learns $L(x)$ items in x hours according to the function

$$L(x) = 40\sqrt{x}, \qquad 0 \le x \le 16$$

(a) Find $L(4)$.   (b) Find $L(9)$.
(c) Find $L(16)$.   (d) Draw the graph of $L(x)$.
(e) Discuss the continuity of $L(x)$.

49. ***Learning Functions.***   Suppose that a learning function is given as

$$L(x) = 30x^{2/3}, \qquad 0 \le x \le 8$$

(a) Draw the graph of $L(x)$.
(b) Discuss the continuity of $L(x)$.

50. ***Voting.***   The registered voting population of a city, $P(t)$, in thousands, is given for the next 8 years as

$$P(t) = 40 + 9t^2 - t^3, \qquad 0 \le t \le 8$$

(a) Find $P(2)$.   (b) Find $P(4)$.
(c) Find $P(6)$.   (d) Draw the graph of $P(t)$.
(e) Discuss the continuity of $P(t)$.

Review Exercises

Find the following limits if they exist.

51. $\lim\limits_{x \to 2} \dfrac{x}{x + 1}$

52. $\lim\limits_{x \to -1} \dfrac{x + 1}{x}$

53. $\lim\limits_{x \to 0} \dfrac{5}{x^2}$

54. $\lim\limits_{x \to 1/3} \left(6 - \dfrac{2}{x} \right)$

55. $\lim\limits_{x \to 2} \dfrac{4}{x - 2}$

56. $\lim\limits_{x \to -1} \dfrac{2}{x + 1}$

2.3 AVERAGE AND INSTANTANEOUS CHANGES

OVERVIEW A person planning to buy a new house either now or in the future is interested in the rate at which prices are changing. Likewise, a manufacturer is interested in changes in manufacturing that could lead to increased profits. As previously stated, calculus is the mathematics used to study change. The concept used is usually the derivative of a function, which we consider in the next section. To introduce this concept we consider **average change** and **instantaneous change**. The average change of a graph over an interval may be interpreted by the slope of the secant line over that interval. The instantaneous change of a graph at a point is the slope of the line that is tangent to the graph at that point. A study of these concepts in this section will motivate a thorough study in the next section on derivatives. In this section we study

- Average rates
- Increments
- Instantaneous rates of change
- Secant lines
- Tangent lines

If you are driving on an interstate at 55 mph, and 4 miles down the road your speed increases to 65 mph, the average change in your speed per mile traveled is

$$\frac{65 - 55}{4} = 2.5$$

That is, in this 4-mile stretch your speed increased by 10 mph, or it has increased 2.5 mph per mile.

In general, the **average rate of change** of any function on an interval is the change in the function values at the endpoints of the interval divided by the change in the independent variable.

TABLE 1

x-Toys	$C(x)$
1	$ 75
2	135
3	185
4	231
5	275

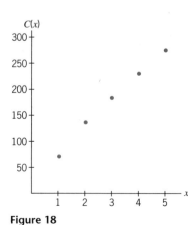

Figure 18

EXAMPLE 24 The cost of manufacturing a given toy includes the setup costs (use of equipment) plus the cost of labor and materials for each toy produced. In general, after a number of toys have been produced, the production cost per toy decreases. We consider a very simplified example of such a change. Table 1 gives the cost, $C(x)$, for the manufacture of x toys. The graph of this function is shown in Figure 18. Find the average rate of change of the cost function $C(x)$ as production increases from

(a) 1 to 2 toys.   (b) 1 to 3 toys.
(c) 1 to 4 toys.   (d) 2 to 5 toys.

SOLUTION

(a) The average rate of change of the cost function between 1 and 2 toys is

$$\frac{C(2) - C(1)}{2 - 1} = \frac{135 - 75}{1} = 60$$

or the average rate of change of the cost is $60. That is, when one toy has been produced, the cost of producing one more toy is $60.

(b)   The average rate of change of the cost function between 1 and 3 toys is

$$\frac{C(3) - C(1)}{3 - 1} = \frac{185 - 75}{2} = 55$$

or the average rate of change of the cost is $55 as x changes from 1 to 3. That is, the average cost of producing each additional toy is $55 as production increases from 1 to 3 toys.

(c)   The average rate of change of the cost function between 1 and 4 toys is

$$\frac{C(4) - C(1)}{4 - 1} = \frac{231 - 75}{3} = 52$$

or the average rate of change of $C(x)$ is $52 as x changes from 1 to 4. That is, the average cost of producing each additional toy is $52 as production increases from 1 to 4 toys.

(d)   The average rate of change of $C(x)$ is

$$\frac{C(5) - C(2)}{5 - 2} = \frac{275 - 135}{3} = 46.67$$

or the average rate of change of $C(x)$ is $46.67 as x changes from 2 to 5. That is, the average cost of producing each extra toy is $46.67 as production increases from 2 to 5 toys.   ■

This example suggests the following formula.

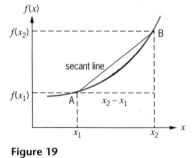

Figure 19

DEFINITION: AVERAGE RATE OF CHANGE

Assuming that the function is defined at x_1 and x_2, the **average rate of change** of $f(x)$ with respect to x as x changes from x_1 to x_2 is the ratio

$$\frac{f(x_2) - f(x_1)}{x_2 - x_1}$$

If we look at the graph of a function such as that in Figure 19, we note that

$$\frac{f(x_2) - f(x_1)}{x_2 - x_1}$$

is the slope of the line through the points $(x_1, f(x_1))$ and $(x_2, f(x_2))$, which are the points A and B in Figure 19. This straight line through A and B is called the **secant line** through A and B of the graph f.

EXAMPLE 25

Use your calculator to graph the function $y = f(x) = x^2 - 1$ and find the average rates of change as x changes from each of the following.

(a) 0 to 2   (b) 0 to 1   (c) 0 to $\frac{1}{2}$

SOLUTION   The graph of the function $y = x^2 - 1$ is shown in Figure 20. Locate the three secants (actually segments).

(a)   The average rate of change from $x = 0$ to $x = 2$ (or the slope of the secant) is

$$\frac{f(x_2) - f(x_1)}{x_2 - x_1} = \frac{f(2) - f(0)}{2 - 0} = \frac{[2^2 - 1] - [0^2 - 1]}{2 - 0} = \frac{3 - (-1)}{2} = 2$$

(b)   The average rate of change from $x = 0$ to $x = 1$ (or the slope of the secant) is

$$\frac{f(x_2) - f(x_1)}{x_2 - x_1} = \frac{f(1) - f(0)}{1 - 0} = \frac{[1^2 - 1] - [0^2 - 1]}{1 - 0} = \frac{0 - (-1)}{1 - 0} = 1$$

(c)   The average rate of change from $x = 0$ to $x = \frac{1}{2}$ (or the slope of the secant) is

$$\frac{f(x_2) - f(x_1)}{x_2 - x_1} = \frac{f(\frac{1}{2}) - f(0)}{\frac{1}{2} - 0} = \frac{[(\frac{1}{2})^2 - 1] - [0^2 - 1]}{\frac{1}{2} - 0}$$

$$= \frac{-\frac{3}{4} - (-1)}{\frac{1}{2}} = \frac{\frac{1}{4}}{\frac{1}{2}} = \frac{1}{2} \qquad \blacksquare$$

Practice Problem 1 Find the average rate of change in the function $f(x) = 2x^2$ as x changes from 1 to 4.

ANSWER $\dfrac{f(4) - f(1)}{4 - 1} = \dfrac{2(4)^2 - 2(1)^2}{4 - 1} = \dfrac{32 - 2}{3} = 10$

Increments

Before we work toward instantaneous rates of change, let's simplify our notation. If $y = f(x)$, the change in x and the corresponding change in y, called **increments** in x and y, can be denoted by h and $f(x + h) - f(x)$, respectively. For example, in Figure 21, for $y = f(x)$, we have

$$h = x_2 - x_1 \quad \text{or} \quad x_2 = x_1 + h$$

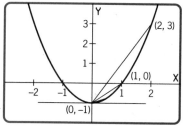

$y = x^2 - 1$ X:[-3, 3] Y:[-2, 4]

Figure 20

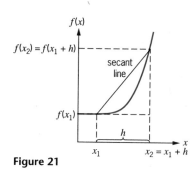

Figure 21

Likewise,

$$f(x_2) - f(x_1) = f(x_1 + h) - f(x_1)$$

This is the change in y corresponding to a change of h in x.

EXAMPLE 26   For the function $y = f(x) = x^2$, find

(a) h and $f(x_1 + h) - f(x_1)$ for $x_1 = 0.6$ and $x_2 = 1$.

(b) $\dfrac{f(x_1 + h) - f(x_1)}{h}$ given $h = 0.5$ and $x_1 = 1$.

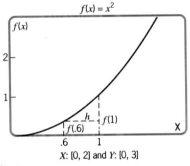

Figure 22

SOLUTION

(a)   As shown in Figure 22, h is the change in x from $x_1 = 0.6$ to $x_2 = 1$, or

$$h = x_2 - x_1 = 1 - 0.6 = 0.4$$

But $f(x_1 + h) - f(x_1)$ is the change in y as x changes from 0.6 to 1, or

$$f(x_1 + h) - f(x_1) = f(x_2) - f(x_1) = f(1) - f(0.6) = 1^2 - (0.6)^2 = 0.64$$

(b)   If we let $h = 0.5$ when $x_1 = 1$, then $x_1 + h = 1 + 0.5 = 1.5$.

$$
\begin{aligned}
\frac{f(x_1 + h) - f(x_1)}{h} &= \frac{f(1.5) - f(1)}{0.5} \\
&= \frac{(1.5)^2 - 1^2}{0.5} \\
&= \frac{2.25 - 1}{0.5} \\
&= 2.5
\end{aligned}
$$

Practice Problem 2 Let $y = f(x) = 2x + 3$. Find

(a) h and $f(x_1 + h) - f(x_1)$ for $x_1 = 2$ and $x_2 = 3$.

(b) $\dfrac{f(x_1 + h) - f(x_1)}{h}$ for $h = 1$ and $x_1 = 1$.

ANSWER

(a) $h = 3 - 2 = 1$ and $f(3) - f(2) = 9 - 7 = 2$

(b) $\dfrac{f(2) - f(1)}{1} = \dfrac{7 - 5}{1} = 2$

Looking at Figure 23, we see that the *slope of the secant line* from x_1 to $x_2 = x_1 + h$ or the *average rate of change* of f from x_1 to x_2 is

$$\frac{f(x_1 + h) - f(x_1)}{h}$$

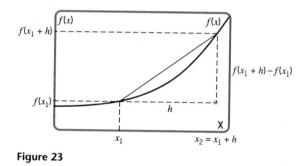

Figure 23

EXAMPLE 27 For $f(x) = x^2 + 2$, find the average rate of change when

(a) $x_1 = 1, h = 0.5.$ (b) $x_1 = 3, h \to 0.$

SOLUTION

(a) The average rate of change is

$$\begin{aligned}
\frac{f(x_1 + h) - f(x_1)}{h} &= \frac{f(1.5) - f(1)}{0.5} \qquad x_1 + h = 1 + 0.5 = 1.5 \\
&= \frac{(1.5)^2 + 2 - (1^2 + 2)}{0.5} \\
&= \frac{4.25 - 3}{0.5} = 2.5
\end{aligned}$$

(b)   The average rate of change is

$$\frac{f(3+h)-f(3)}{h} = \frac{(3+h)^2+2-[(3)^2+2]}{h}$$

$$= \frac{9+6h+h^2+2-11}{h}$$

$$= \frac{6h+h^2}{h} \qquad h \neq 0$$

$$= 6+h$$

As $h \rightarrow 0$, the average rate of change is 6.   ■

These examples suggest the following definitions.

DEFINITION: AVERAGE RATE AND INSTANTANEOUS RATE

If $y = f(x)$, at $x = x_1$ and h is any number (usually small) but not equal to zero, then

$$\textbf{Average rate} = \frac{f(x_1+h)-f(x_1)}{h}$$

$$\textbf{Instantaneous rate} = \lim_{h \to 0}\frac{f(x_1+h)-f(x_1)}{h}$$

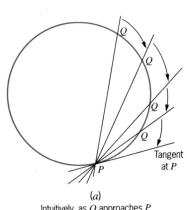

(a)
Intuitively, as Q approaches P along the circle, the secant line rotates into the tangent line.

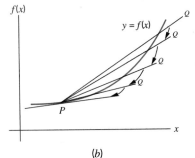

(b)
By definition, as Q approaches P along the graph of $y = f(x)$ from either side of P, the tangent line is the limiting position of the secant line.

Figure 24

Tangent Line to a Graph

What do we mean by the tangent line to the graph of a function at a point? In high school geometry the tangent line to the graph of a circle is defined as the line that intersects the circle in only one point. However, for curves other than a circle, the tangent line to a curve at a point might intersect the curve at other points. Suppose that we define the tangent line to a circle in terms of secant lines. In Figure 24(a), let points P and Q determine a secant line. Now consider point Q as moving toward point P on the circle as shown in the figure. As the point Q moves toward P, the secant rotates into the position of the tangent line of the circle at point P. This is the idea we use to define the tangent line to the graph of any function. [See Figure 24(b).]

Let's write the slope of the secant line in Figure 25 in terms of h. We have

$$\text{Slope of secant line} = \frac{f(x_1+h)-f(x_1)}{h}$$

As we let h approach zero, B approaches A and the secant lines approach a limiting position. The line that the secant lines approach is the line that is **tangent**

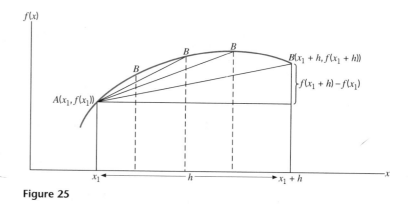

Figure 25

to the curve at point A, and the limit of the slopes of the secant lines is the slope of the tangent line.

TANGENT LINE

For the graph of the function $f(x)$, the line that is **tangent** to $y = f(x)$ at the point $(x_1, f(x_1))$ is the line that passes through this point and has a slope of

$$\lim_{h \to 0} \frac{f(x_1 + h) - f(x_1)}{h}$$

if this limit exists.

Note that the slope of a line that is tangent to a graph at $x = x_1$ is sometimes referred to as the **slope of the curve** (or graph) at $x = x_1$.

EXAMPLE 28 Find the slope of the graph of $y = x^2 - 1$ at $x = 0$.

SOLUTION

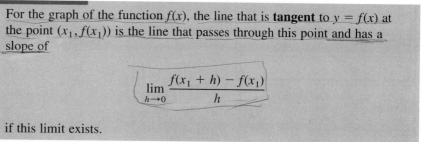

$$\text{Slope of tangent line} = \lim_{h \to 0} \frac{f(x_1 + h) - f(x_1)}{h}$$

$$= \lim_{h \to 0} \frac{f(0 + h) - f(0)}{h} \qquad \text{At } x = 0$$

$$= \lim_{h \to 0} \frac{[(0 + h)^2 - 1] - [0^2 - 1]}{h}$$

$$= \lim_{h \to 0} \frac{(h)^2}{h}$$

$$= \lim_{h \to 0} h = 0 \qquad h \neq 0$$

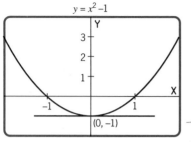

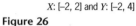

X: [-2, 2] and Y: [-2, 4]

Figure 26

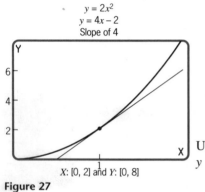

X: [0, 2] and Y: [0, 8]

Figure 27

Note that this answer agrees with the slope of what was thought to be the tangent line as drawn in Figure 20. For convenience, we repeat the pertinent parts of Figure 20 in Figure 26. ■

EXAMPLE 29   Find an equation of the tangent line to the graph of $y = 2x^2$ at (1, 2).

SOLUTION   First we need to find the slope of the tangent line to $y = 2x^2$ at (1, 2) (see Figure 27).

$$\text{Slope} = \lim_{h \to 0} \frac{f(1 + h) - f(1)}{h}$$

$$= \lim_{h \to 0} \frac{2(1 + h)^2 - 2(1)^2}{h}$$

$$= \lim_{h \to 0} 2\,\frac{1 + 2h + h^2 - 1}{h}$$

$$= \lim_{h \to 0} 2\,\frac{2h + h^2}{h} \qquad \text{Divide by } h \neq 0$$

$$= \lim_{h \to 0} 2(2 + h) = 4$$

Using the point-slope equation of a line, $y - y_1 = m(x - x_1)$, we have $y - 2 = 4(x - 1)$ or $y = 4x - 2$. ■

SUMMARY

We learned two important concepts in this section:

$$\lim_{h \to 0} (\text{Average change}) = (\text{Instantaneous change})$$

$$\lim_{h \to 0} (\text{Secant slopes}) = (\text{Tangent slope})$$

We use these concepts in the next three sections.

Exercise Set 2.3

For Exercises 1–7, find each of the following for the function $y = f(x) = 3x^2$.

1. $\dfrac{f(x_2) - f(x_1)}{x_2 - x_1}$ when $x_1 = 2, x_2 = 4$

2. $f(x_1 + h)$ when $x_1 = 2, h = 1$

3. $\dfrac{f(x_1 + h) - f(x_1)}{h}$ when $x_1 = 2, h = 1$

4. The average rate of change when x changes from 3 to 5

5. $\dfrac{f(3 + h) - f(3)}{h}$

6. Find $\displaystyle\lim_{h \to 0} \dfrac{f(3 + h) - f(3)}{h}$ and interpret your answer.

7. Find $\displaystyle\lim_{h \to 0} \dfrac{f(2 + h) - f(2)}{h}$ and interpret your answer.

Find the average rate of change for the following functions.

8. $y = 3x + 4$ between $x = 1$ and $x = 4$

9. $y = x^2 + 2x$ between $x = 0$ and $x = 3$

10. $y = 2x^3 - 6$ between $x = 1$ and $x = 3$

11. $y = \sqrt{x}$ between $x = 1$ and $x = 4$

Find the instantaneous rate of change for the following functions.

12. Exercise 8 at $x = 2$ 13. Exercise 9 at $x = 0$

14. Find the average rate of change of $f(x) = 3x^2 + 2$ from

    (a) $x = 1$ to $x = 2$.   (b) $x = 1$ to $x = 1.5$.

    (c) $x = 1$ to $x = 1.1$.   (d) $x = 1$ to $x = 1.01$.

15. Find the instantaneous rate of change of $f(x) = 3x^2 + 2$ at $x = 1$.

16. Find the average rate of change of $f(x) = -2x^2 + 3$ from

    (a) $x = 2$ to $x = 3$.   (b) $x = 2$ to $x = 2.5$.

    (c) $x = 2$ to $x = 2.1$.   (d) $x = 2$ to $x = 2.01$.

17. Find the instantaneous rate of change of $f(x) = -2x^2 + 3$ at $x = 2$.

18. **Exam.** What is the y-coordinate of the point on the curve $y = 2x^2 - 3x$ at which the slope of the tangent line is the same as that of the slope of the secant line between $x = 1$ and $x = 2$?

    (a) -1 (b) 0   (c) 1

    (d) 3   (e) 9

Applications (Business and Economics)

19. **Consumer Price Index.** The consumer price index (CPI), compiled by the U.S. Department of Labor, gives the price today of $100 worth of food, clothing, housing, fuel, and so on in 1967.

Year	CPI
1982	$289
1983	297
1984	308
1985	320
1986	333
1987	348
1988	363

Find the average change in the consumer price index from

    (a) 1983 to 1985.   (b) 1986 to 1988.

    (c) 1984 to 1988.   (d) 1982 to 1988.

20. **Demand Function.** The price of an article is given as a function of the number of the articles available, x, as

$$p(x) = \frac{10{,}000}{x^2} + \frac{8000}{x}$$

Find the average change in $p(x)$ as

    (a) x changes from 2 to 4.

    (b) x changes from 1 to 2.

    (c) x changes from $\frac{1}{2}$ to 1.

21. **Profit Functions.** A company lists its profit function as

$$P(x) = 8x - 0.02x^2 - 500$$

    (a) Find the average rate of change in profit from $x = 100$ to $x = 200$.

    (b) Find the instantaneous rate of change in profit at $x = 100$.

    (c) Find the instantaneous rate of change in profit at $x = 200$.

    (d) Find the profit at $x = 200$.

22. **Cost Functions.** A cost function is given as

$$C(x) = 40{,}000 - 300x + x^2$$

Graph the function and find the instantaneous rate of change of cost when $x = 100$, when $x = 150$, and when $x = 200$.

Applications (Social and Life Sciences)

23. **Bacteria.** The number of bacteria N, in thousands, present x hours after being treated by an antibiotic is

$$N(x) = x^2 - 6x + 10, \qquad 0 \le x \le 6$$

Find the average rate of change of the number present from $x = 1$ to $x = 3$. Find the instantaneous rate of change at $x = 1$.

24. **Births.** At Sunnyside Hospital the number of births is recorded by year as follows:

Year	1987	1988	1989	1990	1991	1992	1993
Births	1040	1102	1230	1308	1402	1460	1480

Find the average rate of change in the number of births from
(a) 1987 to 1990.   (b) 1990 to 1993.
(c) 1989 to 1993.

Review Exercises

Find the following limits.

25. $\lim\limits_{x \to 2} \dfrac{x^2 - 3x + 2}{x - 2}$

26. $\lim\limits_{x \to -1} \dfrac{x + 1}{x^2 + 3x + 2}$

27. $\lim\limits_{x \to 1} \dfrac{3x^2 + 4x - 1}{5x^2 + 2x + 3}$

28. $\lim\limits_{x \to 2} (x - 2)(x^2 - 2x - 2)$

29. $\lim\limits_{x \to 3} \dfrac{2x^2 - 2x - 12}{x - 3}$

30. $\lim\limits_{x \to 0} \dfrac{5x^3 + 3x}{x}$

2.4 DEFINITION OF THE DERIVATIVE

OVERVIEW   In the last section, we used an expression that we will call the **difference quotient** for function f,

$$\frac{f(x_1 + h) - f(x_1)}{h}$$

which gave the average change in $f(x)$ from x_1 to $x_1 + h$. If the limit of this difference,

$$\lim\limits_{h \to 0} \frac{f(x_1 + h) - f(x_1)}{h}, \qquad h \neq 0$$

exists, it can be interpreted as the *slope of the tangent line* to the graph of the function at $x = x_1$. It can also be interpreted as the *instantaneous rate of change* of $f(x)$ at $x = x_1$. In this section we use this notation to define the **derivative** of $f(x)$, and we investigate some of its many applications. In economics, if $C(x)$ represents a cost function, the derivative of $C(x)$ is the rate of change of cost and is called the **marginal cost** function. The derivative of the profit function $P(x)$ is called the **marginal profit** function (the rate of change of profit); the derivative of the demand function $D(x)$ is the **marginal demand** function; and the derivative of the supply function $S(x)$ is the **marginal supply** function. In medicine, if $P(x)$ represents systolic blood pressure, the derivative of $P(x)$ may be thought of as the sensitivity to a drug. In this section, we

- Define the derivative of a function
- Outline a procedure for finding the derivative
- Illustrate the use of derivatives

In the difference quotient above, let's replace the fixed number x_1 by the more general symbol x to get

$$\frac{f(x + h) - f(x)}{h}, \qquad h \neq 0$$

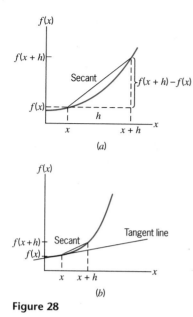

Figure 28

This difference quotient is pictured in Figure 28(a). Geometrically, this quotient gives the slope of the secant line as shown. As $h \rightarrow 0$ in Figure 28(b), the slopes of the secant lines approach the slope of the tangent line at x.

This difference quotient as $h \rightarrow 0$ provides for us one of the basic ideas of calculus, the derivative.

DEFINITION: DERIVATIVE OF A FUNCTION

For $y = f(x)$ and h, any number such that $h \neq 0$, the **derivative** of $f(x)$, which is denoted by $f'(x)$, is defined by

$$f'(x) = \lim_{h \to 0} \frac{f(x + h) - f(x)}{h}$$

provided that this limit exists.

Other symbols that can be used to denote the derivative are

$$y', \quad D_x f, \quad \text{and} \quad \frac{dy}{dx}$$

which can be read as "the derivative of y," or "the derivative of a function of x with respect to x."

NOTE: The dy/dx should not be regarded, at this time, as a quotient of two entities, dy and dx.

EXAMPLE 30   Use the definition of the derivative to find $f'(x)$ for the function $f(x) = x^2 - 1$.

SOLUTION   First, we set up the difference quotient:

$$\frac{f(x + h) - f(x)}{h} = \frac{[(x + h)^2 - 1] - [x^2 - 1]}{h}$$

Second, we write the limit as $h \rightarrow 0$:

$$\lim_{h \to 0} \frac{[(x + h)^2 - 1] - [x^2 - 1]}{h}$$

Third, we simplify this expression as much as possible:

$$\lim_{h \to 0} \frac{x^2 + 2xh + h^2 - 1 - x^2 + 1}{h} = \lim_{h \to 0} \frac{2xh + h^2}{h}$$

$$= \lim_{h \to 0} \frac{h(2x + h)}{h} \qquad h \neq 0$$

$$= \lim_{h \to 0} (2x + h)$$

Fourth, to evaluate the limit as $h \rightarrow 0$, we use the properties of limits that allow us to replace h with 0, since $2x + h$ is a continuous function:

$$\lim_{h \to 0} (2x + h) = 2x + 0 = 2x$$

Therefore, $f'(x) = 2x$ gives the derivative of the function $f(x) = x^2 - 1$ at any real number x. ∎

Procedure for Finding a Derivative

1. Set up the difference quotient.
2. Write as a limit.
3. Simplify this limit so that h is no longer a factor of the denominator.
4. Evaluate the limit by replacing h with zero.

EXAMPLE 31   Given $f(x) = 2x^2 + 3$, compute $f'(x)$.

SOLUTION

1. Set up the difference quotient.

$$\frac{f(x + h) - f(x)}{h} = \frac{[2(x + h)^2 + 3] - [2x^2 + 3]}{h}$$

2. Write as a limit.

$$\lim_{h \to 0} \frac{[2(x + h)^2 + 3] - [2x^2 + 3]}{h}$$

3. Simplify the limit.

$$= \lim_{h \to 0} \frac{2x^2 + 4xh + 2h^2 + 3 - 2x^2 - 3}{h}$$

$$= \lim_{h \to 0} \frac{4xh + 2h^2}{h} = \frac{h(4x + 2h)}{h}$$

$$= \lim_{h \to 0} (4x + 2h) \qquad h \neq 0$$

4. Evaluate the limit.

$$= 4x + 2(0)$$

Thus, we have $f'(x) = 4x$. ∎

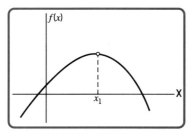

(*a*) Function is not defined at $x = x_1$

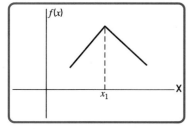

(*b*) Limit of the function does not exist at $x = x_1$

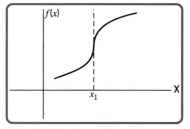

(*c*) No unique tangent line at $x = x_1$

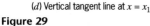

(*d*) Vertical tangent line at $x = x_1$

Figure 29

Calculator Note

Most calculators have a menu for finding the value of the derivative at a point or the value of a function at a point. We illustrate as follows: Press $\boxed{\text{MATH}}$, (select 8), and then insert the expression or function, the variable, and the value of the variable:

$$\text{nDeriv(expression, variable, value)}$$

To find $f'(1)$ in Example 31, we use

$$\text{nDeriv}(2x^2 + 3, x, 1)$$

to obtain a value of 4. From Example 31, $f'(1) = 4 \cdot 1 = 4$.

Practice Problem 1 Find the derivative of $f(x) = 3x^2$.

ANSWER

$$\lim_{h \to 0} \frac{3(x + h)^2 - 3x^2}{h} = \lim_{h \to 0} \frac{6xh + 3h^2}{h} = \lim_{h \to 0} \frac{h(6x + 3h)}{h}$$
$$= \lim_{h \to 0} (6x + 3h) = 6x + 3(0) = 6x$$

Nonexistence of Derivatives

The existence of a derivative at $x = x_1$ depends on the existence of the limit

$$\lim_{h \to 0} \frac{f(x_1 + h) - f(x_1)}{h}$$

If the limit does not exist, we say that the function is **not differentiable** at $x = x_1$ or that $f'(x_1)$ does not exist. A derivative will not exist at a point, x_1, if:

1. The function is not defined at the point x_1.
2. The limit of the function as x approaches x_1 does not exist.
3. There is no unique tangent line at $(x_1, f(x_1))$.
4. There is a vertical tangent line at $x = x_1$.

Examples of these cases are shown in Figure 29.

If the derivative of $f(x)$ exists at $x = a$, what can we say about the continuity of the function at $x = a$? If the derivative of $f(x)$ exists at $x = a$, then $f(x)$ is continuous at $x = a$. The converse of this statement is not true. If a function is continuous at $x = a$, then its derivative at $x = a$ may or may not exist; Figures 29(*c*) and 29(*d*) show examples in which f is continuous at x_1 but $f'(x_1)$ does not exist. As we try to find derivatives of functions, it is important to remember that if a function is not continuous at a point, the function does not have a derivative at that point.

Here is an example of a function that has a derivative everywhere except at $x = 0$, because the function does not even exist at $x = 0$.

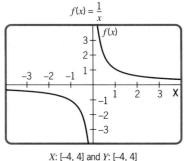

$f(x) = \dfrac{1}{x}$

X: [–4, 4] and Y: [–4, 4]

Figure 30

EXAMPLE 32   Find the derivative of $f(x) = 1/x$, which is shown in Figure 30.

SOLUTION

1.   Set up the difference quotient.

$$\frac{f(x + h) - f(x)}{h} = \frac{\dfrac{1}{x + h} - \dfrac{1}{x}}{h}$$

2.   Write as a limit.

$$\lim_{h \to 0} \frac{\dfrac{1}{x + h} - \dfrac{1}{x}}{h}$$

3.   Simplify the limit by finding a common denominator.

$$= \lim_{h \to 0} \frac{\dfrac{x}{x(x + h)} - \dfrac{(x + h)}{x(x + h)}}{h}$$

$$= \lim_{h \to 0} \frac{x - (x + h)}{x(x + h)h} = \lim_{h \to 0} \frac{-h}{x(x + h)h}$$

$$= \lim_{h \to 0} \frac{-1}{x(x + h)}$$

4.   Evaluate the limit.

$$= \frac{-1}{x(x + 0)}$$

Therefore,

$$f'(x) = \frac{-1}{x^2}$$

Notice that $f'(0)$ does not exist because there is no number $-1/0^2$.

Calculator Note

In the preceding example, $f'(x) = -1/x^2$, so $f'(1) = -1$. Using nDeriv(1/x, x, 1) on your graphing calculator gives an answer of -1.000001. Why is there a discrepancy? The value of the derivative from the calculator is the result of a numerical procedure where the variable changes by increments. The answer is not accurate to the sixth decimal place. Consider the answer as -1 to five decimal places. If we evaluate the derivative at $x = 0$, we get $f'(0) \approx f'(.0010001) = -4,999,750,012$. This is as accurately as the calculator can express such numbers that increase without bound.

EXAMPLE 33   Find the derivative of $f(x) = \sqrt{x}$ and check your answer at $x = 9$ with a calculator.

SOLUTION   To simplify this expression, we multiply both the numerator and denominator by $\sqrt{x + h} + \sqrt{x}$.

$$
\begin{aligned}
f'(x) &= \lim_{h \to 0} \frac{\sqrt{x + h} - \sqrt{x}}{h} \\[2mm]
&= \lim_{h \to 0} \frac{(\sqrt{x + h} - \sqrt{x})(\sqrt{x + h} + \sqrt{x})}{h(\sqrt{x + h} + \sqrt{x})} \\[2mm]
&= \lim_{h \to 0} \frac{(\sqrt{x + h})^2 - (\sqrt{x})^2}{h(\sqrt{x + h} + \sqrt{x})} \\[2mm]
&= \lim_{h \to 0} \frac{h}{h(\sqrt{x + h} + \sqrt{x})} \\[2mm]
&= \lim_{h \to 0} \frac{1}{\sqrt{x + h} + \sqrt{x}} \\[2mm]
&= \frac{1}{\sqrt{x + 0} + \sqrt{x}} \\[2mm]
&= \frac{1}{2\sqrt{x}}
\end{aligned}
$$

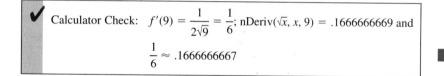

✔ Calculator Check: $f'(9) = \dfrac{1}{2\sqrt{9}} = \dfrac{1}{6}$; nDeriv($\sqrt{x}$, x, 9) = .1666666669 and

$\dfrac{1}{6} \approx .1666666667$

■

Economists make use of the derivative of a function to determine **marginal cost**.

EXAMPLE 34   A small manufacturer of boats (without motors) finds that to produce and sell x units of his product per month, it costs him $10x - 0.1x^2$ hundreds of dollars. That is, his monthly total-cost function is given by $C(x) = 10x - 0.1x^2$ for $0 \le x \le 10$. The graph of this function is shown in Figure 31. Find the marginal cost function.

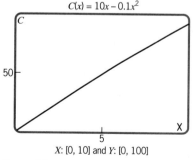

$C(x) = 10x - 0.1x^2$

X: [0, 10] and Y: [0, 100]

Figure 31

SOLUTION   The derivative of $C(x)$ is the marginal cost:

$$\text{Marginal cost} = C'(x)$$

$$= \lim_{h \to 0} \frac{C(x + h) - C(x)}{h}$$

$$= \lim_{h \to 0} \frac{[10(x + h) - 0.1(x + h)^2] - [10x - 0.1x^2]}{h}$$

$$= \lim_{h \to 0} \frac{10h - 0.2xh - 0.1h^2}{h}$$

$$= \lim_{h \to 0} \frac{h(10 - 0.2x - 0.1h)}{h}$$

$$= \lim_{h \to 0} (10 - 0.2x - 0.1h)$$

$$= 10 - 0.2x - 0$$

$$= 10 - 0.2x$$

Practice Problem 2   A revenue function is given as $R(x) = 200x - 0.01x^2$. Find an expression for marginal revenue: $\text{MR} = R'(x)$. What is the approximate additional revenue for one more unit when $x = 10$?

ANSWER   We have $\text{MR} = R'(x) = 200 - 0.02x$ and $R'(10) = 200 - 0.02(10) = 199.80$, so the additional revenue for one more unit when $x = 10$ will be \$199.80.

SUMMARY

The derivative of a function f, at a number x in its domain, is defined to be

$$f'(x) = \lim_{h \to 0} \frac{f(x + h) - f(x)}{h}$$

Since the difference quotient

$$\frac{[f(x + h) - f(x)]}{h}$$

is the slope of the line joining $(x, f(x))$ and $((x + h), f(x + h))$, the derivative may be interpreted as the slope of a tangent line to the curve. Economists often refer to the derivative of a function as the marginal value of that function.

Exercise Set 2.4

Find the derivative of the following functions.

1. $f(x) = 3x$

2. $f(x) = 2x$

3. $f(x) = 3x + 5$

4. $f(x) = -2x + 5$

5. $f(x) = \dfrac{5x - 2}{3}$

6. $f(x) = \dfrac{-3x - 5}{2}$

Find the derivative of each function, and then find $f'(0)$, $f'(1)$, and $f'(-1)$. Check $f'(1)$ using a graphing calculator.

7. $f(x) = 3x^2 + 2$

8. $f(x) = x^2$

9. $f(x) = 7x + x^2$

10. $f(x) = x^2 - 4x$

11. $f(x) = \dfrac{4}{x}$

12. $f(x) = \dfrac{5}{x}$

13. $f(x) = \sqrt{3x}$

14. $f(x) = \sqrt{5x}$

15. $f(x) = \dfrac{1}{\sqrt{x + 1}}$

16. $f(x) = \dfrac{1}{\sqrt{x + 4}}$

17. $f(x) = \dfrac{2}{x^2}$

18. $f(x) = \dfrac{3}{x^2}$

19. $f(x) = x^3$

20. $f(x) = 2x^3$
    [**Hint:** $(x + h)^3 = x^3 + 3x^2h + 3xh^2 + h^3$.]

21. Using your calculator, draw the graph of $y = x^2 - 3$ and evaluate the derivative at $x = 1$ on the graph. (Use ZOOM 4 for your window.) Now, find the derivative of $y = x^2 - 3$ and evaluate at $x = 1$. Are the answers approximately the same?

22. Work Exercise 21 using $y = 8/x$.

23. Find the slope of the tangent line to the graph of $f(x) = 2x^2 + 2x$ at $x = 1$, and then find the equation of the tangent line. [Use $y - y_1 = m(x - x_1)$.]

24. Find the slope of the tangent line to the graph of $f(x) = 3x^2$ at $x = 2$ and then find the equation of the tangent line.

25. For the function $f(x) = x^2 + 1$,
    (a) find the equation for the slope of the tangent lines to this curve.
    (b) find the slope at $x = 2$.
    (c) find an equation of the tangent line drawn to this curve at $x = 2$.
    (d) graph the function and draw the tangent line at $x = 2$.

Applications (Business and Economics)

26. ***Manufacturing.*** On an assembly line the rate at which articles are produced decreases with the number of hours on the line. For example, the number of articles produced at any given time t (t is time at work on the assembly line) is $f(t) = 400t - 3t^2$.
    (a) How many articles have been produced at the end of 2 hours?
    (b) What is the instantaneous rate of producing articles at the end of 2 hours?
    (c) Repeat part (a) at the end of 6 hours.
    (d) Repeat part (b) at the end of 6 hours.
    (e) Compare parts (b) and (d) and see if they agree with the first sentence of the problem.

27. ***Cost Functions.*** A cost function is given as

    $$C(x) = 900 + 300x + x^2$$

    (a) Find the marginal cost function, $C'(x)$.
    (b) Find $C'(2)$. Check with a graphing calculator.
    (c) Find $C'(3)$. Check with a graphing calculator.

28. ***Demand Functions.*** A demand function is given as

    $$p(x) = 144 - x^2, \qquad 1 \le x \le 12$$

    (a) Find the instantaneous rate of change, $p'(x)$.
    (b) Find $p'(2)$. Check with a graphing calculator.
    (c) Find $p'(6)$. Check with a graphing calculator.

29. ***Demand Functions.*** A demand function is given as

    $$p(x) = 169 - x^2, \qquad 1 \le x \le 13$$

    (a) Find the instantaneous rate of change, $p'(x)$.
    (b) Find $p'(3)$. Check with a graphing calculator.
    (c) Find $p'(5)$. Check with a graphing calculator.

Applications (Social and Life Sciences)

30. ***Human Sensitivity.*** The systolic blood pressure of a patient an hour after receiving a drug is given by

    $$P(x) = 136 + 18x - 8x^2, \qquad 0 \le x \le 3$$

    (a) What is the sensitivity dP/dx of the patient to this drug?
    (b) What is the sensitivity when $x = 2$? Check with a graphing calculator.

31. ***Human Sensitivity.*** A doctor administers x milligrams of a drug and records a patient's blood pressure. The systolic pressure is approximated by

$$P(x) = 140 + 10x - 5x^2, \qquad 0 \le x \le 4$$

(a) What is the sensitivity dP/dx of the patient to the drug?

(b) What is the sensitivity when 2 milligrams are administered? Check with a graphing calculator.

(c) What is the sensitivity when 3 milligrams are administered? Check with a graphing calculator.

32. ***Learning Functions.*** A learning function for the number of words learned in a foreign language class after t hours of study is given as

$$N(t) = 15t - t^2, \qquad 0 \le t \le 8$$

(a) Find the rate of change of this function.

(b) Find $N'(2)$ and $N'(4)$. Check with a graphing calculator.

33. ***Bacteria.*** When an antibiotic is introduced into a culture of bacteria, the number of bacteria present after t hours is given by $N(t) = 2000 + 10t - 5t^2$,

where $N(t)$ is the number (in thousands) of bacteria present at the end of t hours.

(a) Find the number of bacteria present after 2 hours.

(b) Find the rate of change in the number of bacteria present at the end of 2 hours. Check with a graphing calculator.

(c) Repeat part (a) at the end of 4 hours. Check with a graphing calculator.

(d) Repeat part (b) at the end of 4 hours. Check with a graphing calculator.

Review Exercises

Find the following limits if they exist.

34. $\displaystyle \lim_{x \to 1/2} \frac{2x - 3}{4x - 2}$

35. $\displaystyle \lim_{x \to -2/3} \frac{3x^2 + 4x - 1}{3x + 2}$

36. Find the average rate of change of $f(x) = x^2 - 4$ from $x = 1$ to $x = 2$.

2.5 TECHNIQUES FOR FINDING DERIVATIVES

OVERVIEW In the preceding section, we found a derivative by using the definition formula

$$f'(x) = \lim_{h \to 0} \frac{f(x + h) - f(x)}{h}$$

In this section and the sections that follow, we discover some rules or formulas from this definition so that we can take the derivative of a very large number of functions without using the definition formula. In fact, these rules or formulas will make the derivatives of many functions easy to obtain. In this section we focus our attention on

- Derivative of x to a power
- Derivative of a constant
- Derivative of a constant times a function
- Derivative of a sum
- Marginal analysis

Geometrically, the graph of $y = f(x) = C$, where C is any constant, is a horizontal line; that is, it has a slope of zero. Intuitively, we know that the derivative of $f(x) = C$ must be zero, since there is no change in the slope. However, we verify this fact with our four-step process given in the preceding section.

1. $\dfrac{f(x + h) - f(x)}{h} = \dfrac{C - C}{h}$

2. $\lim\limits_{h \to 0} \dfrac{C - C}{h}$

3. $= \lim\limits_{h \to 0} \dfrac{0}{h} = \lim\limits_{h \to 0} 0$

4. $= 0$

Thus we have a proof that the derivative of a constant function is zero.

DERIVATIVE OF A CONSTANT FUNCTION

If $f(x) = C$, then $f'(x) = 0$.

EXAMPLE 35

(a) If $f(x) = -5$, then $f'(x) = 0$.
(b) If $f(x) = 3$, then $f'(x) = 0$.

Consider now the derivative of a constant times a function. Suppose that $f(x) = C \cdot g(x)$, where C is a constant and $g'(x)$ exists. Then

1. $\dfrac{f(x + h) - f(x)}{h} = \dfrac{C \cdot g(x + h) - C \cdot g(x)}{h}$

2. $\lim\limits_{h \to 0} \dfrac{C \cdot g(x + h) - C \cdot g(x)}{h}$

3. $= \lim\limits_{h \to 0} C \left[\dfrac{g(x + h) - g(x)}{h} \right] = C \lim\limits_{h \to 0} \left[\dfrac{g(x + h) - g(x)}{h} \right]$

4. $= Cg'(x)$

Therefore we have proved that $f'(x) = Cg'(x)$.

DEFINITION: DERIVATIVE OF A CONSTANT TIMES A FUNCTION

Let C be any real number and $g(x)$ be any function that is differentiable. The derivative of

$$f(x) = Cg(x) \quad \text{is} \quad f'(x) = Cg'(x)$$

That is, the **derivative of a constant times a function** is the constant times the derivative of the function.

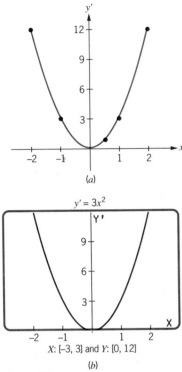

(a)

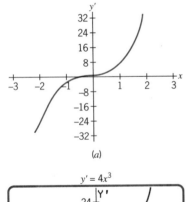

X: [-3, 3] and Y: [0, 12]

(b)

Figure 32

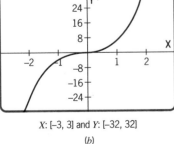

Figure 33

EXAMPLE 36   Find the derivative of $y = 4x^2$.

SOLUTION

Consider y as the product of 4 and x^2. As we found in the exercise set in the last section, the derivative of x^2 is $2x$. Therefore,

$$y' = 4(2x) = 8x$$

Functions of the form $y = x^n$ occur often in application problems. Next we examine some special cases for this type of function and then state the power rule without proof. For example, we have already shown that if $f(x) = x$, then $f'(x) = 1 \cdot x^0 = 1$, and if $f(x) = x^2$, then $f'(x) = 2x^{2-1} = 2x$. It would appear that if $f(x) = x^n$, then $f'(x) = nx^{n-1}$.

Let's continue to explore this conjecture with our calculator.

First, with our calculator, we obtain values of the derivative of $f(x) = x^3$ at a number of points, and then we draw a smooth curve through these points to represent the graph of $f'(x)$; see Figure 32(a). Next, we draw the graph of our conjectured derivative, $f'(x) = 3x^2$; see Figure 32(b). Are the two graphs in agreement? The two graphs appear to coincide. Thus, if $f(x) = x^3$, then $f'(x)$ appears to be $3x^2$.

Similarly, compare the graphs of the derivative values of $f(x) = x^4$ and $f'(x) = 4x^3$ shown in Figure 33. Again, the curves seem to coincide and it appears that if $f(x) = x^4$, then $f'(x) = 4x^3$.

In general, for any positive integer n, it is true that if $f(x) = x^n$, then $f'(x) = nx^{n-1}$. A similar formula can be developed when n is a negative integer. Apply the formula in Example 37, and then show that the answer is correct by using the definition of a derivative.

EXAMPLE 37   Use the definition of a derivative given in the preceding section to verify that if

$$f(x) = x^{-1}, \quad \text{then} \quad f'(x) = -1x^{-1-1} = -x^{-2} = -\frac{1}{x^2}$$

SOLUTION

$$\lim_{h \to 0} \frac{(x + h)^{-1} - x^{-1}}{h} = \lim_{h \to 0} \frac{\dfrac{1}{x + h} - \dfrac{1}{x}}{h}$$

$$= \lim_{h \to 0} \frac{x - (x + h)}{x(x + h)h} \qquad \text{Multiply by } \frac{x(x + h)}{x(x + h)}.$$

$$= \lim_{h \to 0} \frac{-h}{x(x + h)h}$$

$$= \lim_{h \to 0} \frac{-1}{x(x + h)}$$

$$= \frac{-1}{x(x + 0)} = \frac{-1}{x^2}$$

Do you think the formula for the derivative of x^n holds when n is not an integer? Apply the formula in Example 38, and then show that the answer is correct by using the definition of the derivative.

EXAMPLE 38   Use the definition of a derivative in the preceding section to verify that if

$$f(x) = x^{1/2}, \quad \text{then} \quad f'(x) = \frac{1}{2} x^{(1/2)-1} = \frac{1}{2} x^{-1/2}$$

SOLUTION

$$\lim_{h \to 0} \frac{(x+h)^{1/2} - x^{1/2}}{h} = \lim_{h \to 0} \frac{\sqrt{x+h} - \sqrt{x}}{h} \cdot \frac{\sqrt{x+h} + \sqrt{x}}{\sqrt{x+h} + \sqrt{x}}$$

$$= \lim_{h \to 0} \frac{(x+h) - x}{h(\sqrt{x+h} + \sqrt{x})} \qquad \text{Rationalize the numerator.}$$

$$= \lim_{h \to 0} \frac{h}{h(\sqrt{x+h} + \sqrt{x})}$$

$$= \lim_{h \to 0} \frac{1}{\sqrt{x+h} + \sqrt{x}}$$

$$= \frac{1}{\sqrt{x+0} + \sqrt{x}}$$

$$= \frac{1}{2\sqrt{x}}$$

$$= \frac{x^{-1/2}}{2} \qquad\blacksquare$$

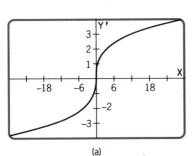

(a)

$$y' = \frac{4}{3} x^{\frac{1}{3}}$$

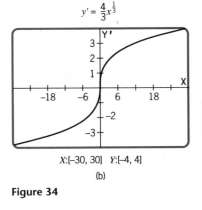

X:[-30, 30]   Y:[-4, 4]

(b)

Figure 34

Calculator Note

Show that if $f(x) = x^{4/3}$, then $f'(x) = \dfrac{4x^{\frac{1}{3}}}{3}$. This time instead of finding the value of the derivative at several points and then drawing the graph, we will accomplish both on the calculator in one step. Press $\boxed{\text{Y =}}$ $\boxed{\text{MATH}}$ (Select 8) and insert nDeriv ((x $\boxed{\wedge}$ 4) $\boxed{\wedge}$ (1/3), x, x) $\boxed{\text{GRAPH}}$. The graph is shown in Figure 34(a). Then the calculator graph of our assumed answer $y' = \dfrac{4}{3} x^{1/3}$ is found in Figure 34(b).

Since the two graphs in Figure 34 seem to be identical, it appears that the derivative of $f(x) = x^n$ is $f'(x) = nx^{n-1}$ when n is a rational number.

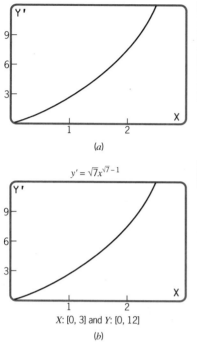

$y' = \sqrt{7}x^{\sqrt{7}-1}$

X: [0, 3] and Y: [0, 12]

(b)

Figure 35

Calculator Note

Using a calculator we demonstrate that the power rule holds for n, an irrational number such as $\sqrt{7}$, consider $f(x) = x^{\sqrt{7}}$. Figure 35(a) shows the graph of y' using the numerical procedure for finding a derivative on a calculator. Press [Y =] [MATH] (select 8) and insert nDeriv($x^{\sqrt{7}}$, x, x). Our conjecture will be that $f'(x) = \sqrt{7}x^{\sqrt{7}-1}$. Figure 35(b) is the graph of our assumed derivative. Note that they are identical.

A formal proof of the following theorem is beyond the scope of this book.

POWER RULE

Let $f(x) = x^n$, where n is any real number. Then $f'(x) = nx^{n-1}$.

EXAMPLE 39

$$\text{If}\quad f(x) = x^4, \quad \text{then}\quad f'(x) = 4x^3 \qquad n = 4; n - 1 = 3$$

EXAMPLE 40

$$\text{If}\quad y = x^{-3}, \quad \text{then}\quad \frac{dy}{dx} = -3x^{-4} \qquad n = -3; n - 1 = -4$$

EXAMPLE 41

$$\text{If}\quad f(x) = x^{1/3}, \quad \text{then}\quad f'(x) = \frac{1}{3}x^{(1/3)-1} = \frac{1}{3}x^{-2/3} \qquad n = \tfrac{1}{3}; n - 1 = -\tfrac{2}{3}$$

The notation

$$\frac{d}{dx}[f(x)]$$

means *find the derivative of the function inside the brackets.*

EXAMPLE 42   If $y = \sqrt[3]{x^2}$, find y'. Recall that $\sqrt[3]{x^2}$ can be written equivalently as $(x^2)^{1/3} = x^{2/3}$.

SOLUTION $y' = \dfrac{d}{dx}[x^{2/3}] = \dfrac{2}{3}x^{(2/3)-1} = \dfrac{2}{3}x^{-1/3} = \dfrac{2}{3\sqrt[3]{x}}$

Practice Problem 1 If $y = \sqrt{x^3}$, find y'.

ANSWER $y' = \dfrac{3\sqrt{x}}{2}$

EXAMPLE 43   Find the derivative of $y = 7x^3$.

SOLUTION   Consider y as a product of 7 and x^3. Then

$$y' = 7\frac{d}{dx}[x^3] \qquad \text{Derivative of constant times a function}$$

$$= 7 \cdot 3x^2 \qquad \text{Power rule}$$

$$= 21x^2$$

■

EXAMPLE 44   Find $\dfrac{d}{dx}\left[\dfrac{x^3}{4}\right]$.

SOLUTION   We can write $x^3/4$ as $\frac{1}{4}x^3$. Thus,

$$\frac{d}{dx}\left[\frac{1}{4}x^3\right] = \frac{1}{4}\frac{d}{dx}[x^3] = \frac{1}{4}[3x^2] = \frac{3x^2}{4}$$

■

EXAMPLE 45   Find the derivative of $f(x) = x^\pi$.

SOLUTION   For this example $n = \pi$, so $f'(x) = \pi x^{\pi-1}$.

■

Practice Problem 2   Find y' for $y = \dfrac{3}{\sqrt{x}}$. (**Hint:** Write $1/\sqrt{x}$ using a negative exponent.)

ANSWER $y' = \dfrac{-3}{2x\sqrt{x}}$

The next rule is a great time-saver, especially when taking the derivative of a polynomial. To differentiate the sum of two functions, differentiate each function individually and add the two derivatives. That is, the derivative of a sum (or difference) is the sum (or difference) of the derivatives.

DERIVATIVE OF A SUM OF TWO FUNCTIONS

Let $f(x)$ and $g(x)$ be two functions whose derivatives exist. The derivative of their sum (or difference) is the sum (or difference) of their derivatives:

$$\frac{d}{dx}[f(x) \pm g(x)] = \frac{d[f(x)]}{dx} \pm \frac{d[g(x)]}{dx} = f'(x) \pm g'(x)$$

To prove this theorem, we apply our four-step procedure. Let

$$s(x) = f(x) + g(x)$$

1. $$\dfrac{s(x + h) - s(x)}{h} = \dfrac{[f(x + h) + g(x + h)] - [f(x) + g(x)]}{h}$$

2. $$s'(x) = \lim_{h \to 0} \dfrac{[f(x + h) + g(x + h)] - [f(x) + g(x)]}{h}$$

3. $$= \lim_{h \to 0} \dfrac{f(x + h) - f(x)}{h} + \lim_{h \to 0} \dfrac{g(x + h) - g(x)}{h}$$

4. $$= f'(x) + g'(x)$$

Therefore, $s'(x) = f'(x) + g'(x)$.

EXAMPLE 46   Find the derivative of $5x^3 + 3x^2$.

SOLUTION   Let $f(x) = 5x^3$ and $g(x) = 3x^2$. Then

$$\frac{d}{dx}[f(x) + g(x)] = \frac{d[f(x)]}{dx} + \frac{d[g(x)]}{dx}$$

$$\frac{d}{dx}[5x^3 + 3x^2] = \frac{d[5x^3]}{dx} + \frac{d[3x^2]}{dx}$$

$$= 5\frac{d(x^3)}{dx} + 3\frac{d(x^2)}{dx} \qquad \text{Constant times a function}$$

$$= 5(3x^2) + 3(2x) \qquad \text{Power rule}$$

$$= 15x^2 + 6x$$

By the repeated application of the sum formula, we see that the derivative of a polynomial is simply the sum of the derivatives of its terms.

EXAMPLE 47   Find the derivative of $f(x) = 7x^4 - 5x^3 + 3x^2 - 2$.

SOLUTION

$$f'(x) = \frac{d}{dx}(7x^4) + \frac{d}{dx}(-5x^3) + \frac{d}{dx}(3x^2) + \frac{d}{dx}(-2) \qquad \text{Derivative of a sum}$$

$$= 7\frac{d}{dx}(x^4) - 5\frac{d}{dx}(x^3) + 3\frac{d}{dx}(x^2) + 0 \qquad \begin{array}{l}\text{Derivative of a constant}\\\text{times a function and}\\\text{derivative of a constant}\end{array}$$

$$= 7(4x^3) - 5(3x^2) + 3(2x) \qquad \text{Power rule}$$

$$= 28x^3 - 15x^2 + 6x$$

EXAMPLE 48   Find the derivative of $f(x) = 5x^3 - \dfrac{2}{3}x^{-2} + x^{1/2} - 3$.

SOLUTION

$$f'(x) = 5 \cdot 3x^{(3-1)} + \left(\frac{-2}{3}\right) \cdot (-2)x^{(-2-1)} + \frac{1}{2}x^{(1/2)-1} + 0$$

$$= 15x^2 + \frac{4}{3}x^{-3} + \frac{1}{2}x^{-1/2}$$

$$= 15x^2 + \frac{4}{3x^3} + \frac{1}{2\sqrt{x}}$$

EXAMPLE 49

(a)   Find an equation of the line tangent to $y = 4x^2 + 2x$ at the point $(1, 6)$.

(b)   Where is $y' = 0$?

SOLUTION

(a)   The slope m is the derivative of the function evaluated at $(1, 6)$:

$$f'(x) = 8x + 2$$

$$f'(1) = 8(1) + 2$$
$$= 10$$

The slope m is 10 at the point $(1, 6)$ and an equation of the line is

$$y - 6 = 10(x - 1) \quad \text{or} \quad y = 10x - 4$$

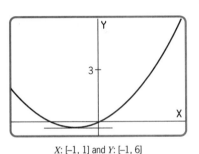

X: [-1, 1] and Y: [-1, 6]

Figure 36

(b) $y' = 0$ wherever $8x + 2 = 0$, that is, where $x = -\frac{1}{4}$ (see Figure 36).

Practice Problem 3 Find an equation of the line that is tangent to the graph of $f(x) = 25 - 3x^2$ at the point $(2, 13)$.

ANSWER   Since $f'(x) = -6x$, $f'(2) = -12$. The tangent line has a slope of -12 and contains the point $(2, 13)$. The tangent line is $y - 13 = -12(x - 2)$ or $y = -12x + 37$.

EXAMPLE 50   If $f(x) = (x + 1)(x + 2)$, find $f'(x)$. (**Hint:** Multiply the two factors.)

SOLUTION $f(x) = x^2 + 3x + 2; f'(x) = 2x + 3$.

EXAMPLE 51   If $f(x) = \dfrac{3x^2 + 3x - 8}{x}$, find $f'(x)$. (**Hint:** Divide by x and write as three terms.)

SOLUTION $f(x) = 3x + 3 - 8x^{-1}$; therefore, $f'(x) = 3 + 0 - 8(-1)x^{-2} = 3 + 8x^{-2}$.

The symbols x and y can represent any variables that have a functional relationship, and $f'(x)$ or dy/dx means the rate of change of y with respect to x. Other letters, such as $h = s(t)$, could be used to represent the same idea. In this case the derivative would be $s'(t)$ or dh/dt and would mean the rate of change of h with respect to t.

Practice Problem 4 The total population of a city at time t is given by $p(t) = 1000 + \frac{1}{30}t^2$. What is the rate of change of the population with respect to time at $t = 45$?

ANSWER $p'(t) = 0 + \dfrac{1}{30} \cdot 2t = \dfrac{t}{15}$ and $p'(45) = 3$ persons per year.

Marginal Analysis

Recall from the preceding section that *marginal* refers to rate of change, that is, to a derivative. (The term *marginal* comes from the Marginalist School of Economic Thought, which originated in Austria with mathematics being applied to economics.) We can think of **marginal cost** as the approximate cost of producing one more item after x items have already been produced. Management must be careful to keep track of marginal costs. If the marginal cost of producing an extra unit exceeds the revenue received from the sale, the company will lose money on that additional unit.

EXAMPLE 52   Suppose that the total cost to produce x hundred cases (24 large bottles per case) of a cola under given conditions is given by

$$C(x) = 600 + 1000x - 2x^2$$

(a)   Find the cost of producing 500 cases.
(b)   Find an expression for the marginal cost.
(c)   Find the marginal cost when $x = 5$.
(d)   When 500 cases have been produced, what is the approximate cost of producing 100 more cases?

SOLUTION

(a)   The cost of producing 500 cases of cola is $C(5)$ or

$$C(5) = 600 + 1000(5) - 2(5)^2 = \$5550$$

(b)   The marginal cost is the derivative with respect to x of the cost function. That is, marginal cost is

$$C'(x) = 1000 - 4x$$

(c) The marginal cost when $x = 5$ is

$$C'(5) = 1000 - 4(5) \quad \text{or} \quad 1000 - 20 = \$980$$

(d) If 500 cases is equivalent to $x = 5$, then 100 more cases is equivalent to a change in x of 1. Since the marginal cost is \$980, the approximate cost of one more x when $x = 5$ is \$980, or the cost of 100 additional cases when the number of cases is 500 is \$980. ■

Calculator Note

Again, we need to take time to emphasize the usefulness of the calculator for problems such as Example 52. The manager of the cola plant has her cost function, which she places in her graphing calculator as $y = 600 + 1000x - 2x^2$. Under the CALC menu she can obtain the cost for any production and the marginal cost for any level of production as she moves the cursor along the graph. Also, she can use TABLE with $Y_1 = 600 + 1000X - 2X^2$ and $Y_2 = 1000 - 4X$ to obtain a table similar to the following.

Production (number of cases of cola)	5	25	50	100
Approximate Cost	\$5550	\$24,350	\$45,600	\$80,600
Approximate Marginal Cost	\$980	\$900	\$800	\$600

Then, if the manager wishes, she can change parts of the cost function and note changes in production costs and marginal costs.

In the preceding example, if we classify the terms of $C(x)$ as

$$C(x) = \underbrace{600}_{\text{Fixed cost}} + \underbrace{100x - 2x^2}_{\text{Variable cost}}$$

then only the part associated with variable cost is found in the derivative (the derivative of 600 is 0). This is in agreement with the economic principle which states that fixed costs of a company have no effect on marginal costs. Generally, marginal cost decreases as more units are produced until it reaches some minimum value, and then it starts to increase. Often this increase is due to the need to add more machinery, more labor, or overtime pay.

It is often appropriate in business to consider average costs and average profits. To take the average of several numerical values, add the values and then divide the total by the number of values. For example, the average of 50, 47, 53, 62, and 58 is computed as

$$\frac{50 + 47 + 53 + 62 + 58}{5} = \frac{270}{5} = 54$$

It is often enlightening to compute and compare the average cost per unit, the average revenue per unit, and the average profit per unit with the marginal functions. These averages per unit are defined as follows.

DEFINITION: AVERAGE COST, REVENUE, PROFIT

If x is the number of units produced, then

$$\textbf{Average cost} = \overline{C}(x) = \frac{C(x)}{x} \qquad \text{Cost per unit}$$

$$\textbf{Average revenue} = \overline{R}(x) = \frac{R(x)}{x} \qquad \text{Revenue per unit}$$

$$\textbf{Average profit} = \overline{P}(x) = \frac{P(x)}{x} \qquad \text{Profit per unit}$$

EXAMPLE 53   For $C(x) = 500 + 20x^2$, find the average cost and the marginal cost and sketch these on the same coordinate system.

SOLUTION   We have

$$\overline{C}(x) = \frac{C(x)}{x} = \frac{500}{x} + 20x \quad \text{and} \quad C'(x) = 40x$$

In Figure 37, $C'(x)$ and $\overline{C}(x)$ appear to intersect at the point where $\overline{C}(x)$ is a minimum.

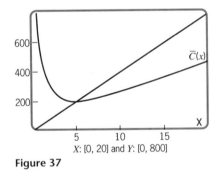

600
400
200

X

5   10   15

X: [0, 20] and Y: [0, 800]

Figure 37

NOTE: In economic theory, it is well known that at the level of production where the average cost is at a minimum, the average cost equals the marginal cost. We will be able to verify this in the next chapter.

Practice Problem 5 Use a graphing calculator for $C(x) = x^3 - 4x^2 + 8x$ to show that the marginal cost function and the average cost function intersect at the minimum value of the average cost function. (Use $0 \le x \le 8$ and $0 \le y \le 10$.)

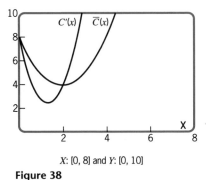

X: [0, 8] and Y: [0, 10]

Figure 38

ANSWER   We have $C'(x) = 3x^2 - 8x + 8$ and $\overline{C}(x) = x^2 - 4x + 8$. As shown in Figure 38, the graph for the marginal cost function appears to intersect the graph for the average cost function at the point (2, 4), which is the point at which there is a minimum value of the average cost function.

SUMMARY

In this section we introduced the following concepts involving derivatives.

1.   The derivative of a constant is zero.

$$\frac{d}{dx} C = 0$$

2.   The derivative of x^n, where n is any real number, is nx^{n-1}.

$$\frac{d}{dx} x^n = nx^{n-1}$$

3.   The derivative of a sum (or difference) is the sum (or difference) of the derivatives.

$$\frac{d}{dx} [f(x) \pm g(x)] = f'(x) \pm g'(x) \quad \text{if } f'(x) \text{ and } g'(x) \text{ exist}$$

4.   The derivative of a constant times a function of x is the constant times the derivative of the function.

$$\text{If } f = cg(x), \quad \text{then} \quad f'(x) = cg'(x)$$

Exercise Set 2.5

Find each of the following.

1. $f'(x)$ for $f(x) = 6$

2. $f'(x)$ for $f(x) = 6x + 2$

3. $\dfrac{dy}{dx}$ for $y = -x + 3$

4. $\dfrac{dy}{dx}$ for $y = \sqrt{5}$

5. $D_x[3x^2 + 5]$ (D_x is d/dx.)

6. $D_x[5x^2 - 3x]$

7. y' for $y = 5x^3 - 3x^2 + 7$

8. y' for $y = -3x^3 + 2x^2 - x$

9. $\dfrac{d}{dx}(2x^{-4})$

10. $\dfrac{d}{dx}\left(\dfrac{4}{x^2}\right)$

11. $f'(x)$ for $f(x) = \dfrac{2}{\sqrt{x}}$

12. $f'(x)$ for $f(x) = \dfrac{2}{\sqrt[3]{x}}$

Find the derivative $f'(x)$, and then find, if they exist, $f'(0)$, $f'(2)$, and $f'(-3)$ for the following functions. Check the values for $f'(-3)$ using a graphing calculator.

13. $f(x) = 6x^4 - 3x$

14. $f(x) = 5x^4 - 3x^3 + 2x^2$

15. $f(x) = \dfrac{3}{x} + 4x$

16. $f(x) = -3x^2 - \dfrac{4}{x}$

17. $f(x) = 4x^2 - \dfrac{4}{x}$

18. $f(x) = \dfrac{-3}{x} + 4x$

19. $f(x) = 4x^{1/2}$

20. $f(x) = 6x^{2/3}$

21. $f(x) = 6x^{2/3} + 4x^{1/2}$

22. $f(x) = 9x^{-2/3} - \dfrac{4}{\sqrt{x}}$

23. $f(x) = 3x^2 - \dfrac{4}{\sqrt{x}}$ 24. $f(x) = 3x^4 - \dfrac{6}{\sqrt[3]{x}}$

25. $f(x) = 7\sqrt{x} + 9\sqrt[3]{x}$ 26. $f(x) = \dfrac{6}{\sqrt{x}} + \dfrac{3}{\sqrt[3]{x}}$

For the following exercises, find the points on the graph of each where the tangent line is horizontal (i.e., $y' = 0$).

27. $y = 3x^2$ 28. $y = 5x^2$

29. $y = 4x^2 - x$

30. $y = 0.01x^2 + 0.4x + 30$

For Exercises 31–34, find the equation of the tangent line to the graph at $x = 1$.

31. $y = 3x^2 + 2$ 32. $y = \dfrac{8}{x} + 3$

33. $y = -0.02x^2 + 0.2x$ 34. $y = \dfrac{1}{3}x^3 + 4$

35. **Exam.**   Reading Company manufactures and sells an industrial-strength cleaning fluid. The equations presented below represent the revenue (R) and cost (C) functions for the company, where x is equal to a thousand gallons of fluid.

$$R(x) = -80 + 26x - 0.05x^2 \quad \text{and}$$
$$C(x) = 40 + 8x + 0.01x^2$$

Which of the following functions represents the marginal cost of 1 gallon of cleaning fluid?
(a) $8x + 0.01x^2$ (b) $8 + 0.01x$
(c) $8 + 0.02x$ (d) $\dfrac{8}{0.02x}$

(e) Some function other than those given above.

36. **Exam.**   For $C(x)$ and $R(x)$ given in Exercise 35, what is the marginal profit?
(a) $R(x) - C(x)$ (b) $18 - 1.2x$
(c) $\dfrac{R'(x)}{C'(x)}$ (d) $18x - 1.2x^2$

(e) Some function other than those given above.

Applications (Business and Economics)

37. **Cost Functions.**   A cost function is given as

$$C(x) = 800 + 400x - x^2$$

(a) Find the marginal cost function, $C'(x)$.
(b) Find $C'(1)$.   (c) Find $C'(2)$.
(d) Find $\bar{C}(2)$.
(e) Check parts (b), (c), and (d) with a graphing calculator.

38. **Demand Functions.**   A demand function is given as

$$p(x) = 138 - x^2, \qquad 1 \le x \le 11$$

(a) Find the instantaneous rate of change, $p'(x)$.
(b) Find $p'(2)$.
(c) Find $p'(5)$.
(d) Check parts (b) and (c) with a graphing calculator.

39. **Revenue Functions.**   A revenue function is given as

$$R(x) = 20x - \dfrac{x^2}{500}, \qquad 0 \le x \le 10{,}000$$

(a) Find the marginal revenue function.
(b) Find the value of x that makes the marginal revenue equal to zero.
(c) Find the value of the revenue at the value of x found in part (b).
(d) Find $\bar{R}(x)$.

40. **Profit Functions.**   A profit curve is given by

$$P(x) = \dfrac{x^2}{2} + 4x$$

(a) Find the derivative.
(b) What values of x make the derivative positive?
(c) What value of x makes the derivative zero?
(d) What values of x make the derivative negative?
(e) Find $\bar{P}(x)$.

41. **Cost, Revenue, Profit.**   Cost and revenue functions are given as

$$C(x) = 6x + 100 \quad \text{and} \quad R(x) = 40x - 0.01x^2$$

(a) Find the marginal cost at $x = 4$. At $x = 10$. Check with a graphing calculator.
(b) What can you say about marginal cost?
(c) Find the marginal revenue at $x = 5$. Check with a graphing calculator.

(d) Find the marginal profit at $x = 4$. Check with a graphing calculator.

(e) What is the profit when $x = 3$?

(f) Find $\overline{P}(x)$.

Applications (Social and Life Sciences)

42. **Pulse Rates.** A doctor has found the relation between the pulse rate y and a person's height x, in inches, to be approximately

$$y = \frac{600}{\sqrt{x}}, \qquad 34 \le x \le 74$$

Find the rate of change of the pulse rate at the following heights.

(a) 36 inches   (b) 49 inches

(c) 64 inches

43. **Rate of Pollution Change.**   Suppose that the concentration of a pollutant in the air is given in parts per million by

$$P(x) = 0.2x^{-2}$$

where x is the distance in miles from a factory.

(a) Find the instantaneous rate of change of pollutant.

(b) Find $P'(1)$ and check.

(c) Find $P'(2)$ and check.

44. **Learning Functions.**   Suppose that a person learns $f(x)$ instructions in x hours according to the function

$$f(x) = 48\sqrt{x}, \qquad 0 \le x \le 10$$

(a) Find the instantaneous rate of change of learning.

(b) Find $f'(1)$.

(c) Find $f'(4)$.

(d) Find $f'(9)$.

(e) Check parts (b), (c), and (d) using a graphing calculator.

45. **Learning Functions.** Suppose that the learning function in Exercise 44 is changed to

$$f(x) = 72 \sqrt[3]{x}, \qquad 0 \le x \le 10$$

(a) Find the instantaneous rate of change of learning.

(b) Find $f'(1)$.

(c) Find the instantaneous rate of change of learning at the end of 8 hours; that is, find $f'(8)$.

(d) Find the number of items learned at the end of 8 hours, $f(8)$.

(e) Check parts (b), (c), and (d) with a graphing calculator.

Review Exercises

Find the following limits if they exist.

46. $\displaystyle\lim_{x \to 1/\sqrt{3}} \frac{6x^3 + 4}{3x^2 - 1}$ 47. $\displaystyle\lim_{x \to 2} \frac{x^2 + x + 4}{x - 2}$

Use the definition of average change for $h = 0.1$ to find the average change in y when $x = 2$ in the following functions.

48. $y = 6x^2 + 5$ 49. $y = 4x^2 + 3x + 2$

Use the definition of a derivative,

$$\lim_{h \to 0} \frac{f(x + h) - f(x)}{h}$$

to find the derivatives of the following functions.

50. $y = 6x^2 + 5$ 51. $y = 4x^2 + 3x + 2$

2.6 PRODUCTS AND QUOTIENTS

OVERVIEW   Since the derivative of the sum of two functions is the sum of their derivatives, if these derivatives exist, we might expect the derivative of the product of two functions to be the product of their derivatives. In this section we show that this statement is not true. We develop formulas for finding the derivative of the product of two functions and the quotients of two functions. When these formulas are combined with formulas already obtained, we can find the derivatives of many new functions. These two formulas significantly increase our ability to compute and apply derivatives. In this section we develop and use

- The product formula
- The quotient formula

There are situations where a function can be considered the product of two other functions.

EXAMPLE 54   For $y = (x^2 + x - 1)(3x + 2)$, find dy/dx.

SOLUTION   One way we can solve this problem is to multiply the two functions and then take the derivative term by term.

$$y = (x^2 + x - 1)(3x + 2)$$
$$= 3x^3 + 5x^2 - x - 2$$

Then we have

$$y' = 9x^2 + 10x - 1 \qquad \blacksquare$$

Now let's consider this example without multiplying the two factors to get a polynomial. Think of the problem as a product of two functions,

$$y = f(x) \cdot g(x)$$

where $f(x) = (x^2 + x - 1)$ and $g(x) = 3x + 2$. We use the following formula for problems of this nature.

PRODUCT FORMULA

Let $f(x)$ and $g(x)$ be two functions whose derivatives with respect to x exist. Then

$$\frac{d}{dx}[f(x)g(x)] = f(x)g'(x) + g(x)f'(x)$$

Sometimes students use a memory scheme for the product formula. Let's write the formula as *First factor times derivative of the Second plus the Second factor times the derivative of the First*, abbreviated as $FS' + SF'$.

EXAMPLE 55   For

$$y = \overbrace{(x^2 + x - 1)}^{F}\overbrace{(3x + 2)}^{S}$$

find dy/dx.

SOLUTION

$$\frac{dy}{dx} = \overbrace{(x^2 + x - 1)}^{F} \cdot \overbrace{(3)}^{S'} + \overbrace{(3x + 2)}^{S} \cdot \overbrace{(2x + 1)}^{F'}$$

$$= 3x^2 + 3x - 3 + 6x^2 + 7x + 2$$
$$= 9x^2 + 10x - 1$$

Note that this is the same answer that we obtained in Example 54.

EXAMPLE 56 For $f(x) = x^2(x + 1)$, find $f'(x)$.

SOLUTION

$$f(x) = \overbrace{x^2}^{F} \cdot \overbrace{(x + 1)}^{S}$$

$$f'(x) = \overbrace{x^2(1)}^{F \cdot S'} + \overbrace{(x + 1)(2x)}^{S \cdot F'}$$
$$= 3x^2 + 2x$$

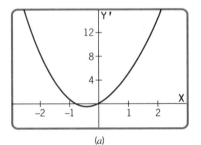

(a)

$y' = 3x^2 + 2x$

(b)

X: [–3, 3] and Y: [–4, 16]

Figure 39

Calculator Note

A formal proof of the product formula is given later in this section. For now, we support the product rule graphically. In Figure 39(a) is shown the graph of the numerical values of the derivative of $f(x) = x^2 (x + 1)$. (See page 95.) Compare this graph with the graph in Figure 39(b) of $y' = 3x^2 + 2x$, obtained using the product formula in the preceding example. Note that the graphs in Figure 39 appear to be identical.

Practice Problem 1 For $y = (3x^2 + 5x)(x^2)$, find y'.

ANSWER $y' = 12x^3 + 15x^2$

EXAMPLE 57 For $y = \dfrac{2x + 1}{x^2}$, find y'.

SOLUTION

$$y = (2x + 1) \cdot \frac{1}{x^2} \qquad \text{Rewrite as a product.}$$

$$= \overbrace{(2x + 1) \cdot x^{-2}}^{F \cdot S}$$

$$\overbrace{y' = (2x + 1)(-2x^{-3}) + x^{-2}(2)}^{F \cdot S' + S \cdot F'}$$

$$= \frac{-2(2x + 1)}{x^3} + \frac{2}{x^2}$$

$$= \frac{-4x - 2 + 2x}{x^3}$$

$$= \frac{-2x - 2}{x^3}$$

For practice, let's find this derivative another way. Write the equation as

$$y = \frac{2x}{x^2} + \frac{1}{x^2} = 2x^{-1} + x^{-2}$$

This results in

$$y' = -2x^{-2} - 2x^{-3} = \frac{-2}{x^2} - \frac{2}{x^3} = \frac{-2x - 2}{x^3}$$

which is the same answer.

Practice Problem 2 If $h(x) = (\sqrt[3]{x} - 3x^2)(9x^{-1})$, find $h'(x)$.

ANSWER $h'(x) = \dfrac{-6\sqrt[3]{x} - 27x^2}{x^2}$

Calculator Note

You can catch careless mistakes by doing a quick check with a calculator. Let's evaluate $h'(x)$ from Practice Problem 2 at $x = 1$.

$$h'(1) = \frac{-6 - 27}{1} = -33$$

Using nDeriv, $h'(1) \approx -33$. Although evaluation at a single point is not proof that an accurate derivative has been obtained, such a check can help you locate mistakes and should increase your confidence level.

We now give a formal proof of the product formula using our four-step procedure for finding a derivative (from the definition of a derivative). Let $y(x) = f(x) \cdot g(x)$. Then

1. $\dfrac{y(x + h) - y(x)}{h} = \dfrac{f(x + h) \cdot g(x + h) - f(x)g(x)}{h}$

2. $y' = \lim\limits_{h \to 0} \dfrac{f(x + h) \cdot g(x + h) - f(x)g(x)}{h}$

Now add and subtract $f(x + h)g(x)$ as needed to get

3. $\lim\limits_{h\to 0} \dfrac{[f(x + h)g(x + h) - f(x + h)g(x)] + [f(x + h)g(x) - f(x)g(x)]}{h}$

$= \lim\limits_{h\to 0} \dfrac{f(x + h)g(x + h) - f(x + h)g(x)}{h} + \lim\limits_{h\to 0} \dfrac{f(x + h)g(x) - f(x)g(x)}{h}$

$= \lim\limits_{h\to 0} f(x + h) \cdot \lim\limits_{h\to 0} \dfrac{g(x + h) - g(x)}{h} + \lim\limits_{h\to 0} g(x) \cdot \lim\limits_{h\to 0} \dfrac{f(x + h) - f(x)}{h}$

4. $= f(x)g'(x) + g(x)f'(x)$

Quotient Formula

Let's now consider situations where the function is the quotient of two functions, $f(x)$ and $g(x)$.

EXAMPLE 58   Find the derivative of $y = \dfrac{x^3 + 4}{x^2}$.

SOLUTION   Without a formula for the derivative of a quotient, we write the function as two fractions.

$$y = \frac{x^3 + 4}{x^2} = \frac{x^3}{x^2} + \frac{4}{x^2} = x + 4x^{-2}$$

Therefore,

$$y' = 1 - 8x^{-3} = 1 - \frac{8}{x^3} = \frac{x^3 - 8}{x^3}$$

Many quotients cannot be written as a sum or difference of two fractions, so we need a formula for the derivative of a quotient of two functions. You might be tempted to think that the derivative of a quotient, $f(x)/g(x)$, is the quotient of the derivatives $f'(x)$ and $g'(x)$. We can see that this is incorrect from the fact that $x^2 = x^5/x^3$, but

$$\frac{d}{dx}x^2 \neq \frac{\dfrac{d}{dx}[x^5]}{\dfrac{d}{dx}[x^3]}$$

We know that they are unequal because

$$\frac{d}{dx}x^2 = 2x \quad \text{and} \quad \frac{\dfrac{d}{dx}x^5}{\dfrac{d}{dx}x^3} = \frac{5x^4}{3x^2} = \frac{5}{3}x^2$$

The correct formula for the derivative of a quotient is as follows.

QUOTIENT FORMULA

Let $f(x)$ and $g(x)$ be two differentiable functions with respect to x. If $g(x) \neq 0$, then

$$\frac{d}{dx}\left[\frac{f(x)}{g(x)}\right] = \frac{g(x)f'(x) - f(x)g'(x)}{[g(x)]^2}$$

In other words, the quotient formula states that the derivative of a quotient is **the denominator times the derivative of the numerator minus the numerator times the derivative of the denominator, all divided by the denominator squared:**

$$\frac{DN' - ND'}{D^2}$$

EXAMPLE 59   Using the quotient formula, find the derivative of

$$y = \frac{x^3 + 4}{x^2}$$

SOLUTION

$$y = \frac{\overbrace{x^3 + 4}^{N}}{\underbrace{x^2}_{D}}$$

$$y' = \frac{x^2 \cdot (3x^2) - (x^3 + 4) \cdot 2x}{(x^2)^2} \qquad \frac{DN' - ND'}{D^2}$$

$$= \frac{3x^4 - 2x^4 - 8x}{x^4}$$

$$= \frac{x^4 - 8x}{x^4}$$

$$= \frac{x^3 - 8}{x^3} \qquad\qquad \text{Divide by } x.$$

Notice that this is the same answer that we found in Example 58.

EXAMPLE 60   If $y = \dfrac{3x - 2}{5x + 3}$, find $\dfrac{dy}{dx}$.

SOLUTION

$$y = \frac{\overbrace{3x - 2}^{N}}{\underbrace{5x + 3}_{D}}$$

$$\frac{dy}{dx} = \frac{(5x + 3)(3) - (3x - 2)(5)}{(5x + 3)^2} \qquad \frac{DN' - ND'}{D^2}$$

$$= \frac{15x + 9 - 15x + 10}{(5x + 3)^2}$$

$$= \frac{19}{(5x + 3)^2} \qquad \blacksquare$$

The formal proof of the quotient formula is very similar to that of the product formula and will not be given at this time. However, we give support graphically to the quotient formula as follows.

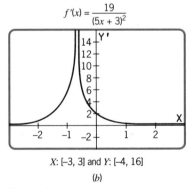

$f'(x) = \dfrac{19}{(5x + 3)^2}$

X: [–3, 3] and Y: [–4, 16]

(b)

Figure 40

Calculator Note

For the function $f(x) = \dfrac{3x - 2}{5x + 3}$, we obtain in Figure 40(a) the graph of the numerical values of $f'(x)$. (See page 95.) Then using $f'(x) = \dfrac{19}{(5x + 3)^2}$ from the preceding example we obtain the graph shown in Figure 40(b). These two graphs appear to be identical.

Practice Problem 3 Find $y' = \dfrac{6x - 1}{3x + 2}$.

ANSWER $y' = \dfrac{15}{(3x + 2)^2}$

EXAMPLE 61   Find the derivative of $y = \dfrac{2x + 1}{\sqrt{x}}$.

SOLUTION

$$y = \frac{\overbrace{2x + 1}^{N}}{\underbrace{\sqrt{x}}_{D}}$$

$$y' = \frac{\sqrt{x}(2) - (2x + 1)\left(\frac{1}{2\sqrt{x}}\right)}{(\sqrt{x})^2} \qquad \frac{DN' - ND'}{D^2}$$

$$= \frac{2\sqrt{x} - \dfrac{x}{\sqrt{x}} - \dfrac{1}{2\sqrt{x}}}{x} \qquad \text{Multiply and simplify.}$$

$$= \frac{4x - 2x - 1}{2x\sqrt{x}} \qquad \text{Multiply numerator and denominator by } 2\sqrt{x}.$$

$$= \frac{2x - 1}{2x\sqrt{x}}$$

Calculator Check: $y'(1) = \dfrac{2 \cdot 1 - 1}{2 \cdot 1\sqrt{1}} = \dfrac{1}{2}$ and nDeriv$((2x + 1)/(\sqrt{x}), x, 1)$

$$= .4999998126.$$

Practice Problem 4 Find the derivative of $y = \dfrac{x^2}{2x + 1}$.

ANSWER $y' = \dfrac{2x^2 + 2x}{(2x + 1)^2}$

Whenever the numerator of a quotient is a constant, the quotient rule may not be the rule that you should use. Although you get the correct answer, often the problem can be worked more easily by rewriting the function and using the power rule. This is demonstrated in the next example.

EXAMPLE 62   Find the derivative of $y = \dfrac{2}{x^2}$.

SOLUTION   Using the quotient rule we have

$$y' = \frac{x^2 \cdot \dfrac{d}{dx}(2) - 2 \cdot \dfrac{d}{dx}(x^2)}{(x^2)^2} = \frac{x^2 \cdot 0 - 4x}{x^4} = \frac{-4}{x^3}$$

If we rewrite the function as $y = 2/x^2 = 2x^{-2}$, we can use the power rule and arrive at the derivative much more quickly.

$$y' = -2(2)x^{-3} = \frac{-4}{x^3}$$

EXAMPLE 63 Find the equation of the tangent line to the graph of $y = x/(x + 1)$ when $x = 1$.

SOLUTION When $x = 1$,

$$y = \frac{1}{1 + 1} = \frac{1}{2}$$

so the equation of the tangent line will be $y - \frac{1}{2} = m(x - 1)$, where $m = y'(1)$.

$$y'(x) = \frac{(x + 1)(1) - x \cdot 1}{(x + 1)^2} = \frac{1}{(x + 1)^2} \qquad \frac{DN' - ND'}{D^2}$$

Evaluating $y'(1) = \frac{1}{4}$, we now have the slope and can find an equation of the tangent line.

$$y - \frac{1}{2} = \frac{1}{4}(x - 1) \quad \text{or} \quad y = \frac{1}{4}x + \frac{1}{4}$$

∎

EXAMPLE 64 If $y = \frac{3x^2 - 1}{2x^2 + 2}$, find $\frac{dy}{dx}$ and check your answer at $x = 1$.

SOLUTION

$$\frac{dy}{dx} = \frac{(2x^2 + 2)(6x) - (3x^2 - 1)(4x)}{(2x^2 + 2)^2} \qquad \frac{DN' - ND'}{D^2}$$

$$= \frac{12x^3 + 12x - 12x^3 + 4x}{(2x^2 + 2)^2}$$

$$= \frac{16x}{(2x^2 + 2)^2}$$

> ✔ Calculator Check: $\dfrac{dy}{dx}$ at $x = 1$ is $\dfrac{16 \cdot 1}{(2 \cdot 1 + 2)^2} = \dfrac{16}{16} = 1$.
>
> $\text{nDeriv}((3x^2 - 1)/(2x^2 + 2), x, 1) = 1$.

∎

Now that we have a quotient formula, we can introduce a new concept in our discussion of marginal analysis. The average cost per item can be found by dividing the total cost by the number of items, $C(x)/x$, and the derivative of this expression is the **marginal average cost**.

DEFINITION: MARGINAL AVERAGE COST

The **marginal average cost** is the derivative of the average cost.

$$\text{Marginal average cost} = \frac{d}{dx}\left[\frac{C(x)}{x}\right] = \frac{d}{dx}[\overline{C}(x)] = \overline{C}'(x)$$

Industry is interested in making the average cost as small as possible. In the next chapter we will learn how to accomplish this by finding the minimum average cost.

EXAMPLE 65

The total cost to manufacture x automobile engines in a week is

$$C(x) = 1000 + 500x - x^3$$

(a) Find the average cost per engine function.
(b) Find the marginal average cost function.

SOLUTION

(a) $\overline{C}(x) = \dfrac{C(x)}{x} = \dfrac{1000 + 500x - x^3}{x}$. By dividing through by x we have

$$\overline{C}(x) = 1000x^{-1} + 500 - x^2$$

(b) The marginal average cost function is

$$\frac{d}{dx}[\overline{C}(x)] = -1000x^{-2} - 2x = \frac{-1000}{x^2} - 2x \qquad ■$$

SUMMARY

1. *Product formula:* The derivative of a product is the first factor times the derivative of the second factor plus the second factor times the derivative of the first factor. If $f'(x)$ and $g'(x)$ exist, then

$$\frac{d}{dx}[f(x)g(x)] = f(x)g'(x) + g(x)f'(x)$$

2. *Quotient formula:* The derivative of a quotient is the denominator times the derivative of the numerator minus the numerator times the derivative of the denominator, all divided by the denominator squared. That is,

$$\frac{d}{dx}\left[\frac{f(x)}{g(x)}\right] = \frac{g(x)f'(x) - f(x)g'(x)}{[g(x)]^2}$$

Exercise Set 2.6

Find dy/dx by the product formula; then check your answer by multiplying the two factors and taking the derivative term by term.

1. $y = (3x + 2)(x - 3)$ 2. $y = (2x - 3)(x + 2)$
3. $y = (x^2 - 1)(x^2 + 1)$
4. $y = (3x^2 - 2)(3x^2 + 2)$

5. $y = (x^2 + x + 1)(x - 1)$
6. $y = (x^2 - x + 1)(x + 1)$

Find the derivatives of the following functions by using the quotient formula.

7. $y = \dfrac{5x}{2x - 3}$ 8. $y = \dfrac{3x - 2}{5x}$

9. $y = \dfrac{3x + 2}{2x - 3}$

10. $y = \dfrac{3x + 4}{5x - 1}$

Find the following and check each answer by finding the value of the derivative at $x = 1$ on a graphing calculator.

11. $\dfrac{d}{dx} [(3x^2 - 2x + 1)(4x + 5)]$

12. $\dfrac{d}{dx} [(4x^2 - 3x + 2)(3x - 2)]$

13. $D_x \left[\dfrac{5}{x^2 + 1} \right]$

14. $D_x \left[\dfrac{7}{2x^2 - 1} \right]$

15. $\dfrac{d}{dx} \left(\dfrac{3x - 2}{x^2 - 3} \right)$

16. $\dfrac{d}{dx} \left(\dfrac{2x + 3}{x^2 - 2} \right)$

17. Find y' for $y = \dfrac{3\sqrt{x}}{x^2 - 4}$.

18. Find y' for $y = \dfrac{4x^{1/2}}{\sqrt{x} - 3}$.

Find dy/dx for the following functions and check with a graphing calculator.

19. $y = 5(x - 2) \left(\dfrac{4x}{x + 1} \right)$

20. $y = (3x + 2)(2x + 7)x^{-1}$

21. $y = (3x - 2)^{-1}(x^3 + 2)$

22. $y = (2x - 3)^{-1}(3x^2 + 2)$

23. $y = 3(2x + 5)^{-1} + \sqrt{x}\,(3x - 5)$

24. $y = \dfrac{2x^2 - 3}{3x + 2} + \dfrac{3}{x}$

25. $y = \dfrac{2x^2 - 1}{x^3 + 1} + \dfrac{1}{\sqrt{x}}$

26. $y = \sqrt{x}\,(x^2 + x)$

27. Find an equation of the tangent line to the graph of $y = 5/(x^2 - 2x)$ at $(-2, \frac{5}{8})$.

Applications (Business and Economics)

28. **Average Cost.** A cost function is given by $C(x) = 7x^2 - 3x + 10$.
   (a) Find the average cost, $\overline{C}(x) = C(x)/x$.
   (b) Find the marginal cost, $C'(x)$.

   (c) Find the marginal average cost, $\overline{C}'(x)$.
   (d) Find $\overline{C}'(5)$.

29. **Demand Function.** A demand function is given by
$$p(x) = \dfrac{4000(25 - x)}{x^2}, \qquad 1 \le x \le 25$$
   (a) Find the marginal demand function.
   (b) Find the marginal demand when $x = 2$. Check with your graphing calculator.
   (c) Find the marginal average demand.

30. **Average Cost.** A cost function is given by $C(x) = 500x^2 + 2$.
   (a) Find the marginal cost function.
   (b) Find the marginal average cost function.

31. **Average Revenue.** A revenue function is given by $R(x) = 1000(x^2 - 3)$.
   (a) Find the marginal revenue function.
   (b) Find the marginal average revenue function.

32. **Average Profit**
   (a) Using Exercises 30 and 31, find the average profit function.
   (b) Find the marginal profit function.
   (c) Find the marginal average profit function.

Applications (Social and Life Sciences)

33. **Learning Functions.** A learning function giving the proportion of a set of activities performed without a mistake after t hours is given as
$$L = \dfrac{80t}{90t + 85}$$
   (a) Find the instantaneous rate of learning.
   (b) Find the rate of learning at $t = 1$ and check with a graphing calculator.
   (c) Find the rate of learning at $t = 8$ and check.

34. **Rate of Change of Drug Concentration.** The concentration of a drug in a person's bloodstream t hours after injection is
$$c(t) = \dfrac{t}{16t^2 + 10t + 63}$$
   (a) Find the rate of change of the concentration.
   (b) What is the concentration after 1 hour?
   (c) What is the rate of change of the concentration after 1 hour? Check with a graphing calculator.

Review Exercises

Find dy/dx for the following functions.

35. $y = 7x^2 - 3x + 2$ 36. $y = 10x^2 - 5x + 3$
37. $y = 3x^2 - x^3 + 2x^4$ 38. $y = 5 - 3x + 6x^2$
39. $y = 5x^3 - 2x^{-1}$ 40. $y = 4x^3 - 6x^{-2}$
41. $y = 6x^4 - 5x^{-2} + 3$ 42. $y = 7x^{-3} + 2x - 3$

Use the definition of the average change to find the average change in y for the following functions when x = 2 and h = 0.1.

43. $y = 3x^2$ 44. $y = x^2 + 4$

Use the definition of the derivative,

$$\lim_{h \to 0} \frac{f(x + h) - f(x)}{h}$$

to find the derivatives of the following functions.

45. $y = 3x^2$ 46. $y = x^2 + 4$

Find f'(x) for the following functions.

47. $f(x) = 2x^{1/2} + 3x^2$ 48. $f(x) = 3x^{1/3} - 5x^3$
49. $f(x) = 2x^{1/2} - 4x^{-1/2}$ 50. $f(x) = 3x^{1/3} - 3x^{-1/3}$
51. $f(x) = 4x^{-1/2} - 6x^{-2/3}$
52. $f(x) = 9x^{-2/3} + 8x^{-3/4}$

2.7 THE CHAIN RULE

OVERVIEW   In many practical situations, the quantity under consideration is described as a function of one variable, which in turn is a function of a second variable. In this section we develop a formula for the derivative of the original quantity with respect to the second variable. The formula we develop for such derivatives is called the **chain rule**. The chain rule is important because it describes how the derivative of a composition of a function is obtained. For example, a company may know

$$\frac{dC}{dP} = \text{Rate of change of cost with respect to production}$$

$$\frac{dP}{dt} = \text{Rate of change of production with respect to time}$$

Since cost also varies with respect to time, we would like to know dC/dt. In this section we learn how to find dC/dt in terms of dC/dP and dP/dt.

A useful way of combining functions $f(x)$ and $g(x)$ is to let the y-coordinates of $g(x)$ be used as the x-coordinates of $f(x)$. This results in a new function $f(g(x))$, a function of a function, which is called the **composite** of $f(x)$ and $g(x)$. In this section we

- Define the composite of two functions
- Introduce the chain rule
- Develop the general power rule

We begin this section by considering functions of functions.

EXAMPLE 66   If $f(x) = x^2 - 7$ and $g(x) = 1/3x$, then what would we mean by $f(g(x))$? Since $f(x)$ is the function that squares its x values and then subtracts 7, then f would square $g(x)$ and subtract 7:

$$f(g(x)) = [g(x)]^2 - 7$$

or

$$f(g(x)) = \left(\frac{1}{3x}\right)^2 - 7 \qquad \text{After substituting for } g(x)$$

$$= \frac{1}{9x^2} - 7$$

Let's call the new function

$$h(x) = \frac{1}{9x^2} - 7$$

The function h is said to be a **composite** of the two functions f and g.

DEFINITION: COMPOSITE FUNCTION

A function h is a **composite** of functions f and g if

$$h(x) = f[g(x)]$$

The domain of h is the set of all numbers x such that x is in the domain of g and $g(x)$ is in the domain of f.

Similarly, to decide what is meant by $g(f(x))$, we note that $g(x)$ is the function that triples its x values and then takes the reciprocal. Therefore, g would triple $f(x)$ and then take the reciprocal:

$$g(f(x)) = \frac{1}{3 \cdot f(x)}$$

or

$$g(f(x)) = \frac{1}{3 \cdot (x^2 - 7)} \qquad \text{After substituting for } f(x)$$

$$= \frac{1}{3x^2 - 21}$$

We observe that $g(f(x)) \neq f(g(x))$, so the order of the composition makes a difference.

EXAMPLE 67   Find $f(g(x))$ if $f(x) = 5x^2 + 3x$ and $g(x) = x - \sqrt{x}$. Each x in the equation for $f(x)$ may be replaced by $g(x)$ to obtain

$$f(g(x)) = 5 \cdot (g(x))^2 + 3 \cdot (g(x))$$
$$= 5 \cdot (x - \sqrt{x})^2 + 3 \cdot (x - \sqrt{x})$$

Similarly,

$$g(f(x)) = f(x) - \sqrt{f(x)}$$
$$= (5x^2 + 3x) - \sqrt{5x^2 + 3x}$$

EXAMPLE 68   If $f(x) = x^2 + 4$ and $g(x) = 3x + 6$, find $f[g(x)]$.

SOLUTION   Replace each x in $f(x)$ by $g(x)$ to get

$$f[g(x)] = [g(x)]^2 + 4 = (3x + 6)^2 + 4 = 9x^2 + 36x + 40$$

Practice Problem 1 If $f(x) = \sqrt{x + 5}$ and $g(x) = x^2$, find $f[g(x)]$. Also find $g[f(x)]$.

ANSWER $f[g(x)] = \sqrt{x^2 + 5}$ and $g[f(x)] = (\sqrt{x + 5})^2 = x + 5$ for $x \geq -5$

Sometimes it is easier to think of a composite function in terms of two variables, u and x. Let's write the problem in Example 68 as $f(u) = u^2 + 4$ and $u = 3x + 6$. Then $f(3x + 6)$ becomes $(3x + 6)^2 + 4$, which is the same answer that we found for $f[g(x)]$.

EXAMPLE 69   Given $y(u) = u^2$ and $u(x) = x^2 + 1$, find $y[u(x)]$.

SOLUTION $y[u(x)] = (x^2 + 1)^2 = x^4 + 2x^2 + 1$

EXAMPLE 70   Let $u(x) = x^2 + 1$ and $y(u) = u^2$. Then $y[u(x)]$ is a composite function, and it is simple enough so that we can find its derivative with methods we already know.

$$y[u(x)] = (x^2 + 1)^2$$
$$= x^4 + 2x^2 + 1$$
$$\frac{d}{dx}[y(u(x))] = 4x^3 + 4x$$

This can be separated into components as follows:

$$\frac{d}{dx}[y(u(x))] = 2(x^2 + 1) \cdot (2x) \qquad \text{Factor } 4x^3 + 4x.$$

$$= 2u \cdot 2x$$

$$= \frac{d}{du}[u^2] \cdot \frac{d}{dx}[x^2 + 1]$$

$$= \frac{dy}{du} \cdot \frac{du}{dx}$$

In the preceding example we can see that the derivative of a composite is equal to the product of the derivatives of its components. This is an example of the general rule for the derivative of a composite function, called the **chain rule**. We state it without proof.

THE CHAIN RULE

If y is a function of u and u is a function of x, whose range is in the domain of y, and if dy/du and du/dx exist, then y is a function of x and

$$\frac{dy}{dx} = \frac{dy}{du} \cdot \frac{du}{dx}$$

NOTE: These derivatives are *not* considered as fractions where du in the numerator and denominator divide out.

Alternative notations often used for the chain rule are

$$\frac{d}{dx} y(u(x)) = \frac{dy}{du} \cdot \frac{du}{dx} \quad \text{and} \quad [y(u)]'(x) = y'(u(x)) \cdot u'(x)$$

EXAMPLE 71   If $y = 5u^2 + 2u - 1$ and $u = 3x + 2$, then

$$\frac{dy}{du} = 10u + 2 \quad \text{and} \quad \frac{du}{dx} = 3$$

So, by the chain rule, $\dfrac{dy}{dx} = \dfrac{dy}{du} \cdot \dfrac{du}{dx} = (10u + 2)(3)$

Therefore, $dy/dx = 30u + 6$. Notice that when $3x + 2$ is substituted for u we obtain

$$\frac{dy}{dx} = 30(3x + 2) + 6 = 90x + 66 \qquad \blacksquare$$

If the function y is found in terms of x in the preceding example, we obtain

$$\begin{aligned}
y(x) &= 5(3x + 2)^2 + 2(3x + 2) - 1 \\
&= 5(9x^2 + 12x + 4) + 6x + 4 - 1 \\
&= 45x^2 + 60x + 20 + 6x + 3 \\
&= 45x^2 + 66x + 23
\end{aligned}$$

Now dy/dx is $\dfrac{dy}{dx} = 90x + 66$

which agrees with the derivative found by using the chain rule.

EXAMPLE 72   If $P = s^{3/4}$ and $s = 2t^3 - t + 1$, find dP/dt.

SOLUTION

$$\frac{dP}{dt} = \frac{dP}{ds} \cdot \frac{ds}{dt}$$

$$= \frac{3}{4} s^{-1/4} (6t^2 - 1)$$

$$= \frac{3(6t^2 - 1)}{4s^{1/4}}$$

$$= \frac{3(6t^2 - 1)}{4(2t^3 - t + 1)^{1/4}} \qquad \text{Substitute } s = 2t^3 - t + 1.$$

$$= \frac{3(6t^2 - 1)}{4\sqrt[4]{2t^3 - t + 1}}$$

■

Practice Problem 2 If $y = u^2 + 4u - 3$ and $u = x + 6$, find dy/dx by the chain rule.

ANSWER $\dfrac{dy}{du} = 2u + 4$ and $\dfrac{du}{dx} = 1$; so

$$\frac{dy}{dx} = \frac{dy}{du} \cdot \frac{du}{dx} = (2u + 4) \cdot 1 = 2(x + 6) + 4 = 2x + 16$$

EXAMPLE 73   Find dy/dx if $y = (3x^2 + 2x - 5)^5$.

SOLUTION   This derivative could be found by expanding the trinomial and then differentiating each term. However, this method can be very time consuming and not practical for exponents greater than 3. Let's use the chain rule instead. We notice that $y = (3x^2 + 2x - 5)^5$ is in the form $y = u^5$, where $u = 3x^2 + 2x - 5$. Now

$$\frac{dy}{du} = 5u^4 \quad \text{and} \quad \frac{du}{dx} = 6x + 2$$

Substituting these into the chain rule form

$$\frac{dy}{dx} = \frac{dy}{du} \cdot \frac{du}{dx}$$

gives us
$$\frac{dy}{dx} = 5u^4 \cdot (6x + 2)$$

$$= 5(3x^2 + 2x - 5)^4 \cdot (6x + 2)$$

■

Suppose that $y(u) = u^n$ and u is a function of x. We have

$$\frac{dy}{du} = n \cdot u^{n-1} \quad \text{and} \quad \frac{du}{dx} = u'(x)$$

Then

$$\frac{dy}{du} \cdot \frac{du}{dx} = n \cdot u^{n-1} \cdot u'(x)$$

Therefore,

$$\frac{d}{dx}[y(u(x))] = n \cdot [u(x)]^{n-1} \cdot u'(x)$$

which is called the **generalized power rule**.

THE GENERALIZED POWER RULE

If u is a function whose derivative exists at x, n is any real number, and

$$y(u(x)) = [u(x)]^n$$

then

$$\frac{d}{dx}[y(u(x))] = n \cdot [u(x)]^{n-1} \cdot u'(x)$$

That is,

$$\frac{dy}{dx} = n[u(x)]^{n-1} \cdot \frac{du}{dx}$$

EXAMPLE 74   If $y = (3x + 2)^4$, find dy/dx.

SOLUTION   First, we recognize that y is of the form u^n, where $u = 3x + 2$. Then

$$\frac{dy}{dx} = nu^{n-1}\frac{du}{dx}$$

$$\frac{dy}{dx} = 4(3x + 2)^3 \cdot 3 \qquad u = 3x + 2; \frac{du}{dx} = 3$$

$$= 12(3x + 2)^3$$

EXAMPLE 75   Find dy/dx for $y = \sqrt{4x^2 + 3}$ and check the answer with a graphing calculator.

SOLUTION   Write $y = \sqrt{4x^2 + 3}$ as $y = (4x^2 + 3)^{1/2}$. Recognize the form $u^{1/2}$, where $u = 4x^2 + 3$. Then

$$\frac{dy}{dx} = nu^{n-1} \frac{du}{dx}$$

$$\frac{dy}{dx} = \frac{1}{2}(4x^2 + 3)^{-1/2}(8x) \qquad u = 4x^2 + 3; \frac{du}{dx} = 8x$$

$$= \frac{4x}{\sqrt{4x^2 + 3}}$$

✓ Calculator Check: $\dfrac{dy}{dx}$ at $x = 1$ is $\dfrac{4}{\sqrt{7}} \approx 1.511858$. Using n Derive, $dy/dx =$ 1.511858.

Now let's work some similar problems without the ''in between'' variable u.

EXAMPLE 76   If $y = (4x^3 - 3x^2 + x - 2)^3$, find dy/dx and check with a graphing calculator at $x = 1$.

SOLUTION   We use the following formula:

$$\text{If} \quad y = [g(x)]^n, \quad \text{then} \quad \frac{dy}{dx} = n[g(x)]^{n-1} g'(x)$$

Now, $g(x) = 4x^3 - 3x^2 + x - 2$, so $g'(x) = 12x^2 - 6x + 1$. Substituting in the formula we now have

$$\frac{d}{dx}[g(x)]^n = n[g(x)]^{n-1} g'(x)$$

$$\frac{dy}{dx} = 3(4x^3 - 3x^2 + x - 2)^2 g'(x) \qquad\qquad n = 3; n - 1 = 2$$

$$= 3(4x^3 - 3x^2 + x - 2)^2(12x^2 - 6x + 1)$$

✓ Calculator Check: $\dfrac{dy}{dx}$ at $x = 1$ is $3(4 - 3 + 1 - 2)^2(12 - 6 + 1) = 0$.
Using nDeriv, we get $.00034 \approx 0$.

EXAMPLE 77   If $y = (3x^2 + 2)^{1/2}$, find dy/dx and check with a calculator at $x = 1$.

SOLUTION   We have $g(x) = 3x^2 + 2$ so $g'(x) = 6x$, and

$$\frac{dy}{dx} = \frac{1}{2}(3x^2 + 2)^{-1/2}(6x) = \frac{3x}{(3x^2 + 2)^{1/2}} \qquad n = \frac{1}{2}$$

✓ Calculator Check   The derivative dy/dx at $x = 1$ is $3/\sqrt{5} \approx 1.34164$, which is the same answer that you get using nDeriv on your calculator.

EXAMPLE 78   If $y = x^2\sqrt{3x^2 + 2}$, find dy/dx and check at $x = 1$ with a graphing calculator.

SOLUTION   First, you must recognize this as a product so that you can use the product formula. If $y = f(x)g(x)$, then $y' = f(x)g'(x) + f'(x)g(x)$. This gives

$$y' = x^2 \frac{d}{dx}[\sqrt{3x^2 + 2}] + 2x\sqrt{3x^2 + 2} \qquad f(x) = x^2; f'(x) = 2x$$

From Example 77, we know that

$$\frac{d}{dx}\sqrt{3x^2 + 2} = \frac{3x}{(3x^2 + 2)^{1/2}}$$

Substituting this value in the expression for y' gives

$$\begin{aligned}
y' &= x^2 \cdot \frac{3x}{\sqrt{3x^2 + 2}} + 2x\sqrt{3x^2 + 2} \\
&= \frac{3x^3 + 2x(3x^2 + 2)}{\sqrt{3x^2 + 2}} \qquad \text{Using a common denominator} \\
&= \frac{9x^3 + 4x}{\sqrt{3x^2 + 2}}
\end{aligned}$$

✓ Calculator Check: At $x = 1$, $\dfrac{dy}{dx} = \dfrac{13}{\sqrt{5}} \approx 5.81378$

Using nDeriv, it is ≈ 5.81378.

Practice Problem 3   If $y = (2x^3 + 4)^5$, find dy/dx.

ANSWER $\dfrac{dy}{dx} = 5(2x^3 + 4)^{5-1} \cdot \dfrac{d}{dx}(2x^3 + 4) = 5(2x^3 + 4)^4(6x^2)$

EXAMPLE 79   If $y = \dfrac{x}{(x^2 + 3)^2}$, find $\dfrac{dy}{dx}$.

SOLUTION   This is in the form of a quotient, so we start with the quotient formula.

$$\frac{dy}{dx} = \frac{(x^2 + 3)^2 \cdot \dfrac{d}{dx}(x) - x \cdot \dfrac{d}{dx}(x^2 + 3)^2}{[(x^2 + 3)^2]^2}$$

$$= \frac{(x^2 + 3)^2 \cdot (1) - x \cdot [2(x^2 + 3)^1 \cdot 2x]}{(x^2 + 3)^4}$$

$$= \frac{(x^2 + 3)^2 - 4x^2(x^2 + 3)}{(x^2 + 3)^4}$$

$$= \frac{(x^2 + 3)(x^2 + 3 - 4x^2)}{(x^2 + 3)^4}$$

$$= \frac{(3 - 3x^2)}{(x^2 + 3)^3}$$

Practice Problem 4 Differentiate $y = \dfrac{\sqrt{x^2 - 5}}{x + 1}$.

ANSWER $y' = \dfrac{x + 5}{(x + 1)^2 \cdot \sqrt{x^2 - 5}}$

COMMON ERROR   Many times students will differentiate like this:

$$y(x) = (3x^2 + 4x)^5, \quad \text{so} \quad y'(x) = 5(3x^2 + 4x)^4$$

What is wrong with this? The answer is $5(3x^2 + 4x)^4(6x + 4)$.

SUMMARY

1. The derivative of a function of x to the nth power is equal to n times the function to the $(n - 1)$st power times the derivative of the function of x with respect to x.

$$\text{If} \quad f(x) = [u(x)]^n, \quad \text{then} \quad f'(x) = n[u(x)]^{n-1} \cdot \frac{du}{dx}$$

2. If $f(x) = y[u(x)]$, then $f'(x) = \dfrac{dy}{du} \cdot \dfrac{du}{dx}$.

Exercise Set 2.7

In each case find $f[g(x)]$ and $g[f(x)]$.

1. $f(x) = x^2 + 3$, $g(x) = x - 1$
2. $f(x) = x^2 - 7$, $g(x) = x + 5$
3. $f(x) = \sqrt{x - 7}$, $g(x) = x^3$
4. $f(x) = x^4$, $g(x) = \dfrac{1}{x}$

Express y in terms of x only.

5. $y = u^2$, $u = x^2 + 1$
6. $y = \sqrt{u + 1}$, $u = x^2 - 1$
7. $y = \dfrac{1}{\sqrt{u + 3}}$, $u = x^4$
8. $y = \dfrac{1}{u^2}$, $u = \sqrt{x + 3}$

Write each of the following functions in the form $y = f(u)$ and $u = g(x)$.

9. $y = (3x + 4)^5$
10. $y = (x + 7)^4$
11. $y = \dfrac{1}{x + 8}$
12. $y = \sqrt[3]{x - 13}$

Use the chain rule to find dy/dx. To check your results, find y as a function of x and then find dy/dx.

13. $y = 3u$, $u = 3x + 5$
14. $y = 4u$, $u = 3x - 2$
15. $y = -4u$, $u = 3x^2$
16. $y = 2u + 4$, $u = 3x^2$
17. $y = 4u^2$, $u = 3x + 5$
18. $y = -5u^2$, $u = 3x - 2$

Find dy/dx using the chain rule for powers:

$$\frac{dy}{dx} = nu^{n-1}\frac{du}{dx}$$

Check your answers using a graphing calculator.

19. $y = (3x + 4)^2$
20. $y = (4x - 7)^5$
21. $y = (6 - 2x^2)^3$
22. $y = \sqrt{4 - x^2}$

Find each of the following and check with a graphing calculator.

23. $D_x\left[\dfrac{1}{2x + 4}\right]$
24. $D_x\sqrt{5x - 1}$
25. $\dfrac{d}{dx}(2x^3 - 1)^{1/3}$
26. $\dfrac{d}{dx}(1 - 8x^2)^{-1/4}$
27. y' for $y = (3x^2 - 2x + 1)^2$
28. y' for $y = (2x^2 - x)^4$
29. $\dfrac{dy}{dx}$ for $y = -5(2x + 1)^3 - 7$
30. $\dfrac{dy}{dx}$ for $y = 3 - 4(3x - 2)^3$
31. $D_x[(x - 1)^3 \sqrt{x}]$
32. $D_x[(x + 5)^2 \sqrt[3]{x}]$
33. $\dfrac{d}{dx}\left(\dfrac{1}{\sqrt{2x^2 - 1}}\right)$
34. $\dfrac{d}{dx}\left[\dfrac{2x + 1}{(2x^2 + 1)^3}\right]$

Find the derivative of each function.

35. $f(x) = (3x + 5)(2x - 3)^2$
36. $f(x) = (3x - 2)^2(2x + 3)$
37. $f(x) = 4(2x^2 + x - 1)^{1/2}$
38. $f(x) = 9(3x^2 - x + 1)^{1/3}$
39. $f(x) = 4x^{1/2}(3x - 1)^2$
40. $f(x) = 6x^{-1/2}(4x - 3)^2$
41. $f(x) = \dfrac{4x^2}{(2x - 3)^2}$
42. $f(x) = \dfrac{3x^2 - 2}{(2x + 3)^2}$
43. $f(x) = [(x + 3)\sqrt{x}]^4$
44. $f(x) = \left(\dfrac{x + 1}{x - 1}\right)^5$
45. $f(x) = \left(\dfrac{x - 3}{x + 2}\right)^2$
46. $f(x) = [(x - 7)\sqrt{x}]^3$

Applications (Business and Economics)

47. **Cost Functions.**   A cost function is found to be $C(x) = (3x - 10)^2 + 24$.
    (a) Find the marginal cost function.
    (b) Find $C'(3)$.
    (c) Find $C'(4)$.
    (d) Find $\overline{C}'(x)$.

48. **Revenue Functions**. A revenue function for the sale of stereo sets is

$$R(x) = 4608 - 372\left(12 - \frac{x}{3}\right) - \left(12 - \frac{x}{3}\right)^2$$

where $0 \le x \le 36$.
(a) Find the marginal revenue function.
(b) Find $R'(9)$.
(c) Find $R'(18)$.
(d) Find $\bar{R}(x)$.

Applications (Social and Life Sciences)

49. **Learning Functions**. For the learning function giving the number of actions learned after x hours of practice,

$$L(x) = 36(2x - 1), \qquad \tfrac{1}{2} \le x \le 15$$

find the rate of learning at the end of (a) 1 hour and (b) 14 hours.

50. **Pollution**. The pollution from a factory is given by

$$P(x) = 0.4(2x + 3)^{-2}$$

where x is measured in miles. Find the rate of change of the pollution. What is the rate of change of pollution at (a) 1 mile and (b) 3 miles?

Review Exercises

Find the derivatives of the following functions.

51. $y = \dfrac{2x + 5}{3x - 2}$

52. $y = \dfrac{4x + 3}{2x - 3}$

53. $y = (4x^2 - 2)(3x + 2)$

54. $y = (5x^2 + 2)(3x + 5)$

55. $y = 3(2x + 5)^3$

56. $y = 4(3x - 2)^2$

Chapter Review

Review the following definitions and concepts to ensure that you understand and can use each of them.

Slope of a function	Derivative of a function
Limit of a function	Average rate of change
Tangent to a graph	Four-step procedure for derivatives
Continuity	Instantaneous rate of change
Marginal cost function	Marginal supply function
Marginal profit function	Average cost function
Marginal demand function	Marginal average cost function
Power rule	Quotient rule
Product rule	General power rule

Make sure you understand and know when and how to use the following formulas.

$$\frac{dy}{dx} = \lim_{h \to 0} \frac{f(x + h) - f(x)}{h} \qquad \frac{d}{dx} x^n = nx^{n-1}$$

$$\text{Average change in } f(x) = \frac{f(x + h) - f(x)}{h}$$

$$\frac{d}{dx} C = 0 \qquad\qquad \frac{d}{dx}[C f(x)] = C \frac{d}{dx}[f(x)]$$

$$\frac{dy}{dx} = \frac{dy}{du} \cdot \frac{du}{dx}$$

$$\frac{d}{dx}[f(x) \pm g(x)] = \frac{d}{dx}[f(x)] \pm \frac{d}{dx}[g(x)]$$

$$\frac{d}{dx}[g(x)]^n = n[g(x)]^{n-1}\,g'(x)$$

$$\frac{d}{dx}[f(x)g(x)] = f(x)g'(x) + g(x)f'(x)$$

$$\frac{d}{dx}\left[\frac{f(x)}{g(x)}\right] = \frac{g(x)f'(x) - f(x)g'(x)}{[g(x)]^2}$$

Chapter Test

1. Define $f'(x)$ in terms of limits.
2. Using the preceding definition find the derivative of $f(x) = 3x^2$.
3. If $y = 2x^2 + 4x + 1$, find the average change in y as x changes from 2 to 3.
4. For $f(x) = 4x - 3x^2$, find $f'(0)$ and $f'(-1)$.
5. For $y = \sqrt{3}$, find y'.
6. Given $y = -4u^2 + 3$ and $u = x^2$, find dy/dx using the chain rule.
7. A cost function is given as $1000 + 20x + x^2$. Find the marginal average cost at $x = 2$.

8. Find
$$\lim_{x \to 0} \frac{3x - 2}{4x + 1}$$

9. Find
$$\lim_{x \to 1} \frac{x^2 - 3x + 3}{x - 1}$$

10. Find $f'(x)$ for $f(x) = (3/x) + 4\sqrt{x}$.
11. A profit function is given as

$$P(x) = \frac{x^2}{2} + 4x$$

Find where $P'(x) > 0$ and where $P'(x) < 0$.

12. Find y' for $y = 2(x^2 - 3)^{3/2}$. Check with a graphing calculator at $x = 4$.
13. Find y' for $y = (3x + 2)^{1/2}(2x - 3)$. Check with a graphing calculator at $x = 1$.

14. Find y' for
$$y = \frac{2x + 5}{\sqrt{x^2 - 1}}$$

Check with a graphing calculator at $x = 2$.

15. A demand function is given as

$$p(x) = \frac{6000(20 - x)}{x^2}, \qquad 1 \leq x \leq 20$$

(a) Find the marginal demand function.
(b) Find the marginal average demand function.

CHAPTER

3

Applications of Differentiation and Additional Derivatives

In this chapter we consider applications that are important in the business world. A CEO desires to optimize profits, but many factors must be considered. What will be the result of increased production? Will the increase in revenue exceed the additional cost? To accomplish this goal of optimizing profits, one of the vice presidents is given the responsibility of minimizing inventory costs. In a like manner, a physician might be interested in maximizing the amount of medicine in the bloodstream after a given time, or an agricultural researcher might be interested in how long it will take a plant to reach its maximum height. In this chapter, we learn to find maximum and minimum values of functions. This study leads to the consideration of graphing functions. In the process of applying calculus concepts to the graphing of functions, we find a need for additional procedures in taking derivatives. Thus, we consider higher-order derivatives.

Throughout this chapter, and those that follow, we will be looking at revenue, cost, and profit functions. These functions are actually models that fit patterns determined from empirical data. In some cases we will state the domain. However, even if a domain is not stated, you should be aware that these functions will usually involve limited domains.

3.1   HOW DERIVATIVES ARE RELATED TO GRAPHS

OVERVIEW   As indicated previously, finding the greatest and least values attained by a function has extensive applications. A business owner wants to find the greatest value of a profit function or perhaps the least value attained by a cost function. The management of a restaurant needs to be able to minimize waste. To locate values that maximize or minimize, we apply principles of calculus to graphs. Recall that the slope of a curve at a point is the slope of the tangent line to the curve at the point. Also remember that this slope can be found by evaluating the derivative of the function at the point. By studying changes in the slope (changes in the sign of the derivative), we obtain interesting information that is relevant to finding maximum and minimum values of the function. In this section we find

- Intervals where a function is increasing
- Intervals where a function is decreasing
- Critical points

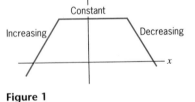

Figure 1

If the graph of a function rises from left to right on an interval [$f(x)$ increases], we say that the function is **increasing** on the interval. If the graph falls [$f(x)$ decreases] from left to right on an interval, we say that the function is **decreasing** on the interval. If the graph of a function is horizontal on an interval, we say that the function is **constant** on the interval. (See Figure 1.) We describe these terms mathematically as follows.

DEFINITION: INCREASING, DECREASING, CONSTANT

1.   A function is **increasing** on an open interval I if, for every a and b in I,

$$f(a) < f(b), \quad \text{when } a < b$$

2.   A function is **decreasing** on an open interval I if, for every a and b in I,

$$f(a) > f(b), \quad \text{when } a < b$$

3.   A function is **constant** on an open interval I if, for every a and b in I,

$$f(a) = f(b), \quad \text{when } a < b$$

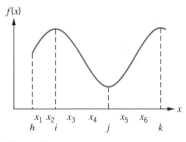

Figure 2

EXAMPLE 1   Using the preceding definition, locate the intervals on an open interval I where $f(x)$ is increasing and the intervals where $f(x)$ is decreasing on the interval (h, k), as shown in Figure 2.

SOLUTION

The function is increasing on (h, i) because $f(x_1) < f(x_2)$ for all x values on this interval where $x_1 < x_2$.

The function is decreasing on (i, j) because $f(x_3) > f(x_4)$ for all x values on this interval where $x_3 < x_4$.

The function is increasing on (j, k) because $f(x_5) < f(x_6)$ for all x values on this interval where $x_5 < x_6$.   ∎

Practice Problem 1 On a graphing calculator, graph the function $f(x) = x^3 - 3x^2$ over the interval $(-1, 4)$. Determine the intervals where the function is increasing and the intervals where the function is decreasing.

ANSWER   By using the Trace function (see page 3 of Appendix A), you can see in Figure 3 that the function increases on the intervals $(-1, 0)$ and $(2, 4)$ and decreases on the interval $(0, 2)$.

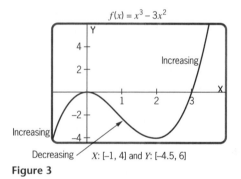

$f(x) = x^3 - 3x^2$

$X: [-1, 4]$ and $Y: [-4.5, 6]$

Figure 3

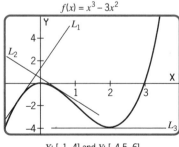

$f(x) = x^3 - 3x^2$

$X: [-1, 4]$ and $Y: [-4.5, 6]$

Figure 4

Now let's examine the graphs of functions and the slopes of tangent lines to the functions we examine. Using the function $f(x) = x^3 - 3x^2$, let's see how the slopes of the tangent lines relate to the curve. Looking at Figure 4, we see that the first tangent line drawn (L_1) is rising from left to right and therefore has a positive slope. The second tangent line drawn (L_2) is going down from left to right and therefore has a negative slope. The third tangent line drawn (L_3) is a horizontal line and therefore has slope 0. What relationship do you see between the slopes of the tangent lines and whether the function is increasing or decreasing?

Calculator Note

1.   You can draw a tangent line with your calculator. For example in Figure 4, to draw the tangent line at $x = -0.5$ use the DRAW DRAW menu as follows. From the home screen, press [2nd] [DRAW] (Select 5) and insert Tangent $(x^3 - 3x^2, -.5)$ [ENTER]. Draw several tangent lines on the graph of $f(x) = x^3 - 3x^2$.

Do these lines confirm what we have just discussed? Are the tangent lines rising from left to right when f is increasing? If you were to draw a tangent line anywhere on the interval $(-\infty, 0)$ would it have a positive or negative slope? What about a tangent line on the interval $(0, 2)$?

2. With your calculator draw the graph of another polynomial, such as $y = 4x^4 - 5x^2 + x$ with X:$[-2, 2]$, Y:$[-3, 3]$ and then draw some tangent lines along the curve. What can you determine about the slope of the tangent lines as they relate to the function?

3. With your graphing calculator draw the graph of $y = 3x^3 - 2x^2 - 1$ and then $y' = 9x^2 - 4x$. Use the intervals X:$[-3, 3]$ and Y:$[-15, 15]$. Can you see the relationship between y and y' when y is increasing? What about when y is decreasing?

It would seem that when the derivative is positive over an interval (the slopes of the tangent lines are positive), the function is increasing on the interval. Likewise, when the derivative is negative over an interval (the slopes of the tangent lines are negative) the function is decreasing on that interval. When the derivative is 0, the function is not increasing or decreasing. We state this property as follows.

POSITIVE AND NEGATIVE SLOPES

(a) If $f'(x) > 0$ for all x in an open interval I, and if $y = f(x)$ is a differentiable function on I, then $y = f(x)$ is increasing for all x in the interval.

(b) If $f'(x) < 0$ for all x in an open interval I, and if $y = f(x)$ is a differentiable function on I, then $y = f(x)$ is decreasing for all x in the interval.

(c) If $f'(x) = 0$ for all x in an open interval I, and if $y = f(x)$ is a differentiable function on I, then $y = f(x)$ is constant for all x in the interval.

A function is increasing (the graph is rising) when its derivative is positive and decreasing (the graph is falling) when its derivative is negative. If the derivative of a function is defined for all points in an interval, to change from positive to negative or from negative to positive, $f'(x)$ must assume the value of 0 on the interval. (Think of a number line. What number divides the positive numbers from the negative numbers? Zero.) That is, if the graph of f changes from rising to falling or falling to rising, at this change the slope of the tangent line is 0 and the tangent line is horizontal, as we saw in Figure 4.

EXAMPLE 2   With your calculator, draw the graph of $f(x) = 5x^4 - 4x^5$ and $f'(x) = 20x^3 - 20x^4$. Determine the intervals where $f(x)$ is increasing and where it is decreasing. Compare these intervals to the intervals where $f'(x) > 0$ and where $f'(x) < 0$.

SOLUTION   The graphs shown in Figure 5 indicate that $f(x)$ is increasing and $f'(x) > 0$ on $(0, 1)$ and $f(x)$ is decreasing and $f'(x) < 0$ on $(-\infty, 0)$, $(1, \infty)$. ∎

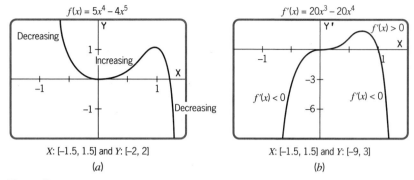

$f(x) = 5x^4 - 4x^5$

$f'(x) = 20x^3 - 20x^4$

X: [-1.5, 1.5] and Y: [-2, 2]

(a)

X: [-1.5, 1.5] and Y: [-9, 3]

(b)

Figure 5

Calculator Note

Have your calculator draw tangent lines at (0, 0) and (1, 0) on the curve $y = 5x^4 - 4x^5$. What do you see? Can you see the first tangent line? Note that both are horizontal.

Tangent lines to the graph of $f(x)$ at the points (0, 0) and (1, 1) in Figure 5(a) are horizontal lines and have slope $= 0$. Note in Figure 5(b) that $f'(x) = 0$ at these two points. When a derivative equals 0 or the derivative is undefined, we classify the points as **critical points**.

DEFINITION: CRITICAL POINT

If c is in the domain of f and if either $f'(c) = 0$ or $f'(c)$ is undefined, then c is called a **critical value** of f. If c is a critical value, then the point $(c, f(c))$ is called a **critical point**.

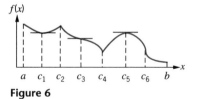

$f(x)$

$a \quad c_1 \quad c_2 \quad c_3 \quad c_4 \quad c_5 \quad c_6 \quad b$

Figure 6

EXAMPLE 3   Find the critical values on (a, b) for the function shown in Figure 6.

SOLUTION   Critical values occur at $x = c_1, c_2, c_3, c_4, c_5, c_6$. At $x = c_1$, $x = c_3$, $x = c_5$, the graph shows horizontal tangents. $f'(x) = 0$ at each of these points. What about the other critical values? Note that even though at c_3 the tangent appears to be horizontal, the function remains decreasing. At c_2 and c_4 it is possible to draw more than one tangent line, and at c_6 there is a vertical tangent; therefore, $f'(x)$ is undefined at c_2, c_4, and c_6. ∎

Steps for Finding Critical Points

1. Determine the domain of the function.
2. Compute the derivative.
3. Factor the derivative, if possible, and determine the points at which the function is defined and the derivative is 0 or is undefined.

NOTE: Even though a critical point is found, it does not necessarily mean that the function changes from increasing to decreasing or from decreasing to increasing. For example, draw a calculator graph of the function $f(x) = x^{1/3}$ and note that the graph of the function reveals that f is always increasing. However, there is a critical value at $x = 0$, where the derivative $f'(x) = 1/(3x^{2/3})$ is undefined. Have your calculator draw the graph of $f'(x) = 1/(3x^{2/3})$ and confirm this.

EXAMPLE 4   Find all critical points for $y = 7x^2 - 14x + 3$.

SOLUTION   Critical points for a function occur where the derivative is 0 or does not exist but the value is within the domain of the function. To find the critical points, we follow the steps listed above. Note that the function is defined for all real values.

1. We have

$$\frac{dy}{dx} = y' = 14x - 14 = 14(x - 1)$$

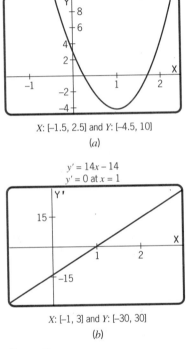

y = 7x² – 14x + 3

X: [–1.5, 2.5] and Y: [–4.5, 10]

(a)

y' = 14x – 14
y' = 0 at x = 1

X: [–1, 3] and Y: [–30, 30]

(b)

Figure 7

(The derivative is defined for all real values of x.)

2. Setting the derivative equal to 0 and solving for x we have

$$14(x - 1) = 0 \quad \text{and} \quad x = 1$$

3. The only critical value occurs at $x = 1$. At $x = 1$ we have

$$y = 7(1)^2 - 14(1) + 3 = -4$$

The only critical point for this function is $(1, -4)$.   ■

Practice Problem 2 Use your calculator to draw the graph of y and y' from Example 4 and verify that $(1, -4)$ is the only critical point.

ANSWER   The graphs are shown in Figure 7. Note that y' changes signs only once, at $(1, 0)$, and y has only the one critical point.

Practice Problem 3 Find the critical points algebraically for the function $f(x) = 2x^3 - 9x^2 + 12x - 10$.

ANSWER $(1, -5)$ and $(2, -6)$

Practice Problem 4 With a calculator, draw the graph of the function $f(x) = 2x^3 - 9x^2 + 12x - 10$ from Practice Problem 3 and $f'(x) = 6x^2 - 18x + 12$ to validate graphically your answer in Practice Problem 3.

ANSWER   From the graphs in Figure 8, we can see that the critical points are at $(1, -5)$ and $(2, -6)$ and that $f'(x)$ changes signs at $x = 1$ and $x = 2$. This confirms the algebraic work done in Practice Problem 3.

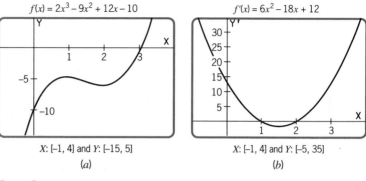

Figure 8

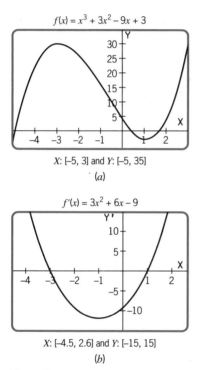

Figure 9

In application problems it is important to know the exact value of the critical points. Therefore, we use algebraic techniques to find the critical points and we then validate our work with a graphical analysis of the function.

EXAMPLE 5   Determine the intervals where the function $f(x) = x^3 + 3x^2 - 9x + 3$ is increasing and where it is decreasing by finding the critical points algebraically and then verifying by using the graphs of f and f'.

SOLUTION   For $f(x) = x^3 + 3x^2 - 9x + 3$, we have $f'(x) = 3x^2 + 6x - 9$. We note here that both f and f' are defined for all real numbers, so all critical points will occur where $f'(x) = 0$. Thus, we examine the equation $f'(x) = 0$.

$$f'(x) = 3x^2 + 6x - 9 = 3(x + 3)(x - 1) = 0$$

From this we determine that $f'(x) = 0$ only when $x = -3$ or $x = 1$; so we have two critical values. Looking at the graphs in Figure 9, we see that f is increasing [and $f'(x) > 0$] over $(-\infty, -3)$, $(1, \infty)$ and f is decreasing [and $f' < 0$] over $(-3, 1)$. ∎

Even though the turning points of a function can be found by looking at the graph of a function or its derivative, these values are often decimal approximations. That is why we usually use both the graphing calculator and algebraic techniques of factoring and reduction to find our answers. It is helpful to support our algebraic work with graphing. Likewise, graphical work can be supported with algebraic work.

EXAMPLE 6   Determine the intervals where $f(x) = \frac{2}{3}x^{3/2} - 2x^{1/2}$ is increasing and where it is decreasing for $x \geq 0$. Use both algebraic techniques and the graphs of f and f'.

SOLUTION   The domain of the function is $[0, \infty)$. For $f(x) = \frac{2}{3}x^{3/2} - 2x^{1/2}$, we have

$$
\begin{aligned}
f'(x) &= \frac{2}{3} \cdot \frac{3}{2} x^{1/2} - 2 \cdot \frac{1}{2} x^{-1/2} \\
&= x^{1/2} - x^{-1/2} \\
&= \sqrt{x} - \frac{1}{\sqrt{x}} \qquad &&\text{Convert to radical form.} \\
&= \frac{x - 1}{\sqrt{x}} \qquad &&\text{Combine the terms.}
\end{aligned}
$$

Notice that $f(0)$ is defined even though $f'(0)$ is undefined and if $x = 1$, $f'(x) = 0$. Therefore, our critical values are $x = 0$ (derivative is undefined) and $x = 1$ (derivative is zero). Looking at the graphs in Figure 10 we see that $f(x)$ is decreasing on $(0, 1)$ and is increasing on $(1, \infty)$. We also see that f' changes from negative to positive at $x = 1$.

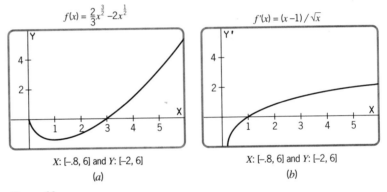

$f(x) = \frac{2}{3}x^{\frac{3}{2}} - 2x^{\frac{1}{2}}$

$f'(x) = (x-1)/\sqrt{x}$

X: [-.8, 6] and Y: [-2, 6]

(a)

X: [-.8, 6] and Y: [-2, 6]

(b)

Figure 10

For some problems in the exercise set, we ask you to locate intervals where a function is increasing and where it is decreasing. If you have critical values at c_1, c_2, and c_3 for a function with $c_1 < c_2 < c_3$ on a closed interval $[a, b]$, consider the closed interval as being divided into the following subintervals.

Then, for a continuous curve between any two of these consecutive points, the function is either decreasing or increasing. That is, find the sign of $f'(x)$ for some x on $a < x < c_1$. If the sign is positive, the function is increasing on (a, c_1). If the sign is negative, the function is decreasing on (a, c_1). This procedure can be generalized for any number of critical points on a closed interval. In the next section, we relate critical points to points associated with relative maxima and relative minima.

Practice Problem 5 For $y = 1/(x - 2)^2$ find the intervals where the function is increasing and where it is decreasing. Using your graphing calculator, graph both y and y' to confirm your answer.

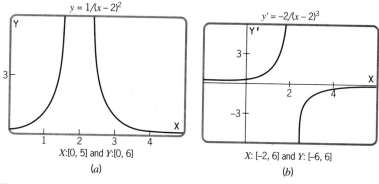

X:[0, 5] and Y:[0, 6]
(a)

X: [-2, 6] and Y: [-6, 6]
(b)

Figure 11

ANSWER $y' = -2/(x - 2)^3$. From Figure 11 we have $y' < 0$ when $x > 2$ and $y' > 0$ when $x < 2$. Therefore, the function is decreasing on $(2, \infty)$ and increasing on $(-\infty, 2)$. Notice that y is not defined for $x = 2$, and therefore, $x = 2$ is not a critical value.

Practice Problem 6 For the function defined by the graph in Figure 12,

(a) find the intervals where the function is increasing over (a, d).
(b) find all the critical values over (a, d).

ANSWER

(a) The function is increasing on (a, b) and (c, d).
(b) Critical values occur at $x = b$ and $x = c$.

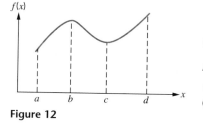

Figure 12

EXAMPLE 7   For $f(x) = x^3(\sqrt{x} + 1)$, find any critical points and the intervals where f is increasing and where it is decreasing. Use f' and the calculator graphs of both f and f'.

SOLUTION   Note that the domain of the function is $[0, \infty)$, and the function will always be positive. We now find the derivative.

$$f(x) = x^3(\sqrt{x} + 1) = x^3(x^{1/2} + 1) = x^{7/2} + x^3 \qquad \text{Rewrite the function to find } f'.$$

$$f'(x) = \frac{7}{2} x^{5/2} + 3x^2 \qquad \text{Find the derivative.}$$

$$= x^2 \left(\frac{7}{2} x^{1/2} + 3 \right) \qquad \text{Factor the derivative.}$$

The only critical value is $x = 0$, since there is no real number that satisfies the equation $(\frac{7}{2}x^{1/2} + 3) = 0$. Therefore, the only critical point is $(0, f(0)) = (0, 0)$ and the function is increasing over $(0, \infty)$. The graphs are shown in Figure 13. ■

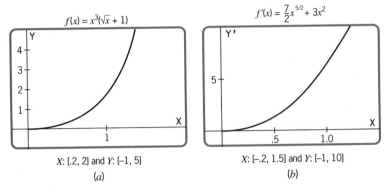

$f(x) = x^3(\sqrt{x} + 1)$

X: [.2, 2] and Y: [-1, 5]

(a)

$f'(x) = \frac{7}{2}x^{5/2} + 3x^2$

X: [-.2, 1.5] and Y: [-1, 10]

(b)

Figure 13

COMMON ERROR   Students sometimes substitute a critical value in the first derivative, $(c, f'(c))$, to get a critical point. Why is this incorrect?

SUMMARY

When the derivative of a function is positive, the function is increasing. Likewise, when the derivative is negative, the function is decreasing. We can determine these intervals by locating the critical points. Remember, a critical value occurs at c if $f(c)$ is defined and if $f'(c) = 0$ or if $f'(c)$ is undefined. An analysis of the graph in conjunction with the knowledge of the values of the critical points enables you to find the intervals where a function is increasing or decreasing.

Exercise Set 3.1

Find all critical points algebraically for the following functions and determine the intervals where the function is increasing.

1. $y = 5$
2. $y = -3x$
3. $f(x) = 3x + 5$
4. $f(x) = -5x + 3$
5. $f(x) = x^2$
6. $f(x) = -3x^2$
7. $y = x^2 + 2x$
8. $y = -3x^2 + 2$
9. $y = 4x^2 + 3x - 2$
10. $y = -5x^2 - 3x + 8$
11. $y = 3x^2 + 12x - 5$
12. $y = 5x^2 - 4x + 3$
13. $y = 12x^{2/3} + x$
14. $y = 4x^{1/2} - 2x$

Determine the intervals where the following functions are increasing or decreasing. The functions continue in the directions indicated with no more critical points.

15.

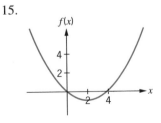

16.

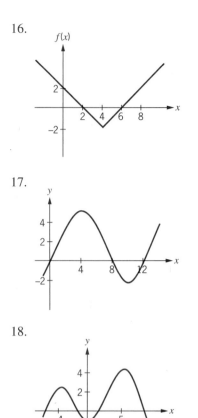

17.

18.

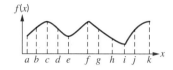

(a) Find all intervals where $f(x)$ is increasing.
(b) Find all intervals where $f(x)$ is decreasing.
(c) Find all the critical values on the interval (a, k).

24. Refer to the function $f(x)$ represented by the following graph to answer each question.

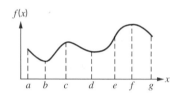

(a) Find all intervals where $f(x)$ is increasing.
(b) Find all intervals where $f(x)$ is decreasing.
(c) Find all the critical values on the interval (a, g).

Find all critical points, intervals where the functions are increasing, and intervals where the functions are decreasing. On a graphing calculator verify your answer from the graphs of the functions and their derivatives.

In Exercises 19–22, assume that the function is continuous. From the given information, sketch the graph of a function that meets the given requirements.

19. The derivative is negative over $(-\infty, 1)$ and positive over $(1, \infty)$. The function contains the points $(0, 3)$, $(1, 1)$, $(2, 2)$, and $(4, 6)$.

20. The derivative is negative over $(-\infty, -1)$ and positive over $(-1, \infty)$. The function contains the points $(-2, 2)$, $(-1, 1)$, $(0, 2)$, and $(1, 5)$.

21. The derivative is positive over $(-\infty, 3)$ and negative over $(3, \infty)$. The function contains the points $(1, 7)$, $(2, 10)$, $(3, 11)$, and $(4, 10)$.

22. The derivative is positive over $(-\infty, -2)$ and negative over $(-2, \infty)$. The function contains the points $(-4, 1)$, $(-2, 5)$, $(0, 1)$, and $(1, -4)$.

23. Refer to the function $f(x)$ represented by the following graph to answer each question.

25. $y = 2x^2$

26. $y = 3x^2$

27. $y = -2x^2 + 3$

28. $y = -2x^2 + 4x$

29. $y = 3x^2 + 3x + 2$

30. $y = -3x^2 - 3x + 2$

31. $y = x^3$

32. $y = -x^3$

33. $y = x^3 + 3x$

34. $y = x^3 - 3x$

35. $y = x^3 + 6x^2$

36. $y = x^3 - 6x^2$

37. $y = 2x^3 - 6x^2$

38. $y = 2x^3 + 6x^2$

39. $y = x^3 - 6x^2 + 9x$

40. $y = x^3 + 6x^2 + 3x - 5$

41. $y = 3x^{1/3} + 8x$

42. $y = 3x^{5/3}$

43. $y = 2\sqrt{x}$

44. $y = \sqrt{x} - \frac{1}{2}x$

45. $f(x) = \dfrac{x^3 - 5x^2 + 3x + 9}{x - 3}$

46. $f(x) = (x + 2)^{3/2} + 4$

Applications (Business and Economics)

47. **Profit Function.**   Profit P is related to selling price S by the formula

$$P = 10,000S - 250S^2$$

For what range of prices is the profit increasing and for what range of prices is it decreasing?

48. *Demand Function.*   A price per unit $p(x)$ for a product is given as a function of the number of units x by

$$p(x) = 800x^{-2} + 400x^{-1}$$

for $x > 0$. Is this function increasing or decreasing?

49. *Cost of Production.*   For $x \geq 1$, find the production values of x for which the cost of production is decreasing and the values of x for which it is increasing given

$$C(x) = x^3 - 6x^2 + 9x + 150$$

50. *Average Cost.*   The cost of producing x units of a given product is given by

$$C(x) = 2000 + 10x + 0.01x^2$$

(a) Find the intervals where the cost function is increasing or decreasing.

(b) Find the intervals where the average cost function $C(x)/x$ is increasing or decreasing.

51. *Marginal Analysis.*   A manufacturer's profit function is given as $P(x) = R(x) - C(x)$, where $R(x)$ is the revenue function and $C(x)$ is the cost function. Suppose that the profit function has a critical value at a production level of $x = c$. Argue that the marginal cost and the marginal revenue at this production level are equal.

Applications (Social and Life Sciences)

52. *Drug Concentration.*   The concentration $c(t)$ of a drug in a patient's bloodstream t hours after a drug is taken is given by

$$c(t) = \frac{0.2t}{t^2 + 4t + 9}$$

For what interval of time will the concentration be increasing and for what interval will it be decreasing?

53. *Population.*   A population study has been made for 8 years in a community and the following population function P has been obtained:

$$P = 1000t^2 - 6000t + 29,000$$

where t is time measured in years from the beginning date of the study. When is the population increasing and when is it decreasing?

3.2   FINDING RELATIVE MAXIMA AND RELATIVE MINIMA

OVERVIEW   When the graph of a continuous function changes from rising to falling, a local high point occurs. In a like manner, when the graph changes from falling to rising, a local low point occurs. In this section we define or name these points and demonstrate how they facilitate curve tracing. The calculus techniques of this section can be applied to a wide variety of application problems (finding maximum velocity or time of peak efficiency of a worker). When we obtain the function that represents a mathematical model of some problem, we can then analyze the model relative to maximum and minimum values. In this section we study

- Relative maxima
- Relative minima
- The first derivative test

When a ''high point'' or ''low point'' occurs on a graph, we refer to that point as an *extremum*. In this chapter, we consider two kinds of *extrema:* **absolute extrema** and **relative extrema**. The greatest value of a function (if one exists) over a specified interval or the entire domain is called the **absolute maximum**. The least value of a function (if one exists) over a specified interval or the entire domain is called the **absolute minimum**. We will discuss these in a later section. In this section we turn our attention to the other kind of extrema, called relative extrema.

DEFINITION: RELATIVE MAXIMUM AND RELATIVE MINIMUM

1. $f(c)$ is a **relative maximum** if there exists an interval (a, b) with $(a < c < b)$ such that $f(x) \leq f(c)$ for all x in (a, b).
2. $f(c)$ is a **relative minimum** if there exists an interval (a, b) with $(a < c < b)$ such that $f(x) \geq f(c)$ for all x in (a, b).

This definition is not as complicated as it may seem. **A relative maximum is simply a high point on the graph where the function changes from *increasing* to *decreasing*. Similarly, a relative minimum is a low point on the graph where the function changes from *decreasing* to *increasing*.** There can be more than one relative maximum or relative minimum on a given interval or over the entire domain of the function.

EXAMPLE 8   Locate points of relative maxima and relative minima in Figure 14.

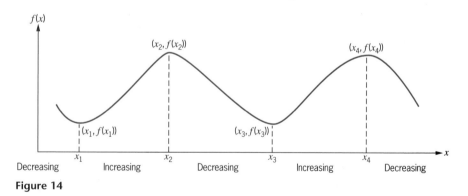

Figure 14

SOLUTION   By inspection we see that

$f(x_1)$ is a relative minimum. The function changes from decreasing to increasing at x_1.

$f(x_2)$ is a relative maximum. The function changes from increasing to decreasing at x_2.

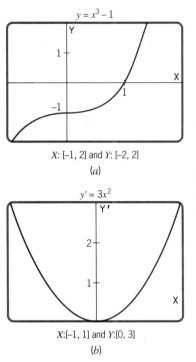

$y = x^3 - 1$

X: [–1, 2] and Y: [–2, 2]

(a)

$y' = 3x^2$

X:[–1, 1] and Y:[0, 3]

(b)

Figure 15

$f(x_3)$ is a relative minimum. The function changes from decreasing to increasing at x_3.

$f(x_4)$ is a relative maximum. The function changes from increasing to decreasing at x_4.

By examining the points where relative maxima and relative minima occur in Figure 14, we note that all points where relative maxima and relative minima occur are critical points. It is possible to prove the following theorem.

THEOREM

If $f(x)$ is continuous on the interval (a, b) and $f(c)$ is a relative maximum or relative minimum where $a < c < b$, then either $f'(c) = 0$ or $f'(c)$ does not exist. That is, c is a critical point.

The preceding theorem states that relative extrema can occur only at critical points. However, be careful, because the existence of a critical point does not always mean that there is an extremum. There are times when a critical point is present but there is no extremum at that point.

With your graphing calculator draw the graphs of $y = x^3 - 1$ and $y' = 3x^2$ shown in Figure 15.

We can determine that $(0, -1)$ is a critical point; however, there is neither a relative maximum nor a relative minimum at $(0, -1)$, since y' is always positive (except when it is 0 at $x = 0$), indicating that the curve is increasing on each side of $x = 0$. What is different about this critical point? (The function does not change from increasing to decreasing.) This difference leads to the following test for relative maxima and relative minima.

FIRST DERIVATIVE TEST FOR RELATIVE EXTREMA

At critical value c:

(a)   The function has a relative maximum at $(c, f(c))$ if the derivative is positive everywhere in an interval just to the left of c and negative in an interval just to the right of c.

(b)   The function has a relative minimum at $(c, f(c))$ if the derivative is negative everywhere in an interval just to the left of c and positive just to the right of c.

(c)   If the derivative does not change signs just to the right and just to the left of c, then $f(c)$ is not a relative extremum.

This is summarized in Figure 16. Notice that in the last two cases listed in the figure we see no extrema, but it does seem that some change occurs. This is called an **inflection point** [f(critical value) = 0, but there is no change of sign of the first derivative] and will be discussed in the next section. Let's now look at the steps we will follow in determining the exact value of a relative extremum. We

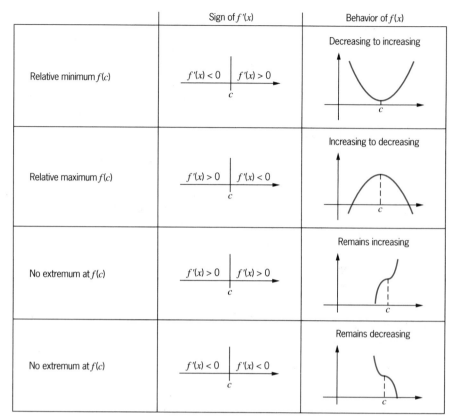

	Sign of $f'(x)$	Behavior of $f(x)$
Relative minimum $f(c)$	$f'(x) < 0$ | $f'(x) > 0$ c	Decreasing to increasing
Relative maximum $f(c)$	$f'(x) > 0$ | $f'(x) < 0$ c	Increasing to decreasing
No extremum at $f(c)$	$f'(x) > 0$ | $f'(x) > 0$ c	Remains increasing
No extremum at $f(c)$	$f'(x) < 0$ | $f'(x) < 0$ c	Remains decreasing

Figure 16

will use both calculus and the graph of the function. The use of graphing calculators along with the knowledge gained from the first derivative help to locate the extremum.

Steps for Determining Relative Extrema of a Function

1. Find the derivative.
2. Factor the derivative, if possible, and determine critical values.
3. Graph the function and its derivative. If the function changes from increasing to decreasing or from decreasing to increasing at a critical point, you have found an extremum. You should graph the derivative as well, to confirm your answer with the first derivative test.
4. Evaluate the function at the critical value to determine the point where the relative extremum is located.

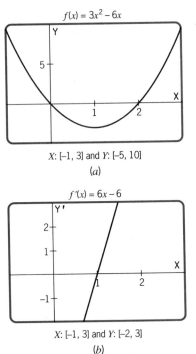

$f(x) = 3x^2 - 6x$

X: [-1, 3] and Y: [-5, 10]

(a)

$f'(x) = 6x - 6$

X: [-1, 3] and Y: [-2, 3]

(b)

Figure 17

EXAMPLE 9   Find any relative extrema for $f(x) = 3x^2 - 6x$ by following the steps listed above.

SOLUTION

1. Find the derivative. The derivative of $f(x) = 3x^2 - 6x$ is $f'(x) = 6x - 6$.
2. Factor the derivative. Factoring f', we have $f'(x) = 6(x - 1)$, and we can determine that the only critical value is $x = 1$, since $f'(1) = 0$.
3. Graph the function and the derivative. The graphs of $f(x) = 3x^2 - 6x$ and $f'(x) = 6x - 6$ are shown in Figure 17.
4. Evaluate the function at the critical values. We can determine from the graph that the function changes from decreasing to increasing and the derivative changes from negative to positive at $x = 1$. Since $f(1) = -3$, there is a relative minimum at the point $(1, -3)$.

EXAMPLE 10   Given the function $f(x) = 2x^3 + 3x^2 - 12x - 4$, find any relative extrema.

SOLUTION   We first determine that $f'(x) = 6x^2 + 6x - 12$ and then solve the equation $f'(x) = 0$. Since $f'(x) = 6x^2 + 6x - 12 = 6(x + 2)(x - 1)$, the critical values are $x = -2$ and $x = 1$. This is confirmed by the graph of f' shown with the graph of f in Figure 18. Looking at the graph, we determine that $f(-2) = 16$ is a relative maximum and $f(1) = -11$ is a relative minimum.

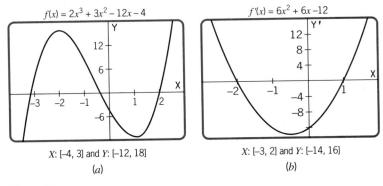

$f(x) = 2x^3 + 3x^2 - 12x - 4$

X: [-4, 3] and Y: [-12, 18]

(a)

$f'(x) = 6x^2 + 6x - 12$

X: [-3, 2] and Y: [-14, 16]

(b)

Figure 18

Practice Problem 1   For the function $f(x) = 2x^3 - 3x^2 - 12x + 8$, locate all relative extrema.

ANSWER   We have $f'(x) = 6x^2 - 6x - 12 = 6(x + 1)(x - 2)$. A relative maximum occurs at $(-1, 15)$ and a relative minimum occurs at $(2, -12)$. This is verified by the graphs in Figure 19.

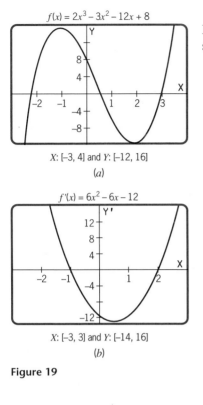

$f(x) = 2x^3 - 3x^2 - 12x + 8$

X: [–3, 4] and Y: [–12, 16]

(a)

$f'(x) = 6x^2 - 6x - 12$

X: [–3, 3] and Y: [–14, 16]

(b)

Figure 19

Let's now consider the four cases where $f'(c)$ is not defined but $f(c)$ is defined. If $f'(x)$ changes signs, then a relative extremum occurs. If $f'(x)$ does not change signs, then there is a vertical tangent. This is summarized in Figure 20.

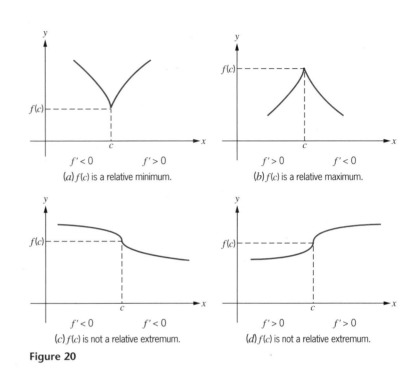

(a) $f(c)$ is a relative minimum.

(b) $f(c)$ is a relative maximum.

(c) $f(c)$ is not a relative extremum.

(d) $f(c)$ is not a relative extremum.

Figure 20

EXAMPLE 11   Draw the graph of $f(x) = 6(x - 1)^{2/3} + 4$ and determine any relative extrema.

SOLUTION   The function is defined for all real numbers, but the derivative,

$$f'(x) = 6 \cdot \tfrac{2}{3}(x - 1)^{-1/3} = \frac{4}{(x - 1)^{1/3}}$$

is not defined at $x = 1$. However, $x = 1$ is a critical value since $f(1)$ is defined. The graph of the function is shown in Figure 21. Be careful about how you enter the function in your calculator. Just using the exponent 2/3 may not work. You will most likely have to enter $\boxed{\text{Y=}}\,6\,\boxed{(}\,\boxed{(}\,\text{X}\,\boxed{-}\,1\,\boxed{)}\,\boxed{\wedge}\,2\,\boxed{)}\,\boxed{\wedge}\,\boxed{(}\,1\,\boxed{\div}\,3\,\boxed{)}\,\boxed{+}\,4.$ We can determine from Figure 21 that a relative minimum occurs at $(1, 4)$.

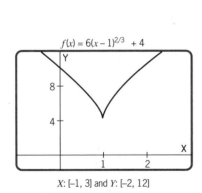

$f(x) = 6(x - 1)^{2/3} + 4$

X: [–1, 3] and Y: [–2, 12]

Figure 21

Practice Problem 2 Find the critical points of each function and draw the graph, then determine where any relative extrema occur.

(a) $f(x) = (x + 2)^{2/3} - 1$ (b) $f(x) = -x^{1/3}$
(c) $f(x) = -(2 - x)^{2/3}$ (d) $f(x) = (x - 2)^{1/3}$

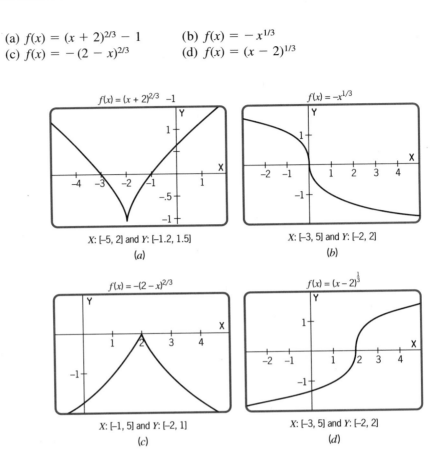

Figure 22

ANSWER

(a) Critical value $x = -2$. There is a relative minimum at $(-2, -1)$; see Figure 22(a).
(b) Critical value $x = 0$. There are no relative extrema; see Figure 22(b).
(c) Critical value $x = 2$. There is a relative maximum at $(2, 0)$; see Figure 22(c).
(d) Critical value $x = 2$. There are no relative extrema; see Figure 22(d).

Now let's turn our attention to a function that has a restricted domain.

EXAMPLE 12   For the function $f(x) = \sqrt{x}\,(x - 4)$, $x > 0$, locate any critical values and relative extrema. Use your graphing calculator to draw the graph and verify the answer.

SOLUTION   We first note that the function is only defined for $x \geq 0$. To find $f'(x)$ for this function, it will be easier to rewrite the function as $f(x) =$

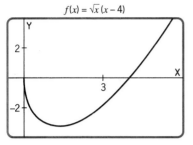

$f(x) = \sqrt{x}\,(x - 4)$

X: [–1.8, 6] and Y: [–4, 4]

Figure 23

$x^{1/2}(x - 4) = x^{3/2} - 4x^{1/2}$. From this we determine that

$$f'(x) = \frac{3}{2}x^{1/2} - 2x^{-1/2}$$

$$= \frac{1}{2}x^{-1/2}(3x - 4)$$

$$= \frac{3x - 4}{2x^{1/2}}$$

We can now determine the only critical value is $x = 4/3$ and from the graph in Figure 23 we see that there is a relative minimum at $(\frac{4}{3}, -3.08)$. ■

Practice Problem 3 On your graphing calculator draw the graphs of each function and determine where any relative extrema occur.

(a) $f(x) = -\sqrt{x}(x - 2)$ (b) $f(x) = |x^2 - 2x - 3|$

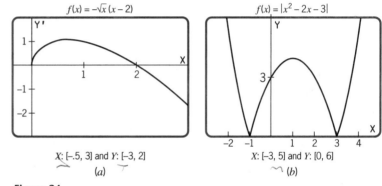

$f(x) = -\sqrt{x}\,(x - 2)$

X: [–.5, 3] and Y: [–3, 2]

(a)

$f(x) = |x^2 - 2x - 3|$

X: [–3, 5] and Y: [0, 6]

(b)

Figure 24

ANSWER   The graphs are shown in Figure 24.

(a) For $x > 0$ there is a relative maximum at $(\frac{2}{3}, 1.09)$.

(b) There are relative minima at $(-1, 0)$ and $(3, 0)$ and there is a relative maximum at $(1, 4)$.

COMMON ERROR   Students sometimes classify $x = 2$ as a critical value of $f(x) = 1/(x - 2)$. Why is this incorrect? (It is incorrect because $x = 2$ is not in the domain of f.)

SUMMARY

In this section we added to our ability to analyze the graphs of functions by using the graphs generated by a graphing calculator and the first derivative test. Sometimes we obtained the graph of the derivative of a function as a means of applying the first derivative test. Remember, if a function is defined at c and either $f'(c) = 0$ or $f'(c)$ is undefined, then a possible extremum has been located.

Exercise Set 3.2

Locate all relative maxima and minima.

1.

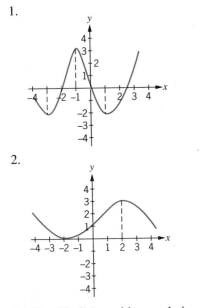

2.

3. Classify $f(c)$ as either a relative maximum, a relative minimum, or neither. Assume the function is continuous over (a, b) and $a < c < b$.

$f'(c)$	$f'(x)$ over (a, c)	$f'(x)$ over (c, b)
(a) 0	+	−
(b) 0	−	−
(c) 0	−	+
(d) 0	+	+
(e) Not defined	−	+
(f) Not defined	−	−
(g) Not defined	+	−
(h) Not defined	+	+

For Exercises 4–6, refer to the function $f(x)$ represented by the following graph. Answer the questions from your observations of the graph.

4. Where are there relative maxima on the interval (a, k)?

5. Where are there relative minima on the interval (a, k)?

6. Name any point(s) where the derivative is undefined on the interval (a, k).

Find the critical points, relative extrema, and intervals over which the function is increasing and decreasing. First find approximate answers with your graphing calculator and then validate or find exact answers with calculus procedures.

7. $y = 2x^2$

8. $y = 3x^2$

9. $f(x) = -2x^2 + 3$

10. $f(x) = -2x^2 + 4x$

11. $f(x) = 3x^2 + 3x + 2$

12. $f(x) = -3x^2 - 3x + 2$

13. $f(x) = x^3$

14. $f(x) = -x^3$

15. $f(x) = x^3 + 4x$

16. $f(x) = x^3 + 4x^2$

17. $f(x) = x^3 - 6x^2$

18. $f(x) = 2x^2 + 4x^4$

19. $f(x) = x(1 - x)^{1/2}$

20. $f(x) = x^2(1 - x)^{1/2}$

21. $f(x) = 4(x + 2)^{2/3}$

22. $f(x) = 2x^{1/3}$

23. $f(x) = 4\sqrt{x}(x - 2)$

24. $f(x) = -2\sqrt{x}(x - 4)$

25. $f(x) = \dfrac{x}{x - 1}$

26. $f(x) = \dfrac{x}{x + 1}$

27. $f(x) = x + \dfrac{4}{x}$

28. $f(x) = x^2 - \dfrac{4}{x}$

29. **Exam.** What values of x produce a relative minimum and a relative maximum, respectively, for $f(x) = 2x^3 + 3x^2 - 12x - 5$?
(a) $-5, 0$ (b) $-2, 1$ (c) $1, -3$
(d) $1, -2$ (e) $2, -1$

30. **Exam.** Vulcan Screw and Bolt Company manufactures and sells screws and bolts. The equations below represent the revenue (R) and cost (C) for a given bolt where x is 1000 cases of bolts.

$$R(x) = 100 + 26x - 0.05x^2 \quad \text{and}$$
$$C(x) = 50 + 8x + 0.01x^2$$

What quantity should be produced to maximize profit?
(a) 520,000 cases   (b) 260,000 cases

(c) 150,000 cases

(d) Some amount other than those given.

31. Find any points at which there are relative extrema for $y = |x - 1|^3$.

Applications (Business and Economics)

All application problems have been selected so that the maxima and minima requested are at relative maxima and minima.

32. **Cost Function.** A cost function is given by $C(x) = 400 - 12x + x^2$, where x is the number of units produced. Find the minimum cost.

33. **Cost Function.** A cost function is given by $C(x) = 2x^2 - 12x + 200$, where x is the number of units produced. Find the minimum cost.

34. **Demand Function.** A unit price $p(x)$ for a product is given as a function of the number of units for $x > 0$ and is defined as

$$p(x) = \frac{60,000 - x}{20,000}$$

Is there a maximum or minimum price?

35. **Profit Function.** Profit P is related to selling price S by the formula

$$P(S) = 10,000S - 250S^2$$

What selling price provides a maximum profit?

36. **Maximum Profit.** The cost and revenue from sales curves are given by $C(x) = 2x^2 - x$ and $R(x) = x^2 + 5x$. Find the number of units for which the profit (in hundreds) will be maximum. What is the maximum profit?

Applications (Social and Life Sciences)

37. **Population.** A population study has been made for 8 years in a community, and the following population function P has been obtained.

$$P(t) = 1000t^2 - 6000t + 29,000$$

where t is time measured in years from the beginning date of the study. What is the minimum population?

38. **Drug Concentration.** The concentration $c(t)$ of a drug in a patient's bloodstream t hours after a drug is taken is given by

$$c(t) = \frac{0.2t}{t^2 + 4t + 9}$$

(a) When will the concentration be a maximum?

(b) What is the maximum concentration?

Review Exercises

Find all critical points and the intervals where the functions are increasing and decreasing. Draw each graph with your graphing calculator to check your answers.

39. $y = x^2 - 2x + 3$

40. $y = x^2 - 4x + 3$

41. $y = 2x^2 - 4x + 3$

42. $y = 3x^2 - 3x + 2$

3.3   HIGHER-ORDER DERIVATIVES AND CONCAVITY

OVERVIEW   Consider a function such as $f(x) = x^3 + 2x^2$. The derivative of this function is $f'(x) = 3x^2 + 4x$. Since $f'(x)$ is also a function of x, there is no reason we cannot take the derivative of it, which we denote as $f''(x) = 6x + 4$. In this section we take derivatives of derivatives. The derivative of a derivative is useful in discussing concavity, which is another tool we can use to analyze and sketch the graph of a function. In this section we study

- Higher-order derivatives
- Concavity
- Points of inflection

Higher-Order Derivatives

The derivative of the derivative of a function is called the second derivative of the function. The derivative of the second derivative gives the third derivative. We could continue this process getting fourth derivatives, fifth derivatives, and so on. These are called **higher-order derivatives**. Many different notations are used to indicate derivatives and higher-order derivatives. Some of the notations are given in Table 1.

TABLE 1

Function	First Derivative	Second Derivative	Third Derivative	nth Derivative
$y = f(x)$	y'	y''	y'''	$y^{(n)}$
$y = f(x)$	$f'(x)$	$f''(x)$	$f'''(x)$	$f^{(n)}(x)$
$y = f(x)$	$D_x f(x)$	$D_x^2 f(x)$	$D_x^3 f(x)$	$D_x^n f(x)$
$y = f(x)$	$\dfrac{dy}{dx}$	$\dfrac{d^2y}{dx^2}$	$\dfrac{d^3y}{dx^3}$	$\dfrac{d^ny}{dx^n}$
$y = f(x)$	$\dfrac{d}{dx}[f(x)]$	$\dfrac{d^2}{dx^2}[f(x)]$	$\dfrac{d^3}{dx^3}[f(x)]$	$\dfrac{d^n}{dx^n}[f(x)]$

EXAMPLE 13   Given $y = 3x^4 + 2x^3 + 5x^2 - x + 2$, find each of the following.

(a) $\dfrac{dy}{dx}$ (b) $\dfrac{d^2y}{dx^2}$ (c) $\dfrac{d^3y}{dx^3}$ (d) $\dfrac{d^4y}{dx^4}$ (e) $\dfrac{d^5y}{dx^5}$

SOLUTION

(a) $\dfrac{dy}{dx} = 12x^3 + 6x^2 + 10x - 1$ First derivative

(b) $\dfrac{d^2y}{dx^2} = 36x^2 + 12x + 10$ Second derivative (derivative of the first derivative)

(c) $\dfrac{d^3y}{dx^3} = 72x + 12$ Third derivative (derivative of the second derivative)

(d) $\dfrac{d^4y}{dx^4} = 72$ Fourth derivative (derivative of the third derivative)

(e) $\dfrac{d^5y}{dx^5} = 0$ All other derivatives will be zero.

EXAMPLE 14   Given $f(x) = 3x^{2/3}$, find   (a) $f'(x)$,   (b) $f''(x)$, and   (c) $f'''(x)$.

SOLUTION

(a) $f'(x) = 3 \cdot \frac{2}{3}x^{-1/3} = 2x^{-1/3}$ First derivative

(b) $f''(x) = 2 \cdot (-\frac{1}{3})x^{-4/3} = (-\frac{2}{3})x^{-4/3}$ Second derivative (derivative of the first derivative)

(c) $f'''(x) = (-\frac{2}{3}) \cdot (-\frac{4}{3})x^{-7/3} = \frac{8}{9}x^{-7/3}$ Third derivative (derivative of the second derivative)

Practice Problem 1 For $y = x^3 + \dfrac{1}{x}$, find $y'''(x)$.

ANSWER $y'''(x) = 6 - 6x^{-4}$

For functions that involve products, such as $f(x) = x(x - 2)^2$, there are two approaches to finding higher-order derivatives. One approach would be to expand the function and then differentiate. However, for a function that is not easily expanded, such as $f(x) = \sqrt{x}(x - 4)^5$, this is not a reasonable approach. For functions such as these, we use the product rule.

EXAMPLE 15   Find $f''(x)$ if $f(x) = x(x - 2)^2$.

SOLUTION   To find $f'(x)$ we use the product formula:

$$f'(x) = x \cdot 2(x - 2) + (x - 2)^2 \cdot 1$$
$$= (x - 2)(2x + x - 2) = (x - 2)(3x - 2) \qquad FS' + SF'$$

Now find the derivative of $f'(x)$ by using the product rule again.

$$f''(x) = (x - 2)(3) + (3x - 2)(1) = 3x - 6 + 3x - 2 = 6x - 8$$

For functions that involve quotients, we again have a choice of techniques. For example, $f(x) = x/(x - 1)$ can be written as $f(x) = x(x - 1)^{-1}$. Therefore, we can use the quotient rule or the product rule. The choice of techniques will usually depend on the complexity of the function. There are times when both techniques can be used. Often, after a first derivative is taken, this derivative can be rewritten and the product rule or the general power rule can be used. This is advisable when the numerator is a constant, as illustrated in the next example.

EXAMPLE 16   Find $f''(x)$ for $f(x) = \dfrac{x}{x - 1}$.

SOLUTION   We use the quotient formula to find the first derivative:

$$f'(x) = \frac{1(x - 1) - x(1)}{(x - 1)^2} \qquad \frac{N'D - ND'}{D^2}$$
$$= \frac{x - 1 - x}{(x - 1)^2} = \frac{-1}{(x - 1)^2}$$

To find the second derivative, we rewrite the first derivative and use the general power rule.

$$f'(x) = -(x - 1)^{-2}$$

$$f''(x) = 2(x - 1)^{-3} \cdot 1 = \frac{2}{(x - 1)^3}$$

Practice Problem 2 Find y'' for each function.

(a) $y = \sqrt{x}(x - 2)$

(b) $y = x^2(x - 3)^2$

(c) $y = \dfrac{2x}{3x - 2}$

ANSWER

(a) $y'' = \frac{3}{4}x^{-1/2} + \frac{1}{2}x^{-3/2}$

(b) $y'' = 12x^2 - 36x + 18$

(c) $y'' = 24(3x - 2)^{-3}$

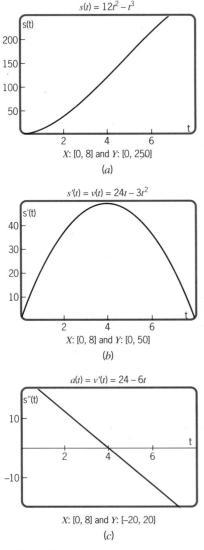

$s(t) = 12t^2 - t^3$

s(t)

X: [0, 8] and Y: [0, 250]

(a)

$s'(t) = v(t) = 24t - 3t^2$

s'(t)

X: [0, 8] and Y: [0, 50]

(b)

$a(t) = v'(t) = 24 - 6t$

s''(t)

X: [0, 8] and Y: [−20, 20]

(c)

Figure 25

Applications

A derivative measures the rate of change of a function. Therefore, f'' measures the rate of change of f' in the same way that f' measures the rate of change of f at some particular point. For example, if $s(t) = 12t^2 - t^3$ is a position function for an object at time t for $0 \le t \le 8$, then $s'(t) = v(t)$ is the *velocity* of the object at time t in feet per second as defined by $v(t) = 24t - 3t^2$. Likewise, $v'(t) = a(t) = 24 - 6t$ is the *acceleration* of the object at time t. We see then that at $t = 2$, the velocity of the object is $v(2) = 36$ and $a(2) = 12$, which tells us that at 2 seconds the object's velocity is still increasing. The object will have increasing velocity when the derivative of the velocity function (acceleration) is positive.

EXAMPLE 17 Using the position function $s(t) = 12t^2 - t^3, 0 \le t \le 8$, find the interval(s) of time when the velocity is increasing. The object will have increasing velocity when the derivative of the velocity function (acceleration) is positive.

SOLUTION We found above that $v(t) = 24t - 3t^2$ and $v'(t) = a(t) = 24 - 6t$. The critical value for v occurs at $t = 4$. Over the interval $(0, 4)$, $a(t) > 0$ and over the interval $(4, 8)$, $a(t) < 0$. Therefore, velocity increases over the time interval 0 seconds to 4 seconds. This is verified by looking at the graphs in Figure 25.

Another useful application of higher-order derivatives is determining the **concavity** of the graph of a function. Look at the two graphs in Figure 26. The graph of $f_1(x)$ can be described as turning upward (concave up) and the graph of $f_2(x)$ turns downward (concave down).

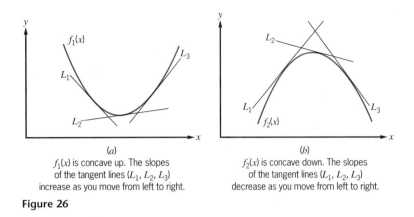

(a)

$f_1(x)$ is concave up. The slopes
of the tangent lines (L_1, L_2, L_3)
increase as you move from left to right.

(b)

$f_2(x)$ is concave down. The slopes
of the tangent lines (L_1, L_2, L_3)
decrease as you move from left to right.

Figure 26

In Figure 26 notice that the tangent lines lie below the graph of $f_1(x)$ and the slopes of the tangent lines of $f_1(x)$ increase from left to right (starting as negative and ending as positive). This means that $f_1'(x)$ is increasing and, therefore, its derivative is positive, $f_1''(x) > 0$. Likewise, the tangent lines lie above the graph of $f_2(x)$ and the slopes of the tangent lines of $f_2(x)$ are decreasing from left to right (starting as positive and ending as negative). This means that the derivative of $f_2'(x)$ must be negative, $f_2''(x) < 0$. This discussion is summarized as follows.

DEFINITION: CONCAVITY OF A FUNCTION

1. If $f''(x) > 0$ on an open interval I, then $f(x)$ is turning upward on I and the graph is said to be **concave upward** over I.
2. If $f''(x) < 0$ on an open interval I, then $f(x)$ is turning downward on I and the graph is said to be **concave downward** over I.

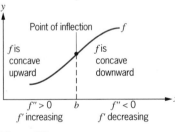

Figure 27

Notice that where the graph is concave upward, $f''(x) > 0$ so that $f'(x)$ is increasing, and where the graph is concave downward, $f''(x) < 0$ so that $f'(x)$ is decreasing. Now, what happens at the point where a graph changes from concave upward to concave downward (Figure 27)? Remember that $f''(x)$ must either assume a value of 0 or it does not exist at this point. The point $(b, f(b))$ where concavity changes is called a **point of inflection**.

DEFINITION: INFLECTION POINT

An **inflection point** is a point on the graph of a function where concavity changes from upward to downward or from downward to upward. At this point, $f''(x) = 0$ or it does not exist.

NOTE: Where $f''(x) = 0$ or where it does not exist a *possible* inflection point occurs. However, if $f''(x)$ does not change sign, there is no inflection point. The graph of the second derivative will help you see what is occurring at the point.

EXAMPLE 18   Determine where the graph of the function $f(x) = x^3 - 6x^2 + 9x + 4$ is concave up, is concave down, and has points of inflection. Also, determine where f is increasing, where it is decreasing, and the values of any relative extrema. Use critical values, possible inflection points, and the graphs of f, f', and f'' to find the information.

SOLUTION

1. Find the first derivative:

$$f'(x) = 3x^2 - 12x + 9 = 3(x^2 - 4x + 3) = 3(x - 3)(x - 1)$$

From this we see that the critical values are $x = 1$ and $x = 3$.

2. Find the second derivative:

$$f''(x) = 6x - 12 = 6(x - 2)$$

From this we determine that the only possible inflection point is at $x = 2$.

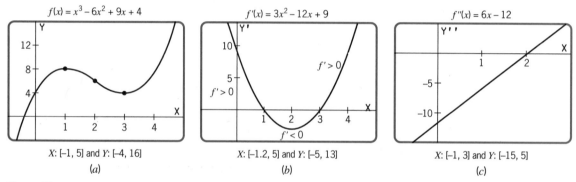

$f(x) = x^3 - 6x^2 + 9x + 4$

X: [−1, 5] and Y: [−4, 16]

(a)

$f'(x) = 3x^2 - 12x + 9$

X: [−1.2, 5] and Y: [−5, 13]

(b)

$f''(x) = 6x - 12$

X: [−1, 3] and Y: [−15, 5]

(c)

Figure 28

3. Using the information found in (1) and (2), we look at the graphs of the function and the derivatives in Figure 28 to obtain the following:

Increasing on $(-\infty, 1)$, $(3, \infty)$ where $f' > 0$.
Decreasing on $(1, 3)$ where $f' < 0$.
Relative maximum at $(1, 8)$ where $f' = 0$ and changes from positive to negative.
Relative minimum at $(3, 4)$ where $f' = 0$ and changes from negative to positive.
Concave down on $(-\infty, 2)$ where $f'' < 0$.
Concave up on $(2, \infty)$ where $f'' > 0$.
Inflection point at $(2, 6)$ where $f'' = 0$ and concavity changes. ■

Practice Problem 3 Draw the graph of $f(x) = x^3 - 12x^2$. Find all extrema and points of inflection.

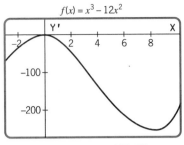

$f(x) = x^3 - 12x^2$

X: [-3, 10] and Y: [-280, 40]

Figure 29

ANSWER The graph is shown in Figure 29.

$f'(x) = 3x^2 - 24x = 3x(x - 8)$. Relative maximum at $(0, 0)$ and relative minimum at $(8, -256)$.

$f''(x) = 6x - 24$. Inflection point at $(4, -128)$.

EXAMPLE 19   For $f(x) = x^3 - 3x^2 + 9x - 3$:

(a) Find the intervals where the graph is concave up and the intervals where it is concave down.

(b) Find the slope of the tangent line at the point of inflection.

SOLUTION

(a) We must first find the second derivative.

$$f(x) = x^3 - 3x^2 + 9x - 3$$

$$f'(x) = 3x^2 - 6x + 9$$

$$f''(x) = 6x - 6 = 6(x - 1)$$

We see from this work that the only possible inflection point occurs at $x = 1$. By drawing the graph of $f''(x)$, we can see that it does indeed change from negative to positive and, therefore, the function is concave down on the interval $(-\infty, 1)$ and concave up on the interval $(1, \infty)$. The graphs of $f''(x)$ and $f(x)$ are shown in Figure 30.

(b) To find the slope of the tangent line at the inflection point, we evaluate $f'(1)$ and determine the slope as $f'(1) = 6$ at $(1, 4)$ ■

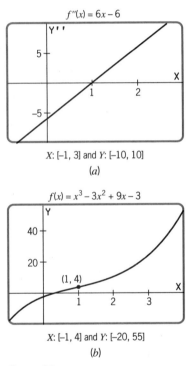

$f''(x) = 6x - 6$

X: [-1, 3] and Y: [-10, 10]

(a)

$f(x) = x^3 - 3x^2 + 9x - 3$

(1, 4)

X: [-1, 4] and Y: [-20, 55]

(b)

Figure 30

EXAMPLE 20   For $f(x) = x^4 - 2x^3$, find any relative extrema and inflection points using the first and second derivatives and the graphs of the functions.

SOLUTION   First, we find the derivatives.

$$f(x) = x^4 - 2x^3$$

$$f'(x) = 4x^3 - 6x^2 = 2x^2(2x - 3)$$ Critical values are $x = 3/2$ and $x = 0$.

$$f''(x) = 12x^2 - 12x = 12x(x - 1)$$ Possible inflection points are $x = 0$ and $x = 1$.

Using the information just found together with the graphs shown in Figure 31, we can determine that there is a relative minimum at $(\frac{3}{2}, -1.7)$ and inflection points at $(0, 0)$ and $(1, -1)$. Notice that even though $f'(0) = 0$ and is a critical value, the derivative does not change signs and there are no relative extrema at that point. ■

Now, let's use our knowledge of the relationship between a function and its first and second derivative to sketch a graph of a function without knowing the definition of the function or its derivatives.

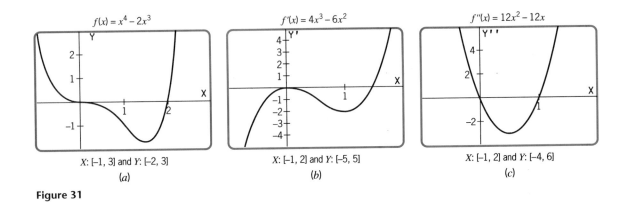

X: [–1, 3] and Y: [–2, 3]

(a)

X: [–1, 2] and Y: [–5, 5]

(b)

X: [–1, 2] and Y: [–4, 6]

(c)

Figure 31

EXAMPLE 21   The graphs of f' and f'' are shown in Figure 32 along with the value of the function f at several points. Use the graphs, what you know about a function and its derivatives, and the values given to sketch the graph of a function f that fits the given information.

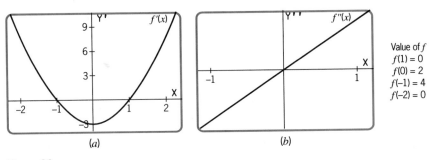

(a)   (b)

Value of f
$f(1) = 0$
$f(0) = 2$
$f(-1) = 4$
$f(-2) = 0$

Figure 32

SOLUTION   Since $f'(x) > 0$ on $(-\infty, -1)$, $(1, \infty)$, we know that f is increasing on those intervals.

Since $f'(x) < 0$ on $(-1, 1)$, we know that f is decreasing on that interval. Also, f' changes from positive to negative at $x = -1$ and f changes from increasing to decreasing; therefore, there is a relative maximum at $x = -1$. Likewise, we can see that a relative minimum occurs at $x = 1$ because f' changes from negative to positive and f changes from decreasing to increasing.

The graph of f'' tells us that f is concave down on the interval $(-\infty, 0)$, since $f'' < 0$ there. Also, f is concave up where $f'' > 0$, which is on the interval $(0, \infty)$. Therefore, the only inflection point occurs at $x = 0$.

We now use the preceding information to sketch a graph of a function meeting the above requirements (Figure 33).

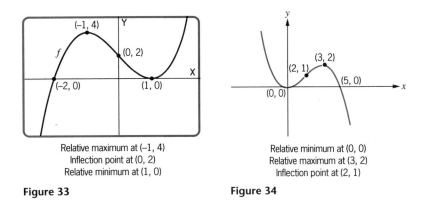

Relative maximum at (–1, 4)
Inflection point at (0, 2)
Relative minimum at (1, 0)

Figure 33

Relative minimum at (0, 0)
Relative maximum at (3, 2)
Inflection point at (2, 1)

Figure 34

Practice Problem 4   Given the graph of a function in Figure 34, sketch the graphs of f' and f''.

ANSWER   The graphs are shown in Figure 35.

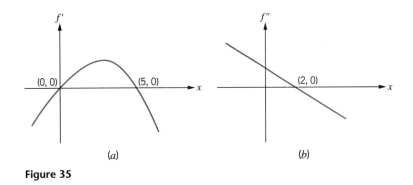

(a)

(b)

Figure 35

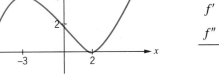

Figure 36

Practice Problem 5   Use the given information on $f'(x)$ and $f''(x)$ and the ordered pairs to the right to sketch a rough graph of the function. Assume that the function is continuous.

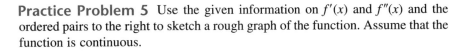

$f' > 0$	$f' < 0$	$f' < 0$	$f' > 0$	f contains the points
$f'' < 0$	$f'' < 0$	$f'' > 0$	$f'' > 0$	$(-3, 4)$, $(0, 2)$ and $(2, 0)$

ANSWER   The graph is shown in Figure 36.

SUMMARY

In this section we learned that derivatives can have derivatives. These higher-order derivatives are useful in many ways. The use that we concentrated on most in this section was finding the concavity of a graph. Discontinuities may occur in derivatives even though a function is continuous. This fact does not affect the usefulness of the derivative in determining possible inflection points and concavity. The conclusions are summarized in Table 2.

TABLE 2

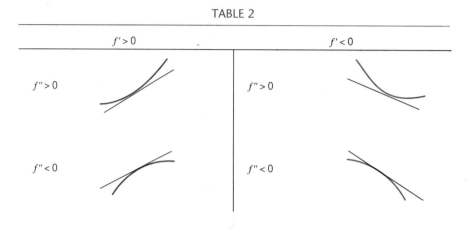

	$f' > 0$	$f' < 0$
$f'' > 0$		$f'' > 0$
$f'' < 0$		$f'' < 0$

Exercise Set 3.3

Find

$$\frac{dy}{dx}, \quad \frac{d^2y}{dx^2}, \quad and \quad \frac{d^3y}{dx^3}$$

for each function.

1. $y = 25$
2. $y = -5$
3. $y = 3x + 5$
4. $y = 3x^2 - 5$
5. $y = 35x^2 - 27x + 5$
6. $y = -3x^3 + 4x + 2$
7. $y = 2x^3 + 3x^2 - x + 2$
8. $y = 4x^3 - 3x^2 + 5x - 3$
9. $y = 2x^6 + 3x^3 - 2x + 7$
10. $y = -3x^8 + 2x^5 - 3x^2$
11. $y = \frac{1}{2}x^4 + \frac{1}{3}x^3$
12. $y = -\frac{2}{3}x^3 + \frac{1}{2}x^2$
13. $y = 3x^{1/3}$
14. $y = 3x^{2/3}$
15. $y = x\sqrt[3]{x}$
16. $y = \sqrt{16 - x}$
17. $y = \frac{3}{x - 1}$
18. $y = \frac{4}{(x - 1)^4}$

Use the figure to answer Exercises 19–21.

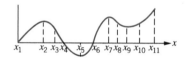

19. For the points named on the figure, list all x_i's over the interval (x_1, x_{11}) for each concept.
    (a) Critical point   (b) Inflection point
    (c) x-intercept   (d) Relative maximum
    (e) Relative minimum

20. Does it appear that there are points in the interval (x_1, x_{11}) where $f'(x)$ is undefined?

21. For each blank, write one of the following: positive, negative, or zero.
    (a) At x_2, f is _____, f' is _____, and f'' is _____.
    (b) At x_5, f is _____, f' is _____, and f'' is _____.
    (c) At x_6, f is _____, f' is _____, and f'' is _____.

(d) At x_{10}, f is ———, f' is ———, and f'' is ———.

(e) At x_8, f is ———, f' is ———, and f'' is ———.

(f) On the interval (x_3, x_6), f'' is ———.
(g) On the interval (x_8, x_{10}), f'' is ———.
(h) Wherever f is increasing, f' is ———.
(i) Wherever f is concave up, f'' is ———.
(j) List the intervals where f is decreasing.
(k) List the intervals where f is concave down over the interval (x_1, x_{11}).

Discuss whether the function described below is increasing or decreasing, is concave upward or concave downward, and draw a small section of the curve in some interval containing c.

22. $f'(c) > 0$, $f''(c) > 0$
23. $f'(c) > 0$, $f''(c) < 0$
24. $f'(c) < 0$, $f''(c) > 0$
25. $f'(c) < 0$, $f''(c) < 0$
26. $f'(c) = 0$, $f''(c) > 0$
27. $f'(c) = 0$, $f''(c) < 0$
28. Find an equation of the tangent line at the point of inflection for $y = x^3 + 3x^2 + 2$.

In Exercises 29, 30, and 31, use the given information on $f'(x)$ and $f''(x)$ and the table of values to sketch a rough graph of the function. Assume that the function is continuous.

29.

| $f' > 0$ | $f' < 0$ | $f' < 0$ | $f' > 0$ |
$f'' < 0$	$f'' < 0$	$f'' > 0$	$f'' > 0$
-1	0	1	

x	-2	-1	0	1	2
$f(x)$	-2	2	0	-2	2

30.

| $f' < 0$ | $f' > 0$ |
$f'' > 0$	$f'' > 0$
1	

x	0	1	2
$f(x)$	0	-3	0

31.

| $f' < 0$ | $f' > 0$ | $f' > 0$ | $f' < 0$ |
$f'' > 0$	$f'' > 0$	$f'' < 0$	$f'' < 0$
-2	-1	0	

x	-2	-1	0
$f(x)$	-16	-5	0

Find $f''(x)$ for each of the following functions

32. $f(x) = (x^2 - 3)(x - 1)$
33. $f(x) = (x - 3)(x + 4)^2$
34. $f(x) = \dfrac{x + 1}{x + 3}$
35. $f(x) = \dfrac{x}{(x + 3)^2}$
36. $f(x) = (x^2 + 4)^3$
37. $f(x) = \dfrac{1}{\sqrt{x^2 + 4}}$

Find the intervals over which the graph of the given function is concave downward and concave upward, the inflection points, and all relative extrema. Use calculator graphs of the function and its first and second derivatives.

38. $y = x^2$
39. $y = -x^2$
40. $y = 3x^2$
41. $y = -2x^2$
42. $y = 4x^2 + 3x$
43. $y = -3x^2 + 2x$
44. $y = x^3$
45. $y = -x^3$
46. $y = x^3 - 3x^2$
47. $y = x^3 + 4x^2$
48. $y = x^4 - 2x^2$
49. $y = x^4 + x^2$
50. $y = 3x^{5/3} - 15x^{2/3}$
51. $y = x^{5/3} + 5x^{2/3}$

Given the graph of f, make a reasonable sketch of f' and f".

52.

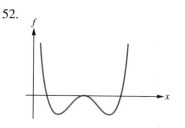

x-intercepts: 1, 4, 7
Relative minima at $x = 2$, $x = 6$
Relative maximum at $x = 4$
Inflection points at $x = 3$, $x = 5$

53.

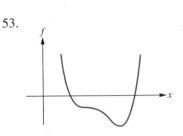

x-intercepts: 4, 14
Relative minima at $x = 6$, $x = 12$
Relative maximum at $x = 8$
Inflection points at $x = 7$, $x = 10$

Given the graphs of f' and f'' along with values of the function in the table, make a reasonable sketch of a function fitting the given requirements.

54.

x	-1	0	0.5	2
$f(x)$	-5	2	8.5	22

55.

x	-2	-1	0	0.5	1
$f(x)$	-4	-2	1	0	-1

56. A helicopter rises vertically, and after t seconds its height $s(t)$ above the ground is given by the formula $s(t) = 6t^2 - t^3$ feet for $0 \le t \le 6$.
 (a) Find the velocity function and the acceleration function.

 (b) Find the velocity and acceleration after 2 seconds.

57. A missile rises to a height of $s(t) = t^3 + 0.5t^2$ feet in t seconds.
 (a) Find the velocity function and the acceleration function.
 (b) Find the velocity and acceleration after 8 seconds.

Applications (Business and Economics)

All application problems have been selected so that the maxima and minima requested are at relative maxima and minima. Use your graphing calculator to find approximate values of relative extrema and then use calculus techniques to find exact values.

58. Cost Function. A cost function is given by $C(x) = (3x - 10)^2 + 24$ for $x \ge 3$. Find the minimum value of the cost function.

59. Maximum Profit. Profit P is related to selling price S by the formula $P = 15{,}000S - 500S^2$ for $0 \le S \le 30$.
 (a) Describe the rate of change of P at $S = 7$.
 (b) For what range of values of S is the profit increasing?
 (c) At what selling price is the profit a maximum?
 (d) What is the maximum profit?

60. Maximum Profit. The cost and revenue from sales curves are given by $C = 2x^2 - 8x$ and $R = x^2 - 6x$, where x is given in millions.
 (a) For what production is the revenue function a minimum?
 (b) For what production is the cost function a minimum? Discuss.
 (c) Find the number of units for which the profit will be a maximum.
 (d) What is the maximum profit?

61. A revenue function for the sale of x stereo sets per year is

$$R(x) = 460x - 372\left(12 - \frac{x}{3}\right) - \left(12 - \frac{x}{3}\right)^2$$

for $0 \le x \le 5000$.

(a) For what production do you have maximum revenue?

(b) Find the maximum revenue.

Applications (Social and Life Sciences)

62. **Drug Concentration**.   The concentration of a drug in a person's blood x hours after administration is

$$c = \frac{x}{20(x^2 + 6)}$$

(a) When will the concentration be a maximum?

(b) What is the maximum concentration?

63. **Bacteria Count**.   Assume that the number of bacteria, in millions, present in a culture at time t is given by $B(t) = t^2(t - 8) + 400$. Find the minimum population.

64. **Contagious Disease**.   Suppose that the number of people who are affected by the introduction of a flu virus in a state is given by $N(t) = 40t + 12 - t^2$, where t is the number of days after the virus is detected. On what day will the maximum number of people be affected, and how many people will be affected?

Review Exercises

Find the critical points, the intervals where the graphs are increasing, and the intervals where the graphs are decreasing. Draw each graph by hand and then check your work with a graphing calculator.

65. $y = x^2 + 3x$

66. $y = x^3 - 3x^2$

67. $y = x^3 - 3x + 2$

68. $y = x^3 + x - 1$

69. $y = x^4 - 4x + 1$

70. $y = x^4 - 2x^2 + 1$

3.4   THE SECOND DERIVATIVE TEST AND ABSOLUTE EXTREMA

OVERVIEW   Our study of concavity leads naturally into another test that helps determine whether a critical point is a relative maximum or relative minimum. A relative maximum may not be an absolute maximum; likewise, a relative minimum may not be an absolute minimum. We consider the differences in absolute and relative extrema. In this section we study

- The second derivative test
- Absolute maximum
- Absolute minimum

In Section 9.2, the first derivative test was given for relative maxima or relative minima (relative extrema). Another test for relative extrema uses the second derivative; this test is used when the second derivative is easy to obtain. Just as the first derivative gives the slope of the tangent line to a curve, the second derivative indicates the concavity of the curve. Suppose that a function is concave down in an open interval and $f'(c) = 0$ for some point c in this interval. Intuitively, a relative maximum exists at $(c, f(c))$; see Figure 37(a). Likewise, if a function is concave up in an open interval and $f'(c) = 0$ for some c in this interval, then a relative minimum exists at $(c, f(c))$; see Figure 37(b). This leads to the following test.

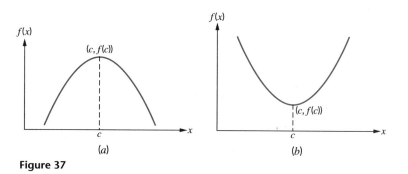

Figure 37

TABLE 3

$f'(c)$	$f''(c)$	$f(c)$
0	$-$	Relative maximum
0	$+$	Relative minimum
0	0	Test fails; use first derivative test or graph

SECOND DERIVATIVE TEST

Let $f'(c) = 0$ and suppose that $f''(x)$ exists in an open interval containing c.

(a) If $f''(c) < 0$, then $f(c)$ is a relative maximum.
(b) If $f''(c) > 0$, then $f(c)$ is a relative minimum.
(c) If $f''(c) = 0$, then the test gives no information.

The second derivative test is summarized in Table 3.

EXAMPLE 22   Use the second derivative test to find the points on $f(x) = x^3 - 3x^2 - 9x + 5$ that are relative extrema.

SOLUTION   We have

$$f'(x) = 3x^2 - 6x - 9 = 3(x^2 - 2x - 3) = 3(x - 3)(x + 1)$$

The critical values of the function are $x = -1$ and $x = 3$, since those are the values that make $f'(x) = 0$. Let's test these critical values in the second derivative, $f''(x) = 6x - 6$.

$$f''(-1) = -6 - 6 = -12 < 0 \qquad \text{So } f(-1) = 10 \text{ is a relative maximum.}$$
$$f''(3) = 18 - 6 = 12 > 0 \qquad \text{So } f(3) = -22 \text{ is a relative minimum.}$$

This is confirmed by the graph of f shown in Figure 38.   ■

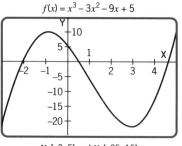

$f(x) = x^3 - 3x^2 - 9x + 5$

X: [-3, 5] and Y: [-25, 15]

Figure 38

EXAMPLE 23   Use the second derivative test to test $f(x) = 2x^3 - 12x^2 + 24x + 12$ for relative extrema.

SOLUTION   We have

$$f'(x) = 6x^2 - 24x + 24 = 6(x^2 - 4x + 4) = 6(x - 2)^2$$

We can see that the only critical value is $x = 2$. Next we apply the second derivative test.

$$f''(x) = 12x - 24 \quad \text{and} \quad f''(2) = 12(2) - 24 = 0$$

The second derivative test fails here, so we must use the first derivative test or a graph. We can see by looking at the graph of $f(x)$ in Figure 39 that the function appears to be increasing for all x. We can confirm this by looking at the graph of f', which is nonnegative for all x; hence, $f(x)$ is increasing for all x. If we look at the graph of f'', we can see that there is an inflection point at $(2, 28)$. ■

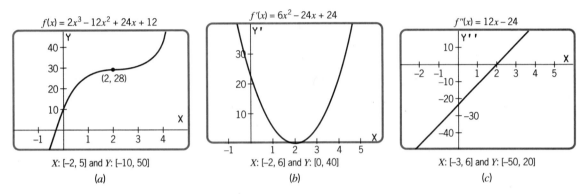

Figure 39

Practice Problem 1 The critical values for the function $f(x) = 2x^3 + 3x^2 - 12x + 8$ are $x = -2$ and $x = 1$. Use the second derivative test to classify as relative maxima, relative minima, or no information.

ANSWER $f''(x) = 12x + 6$. A relative maximum occurs at $x = -2$ ($f''(-2) < 0$) and a relative minimum at $x = 1$ ($f''(1) > 0$).

EXAMPLE 24 Determine where the graph of the function $f(x) = x^3 - 3x^2$:

(a)   Is concave upward.
(b)   Is concave downward.
(c)   Has a point of inflection.
(d)   Is Increasing.
(e)   Is Decreasing.
(f)   Has a relative maximum.
(g)   Has a relative minimum.
(h)   Find the slope of the tangent line at the point of inflection.

SOLUTION

$$f'(x) = 3x^2 - 6x = 3x(x - 2) \qquad \text{Critical values are } x = 0, x = 2.$$

$$f''(x) = 6x - 6 = 6(x - 1) \qquad \text{Possible inflection point at } x = 1$$

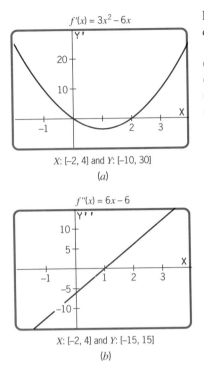

$f'(x) = 3x^2 - 6x$

X: [-2, 4] and Y: [-10, 30]

(a)

$f''(x) = 6x - 6$

X: [-2, 4] and Y: [-15, 15]

(b)

Figure 40

From the information above and the graphs shown in Figure 40 we can conclude the following.

(a) f is concave upward on $(1, \infty)$.
(b) f is concave downward on $(-\infty, 1)$.
(c) There is a point of inflection at $(1, -2)$.
(d) f is increasing on $(-\infty, 0)$ and $(2, \infty)$.
(e) f is decreasing on $(0, 2)$.
(f) $f'(0) = 0$ and $f''(0) = -6 < 0$; therefore, there is a relative maximum at $(0, 0)$.
(g) $f'(2) = 0$ and $f''(2) = 6 > 0$; therefore, there is a relative minimum at $(2, -4)$.
(h) The point of inflection occurs at $x = 1$. The slope of the tangent at this point is $f'(1) = -3$.

All of the preceding information is summarized in Figure 41.   ■

Up until this point, we have discussed relative extrema. Now let's turn our attention to **absolute extrema**. The **absolute maximum** of a function is the greatest value that the function attains in a closed interval, and the **absolute minimum** is the least value attained in a closed interval. A function that is continuous on a closed interval must have an absolute maximum and absolute minimum on that interval.

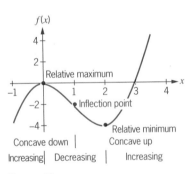

$f(x)$

Relative maximum

Inflection point

Relative minimum

Concave down | Concave up

Increasing| Decreasing | Increasing

Figure 41

DEFINITION: ABSOLUTE MAXIMUM AND ABSOLUTE MINIMUM

(a) A function f has an absolute maximum at $x = c$ if $f(c) \geq f(x)$ for all x in the selected domain of f.
(b) A function f has an absolute minimum at $x = c$ if $f(c) \leq f(x)$ for all x in the selected domain of f.

To determine where the absolute maximum and the absolute minimum occur, study Figure 42. Note that for values in the intervals between the endpoints and the critical points, a continuous function is either increasing or decreasing. So, to find the absolute maximum and the absolute minimum of a continuous function that is not constant on a closed interval, we evaluate the function at each endpoint and at each critical point *within the interval* and simply choose the greatest value for the absolute maximum and the least value for the absolute minimum.

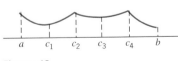

$a \quad c_1 \quad c_2 \quad c_3 \quad c_4 \quad b$

Figure 42

Procedure for Finding Absolute Maximum and Absolute Minimum

For $f(x)$ defined and continuous on $[a, b]$:

Procedure for Finding Absolute Maximum and Absolute Minimum (Continued)

1. Find all critical values $c_1, c_2, \ldots, c_n$ on (a, b).
2. Evaluate $f(a), f(b), f(c_1), f(c_2), \ldots, f(c_n)$.
3. The greatest value found in step 2 is the absolute maximum on $[a, b]$. The least value found in step 2 is the absolute minimum on $[a, b]$.

EXAMPLE 25 Draw the graph of $f(x) = x^3 - 6x^2 + 8$ and find the absolute maximum and the absolute minimum on each given interval.

(a) $[-1, 3]$ (b) $[-1, 5]$ (c) $[-3, 7]$

SOLUTION The graph of $f(x) = x^3 - 6x^2 + 8$ is given in Figure 43. Since $f'(x) = 3x^2 - 12x = 3x(x - 4)$, the critical values are at $x = 0$ and $x = 4$.

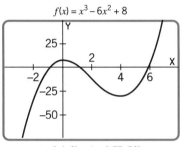

$f(x) = x^3 - 6x^2 + 8$

X: [-4, 8] and Y: [-75, 50]

Figure 43

(a) For the interval $[-1, 3]$ we have $a = -1$ and $b = 3$. The critical value in the interval is $x = 0$.

$$f(-1) = (-1)^3 - 6(-1)^2 + 8 = 1 \qquad \text{Endpoint}$$
$$f(0) = (0)^3 - 6(0)^2 + 8 = 8 \qquad \text{Critical point}$$
$$f(3) = (3)^3 - 6(3)^2 + 8 = -19 \qquad \text{Endpoint}$$

Therefore, for the interval $[-1, 3]$ the absolute maximum occurs at $x = 0$ and is 8. The absolute minimum occurs at $x = 3$ and is -19.

(b) For the interval $[-1, 5]$ we have $a = -1$ and $b = 5$. The critical values are $x = 0$ and $x = 4$.

$$f(-1) = (-1)^3 - 6(-1)^2 + 8 = 1 \qquad \text{Endpoint}$$
$$f(0) = (0)^3 - 6(0)^2 + 8 = 8 \qquad \text{Critical point}$$
$$f(4) = (4)^3 - 6(4)^2 + 8 = -24 \qquad \text{Critical point}$$
$$f(5) = (5)^3 - 6(5)^2 + 8 = -17 \qquad \text{Endpoint}$$

For the interval $[-1, 5]$ the absolute maximum occurs at $x = 0$ and is 8. The absolute minimum occurs at $x = 4$ and is -24.

(c) For the interval $[-3, 7]$ we have $a = -3$ and $b = 7$. The critical values are $x = 0$ and $x = 4$.

$$f(-3) = (-3)^3 - 6(-3)^2 + 8 = -73 \qquad \text{Endpoint}$$
$$f(0) = (0)^3 - 6(0)^2 + 8 = 8 \qquad \text{Critical point}$$
$$f(4) = (4)^3 - 6(4)^2 + 8 = -24 \qquad \text{Critical point}$$
$$f(7) = (7)^3 - 6(7)^2 + 8 = 57 \qquad \text{Endpoint}$$

For the interval $[-3, 7]$, the absolute maximum occurs at $x = 7$ and is 57. The absolute minimum occurs at $x = -3$ and is -73.

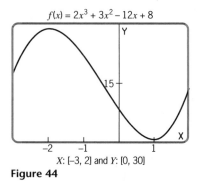

$f(x) = 2x^3 + 3x^2 - 12x + 8$

X: [-3, 2] and Y: [0, 30]

Figure 44

Practice Problem 2 For $f(x) = 2x^3 + 3x^2 - 12x + 8$, find the absolute maximum and absolute minimum on the interval $[-3, 2]$. Verify your answer with a graph.

ANSWER We have

$$f'(x) = 6x^2 + 6x - 12 \quad \text{and} \quad f''(x) = 12x + 6$$

The absolute maximum on $[-3, 2]$ is 28 at $x = -2$ and the absolute minimum is 1 at $x = 1$. The graph is shown in Figure 44.

If we look at the function in Practice Problem 2 as it is defined for all real numbers, there is no absolute maximum and no absolute minimum because the function can become infinitely large in magnitude in both directions. This illustrates the importance of the interval for which an absolute maximum or an absolute minimum is desired.

When there is only one critical value of f on (a, b), and the second derivative is negative at the critical value, the absolute maximum of f on $[a, b]$ occurs at the relative maximum. Likewise, when there is only one critical value of f on (a, b), and the second derivative is positive at the critical value, the absolute minimum of f on $[a, b]$ occurs at the relative minimum. This is summarized in Table 4.

TABLE 4

Test for Absolute Maximum or Minimum			
Let $f(x)$ be continuous on $a < x < b$, and let there be only one critical value c on this interval.			
$f'(c)$	$f''(c)$	$f(c)$	Graph
0	−	Absolute maximum	
0	+	Absolute minimum	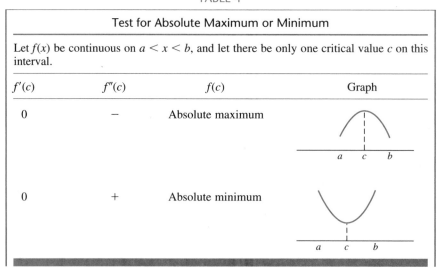

EXAMPLE 26 Find the absolute maximum and the absolute minimum of $f(x) = 3x^2 - 6x$ on $[0, 4]$.

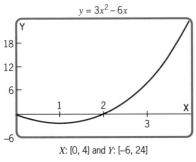

$y = 3x^2 - 6x$

X: [0, 4] and Y: [−6, 24]

Figure 45

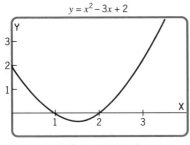

$y = x^2 - 3x + 2$

X: [0, 4] and Y: [−1, 4]

Figure 46

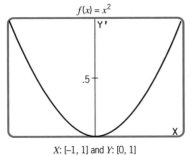

$f(x) = x^2$

X: [−1, 1] and Y: [0, 1]

Figure 47

SOLUTION We have $f'(x) = 6x - 6 = 6(x - 1)$; therefore, the only critical value is $x = 1$, with $f'(1) = 0$. Since $x = 1$ is the only critical value on [0, 4], the second derivative test can be used.

$$f''(x) = 6 \quad \text{and} \quad f''(1) = 6$$

Since $f''(x)$ is positive, the absolute minimum of $f(x)$ occurs at $x = 1$ and is $f(1) = 3(1)^2 - 6(1) = -3$. The absolute maximum is found by evaluating $f(x)$ at the two endpoints of the interval $x = 0$ and $x = 4$.

$$f(0) = 3(0)^2 - 6(0) = 0 \quad \text{and} \quad f(4) = 3(4)^2 - 6(4) = 48 - 24 = 24$$

Thus, the absolute maximum of f occurs at $x = 4$ and is 24. ∎

Practice Problem 3 With your graphing calculator, draw the graph of $y = 3x^2 - 6x$ on the intervals $0 \le x \le 4$ and $-6 \le y \le 24$. Then from the graph find x for the absolute maximum and the absolute minimum. Do your answers agree with those in Example 26?

ANSWER The graph is shown in Figure 45. The absolute maximum on [0, 4] is 24 and the absolute minimum is -3, which agree with Example 26.

Practice Problem 4 Using your graphing calculator, draw the graph and find the absolute minimum for $y = x^2 - 3x + 2$ on [0, 4].

ANSWER The graph is shown in Figure 46. The absolute minimum is $-\frac{1}{4}$ at $x = \frac{3}{2}$.

The absolute extrema can occur at more than one point on a closed interval. For example, $f(x) = x^2$ on $[-1, 1]$ has an absolute maximum at $x = -1$ and $x = 1$. This is demonstrated in Figure 47.

SUMMARY

The second derivative test gives us a method of determining relative extrema without having to see a graph. This is useful in applications, as we will see in Section 9.6. Over a closed interval, a continuous function that is not constant has an absolute maximum and absolute minimum. These occur at the endpoints of the interval or at critical points that are within the closed interval.

Exercise Set 3.4

Find the location of all absolute maxima and absolute minima (if any) for the following functions.

1.

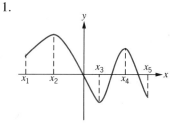

2.

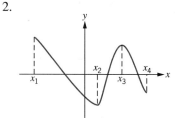

3.

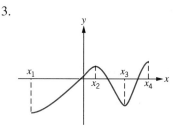

4.

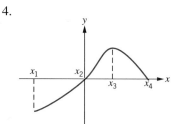

Find the absolute maximum and the absolute minimum on the given intervals.

5. $y = x^2 - x$
   (a) $[0, 1]$ (b) $[0, 3]$ (c) $[-2, 3]$

6. $y = x^2 + 2x$
   (a) $[-2, 0]$ (b) $[-3, 1]$ (c) $[-4, 4]$

7. $y = 2x^3$
   (a) $[-1, 1]$ (b) $[-2, 2]$ (c) $[-3, 3]$

8. $y = x^3 - 3x^2$
   (a) $[-1, 3]$ (b) $[-2, 4]$ (c) $[-3, 6]$

9. $y = x^4 + 2x^2$
   (a) $[-1, 1]$ (b) $[-2, 2]$ (c) $[-3, 2]$

10. $y = 2x + \dfrac{1}{x}$
   (a) $[1, 2]$ (b) $[-2, -1]$ (c) $[1, 3]$

Using the second derivative test (if possible), find relative maxima and relative minima for each function. Verify your answers with your graphing calculator.

11. $y = 6x^2 + 4x^3$ 12. $y = 9x^4 - 4x^2$

13. $y = 3x^4 - 4x^3$ 14. $y = 3x^4 - 6x^3$

15. $y = x^3 + \dfrac{19}{2}x^2 + 20x$ 16. $y = 6(x - 2)^{2/3}$

Find the critical points, relative extrema, absolute extrema, and inflection points of the following functions. Use calculator graphs to verify your work.

17. $y = 4x^2 - x$ 18. $y = x^2 + 2x$

19. $y = 2x^3$ 20. $y = -2x^3$

21. $y = x^3 - 3x^2$ 22. $y = x^3 + 3x^2$

23. $y = x^4 + 2x^2$ 24. $y = 3x - x^{-1}$

25. $y = 2x + x^{-1}$

Find the intervals where the function is increasing or decreasing, and where the graph is concave upward or downward. Then find all points of relative maxima, relative minima, and points of inflection. First find all of these using a graphing calculator. Then, using calculus techniques, indicate all of these on a coordinate system as you graph the function.

26. $f(x) = x^2(x - 1)$ 27. $f(x) = \sqrt{x}$

28. $f(x) = \sqrt[3]{x}$ 29. $f(x) = x(x - 1)^2$

30. $f(x) = x - 3x^{1/3}$ 31. $f(x) = x^{4/3} + 4x^{1/3}$

32. **Exam.** The five graphs below have equal scales on the x- and y-axis. Which one could be the graph of $y = 3x^4 - 4x^3$?

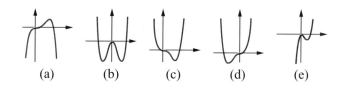

(a)   (b)   (c)   (d)   (e)

33. **Exam.**   On which of the following open intervals is the function $f(x) = (x^3/6) - 2x$ decreasing and concave upward?

(a) $(-2, 2)$ (b) $(-2, 0)$ (c) $(0, 2)$
(d) $(2, 3)$ (e) $(-2, 4)$

34. **Exam.**   What is the maximum value of a function defined by $f(x) = x^3 + x^2$ over $[-2, 1]$?

(a) -2 (b) 0 (c) $\dfrac{4}{27}$ (d) $\dfrac{20}{27}$

(e) 2

35. **Exam.**   To find the point at which a relative minimum cost occurs given a total cost equation, the steps are as follows: (1) Find the first derivative, (2) set the first derivative equal to zero, and (3) solve the equation. Using the solution(s) so derived, which additional steps must be taken to determine if a minimum exists?

(a) Substitute the solution(s) in the first derivative of the cost function, and a negative result indicates a minimum.

(b) Substitute the solution(s) in the first derivative of the cost function, and a positive result indicates a minimum.

(c) Substitute the solution(s) in the second derivative of the cost function, and a negative result indicates a minimum.

(d) Substitute the solution(s) in the second derivative of the cost function, and a positive result indicates a minimum.

36. Use a graphing calculator to find the absolute minimum of

$$y = \frac{4}{|x^2 + 2x - 3|}$$

over the interval $[-2, 0]$.

Applications (Business and Economics)

37. **Average Cost Function.**   If a cost function is given by $C(x) = 2x^2 + 50$, for $0 \le x \le 100$, find the minimum value of the average cost function,

$$\overline{C}(x) = \frac{C(x)}{x}$$

38. **Average Revenue.**   A revenue function is given by $R(x) = 3x^2 - 10x + 75$, for $0 \le x \le 100$. Find the minimum value of the average revenue function,

$$\overline{R}(x) = \frac{R(x)}{x}$$

39. **Average Profit.**   Using Exercises 37 and 38, find the minimum value of the average profit function,

$$\overline{P}(x) = \frac{P(x)}{x}$$

40. **Average Cost Function.**   Given that the cost function of producing x items of a given product is

$$C(x) = 40 + 10x + \frac{x^3}{200}$$

find the maximum value of the average cost function,

$$\overline{C}(x) = \frac{C(x)}{x}$$

and compare the value of $\overline{C}(x)$ with $C'(x)$, the marginal cost at the value of x where $\overline{C}(x)$ is a minimum.

41. **Revenue Function.**   The cost and revenue from sales are

$$C(x) = x^3 - 15x^2 + 72x + 25$$
$$\text{and}\quad R(x) = 51x - 3x^2$$

Draw the cost function and the revenue function at the same time. Where does profit occur and where does loss occur? Find the point at which there is maximum profit.

Applications (Social and Life Sciences)

42. **Maximum Work.**   A psychiatrist has found that for $t \ge 0$, the equation $y(t) = c_0 + c_1 t + c_2 t^2 + c_3 t^3$, where c_0, c_1, c_2, and c_3 are constants, describes the relationship between amount of work output $y(t)$ and elapsed time t for $0 \le t \le 5$. Find any relative max-

ima or relative minima, and sketch the curve when $c_0 = 0$, $c_1 = 92$, $c_2 = 10$, and $c_3 = -6$.

Review Exercises

Find the critical points, intervals where the graph is increasing and where it is decreasing, relative maxima and minima, points of inflection, intervals where the curve *is concave upward and concave downward, and sketch the graphs of the following functions.*

43. $y = x^{-1} + x^{-2}$

44. $y = x + x^{2/3}$

45. $y = \dfrac{x^3}{9(x + 2)}$

46. $y = \left(\dfrac{1}{7}\right)(2x^3 - 9x^2 + 12x + 3)$

47. $y = x^{5/3} + 5x^{2/3}$

3.5 INFINITE LIMITS; LIMITS AT INFINITY; CURVE TRACING

OVERVIEW   The real number system is unbounded. There is no *largest* real number and no *smallest* real number. Geometrically, this means that the real number line extends indefinitely to the right and to the left (Figure 48). We have already used the symbols ∞ and $-\infty$ (read "infinity" and "negative infinity") to denote this unboundedness. The symbol ∞ was introduced by the English mathematician John Wallis (1616–1703). The symbol does not represent a number and cannot be manipulated algebraically. It has meaning primarily in connection with limits. In this section we give additional attention to limits that do not exist. That is, we consider functions that increase without bound. In addition, we sometimes need to determine limits when the independent variable is getting larger and larger. This new concept of a limit will be very useful in graphical analysis such as that used in studying growth of bacteria or the average cost function in business. In this section we

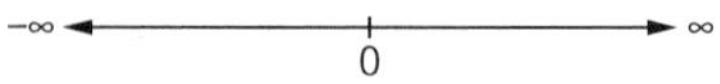

Figure 48

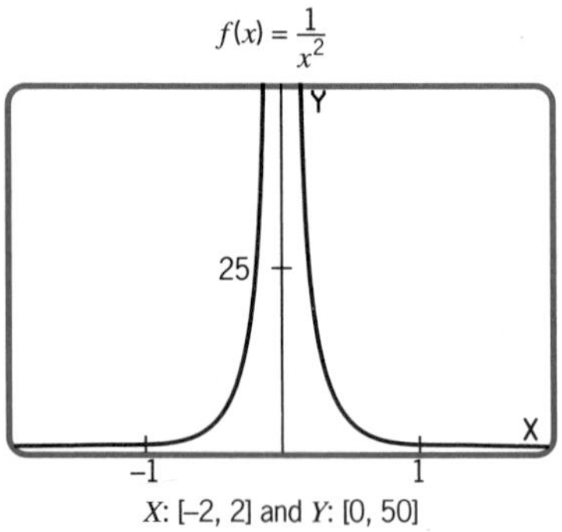

- Define vertical and horizontal asymptotes
- Evaluate limits at infinity
- Analyze graphs of rational functions

$f(x) = \dfrac{1}{x^2}$

Figure 49

X: [-2, 2] and Y: [0, 50]

Consider the graph of $f(x) = 1/x^2$ shown in Figure 49. Consider the function and $\lim\limits_{x \to 0} f(x)$. We see that $f(x)$ gets larger and larger as x gets closer and closer to 0. We say that the limit of the function increases without bound as x approaches 0, or

$$\lim_{x \to 0} \frac{1}{x^2} \to \infty$$

The notation $\to \infty$ is read "approaches infinity," which means the function is increasing without bound as x approaches 0.

Generally, if $f(x)$ increases without bound as x gets closer and closer to some number a from both sides of a, then $\lim\limits_{x \to a} f(x) \to \infty$.

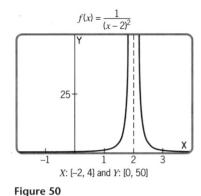

$$f(x) = \frac{1}{(x-2)^2}$$

X: [-2, 4] and Y: [0, 50]

Figure 50

EXAMPLE 27   Find $\lim\limits_{x \to 2} \dfrac{1}{(x-2)^2}$.

SOLUTION   Study the following table along with the graph in Figure 50.

x	1.9	1.99	1.999	→ 2 ←	2.001	2.01	2.1
$\dfrac{1}{(x-2)^2}$	100	10,000	1,000,000	→ ∞ ←	1,000,000	10,000	100

It seems that as $x \to 2$, $1/(x-2)^2$ gets larger and larger, or, increases without bound. That is,

$$\lim_{x \to 2} \frac{1}{(x-2)^2} \to \infty$$

This is verified in the graph shown in Figure 50.

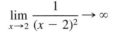

Calculator Note

In order to get a look at how the values of the function are changing as x gets closer and closer to 2, you can use the $\boxed{\text{TRACE}}$ key. However, you can also have your calculator evaluate the function at values of x very close to $x = 2$. Unlike the $\boxed{\text{TRACE}}$ key, the evaluation function allows you to select the values of x.

Practice Problem 1   Draw the graph of

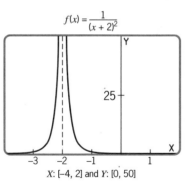

$$f(x) = \frac{1}{(x+2)^2}$$

X: [-4, 2] and Y: [0, 50]

Figure 51

$$f(x) = \frac{1}{(x+2)^2}$$

and determine what happens as x gets closer and closer to -2.

ANSWER   The graph is shown in Figure 51. As x gets closer and closer to -2, the function becomes increasingly large. We can say that

$$\lim_{x \to -2} \frac{1}{(x+2)^2} \to \infty$$

Sometimes a curve is unbounded in a negative direction. Sometimes a curve is unbounded in a positive direction as x approaches a number from one side and is unbounded in a negative direction as x approaches the number from the other side. We use the following terminology for infinite limits.

INFINITE LIMITS

(a) If as $x \to a^-$, $f(x)$ increases without bound, we indicate this behavior by

$$\lim_{x \to a^-} f(x) \to \infty$$

(b) If as $x \to a^-$, $f(x)$ decreases without bound, we indicate this behavior by

$$\lim_{x \to a^-} f(x) \to -\infty$$

(c) If as $x \to a^+$, $f(x)$ increases without bound, we indicate this behavior by

$$\lim_{x \to a^+} f(x) \to \infty$$

(d) If as $x \to a^+$, $f(x)$ decreases without bound, we indicate this behavior by

$$\lim_{x \to a^+} f(x) \to -\infty$$

It should be emphasized that ∞ is not a number. When we use the notation $\lim_{x \to a} f(x) \to \infty$ or $\lim_{x \to a} f(x) \to -\infty$, the limit *does not exist*. We use ∞ or $-\infty$ to give us an indication of the unbounded behavior of the function.

EXAMPLE 28   Find $\lim_{x \to 3} \dfrac{x}{x - 3}$

as x approaches 3 from the left and from the right.

SOLUTION

x	2.9	2.99	2.999	$\to$	3	$\leftarrow$	3.001	3.01	3.1
$\dfrac{x}{x - 3}$	-29	-299	-2999	$\to -\infty$		$\infty \leftarrow$	3001	301	31

As x approaches 3 from the left, the function approaches $-\infty$ (gets smaller and smaller or is unbounded in a negative direction). We say that

$$\lim_{x \to 3^-} \frac{x}{x - 3} \to -\infty$$

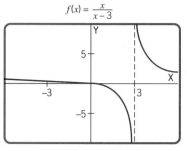

$$f(x) = \frac{x}{x-3}$$

X: [–6, 6] and Y: [–10, 10]

Figure 52

Also, as x approaches 3 from the right, the function approaches ∞ (gets larger and larger or is unbounded in a positive direction). We say that

$$\lim_{x \to 3^+} \frac{x}{x-3} \to \infty$$

The graph is shown in Figure 52 and confirms the behavior we saw indicated in the table of values. The line $x = 3$ is a vertical asymptote.

We have been looking at functions that exhibit **asymptotic behavior**. A function exhibits asymptotic behavior around a vertical asymptote $x = a$ when it becomes arbitrarily close to the asymptote as x approaches a. This leads us to the following definition.

DEFINITION: VERTICAL ASYMPTOTES

If $\lim_{x \to a^+} f(x) \to \pm\infty$ and/or $\lim_{x \to a^-} f(x) \to \pm\infty$, then the line $x = a$ is called a **vertical asymptote**. For a rational function, a vertical asymptote can be found by locating any value of x that will make the denominator (but not the numerator) equal to 0. If the numerator is also 0 at that value, then $x = a$ may not be a vertical asymptote.

In Example 27, $x = 2$ is a vertical asymptote of $f(x) = 1/(x-2)^2$. In Example 28, $x = 3$ is a vertical asymptote of $f(x) = x/(x-3)$. A function such as

$$f(x) = \frac{x(x-1)}{x(x+1)}$$

will have a "hole" in the graph at $x = 0$, but it will not have a vertical asymptote at $x = 0$ because x is a factor of both the numerator and the denominator.

EXAMPLE 29   Find the vertical asymptotes and draw the graph of $y = 3/(x-2)$.

SOLUTION

$$\lim_{x \to 2^-} \frac{3}{x-2} \to -\infty \quad \text{and} \quad \lim_{x \to 2^+} \frac{3}{x-2} \to \infty$$

Thus $x = 2$ is a vertical asymptote. We note that y increases without bound as x approaches this asymptote from the right. As $x \to 2$ from the left $y \to -\infty$. Note this on the graph in Figure 53. With your graphing calculator, duplicate this graph and use the Trace feature to note how the values change rather dramatically as $x \to 2^+$ and $x \to 2^-$.

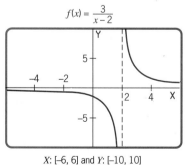

$$f(x) = \frac{3}{x-2}$$

X: [–6, 6] and Y: [–10, 10]

Figure 53

You can determine the behavior of a function as it approaches a vertical asymptote by analyzing the sign of the function very close to that value. It is not necessary to actually calculate the function, but only whether it will be positive or negative. When a function demonstrates asymptotic behavior, that behavior will remain consistent very close to the asymptote. The function will not suddenly change directions. Notice that if $x = 2.1$, then $3/(2.1 - 2) > 0$; therefore,

$$\lim_{x \to 2^+} \frac{3}{x - 2} \to \infty$$

Likewise, if $x = 1.9$, then $3/(1.9 - 2) < 0$ and we have

$$\lim_{x \to 2^-} \frac{3}{x - 2} \to -\infty$$

Calculator Note

As we learned earlier, the evaluation function on your calculator allows you to determine the value of the function at any value for x within the domain of x entered into the calculator. Using this function, you can quickly determine the sign of the function very close to the vertical asymptote. You can also use the Table feature to examine the behavior of the function close to the vertical asymptote.

Limit as $x \to \infty$

In addition to being interested in what happens to function values as the independent variable approaches a particular number, we are often concerned about what happens as the independent variable gets larger and larger or increases without bound. For example, consider the function

$$f(x) = \frac{x - 1}{x} = 1 - \frac{1}{x}$$

and determine what happens as x gets larger and larger. Values for $f(x)$ are tabulated in Table 5. It appears that $f(x) = 1 - 1/x$ approaches 1 as x gets larger and larger. To indicate this, we write

$$\lim_{x \to \infty} \left(1 - \frac{1}{x} \right) = 1$$

where $x \to \infty$ (read "x approaches infinity") means that x is increasing without bound.

TABLE 5

x	1	2	3	5	10	100	1000	100,000	$\to$	∞
$f(x)$	0	$\frac{1}{2}$	$\frac{2}{3}$	$\frac{4}{5}$	0.9	0.99	0.999	0.99999	$\to$	1

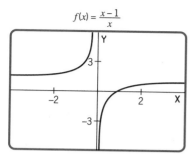

$$f(x) = \frac{x-1}{x}$$

X: [-4, 4] and Y: [-6, 6]

Figure 54

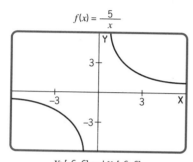

$$f(x) = \frac{5}{x}$$

X: [-6, 6] and Y: [-6, 6]

Figure 55

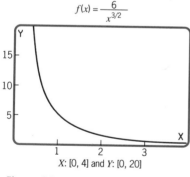

$$f(x) = \frac{6}{x^{3/2}}$$

X: [0, 4] and Y: [0, 20]

Figure 56

Similarly, as x becomes more and more negative, the function also approaches 1. We indicate this by writing $\lim\limits_{x \to -\infty} f(x) = 1$. Note the graph in Figure 54.

For real numbers L and M, the notation $\lim\limits_{x \to \infty} f(x) = L$ means that $f(x)$ approaches L as x gets larger and larger; the notation $\lim\limits_{x \to -\infty} f(x) = M$ means that $f(x)$ approaches M as x gets smaller and smaller. (The number is negative but the absolute value is large.)

All the limit properties listed previously hold for $\lim\limits_{x \to \infty} f(x)$, with the exception of the limit of a quotient. The limit of a quotient is replaced by the following property.

SPECIAL LIMIT

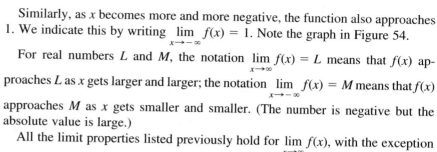

For any constant c,

$$\lim_{x \to \infty} \frac{c}{x^n} = 0 \quad \text{and} \quad \lim_{x \to -\infty} \frac{c}{x^n} = 0 \quad \text{for } n > 0$$

EXAMPLE 30

$$\lim_{x \to \infty} \frac{5}{x} = 0$$

The graph is shown in Figure 55. We see also that

$$\lim_{x \to -\infty} \frac{5}{x} = 0$$

EXAMPLE 31   Find $\lim\limits_{x \to \infty} \dfrac{6}{x^{3/2}}$.

SOLUTION   Since $n = 3/2 > 0$, we can say that

$$\lim_{x \to \infty} \frac{6}{x^{3/2}} = 0$$

See the graph in Figure 56. Note that the function is not defined for $x \leq 0$. ■

EXAMPLE 32   Find $\lim\limits_{x \to \infty} \dfrac{3x^2 - 5}{2x^2 + x - 3}$.

SOLUTION   This function is a quotient of two other functions (both the numerator and denominator are second-degree functions). However, we cannot use the formula for the limit of a quotient because

$$\lim_{x \to \infty} (3x^2 - 5) \to \infty \quad \text{and} \quad \lim_{x \to \infty} (2x^2 + x - 3) \to \infty$$

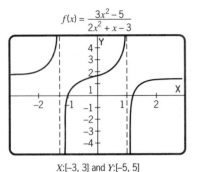

$$f(x) = \frac{3x^2 - 5}{2x^2 + x - 3}$$

X:[-3, 3] and Y:[-5, 5]

Figure 57

In this situation we use algebra to transform the quotient into a form where the numerator and denominator do have limits. The algebraic manipulation is to divide the numerator and the denominator by x raised to the largest exponent that occurs in the expression. In this case, we divide each term in the expression by x^2 and then use the special limit property for a quotient (see Figure 57).

$$\lim_{x \to \infty} \frac{3x^2 - 5}{2x^2 + x - 3} = \lim_{x \to \infty} \frac{3 - \dfrac{5}{x^2}}{2 + \dfrac{1}{x} - \dfrac{3}{x^2}}$$

$$= \frac{\lim_{x \to \infty} \left(3 - \dfrac{5}{x^2}\right)}{\lim_{x \to \infty} \left(2 + \dfrac{1}{x} - \dfrac{3}{x^2}\right)} = \frac{3 - 0}{2 + 0 - 0} = \frac{3}{2}$$

We arrived at the above answer because

$$\lim_{x \to \infty} \frac{-5}{x^2} = 0, \quad \lim_{x \to \infty} \frac{1}{x} = 0, \quad \text{and} \quad \lim_{x \to \infty} \frac{-3}{x^2} = 0$$

Calculator Note

You must use some care when graphing the function

$$f(x) = \frac{3x^2 - 5}{2x^2 + x - 3}$$

If you do not use **dot mode**, you may get a rather strange looking graph if the window is too large. Press the MODE key and moved the cursor down to the line that has ''Connected Dot.'' Move the cursor to the right so that it is flashing over the word ''Dot.'' Press ENTER. The calculator will not now try to connect parts of the graph that should not be connected. Use the ZOOM menu and you will get a good picture of the function.

Practice Problem 2 Find $\lim_{x \to \infty} \dfrac{5x^2 + 2}{3x^2 + 2x}$.

ANSWER $\dfrac{5}{3}$

EXAMPLE 33   Find $\lim_{x \to \infty} \dfrac{3x^2 + x - 5}{2x + 5}$.

SOLUTION   Divide the numerator and denominator by x^2, since the largest exponent in the expression is 2.

$$\lim_{x\to\infty}\frac{3x^2+x-5}{2x+5}=\lim_{x\to\infty}\frac{3+\dfrac{1}{x}-\dfrac{5}{x^2}}{\dfrac{2}{x}+\dfrac{5}{x^2}}$$

The limit of the numerator is 3, and the limit of the denominator is 0; consequently, the quotient becomes very large and the limit as x approaches infinity increases without bound. This is indicated by

$$\lim_{x\to\infty}\frac{3x^2+x-5}{2x+5}\to\infty$$

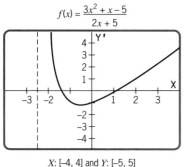

$f(x)=\dfrac{3x^2+x-5}{2x+5}$

X: [–4, 4] and Y: [–5, 5]

Figure 58

Figure 58 shows the graph of the function

$$f(x)=\frac{3x^2+x-5}{2x+5}$$

NOTE: In Examples 30, 31 and 32, the limits exist. In Example 33, the limit fails to exist. With this in mind, we state the following:

1. If the degree of the polynomial in the numerator of a rational function is greater than the degree of the denominator, the limit of the fraction as $x\to\infty$ is $\pm\infty$. (That is, it fails to exist and exhibits unbounded behavior as in Example 33.)
2. If the degree of the numerator and denominator are the same, the limit as $x\to\pm\infty$ is a number that is the quotient of the coefficients of the highest-degree terms in the numerator and in the denominator. (See Example 32.)
3. If the degree of the denominator is greater than the degree of the numerator, the limit of the function as $x\to\pm\infty$ is 0. (See Examples 30 and 31.)

As $|x|$ gets very large, we see in Examples 30 and 31 that the function gets very close to the horizontal line $y=0$ but does not reach it. In Example 32, the function gets very close to the line $y=\frac{3}{2}$ as $|x|$ gets very large, but it does not reach the line. The line $y=3/2$ is called a **horizontal asymptote**.

DEFINITION: HORIZONTAL ASYMPTOTE

If $\lim_{x\to\infty} f(x)=c$ and/or $\lim_{x\to-\infty} f(x)=c$, then the line $y=c$ is called a **horizontal asymptote**.

Be aware of the fact that we are dealing with very large values of $x, x \to \pm\infty$. A function may cross a horizontal asymptote when x is relatively small, but it will not cross a horizontal asymptote as $x \to \pm\infty$ (see Example 32) for these functions.

EXAMPLE 34　Find the horizontal and vertical asymptotes for

$$y = \frac{2x + 1}{x - 1}$$

SOLUTION

$$\lim_{x \to \infty} \frac{2x + 1}{x - 1} = \lim_{x \to \infty} \frac{2 + \dfrac{1}{x}}{1 - \dfrac{1}{x}} = 2 \qquad \text{Remember that } \lim_{x \to \infty} \frac{1}{x} = 0.$$

The line $y = 2$ is a horizontal asymptote.

By inspection, we see that $x = 1$ makes the denominator zero and does not make the numerator zero. Thus, $x = 1$ is a vertical asymptote. Looking at the first derivative we have

$$y' = \frac{2(x - 1) - (2x + 1) \cdot 1}{(x - 1)^2} = \frac{-3}{(x - 1)^2} \qquad \frac{N'D - ND'}{D^2}$$

We see that $y' < 0$ for all values of x, the function is always decreasing. (Draw a graph of y' to confirm this.)

Likewise,

$$y'' = \frac{6}{(x - 1)^3}$$

When $x < 1$, we see that $y'' < 0$ and the graph will be concave downward. When $x > 1$, then $y'' > 0$ and the graph will be concave upward. (Draw the graph of y'' to confirm this.) These conditions are satisfied in the graph of the function in Figure 59.

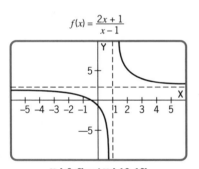

$f(x) = \dfrac{2x + 1}{x - 1}$

X: [–6, 6] and Y: [–10, 10]

Figure 59

Practice Problem 3　Find the horizontal and vertical asymptotes of

$$y = \frac{3x - 1}{1 - x}$$

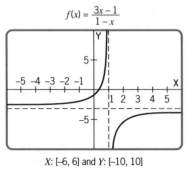

$f(x) = \dfrac{3x - 1}{1 - x}$

X: [–6, 6] and Y: [–10, 10]

Figure 60

ANSWER　There is a vertical asymptote at $x = 1$ and a horizontal asymptote at $y = -3$. The graph is shown in Figure 60.

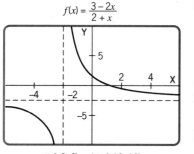

$f(x) = \dfrac{3-2x}{2+x}$

X: [-6, 6] and Y: [-10, 10]

Figure 61

Practice Problem 4 Find the horizontal and vertical asymptotes of

$$y = \frac{3-2x}{2+x}$$

ANSWER There is a vertical asymptote at $x = -2$ and a horizontal asymptote at $y = -2$ (see Figure 61).

EXAMPLE 35 Discuss the graph of the function

$$y = f(x) = \frac{1}{x^2-1}$$

relative to each of the following.

(a)   Vertical asymptotes
(b)   Horizontal asymptotes
(c)   Intervals where the function is increasing and where it is decreasing
(d)   Relative extrema
(e)   Concavity

SOLUTION

(a)   Vertical asymptotes: Since

$$f(x) = \frac{1}{x^2-1} = \frac{1}{(x-1)(x+1)}$$

we see that $x = -1$ and $x = 1$ are vertical asymptotes.

(b)   Horizontal asymptotes: Since

$$\lim_{x\to\infty} \frac{1}{x^2-1} = \lim_{x\to\infty} \frac{\frac{1}{x^2}}{1-\frac{1}{x^2}} = 0$$

we know that $y = 0$ (the x-axis) is a horizontal asymptote.

(c)   Increasing and decreasing intervals: We have

$$f'(x) = \frac{-2x}{(x^2-1)^2} = \frac{-2x}{(x-1)^2(x+1)^2}$$

The only critical value is $x = 0$, since the function is not defined for $x = 1$ and $x = -1$. Look at the graph in Figure 62. From the graph of f', we see that $f' > 0$ over $(-\infty, -1)$, $(-1, 0)$. Thus, $f(x)$ is increasing there.

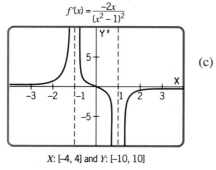

$f'(x) = \dfrac{-2x}{(x^2-1)^2}$

X: [-4, 4] and Y: [-10, 10]

Figure 62

$$f''(x) = \frac{6x^2 + 2}{(x^2 - 1)^3}$$

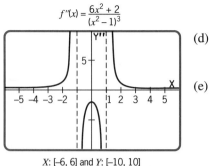

X: [-6, 6] and Y: [-10, 10]

Figure 63

$$f(x) = \frac{1}{x^2 - 1}$$

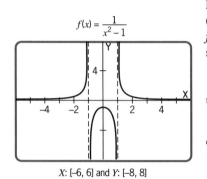

X: [-6, 6] and Y: [-8, 8]

Figure 64

Likewise, $f' < 0$ over $(0, 1)$, $(1, \infty)$; therefore, f is decreasing on those intervals.

(d) Relative extrema: Since $x = 0$ is the only critical point, we can see from the graph that $f'(0) = 0$ and f' changes from positive to negative; therefore, there is a relative maximum at $(0, -1)$ [i.e., $f(0) = -1$].

(e) Concavity: To determine concavity, we must find $f''(x)$.

$$f''(x) = \frac{-2(x^2 - 1)^2 - (-2x) \cdot 2(x^2 - 1)2x}{(x^2 - 1)^4}$$

$$= \frac{(x^2 - 1)(-2x^2 + 2 + 8x^2)}{(x^2 - 1)^4} = \frac{6x^2 + 2}{(x^2 - 1)^3}$$

Looking at the graph of f'' in Figure 63, we determine that $f'' > 0$ on $(-\infty, -1)$ and $(1, \infty)$; therefore, f is concave up on those intervals. Likewise, $f'' < 0$ on $(-1, 1)$, and f is concave down on that interval. The graph of f is shown in Figure 64.

When finding horizontal asymptotes, instead of evaluating the limits you can use the following guidelines derived from taking $\lim\limits_{x \to \infty} [f(x)/g(x)]$.

For a rational function $y = f(x)/g(x)$, if the degree of $f(x) = m$ and the degree of $g(x) = n$, then we have the following:

$$\lim_{x \to \pm\infty} \frac{f(x)}{g(x)} = \begin{cases} 0, & \text{if } m < n \\ \dfrac{\text{Coefficient of } x^m}{\text{Coefficient of } x^n}, & \text{if } m = n \\ \pm\infty & \text{if } m > n \end{cases}$$

EXAMPLE 36 Find each of the following limits using the properties above.

(a) $\lim\limits_{x \to \infty} \dfrac{3x}{x^2 + 2}$

(b) $\lim\limits_{x \to \infty} \dfrac{x^2 - 3}{4x^2 + 2}$

(c) $\lim\limits_{x \to \infty} \dfrac{x^4 + 1}{x^2 + 2x + 1}$

(d) $\lim\limits_{x \to -\infty} \dfrac{x^2}{x + 1}$

SOLUTION

(a) $\lim\limits_{x \to \infty} \dfrac{3x}{x^2 + 2} = 0$ Degree of numerator < Degree of denominator

(b) $\lim\limits_{x \to \infty} \dfrac{x^2 - 3}{4x^2 + 2} = \dfrac{1}{4}$ Degree of numerator = Degree of denominator

(c) $\lim\limits_{x \to \infty} \dfrac{x^4 + 1}{x^2 + 2x + 1} \to \infty$ Degree of numerator > Degree of denominator

(d) $\lim\limits_{x \to -\infty} \dfrac{x^2}{x + 1} \to -\infty$ Degree of numerator > Degree of denominator

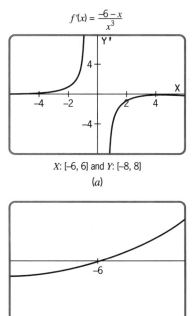

$f'(x) = \dfrac{-6 - x}{x^3}$

X: [–6, 6] and Y: [–8, 8]

(a)

X: [–7, –5] and Y: [–.01, .01]

(b)

Figure 65

Remember, a function may cross a horizontal asymptote, but a function will not cross a vertical asymptote. Be careful when analyzing a graph drawn on your graphing calculator, because inflection points may not be apparent when you are looking at the screen as we will discover in Example 37.

EXAMPLE 37   Analyze the graph of $f(x) = \dfrac{x + 3}{x^2}$.

SOLUTION

1. By examining the function, we can see that it has a vertical asymptote at $x = 0$ and a horizontal asymptote at $y = 0$.
2. We have

$$f'(x) = \frac{-6 - x}{x^3}$$

so the only critical value is $x = -6$. Note that $x = 0$ is not a critical value because the function is not defined there (Figure 65).

3. We have

$$f''(x) = \frac{18 + 2x}{x^4}$$

so there is a possible inflection point at $x = -9$ (Figure 66).

$f''(x) = \dfrac{18 + 2x}{x^4}$

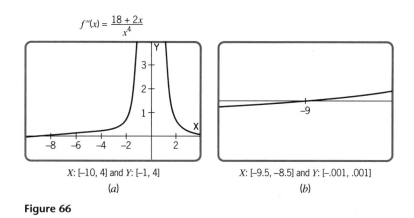

X: [–10, 4] and Y: [–1, 4]

(a)

X: [–9.5, –8.5] and Y: [–.001, .001]

(b)

Figure 66

4. The graphs in Figure 65 will help, but you must use care when using a graphing calculator. There is a sign change in f' at $x = -6$, but it may not be readily apparent on your calculator screen. Look at the graph in Figure 65(b) with a much more narrow range. When a sign change, or lack of change, is not apparent at a critical value or a possible inflection point, you must investigate by using either the Zoom feature or a more restricted range.

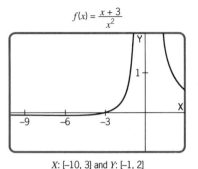

$$f(x) = \frac{x+3}{x^2}$$

X: [-10, 3] and Y: [-1, 2]

Figure 67

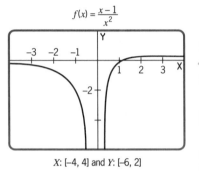

$$f(x) = \frac{x-1}{x^2}$$

X: [-4, 4] and Y: [-6, 2]

Figure 68

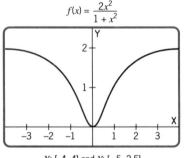

$$f(x) = \frac{2x^2}{1+x^2}$$

X: [-4, 4] and Y: [-.5, 2.5]

Figure 69

5.   We see that there is a relative minimum at $(-6, -\frac{1}{12})$ and an inflection point at $(-9, -\frac{2}{27})$ and that the function is concave down on the interval $(-\infty, -9)$ and concave up on the intervals $(-9, 0)$ and $(0, \infty)$.
6.   The graph of f is shown in Figure 67. Note that the point of inflection is very subtle and is hard to see, so you must examine that part of the graph very carefully.   ◼

Practice Problem 5   For $f(x) = (x - 1)/x^2$ find all extrema, asymptotes, and inflection points.

ANSWER   We have

$$f'(x) = \frac{2-x}{x^3} \quad \text{and} \quad f''(x) = \frac{2x-6}{x^4}$$

From Figure 68 we see that there is a vertical asymptote at $x = 0$ and a horizontal asymptote at $y = 0$. There is a relative maximum at $(2, \frac{1}{4})$ and an inflection point at $(3, \frac{2}{9})$.

Practice Problem 6   Analyze the graph of $f(x) = \dfrac{2x^2}{1+x^2}$.

ANSWER   Looking at the graph in Figure 69, we see that there is a horizontal asymptote at $y = 2$ and there are no vertical asymptotes. By using the first and second derivatives, we find that there is a relative minimum at $(0, 0)$ and there are inflection points at $(0.577, 0.5)$ and $(-0.577, 0.5)$.

EXAMPLE 38   Analyze the graph of

$$f(x) = \frac{2x}{\sqrt{x^2+2}}$$

by using the Trace key on your calculator.

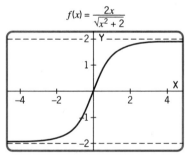

$$f(x) = \frac{2x}{\sqrt{x^2+2}}$$

X: [-4.5, 4.5] and Y: [-2.2, 2.2]

Figure 70

SOLUTION   The graph is shown in Figure 70. This is an example of a function that has two horizontal asymptotes.

$$\lim_{x \to +\infty} f(x) = 2 \quad \text{and} \quad \lim_{x \to -\infty} f(x) = -2$$

Draw the curve with the lines $y = 2$ and $y = -2$ to demonstrate this. ■

SUMMARY

When the limit of a function fails to exist as x approaches some number a either from the left or the right, we use the notation $\lim_{x \to a} f(x) \to \pm\infty$ to indicate this unbounded behavior. Remember that ∞ is not a number and cannot be manipulated like one; it is an indication of a behavior. If, for very large values of x, a function approaches a limiting value L, we use the notation $\lim_{x \to \infty} f(x) = L$. The line $y = L$ is a horizontal asymptote. If there is a value of the variable, c, that makes the denominator of a rational function zero, but does not make the numerator zero, then the line $x = c$ may be a vertical asymptote.

Exercise Set 3.5

Use a calculator to evaluate each function at $x = 10$, $x = 100$, $x = 1000$, and $x = 10,000$. Estimate $\lim_{x \to \infty} f(x)$ for each of the following functions.

1. $f(x) = \dfrac{4}{x + 2}$

2. $f(x) = \dfrac{x + 2}{2x - 2}$

3. $f(x) = \dfrac{x^2}{x + 2}$

4. $f(x) = \dfrac{x + 2}{x}$

Use a calculator to evaluate each function at $x = -10$, $x = -100$, $x = -1000$, and $x = -10,000$. Estimate $\lim_{x \to -\infty} f(x)$ for each of the following functions.

5. $f(x) = \dfrac{4}{x + 2}$

6. $f(x) = \dfrac{x + 2}{2x - 2}$

7. $f(x) = \dfrac{x^2}{x + 2}$

8. $f(x) = \dfrac{x + 2}{x}$

Use a calculator to evaluate each function at $x = 0.9$, $x = 0.99$, $x = 0.999$, and $x = 0.9999$. Estimate $\lim_{x \to 1^-} f(x)$ for each of the following functions from your calculations.

9. $f(x) = \dfrac{2}{x - 1}$

10. $f(x) = \dfrac{2x}{x - 1}$

11. $f(x) = \dfrac{1}{|x - 1|}$

12. $f(x) = \dfrac{2}{(x - 1)^2}$

Use a calculator to evaluate each function at $x = 1.1$, $x = 1.01$, $x = 1.001$, and $x = 1.0001$. Estimate $\lim_{x \to 1^+} f(x)$ for each of the following functions from your calculations.

13. $f(x) = \dfrac{2}{x - 1}$

14. $f(x) = \dfrac{2x}{x - 1}$

15. $f(x) = \dfrac{1}{|x - 1|}$

16. $f(x) = \dfrac{2}{(x - 1)^2}$

Study the following graphs and identify each of the following, if they exist.

(a) *Vertical asymptotes*
(b) *Horizontal asymptotes*

(c) $\lim\limits_{x\to\infty} f(x)$

(d) $\lim\limits_{x\to -\infty} f(x)$

17.

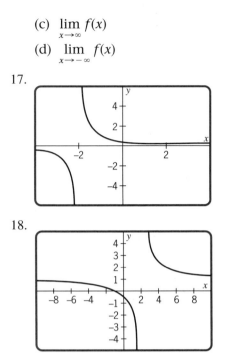

18.

and sketch the graph. Use both your graphing calculator and calculus techniques.

33. $f(x) = \dfrac{x - 3}{x + 2}$

34. $f(x) = \dfrac{1}{(x - 2)^3}$

35. $f(x) = \dfrac{2x}{1 + x}$

36. $f(x) = \dfrac{x}{x^2 - 1}$

37. $f(x) = \dfrac{2x + 3}{x - 1}$

38. $f(x) = \dfrac{5 - 2x}{3 + x}$

39. $f(x) = \dfrac{2x^2 - 6}{(x - 1)^2}$

40. $f(x) = 1 + \dfrac{1}{x} + \dfrac{1}{x^2}$

41. $f(x) = \dfrac{6 - x}{x^2}$

42. $f(x) = \dfrac{x^2}{(x + 2)^2}$

43. $f(x) = \dfrac{1 - 2x}{x^2}$

44. $f(x) = \dfrac{x + 1}{x^2}$

Find any horizontal and vertical asymptotes.

45. $y = \dfrac{x^3}{x^2 - 1}$

46. $y = \dfrac{-4x}{x^2 + 4}$

47. $y = 2 + \dfrac{x^2}{x^4 + 1}$

48. $y = \dfrac{2x^2 - 3x + 5}{x^2 + 1}$

49. Find $\lim\limits_{x\to 0} \sqrt{\dfrac{1 + x}{x^2} - \dfrac{1}{x}}$

*Find the following limits, if they exist. Use both the
properties of limits and graphs of the functions.*

19. $\lim\limits_{x\to\infty} \dfrac{3}{x}$

20. $\lim\limits_{x\to\infty} \dfrac{5}{x}$

21. $\lim\limits_{x\to\infty} \left(7 - \dfrac{5}{x}\right)$

22. $\lim\limits_{x\to\infty} \left(\dfrac{2}{x^2} - \dfrac{3}{x} + 5\right)$

23. $\lim\limits_{x\to\infty} \dfrac{3x - 2}{3x + 5}$

24. $\lim\limits_{x\to\infty} \dfrac{4x^2 - 2}{4x + 2}$

25. $\lim\limits_{x\to\infty} \dfrac{3x^2 - 5x + 7}{4x^2 + x + 1}$

26. $\lim\limits_{x\to\infty} \dfrac{5x^2 + 3x - 7}{6x^2 + 3x + 2}$

27. $\lim\limits_{x\to\infty} \dfrac{3x^3 - 2x + 5}{4x + 5}$

28. $\lim\limits_{x\to\infty} \dfrac{4x + 5}{3x^2 - 2x + 5}$

29. $\lim\limits_{x\to\infty} \dfrac{x^3 + 2x}{x^2 + 4}$

30. $\lim\limits_{x\to\infty} \dfrac{4x^4}{x^2 + 1}$

31. $\lim\limits_{x\to\infty} \dfrac{1}{\sqrt{1 + x}}$

32. $\lim\limits_{x\to\infty} \dfrac{x}{\sqrt{x - 2}}$

*For each function below, find the horizontal and ver-
tical asymptotes, relative extrema, inflection points,*

Applications (Business and Economics)

50. *Average Cost.* The cost function for manufacturing
x hair brushes is $C(x) = 4000 + 2.50x$. Given that
the average cost function is

$$\overline{C}(x) = \dfrac{C(x)}{x}$$

find $\lim\limits_{x\to\infty} \overline{C}(x)$.

51. *Average Profit.* If the hair brushes in Exercise 50
sell for $6 each and the average profit function is

$$\overline{P}(x) = \dfrac{P(x)}{x}$$

find $\lim\limits_{x\to\infty} \overline{P}(x)$.

52. **Supply Function.**   A supply function is given as

$$p(x) = \frac{900}{x} - \frac{800}{x^2}, \qquad x \geq 1$$

where x is the number of available items. Find $\lim_{x \to \infty} p(x)$.

53. **Cost Function.**   A cost function is given as

$$C(x) = 5000 + 10x + \frac{40}{x}$$

where x is the number of items produced. Find $\lim_{x \to \infty} C(x)$.

54. **Profit.**   A company finds that the cost of manufacturing x electric saws in a week is given by

$$C(x) = 0.16x^2 - 57.8x + 6220$$

The company is able to sell the saws at $20 each minus $x/1000$ (discount for large purchases).
(a) Find the revenue function.
(b) Find the maximum profit.
(c) Draw graphs showing $C(x)$ and $R(x)$ and locate the regions for profit and loss.
(d) Mark where the maximum profit occurs on the graph in part (c).

Applications (Social and Life Sciences)

55. **Voting Trends.**   Suppose that the probability P that a person will vote yes as the nth voter on an issue is

$$P = \frac{1}{3} + \frac{1}{10}\left(-\frac{1}{2}\right)^n$$

Find the value of $\lim_{n \to \infty} P$.

56. **Growth.**   A growth function is given by

$$N(t) = \frac{80,000t}{100 + t}$$

where t is the time in minutes. Find the value of $\lim_{t \to \infty} N(t)$.

57. **Bacteria.**   When bacteria are introduced into a medium, they often increase according to the graph shown below. Explain the significance of each of the following.
(a) $x = a$ (b) $(a, f(a))$ (c) $y = b$

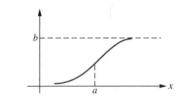

Review Exercises

Find points of absolute maxima and absolute minima on the given closed intervals.

58. $f(x) = x^2(x - 3)$ on $[-3, 3]$
59. $f(x) = x(x - 2)^2$ on $[-3, 3]$

3.6 OPTIMIZATION

OVERVIEW   Many practical problems require determining maximum or minimum values. For example, businesspeople wish to maximize profit and minimize cost. Builders wish to maximize the strength of their structures. The government has to be concerned about maximizing tax revenue, and the retailer wants to minimize inventory cost. Such problems are called optimization problems. They require the determination of absolute maxima or absolute minima. In this section we apply the theory developed in the preceding sections to help determine the absolute extrema.

Although it is impossible to describe mathematical procedures that can solve all optimization problems, it is possible to state some general rules.

Steps in Optimization

1. Use any or all of the techniques and strategies for solving problems you have learned in previous chapters.
2. Work toward a functional relationship between variables to be optimized and one other variable. (If two or more independent variables arise, it will be necessary in this chapter to eliminate all but one.)
3. Find critical values and locate absolute maxima and minima.

For the problems in this section, you should check your work by drawing the graph of the function being optimized. The first two examples should give some understanding of optimization before we consider the more important applications of this section.

EXAMPLE 39   The sum of two positive numbers is 220. Find the two numbers such that the product of the two numbers is a maximum.

SOLUTION   Since the sum of the two numbers is 220, if one number is 100, the other number is $220 - 100 = 120$. The product of the two numbers is $100(120) = 12,000$. This is probably not the maximum value, but we can use this as we proceed. Instead of 100, let the first of the two numbers be x. Then the second number is $220 - x$. The product of these two numbers is

$$P = x(220 - x) = 220x - x^2$$

Furthermore, since the numbers must be positive, we look only at the interval $[0, 220]$. We want P (the product) to be a maximum, so we take the first derivative to locate the critical value of x.

$$\frac{dP}{dx} = 220 - 2x$$

Setting $220 - 2x$ equal to zero yields

$$220 - 2x = 0$$
$$2x = 220$$
$$x = 110 \qquad \text{Critical value}$$
$$\frac{d^2P}{dx^2} = -2 \qquad \text{Second derivative}$$

The second derivative is negative and we now know that there is a relative maximum at $x = 110$. At $x = 0$ and at $x = 220$, the endpoints of the interval, the value of the product is 0. Therefore, at $x = 110$ we see that

$$P(110) = 110(220 - 110) = 12,100$$

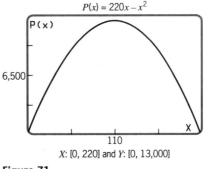

$$P(x) = 220x - x^2$$

X: [0, 220] and Y: [0, 13,000]

Figure 71

is the maximum product. So the two numbers are 110 and 110. The graph of the product function over [0, 220] is shown in Figure 71. ■

Practice Problem 1 Find two numbers whose sum is 72 and whose product is a maximum.

ANSWER   36 and 36

EXAMPLE 40   Using 120 feet of fencing, a farmer wishes to contain a cow in a rectangular plot of land that has one side along the bank of a river. If no fencing is needed along the river, what should be the dimensions of the rectangular field to provide the cow with maximum grazing area?

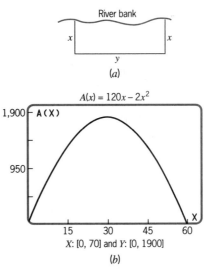

$$A(x) = 120x - 2x^2$$

X: [0, 70] and Y: [0, 1900]

(b)

Figure 72

SOLUTION   Whenever possible, it is a good idea to draw a figure to help set up a problem. The situation is shown in Figure 72(a). By looking at the figure, let's first make a guess of the dimensions we will select. If we were to choose $x = 10$, then we would have $y = 120 - 2(10) = 100$, since a total of 120 feet is available. With these dimensions, the area would be, $A = 100 \cdot 10 = 1000$. Although this may seem like a good guess, we will need a function to maximize to find the dimensions for maximum area. Let's use our guess to help us set up an area function.

First, we must establish a relationship between the two variables. Since we know the amount of fence available, we let y be the length of the side that is parallel to the river and x be the length of the other two sides, resulting in the equation $2x + y = 120$. However, we want the area function to be in terms of one variable so that we can maximize it. By solving for y, we have $y = 120 - 2x$ and now we have a relationship established between the two variables. This will enable us to substitute one for the other as needed.

We now write the area of our rectangular field as a function of x, find the derivative and critical value(s), and then find the second derivative to use the second derivative test for extrema.

$$\text{Area} = x \cdot y \qquad \text{Now substitute } y = 120 - 2x.$$

$$A(x) = x(120 - 2x) = 120x - 2x^2 \qquad \text{This is area as a function of } x \text{ and only } x.$$

$$A'(x) = 120 - 4x \qquad \text{First derivative}$$

$$120 - 4x = 0 \qquad \text{Set the derivative to zero and solve.}$$

$$x = 30 \qquad \text{This is the critical value.}$$

$$A''(x) = -4 \qquad \text{Second derivative is negative.}$$

The second derivative is negative; therefore, we have found a relative maximum. For $x = 30$, we have $y = 120 - 2(30) = 60$. The dimensions should be 30 by 60 feet to produce the maximum area of 1800 square feet. The graph of the area function is shown in Figure 72(b). ■

Practice Problem 2 A farmer wants to fence a rectangular area next to a mountain. Find the dimensions with the largest area possible using 3000 feet of fence if no fence is needed on the mountain side. (Use a figure similar to Figure 72.)

ANSWER   1500 feet by 750 feet

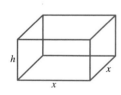

Figure 73

EXAMPLE 41   Design a box with a square base and maximum volume using 216 square inches of cardboard. The box will have a top and bottom.

SOLUTION   We will let the dimensions of the base of the box be x by x and the height be h (see Figure 73). The volume of this box is $V = x^2h$. Before we try to work with variables, let's make a guess of some measurements to help us see how to proceed with the problem. If we let $x = 10$, then the base and top will each measure $10 \cdot 10 = 100$ square inches. That would be a total of 200 square inches, leaving only 16 square inches for the sides! This is quite obviously not a good guess, but we now have an idea of how to continue.

Note that the volume V is a function of two variables, x and h (length or width and height). Just as in Example 40, we will solve for one of the variables in terms of the other. The 216 square inches of cardboard are utilized in the four sides and two ends as follows:

$$\text{Area of four sides} = 4xh \qquad \text{Area of each side is } xh.$$

$$\text{Area of top} + \text{Area of bottom} = 2x^2$$

$$4xh + 2x^2 = 216 \qquad \text{Total area}$$

$$h = \frac{216 - 2x^2}{4x} \qquad \text{Solve for } h.$$

Substituting for h in the expression for V we obtain

$$V = x^2 h$$

$$V = x^2 \left[\frac{216 - 2x^2}{4x}\right]$$

$$V = 54x - \frac{1}{2}x^3$$

$$\frac{dV}{dx} = 54 - \frac{3}{2}x^2 \qquad \text{Differentiate volume function.}$$

Now find the critical value(s).

$$54 - \frac{3}{2}x^2 = 0$$

$$\frac{3}{2}x^2 = 54$$

$$x^2 = 36 \qquad \text{Dimensions of square base}$$

$$x = 6 \qquad \begin{array}{l} -6 \text{ cannot be a critical value, since } x \\ \text{cannot be negative.} \end{array}$$

$$h = \frac{216 - 2(6)^2}{4 \cdot 6}$$

$$= \frac{216 - 72}{24}$$

$$= 6 \text{ inches} \qquad \text{This is the height.}$$

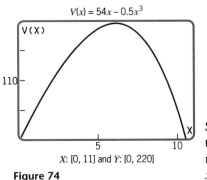

$V(x) = 54x - 0.5x^3$

110

5   10

X: [0, 11] and Y: [0, 220]

Figure 74

Since $d^2V/dx^2 = -3x$ and at $x = 6$ we have $d^2V/dx^2 = -18$, there is a relative maximum at the critical value. Looking at the graph of V in Figure 74, note that the relative maximum is also an absolute maximum. Maximum $V = x^2h = 6^2 \cdot 6 = 216$ cubic inches. ∎

Practice Problem 3 A box with no top is to be made from a piece of metal that measures 20 feet by 20 feet. It is to be made by cutting out squares from each corner of the metal and then folding up the sides. What should the dimensions be to create a box with the maximum volume possible? What is the maximum volume?

ANSWER   A square of width $\frac{10}{3} = 3\frac{1}{3}$ feet should be cut out (Figure 75). The dimension of the box would be $13\frac{1}{3}$ by $13\frac{1}{3}$ by $3\frac{1}{3}$ feet, yielding a volume of 16,000/27 cubic feet.

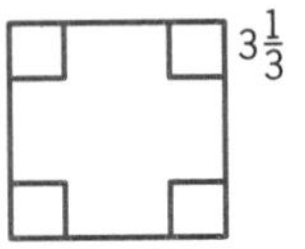

Figure 75

Maximum Profit

The next five examples in this section involve methods of optimization of business problems. Before we continue, make note of the following:

Marginal cost, $C'(x)$: Rate of change of cost
Marginal revenue, $R'(x)$: Rate of change of revenue
Marginal profit, $P'(x)$: Rate of change of profit

Furthermore, we have

$$\text{Profit} = \text{Revenue} - \text{Cost} \qquad P = R - C$$

and

$$\text{Marginal Profit} = \text{Marginal Revenue} - \text{Marginal Cost} \qquad P' = R' - C'$$

Since $P' = 0$ when $R' = C'$, **maximum profit** occurs when

$$\text{Marginal cost} = \text{Marginal revenue}$$

EXAMPLE 42   The revenue function and the cost function for an item are given by

$$R(x) = 12x - \frac{x^2}{12} - \frac{x^3}{6} \quad \text{and} \quad C(x) = 2 + 4x - \frac{x^2}{12}$$

where $0 \le x \le 6$ and x is in thousands. For what value of x is the profit a maximum?

SOLUTION   To maximize profit, we must find the derivative and critical values, and then test with the second derivative.

$$P(x) = R(x) - C(x)$$
$$= 12x - \frac{x^2}{12} - \frac{x^3}{6} - \left[2 + 4x - \frac{x^2}{12} \right]$$

$$P(x) = \frac{-x^3}{6} + 8x - 2 \qquad\qquad \text{Profit function}$$

$$P'(x) = \frac{-x^2}{2} + 8 \qquad\qquad \text{Marginal profit function}$$

$$\frac{-x^2}{2} + 8 = 0$$ Find critical values.

$$16 - x^2 = 0$$

$$x = 4$$ Reject -4, since $x \geq 0$.

$$P(4) = \frac{58}{3}$$ Evaluate the profit at $x = 4$.

$$P''(x) = -x$$ Second derivative

$$P''(4) < 0$$ By the second derivative test, it is a maximum.

By looking at the graph in Figure 76, we see that $x = 4$ is the absolute maximum. Thus, the profit is a maximum at $x = 4$ (or 4000 items). ■

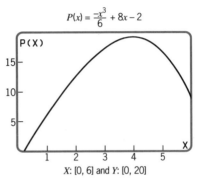

$P(x) = \frac{-x^3}{6} + 8x - 2$

X: [0, 6] and Y: [0, 20]

Figure 76

Profit can be maximized by solving the equation $MR = MC$. A producer can estimate marginal revenue and marginal cost at a given level of production and can raise production if $MR > MC$ or reduce production if $MR < MC$.

Practice Problem 4 A company makes and sells purses. The daily cost function is $C(x) = 80 + 8x - 1.65x^2 + 0.1x^3$ and the revenue function is $R(x) = 32x$, both in dollars. Assuming that the company can produce at most 30 bags per day, what production level x will yield the maximum daily profit?

ANSWER We have

$$P(x) = -80 + 24x + 1.65x^2 - 0.1x^3 \quad \text{and} \quad P'(x) = 24 + 3.3x - 0.3x^2$$

The critical value is $x = 16$, and a production level of 16 bags will yield the maximum daily profit.

EXAMPLE 43 If $C = 0.01x^2 + 5x + 100$ is a cost function, find the average cost function and the level of production x where average cost is a minimum.

SOLUTION   Average cost is

$$\overline{C}(x) = \frac{C(x)}{x}$$

This is the function that we will minimize.

$$\overline{C}(x) = 0.01x + 5 + \frac{100}{x} \qquad \text{Find the average cost function.}$$

$$\overline{C}'(x) = 0.01 - \frac{100}{x^2} \qquad \text{Differentiate.}$$

$$0 = 0.01 - \frac{100}{x^2} \qquad \text{Find the critical values.}$$

$$0.01x^2 = 100$$

$$x^2 = 10{,}000$$

$$x = 100 \qquad \text{Use } x = 100 \text{ and reject } x = -100.$$

$$\overline{C}''(x) = \frac{200}{x^3} \qquad \text{Find the second derivative.}$$

$$\overline{C}''(100) > 0 \qquad \text{Minimum by the second derivative test}$$

Therefore, $x = 100$ is the level of production that yields the minimum average cost (see Figure 77).   ■

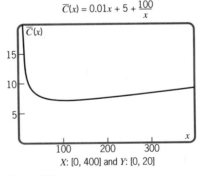

$$\overline{C}(x) = 0.01x + 5 + \frac{100}{x}$$

X: [0, 400] and Y: [0, 20]

Figure 77

Practice Problem 5 If $C(x) = x^2 + 160x + 6400$ is a cost function, find the value of x that gives minimum average cost. Graph the average cost function on your graphing calculator to verify the result.

ANSWER   We have

$$\overline{C}(x) = x + 160 + \frac{6400}{x}$$

so $x = 80$ is the minimum average cost. The graph is shown in Figure 78.

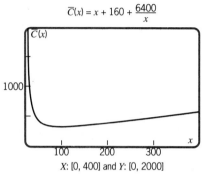

$$\overline{C}(x) = x + 160 + \frac{6400}{x}$$

X: [0, 400] and Y: [0, 2000]

Figure 78

EXAMPLE 44   A company sells a product for $1000 per set of 100 units. The cost, in dollars, of making x sets of 100 units in 1 year is $C(x) = 6 + 2x + 0.01x^2$. Write an expression for profit in terms of x. Find the number of sets of 100 units that would give a maximum profit.

SOLUTION   The revenue for x sets of 100 units would be $R(x) = xp = 1000x$. Since Profit = Revenue − Cost, the profit on x sets of 100 units would be

$$P = R - C$$
$$P = 1000x - (6 + 2x + 0.01x^2)$$
$$= 1000x - 6 - 2x - 0.01x^2$$
$$= 998x - 6 - 0.01x^2$$

The first-derivative condition for a maximum profit is $dP/dx = 998 - 0.02x = 0$. We solve this and find that $x = 49,900$. Likewise, $d^2P/dx^2 = -0.02$, which shows that the function is concave down and is therefore a maximum by the second derivative test. Thus 49,900 sets of 100 units would yield a maximum profit. The maximum profit would be $24,900,094. The graph is shown in Figure 79. ∎

NOTE: The first derivative of C with respect to x, called marginal cost, is the rate of change of cost when one more set of 100 units is produced. For the given example, $dC/dx = 2 + 0.02x$. In like manner, the rate of change of revenue with

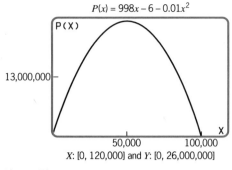

$$P(x) = 998x - 6 - 0.01x^2$$

X: [0, 120,000] and Y: [0, 26,000,000]

Figure 79

respect to the number of sets of units produced is called marginal revenue, and for this example is $dR/dx = 1000$. We see that the marginal cost increases with increasing x, whereas the marginal revenue is a constant. If production is continued until marginal cost equals marginal revenue, the number of sets of 100 units would be obtained from

$$1000 = 2 + 0.02x$$

$$x = 49,900$$

which is the same answer that we obtained in Example 44.

Practice Problem 6 A firm producing baseball gloves determines that in order to sell x gloves per week, the price per glove must be set at $p = 150 - 0.5x$. It is also determined that the cost of producing x number of gloves will be $C(x) = 2500 + 0.25x^2$. How many gloves must be sold to maximize profit each week?

ANSWER   We have $R(x) = xp$ and $P'(x) = R'(x) - C'(x) = 150 - 1.5x$. The critical value is $x = 100$, which is value of x that produces maximum profit.

EXAMPLE 45   A 100-room budget motel is filled to capacity every night at $20 per room. For each $1 increase in rent, two fewer rooms are rented. If each rented room costs $4 to service each day, how much should the management charge for each room to maximize profit?

SOLUTION   The change in the number of rooms rented depends on each $1 increase in rate. Therefore, we will let x = the number of $1 increases in the rent. The remaining functions are then found as follows:

New number of rooms rented = $100 - 2x$, since the number decreases by 2 for each $1 increase.
New rent = $20 + x$.

Revenue = (Number of rooms rented)(Rent per room) = $(100 - 2x)(20 + x)$.

Cost = $4 · Number of rooms rented = $4(100 - 2x) = 400 - 8x$.

Profit = Revenue − Cost = $(100 - 2x)(20 + x) - 4(100 - 2x) = 1600 + 68x - 2x^2$.

Marginal profit = $P'(x) = 68 - 4x$.

From $P'(x) = 68 - 4x = 0$, we find that $x = 17$ is the only critical value. Since $P''(x) = -4 < 0$, the function is concave down and we have found that at $x = 17$ a rent of $20 + $17 = 37 should be charged to obtain maximum profit. The graph is shown in Figure 80.

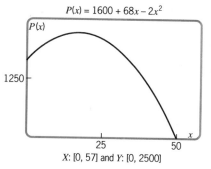

$P(x) = 1600 + 68x - 2x^2$

$P(x)$

1250

25   50

x

X: [0, 57] and Y: [0, 2500]

Figure 80

Practice Problem 7 An electronics supplier sells 50 CD players at $260 each. For every player over 50, the price of the player is reduced $2. If the supplier buys the CD players from the manufacturer for $120 each, how many CD players must be sold for the supplier to reach maximum profit each day?

ANSWER 60 players

EXAMPLE 46 A fruit grower has been planting 30 peach trees per acre in his orchard. The average yield is 400 peaches per tree. It is predicted that for each additional tree planted per acre, the yield will be reduced by 10 peaches per tree. How many trees should be planted per acre in order to have maximum yield?

SOLUTION To understand the problem, we summarize the data as follows. In the past, 30 trees have been planted per acre. The yield has been 400 peaches per tree. Each additional tree planted will reduce the yield per tree by 10 peaches. We can summarize the problem as follows by letting x = the number of trees planted.

New number of trees per acre = $30 + x$

Decrease in yield per tree = $10x$

New yield per tree $= 400 - 10x$

New total yield $= Y(x) = (30 + x)(400 - 10x) = 12,000 + 100x - 10x^2$

Marginal yield $= Y'(x) = 100 - 20x$

By setting $Y'(x) = 0$, we have $100 - 20x = 0$ or $x = 5$. Since $Y''(x) = -20 < 0$, $x = 5$ produces a relative maximum. By planting 5 additional trees production will be

$$Y(5) = 12,000 + 100(5) - 10(5^2) = 12,250$$

peaches per acre [the absolute maximum yield since $Y(x)$ is always concave down]. Note that we have worked this problem as if the function is continuous. We need to realize that if the answer had been 5.3 trees, we could not have 0.3 of a tree. The graph is shown in Figure 81.

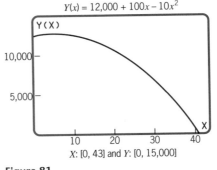

Figure 81

Practice Problem 8 A fruit grower has been planting 25 trees per acre in his orchard. For this design, the average yield is 495 apples per tree. It is predicted that for each additional tree planted per acre the yield will be reduced by 15 apples per tree. How many additional trees should be planted to maximize the yield?

ANSWER 4 trees

Inventories

Every large retail outlet is concerned about the cost of storing the inventory. For example, suppose that a department store sells 3650 refrigerators a year. Of course, it could order all of these at one time, but then the store would have to pay very high **storage costs** (for space, insurance, etc.). If the store makes too many small orders, then costs can increase due to delivery charges, office records, manpower, and so on. We classify these as **reorder costs**. Inventory cost is calculated by adding the storage cost and the reorder cost. We will be trying to minimize

$$\text{Inventory costs} = \text{Storage costs} + \text{Reorder costs}$$

To obtain the function representing inventory costs, we simplify the problem by assuming that the number demanded is uniform per day. That is, 10 refrigerators will be sold each day for 365 days. Now let x be the number ordered originally and at each reordering period. A reorder is made so that the next order arrives just as the inventory is depleted. Thus, we have x refrigerators on hand at the beginning of a period and 0 at the end of the period. Hence the average number in storage over a period is

$$\frac{x + 0}{2} = \frac{x}{2}$$

Since this average is the same for each reorder period, we say that $x/2$ is the average number in storage over the year (see Figure 82).

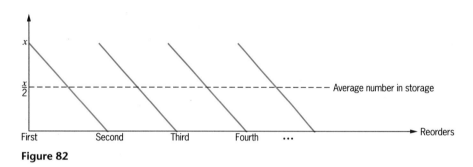

Figure 82

The yearly storage cost is usually found by multiplying the cost of storing 1 unit times $x/2$ (average number in storage). The number of reorders would be the total number needed in a year divided by x; that is, the number of reorders for refrigerators is $3650/x$. Usually, some expression can be found in terms of x for the cost of each order. Therefore, we can now express inventory costs as follows.

DEFINITION: INVENTORY COSTS

$$\text{Inventory costs} = (\text{Cost of storing 1 unit})\left(\frac{x}{2}\right) + (\text{Cost of each reorder})\left(\frac{N}{x}\right)$$

where N is the number of items needed per year and x is the number obtained in each order.

EXAMPLE 47 A department store sells 3600 refrigerators per year. It costs $20 per unit to store a refrigerator for 1 year. The cost of reorders has been estimated to be $10 plus $5 for each refrigerator. How many orders should be placed per year (or how many should be ordered per order) to minimize inventory costs?

SOLUTION   Let x be the number of refrigerators ordered in each reorder. Then, by the preceding discussion, $x/2$ is the average number in storage and $3600/x$ is the number of orders.

$$\text{Inventory costs} = C(x)$$

$$C(x) = 20\left(\frac{x}{2}\right) + (10 + 5x)\left(\frac{3600}{x}\right)$$

$$C(x) = 10x + \frac{36{,}000}{x} + 18{,}000$$

$$C'(x) = 10 - \frac{36{,}000}{x^2}$$

$$C''(x) = \frac{72{,}000}{x^3}$$

Setting $C'(x)$ equal to 0 yields $x = 60$ as a positive critical point. Also, $C''(x)$ is positive for positive values of x; therefore, there is a relative minimum at $x = 60$. Since $C''(x)$ is positive for positive values of x, $x = 60$ provides an absolute minimum. Hence, $3600/60 = 60$ orders should be placed each year and the minimum inventory cost is

$$C(60) = 10 \cdot 60 + \frac{36{,}000}{60} + 18{,}000 = \$19{,}200 \qquad \blacksquare$$

Practice Problem 9   An automobile dealership sells 2000 automobiles each year. It costs $200 to store a car for one year. The cost of reorders is estimated at $125. How many orders should be placed each year to minimize inventory costs?

ANSWER   40 orders of 50 each

EXAMPLE 48   Suppose that it costs $1000 to prepare a factory to produce a batch or run a certain item. After the preparation is complete, it costs $40 to produce each item. After the items are produced, it costs an average of $10 per year for each item held in inventory. If a company needs 5000 items per year, how many units should be run off in each batch to minimize cost?

SOLUTION   Let $x \geq 0$ be the number of items produced in each run or batch. Then

$$\text{Number of runs per year} = \frac{5000}{x}$$

$$\text{Cost to prepare factory for production} = \frac{1000(5000)}{x}$$

$$\text{Cost for 5000 items} = 40(5000)$$

We assume that all the production, or run, is put in inventory, the items are withdrawn at a uniform rate, and the inventory is depleted before a second run is made. Thus, the average number in inventory would be $x/2$. Cost for inventory storage is then $10 \cdot (x/2)$. The total cost for 5000 items is

$$C = 1000 \left(\frac{5000}{x} \right) + 40(5000) + \frac{10x}{2} \qquad \text{Cost function}$$

$$\frac{dC}{dx} = \frac{-5,000,000}{x^2} + 5 \qquad \text{Differentiate the cost function.}$$

$$0 = \frac{-5,000,000}{x^2} + 5 \qquad \text{Find the critical values.}$$

$$= \frac{-5,000,000 + 5x^2}{x^2}$$

$$= \frac{5(-1,000,000 + x^2)}{x^2}$$

$$x = 1000 \qquad \text{Since } x \text{ must be positive, this is the only solution.}$$

$$\frac{dC^2}{dx^2} = \frac{10,000,000}{x^3} \qquad \text{Second derivative will be positive for positive } x.$$

Thus, the minimum cost occurs when 1000 items are produced in each batch. This minimum cost is $210,000. ■

SUMMARY

Solving optimization problems requires practice and development of reasoning skills. In each problem you must identify each function that is to be maximized or minimized. This involves finding the derivative and critical value(s) along with function values at any endpoints, and then using the second derivative to test that you have found an absolute maximum or minimum. It is wise to let your variable be the amount around which the problem changes. For example, let $x =$ the number of $5 increases in a rent problem, or let $x =$ the additional number of trees in an orchard. If you have to develop the function to be optimized, then it is important to break the problem down into small units and see how the variable is used in each unit. Then put the units together to form the proper function. You can use a graphing calculator to sketch the graphs of the functions you have found to see if they seem to be reasonable functions. You should always make sure that your answer is reasonable within the context of the problem.

Exercise Set 3.6

1. Find two numbers whose sum is 52 and whose product is a maximum.

2. The product of two positive numbers is 36. Find the numbers that will make their sum a minimum.

3. The product of two positive numbers is 49. Find the numbers that will make their sum a minimum.

4. Find the dimensions of the rectangle of area 16 for which the perimeter is a minimum.

5. A box with square ends and rectangular sides is to be made from 600 square inches of cardboard. What is the maximum volume of the box? The box will have a top.

6. Find the dimensions of a box with a square top and bottom and rectangular sides with a volume of 27 cubic inches that can be constructed with the minimum amount of cardboard.

7. A rectangular plot of ground containing 576 square feet is to be fenced. Find the dimensions that require the least amount of fence.

8. Work Exercise 7 for a rectangle containing 100 square feet.

9. A rectangular plot of ground containing 1350 square feet is to be fenced, and an additional fence is to be constructed in the middle of the longest side to divide the plot into two equal parts. Find the dimensions that require the least amount of fence.

10. Suppose that the fence used to enclose the plot in Exercise 9 costs $12 per foot, and the fence used to divide the plot costs $6 per foot. Find the dimensions that make the cost a minimum.

11. From a piece of cardboard 60 centimeters by 60 centimeters, square corners are cut out so that the sides can be folded up to form a box. What size squares should be cut out to give a box of maximum volume? What is the maximum volume?

12. Redo Exercise 11 for a piece of cardboard 50 centimeters by 50 centimeters.

13. A company has set aside $3000 to fence in a rectangular portion of land adjacent to their building, using the building as one side of the enclosed area. The cost of the fencing running parallel to the building is $5 per foot installed, and the fencing for the remaining 2 sides will cost $3 per foot installed. Find the dimensions of the maximum enclosed area.

14. **_Exam_**.   A furniture company uses the following model for the replenishment of square feet of white pine lumber used in the manufacture of furniture. If the monthly cost of purchasing, carrying costs, and storage is given by

$$y = \frac{200{,}000}{x} + 0.05x$$

where x is the number of square feet in each order, the quantity ordered that minimized this cost is
(a) 633 square feet   (b) 2000 square feet
(c) 6325 square feet   (d) 4000 square feet
(e) Some number other than those given.

Applications (Business and Economics)

15. **_Cost Function_**.   A cost function is given as $C(x) = 10x + 30 + 0.01x^2$.
    (a) Find the marginal cost, $C'(x)$.
    (b) Find the average cost,

$$\overline{C}(x) = \frac{C(x)}{x}$$

    (c) Find the marginal average cost, $\overline{C}'(x)$.
    (d) Show that the average cost $\overline{C}(x)$ is a minimum when the marginal cost is the same as the average cost.

16. **_Maximum Revenue_**.   For a charter boat to operate, a minimum of 75 people paying $125 each is necessary. For each person in excess of 75, the fare is reduced $1 per person. Find the number of people that will make the revenue a maximum. What is the maximum revenue?

17. **_Cost and Revenue Functions_**.   The cost and revenue from sales are

$$C = x^3 - 15x^2 + 76x + 25 \quad \text{and}$$
$$R = 55x - 3x^2$$

Find the number of units for which the profit will be a maximum.

18. Verify for Exercise 17 that the marginal cost and marginal revenue are equal when the profit is a maximum.

19. ***Maximum Profit.*** A furniture maker can produce 25 tables per week. If $p = 110 - 2x$ is the demand equation for x tables at price p and the cost of producing the tables is $C(x) = 600 + 10x + x^2$ dollars, how many tables should be made each week to give the largest profit? What is the largest profit?
Hint: $R(x) = xp$.

20. ***Maximum Revenue.*** If $R(x) = x\sqrt{800 - x^2}$ is the revenue function for the sale of x tennis rackets per day, how many tennis rackets should be produced and sold to maximize revenue?

21. ***Maximum Profit.*** A company manufactures and sells x CD players per month. The monthly cost and demand equations are given as

$$C(x) = 30,000 + 30x \quad \text{and} \quad p = 300 - \frac{x}{30}$$

where $0 \le x \le 9000$. Find the production level that will realize the maximum profit. What is the maximum profit? ***Hint:*** $R(x) = xp$.

22. ***Minimum Cost.*** For safety reasons, a company plans to fence in a 10,800 square-foot rectangular employee parking lot adjacent to the building by using the building as one side of the enclosed area. The fencing parallel to the building faces a highway and will cost $3 per foot installed and the fencing on the other two sides will cost $2 per foot installed. Find the amount of each type of fence so that the total cost of the fence will be a minimum. What is the minimum cost?

23. ***Maximum Profit.*** A restaurant is being planned on the basis of the following information. For a seating capacity of 50 to 100 persons, the weekly profit is approximately $6 per seat. As the seating capacity increases beyond 100 chairs, the weekly profit on each chair in the restaurant decreases by $0.05 times the excess above 100 chairs. What seating capacity would yield the maximum profit?

24. ***Maximum Income.*** A rental agency has a problem in determining the rent to charge for each of 100 apartments to obtain a maximum income. It is estimated that if the rent is set at $100 per month, all units will be occupied. On the average, one unit will remain vacant for each $5-per-month increase in rent. What should the rent be in order to maximize the income?

25. ***Inventory.*** A company has a contract to supply 500 units per month at a uniform daily rate. Since it costs $100 to start production and the production cost is $5 per unit, it is decided to produce a large quantity at one time, storing the excess units until time for delivery. Storage costs run $10 per item per month. How many items should be made per run to minimize cost?

26. ***Advertising.*** ABC Auto has determined that profits are related to x thousands of dollars spent on advertising by

$$P(x) = 8 + 30x - \frac{x^2}{2}$$

What amount should be spent on advertising to attain maximum profit?

27. ***Inventory.*** ABC Auto sells 400 new automobiles a year. It costs $100 to store one automobile for a year. To reorder automobiles there is a fixed cost of $60 and a cost per automobile of $50. How many automobiles must be ordered each time to minimize inventory costs?

28. ***Inventory.*** An appliance dealer sells 3000 VCRs a year. It costs $12 to store one VCR for a year. To reorder there is a fixed fee of $10 plus $6 for each VCR. How many times a year should the store reorder VCRs, and what should be the size of the order so as to minimize costs?

29. ***Maximum Income.*** If a crop of oranges is harvested now, the average yield of 80 pounds per tree can be sold at $0.40 per pound. From past experience, the owners expect the crop yield to increase at a rate of 10 pounds per week per tree and the price to decrease at a rate of $0.02 per pound per week. When should the oranges be picked to attain maximum sale?

30. ***Maximum Profit.*** An apartment complex has 24 units. Upkeep and utilities come to $900 per month. The manager estimates that he can keep all the apartments rented at $150 per month rent, but he has one vacancy for each $20 per month added to the rent. However, for each vacant apartment, he would save

$30 per month out of the $900 expenses. What should the rent be to maximize the profit? How many apartments will he have rented at this rent? What will be the maximum profit?

31. **Maximum Yield.** If 25 pear trees are planted per acre in an orchard, the yield will be 525 pears per tree. For each additional tree planted per acre, the yield per tree will be reduced by 15 pears. What is the optimal number of trees to plant per acre?

32. **Maximum Revenue.** A company handles an apartment building with 50 units. Experience has shown that if the rent for each of the units is $160 per month, all of the units will be filled, but one unit will become vacant for each $5 increase in the monthly rent. What rent should be charged to maximize the total revenue from the building?

33. **Maximum Revenue.** A bus company charges $10 for a trip to an historical landmark if 30 people travel in a group. But for each person above the 30, the charge will be reduced by $0.20. How many people in a group will maximize the total revenue for the bus company?

34. **Maximum Revenue.** A consulting company will hold a workshop if at least 30 people sign up at a cost of $50 per person. The company will agree to a reduction of $1.25 per person for each person over the 30 minimum. (All attending receive the discount.) In order for the company to maximize its revenue, how many people should attend the workshop? Assume the maximum number allowed is 40.

35. **Maximum Yield.** A pecan grower has found that if 20 trees are planted per acre, each tree will produce an average of 60 pounds of pecans per year. If, for each additional tree planted per acre (up to 15 additional trees) the average yield per tree will drop by 2 pounds, how many additional trees (if any) should be planted to maximize the yield?

36. **Maximum Revenue.** When a travel agency charges $600 for a fantasy baseball weekend with a professional baseball team, it attracts 1000 people. For each $20 decrease in the charge, an additional 100 people will sign up. What price should be charged to maximize the revenue?

Applications (Social and Life Sciences)

37. **Mosquito Population.** Assume that the number of mosquitoes $N(x)$, in thousands, depends on the rainfall, in inches, according to the function

$$N(x) = 60 - 45x + 12x^2 - x^3, \qquad 0 \le x \le 6.4$$

(a) Find the amount of rainfall that will produce the minimum number of mosquitoes.
(b) Find the amount of rainfall that will produce the maximum number of mosquitoes.

38. **Bacteria Population.** Suppose that the bacteria count t days after a treatment is given by

$$C(t) = 30t^2 - 180t + 700$$

(a) When will the count be minimum?
(b) What is the minimum count?

39. **Voter Registration.** The number of registered voters, in thousands, is estimated to grow according to the function for the time in years as

$$N(t) = 12 + 3t - t^3, \qquad 0 \le t \le 3$$

(a) Find the rate of increase.
(b) When is the number a maximum?

Review Exercises

Draw the graph of each function and find the critical points, the intervals where the graph is increasing and where decreasing, the relative maxima and minima, points of inflection, the intervals where the graph is concave upward and where concave downward, and the horizontal and vertical asymptotes if they exist.

40. $y = \dfrac{4x}{x - 2}$

41. $y = \dfrac{4}{x - 2}$

42. $y = (4 - x^{2/3})^{3/2}$

43. $y = \dfrac{x}{2}(16 - x^2)^{1/2}$

3.7 ELASTICITY AS IT AFFECTS BUSINESS DECISIONS (Optional)

OVERVIEW   In this section we are interested in relative change. By relative change, we mean a comparison of change and the value of a variable before the change takes place. For example, a salary increase of $2000 per year for a person whose salary is $10,000 is a 20% increase. A $2000-a-year increase for a person whose salary is $200,000 is a 1% increase. In this section we

- Introduce average elasticity (the rate of the relative changes in two variables)
- Introduce point elasticity, which is the limit of average elasticity as the change approaches 0
- Use these concepts to study revenue, cost, demand, other business applications, and applications in the social and life sciences

Relative Change

Consider the function $y = f(x)$ as x changes by Δx (the change in the value of x). We will define the change in y to be

$$\Delta y = f(x + \Delta x) - f(x)$$

Now let's define the relative change in x and y.

Relative change in x at $x = x_0$ is $\dfrac{\Delta x}{x_0}$ (x_0 is the initial value of x).

Relative change in $y = y_0$ is $\dfrac{\Delta y}{y_0}$ (y_0 is the initial value of y).

EXAMPLE 49

(a) For the function $y = x^2$, find the relative changes in x and y when x changes from 4 to 5.

(b) For the function $y = x$, find the relative changes in x and y as x changes from 4 to 5.

SOLUTION

(a) For x, we have

$$\frac{\Delta x}{x_0} = \frac{5 - 4}{4} = \frac{1}{4} = 0.25 \quad \text{or} \quad 25\%$$

For y, we have

$$\Delta y = f(5) - f(4) = 5^2 - 4^2 = 25 - 16 = 9$$

Therefore,

$$\frac{\Delta y}{y_0} = \frac{5^2 - 4^2}{4^2} = \frac{9}{16} = 0.5625 \quad \text{or} \quad 56\tfrac{1}{4}\%$$

(b) We have

$$\text{Relative change in } x = \frac{5 - 4}{4} = \frac{1}{4} = 25\%$$

$$\text{Relative change in } y = \frac{5 - 4}{4} = \frac{1}{4} = 25\%$$

This time the relative change in y is the same as the relative change in x.

In the preceding example, we say that the function y in part (a) is more sensitive or more responsive to a change in x than the function in part (b). We define average elasticity as a measure of this responsiveness. In part (a) the average elasticity is calculated as

$$\frac{\text{Relative change in } y}{\text{Relative change in } x} = \frac{\dfrac{\Delta y}{y}}{\dfrac{\Delta x}{x}} = \frac{0.5625}{0.25} = 2.25$$

and in part (b) the average elasticity is

$$\frac{\dfrac{\Delta y}{y}}{\dfrac{\Delta x}{x}} = \frac{0.25}{0.25} = 1$$

The average elasticity of 2.25 in part (a) is greater than the average elasticity of 1 in part (b) and is, therefore, more sensitive or more responsive to a change in x.

DEFINITION: AVERAGE ELASTICITY

For a function $y = f(x)$ the **average elasticity** of $f(x)$ relative to x is defined to be

$$\frac{\dfrac{\Delta y}{y}}{\dfrac{\Delta x}{x}} = \frac{\dfrac{f(x + \Delta x) - f(x)}{y}}{\dfrac{\Delta x}{x}} \qquad \text{(Relative change in } y)/(\text{Relative change in } x)$$

$$= \frac{x}{y}\left[\frac{f(x + \Delta x) - f(x)}{\Delta x}\right]$$

EXAMPLE 50   Find the average elasticity of $y = f(x) = x^3$ as x changes from 2 to 3; that is, $\Delta x = 1$.

SOLUTION

$x = 2$ therefore; $\Delta x = 3 - 2 = 1$

$y = f(2) = 2^3 = 8$ therefore; $\Delta y = f(x + \Delta x) - f(x) = f(2 + 1) - f(2) =$
$$3^3 - 2^3 = 27 - 8 = 19$$

$$\frac{\dfrac{\Delta y}{y}}{\dfrac{\Delta x}{x}} = \frac{\dfrac{19}{8}}{\dfrac{1}{2}} = \frac{19}{4}$$

The average elasticity of this function when $x = 2$ and $\Delta x = 1$ is 19/4.   ■

Recall that at the beginning of our study of calculus, we went from average slope to instantaneous (or point) slope by taking the limit as the change in x (Δx) approaches 0. We apply the same reasoning to average elasticity.

$$\lim_{\Delta x \to 0} \frac{\dfrac{\Delta y}{y}}{\dfrac{\Delta x}{x}} = \lim_{\Delta x \to 0} \frac{x}{y} \left[\frac{f(x + \Delta x) - f(x)}{\Delta x} \right]$$

$$= \frac{x}{y} \lim_{\Delta x \to 0} \frac{f(x + \Delta x) - f(x)}{\Delta x} \qquad x/y \text{ is constant relative to } \Delta x.$$

$$= \frac{x}{y} f'(x) = \frac{x f'(x)}{f(x)} \qquad \text{Definition of derivative}$$

Thus, point elasticity (simply called elasticity) is the limit of average elasticity if the limit exists.

DEFINITION: ELASTICITY

For the function $y = f(x)$ the **elasticity** at x is

$$E(x) = \frac{x f'(x)}{f(x)}$$

provided that $f'(x)$ exists and $f(x) \neq 0$.

If $|E(x)| > 1$, then y is elastic relative to x.
If $|E(x)| = 1$, then y has unit elasticity with respect to x.
If $|E(x)| < 1$, then y is inelastic relative to x.

When a function is elastic, the relative change in output is greater than the relative change in input. When a function is inelastic, a relative change in output is less than the relative change in input.

In the business world, Alfred Marshall, a British economist, used the preceding ideas when he introduced what he called price elasticity of demand. This concept provided a measure of the relative amount by which demand would change in response to a change in price.

Suppose that a demand D is given as a function of price p.

$$D = f(p)$$

For most products, demand is a decreasing function of price. As the price (input) increases, the demand (output) decreases. Substituting in the general elasticity formula,

$$E(x) = \frac{xf'(x)}{f(x)} \quad \text{becomes} \quad E(p) = \frac{pf'(p)}{f(p)}$$

Note that $E(p)$ will always be negative. Most management books define $E(p)$ to be

$$\frac{-pf'(p)}{f(p)}$$

so that E will be easier to work with because it will always be positive.

DEFINITION: ELASTICITY OF DEMAND

Suppose that the demand and price are related by

$$D = f(p)$$

The **point elasticity of demand** is

$$E(p) = \frac{-pf'(p)}{f(p)}$$

1. If $0 \leq E(p) < 1$, demand is inelastic.
   (a) Increase in unit price $\Rightarrow$ Increase in revenue.
   (b) Decrease in unit price $\Rightarrow$ Decrease in revenue.
2. If $E(p) > 1$, demand is elastic.
   (a) Increase in unit price $\Rightarrow$ Decrease in revenue.
   (b) Decrease in unit price $\Rightarrow$ Increase in revenue.
3. If $E(p) = 1$, demand has unit elasticity.
   (a) Increase in price $\Rightarrow$ Revenue remains the same.

Note that the absolute value sign on $|E(x)|$ was removed for $E(p)$, since it is assumed that $f(p)$ is a decreasing function; that is, $f'(p)$ is negative at the same time p and $f(p)$ are positive. Thus $E(p)$ is positive.

EXAMPLE 51   If $D = f(p) = 1600 - 80p$, with $0 < p < 20$, find $E(p)$ and give one interpretation at

(a) $p = \$2$ (b) $p = \$10$ (c) $p = \$12$

SOLUTION

$$E(p) = \frac{-pf'(p)}{f(p)} = \frac{p \cdot (80)}{1600 - 80p}, \qquad f'(p) = -80$$

(a)   At $p = 2$, we substitute $p = 2$ into $E(p)$ and get

$$E(2) = \frac{2(80)}{1600 - 80 \cdot 2} = \frac{1}{9} \approx 0.111, \qquad 0 \leq E(p) < 1$$

The demand is inelastic (i.e., the demand is not overly sensitive to a change in price). Actually, a price increase of 10% would result in a

$$0.111(10\%) = 1.1\%$$

increase in demand.

(b)   At $p = 10$, we substitute $p = 10$ into $E(p)$ and get

$$E(10) = \frac{10(80)}{1600 - 80(10)} = 1$$

In this case, a percentage change in price will result in approximately the same percentage change in demand. When $E(p) = 1$, the demand has unit elasticity.

(c)   At $p = 12$, we substitute $p = 12$ into $E(p)$ and get

$$E(12) = \frac{12(80)}{1600 - 80(12)} = 1.5$$

Since $E(p) > 1$, demand is elastic (i.e., there is a change in demand when price changes.) For example, a 10% change in price would result in a

$$1.5(10\%) = 15\%$$

change in demand.

EXAMPLE 52   Given $D = f(p) = 3840 - 20p^2$ for $0 < p < \sqrt{192}$.

(a)   Determine the interval on p where the demand is inelastic and the interval where it is elastic.

(b)   Interpret the result for a 10% increase in price at $p = \$6$.

(c)   Interpret the result for a 10% increase in price at $p = \$10$.

SOLUTION

(a)   We have

$$E(p) = -\frac{p \cdot (-40p)}{3840 - 20p^2} \qquad \frac{dD}{dp} = -40p$$

$$= \frac{2p^2}{192 - p^2} \qquad \text{Factor 20 from numerator and denominator and reduce.}$$

Now, if $E(p) = 1$, then

$$\frac{2p^2}{192 - p^2} = 1$$

$$2p^2 = 192 - p^2$$

$$3p^2 = 192$$

$$p^2 = 64$$

$$p = 8 \qquad \text{p cannot be negative.}$$

When $p = 8$ the demand has unit elasticity. In a similar manner, if $E(p) < 1$, then

$$\frac{2p^2}{192 - p^2} < 1$$

$$2p^2 < 192 - p^2 \qquad 192 - p^2 > 0$$

$$3p^2 < 192$$

$$p^2 < 64$$

$$p < 8 \qquad \text{p cannot be negative.}$$

For $0 < p < 8$, the demand is inelastic. If $E(p) > 1$, then

$$\frac{2p^2}{192 - p^2} > 1$$

$$2p^2 > 192 - p^2 \qquad 192 - p^2 > 0$$

$$3p^2 > 192$$

$$p^2 > 64$$

$$p > 8 \qquad \text{p cannot be negative.}$$

For $8 < p < \sqrt{192}$ demand is elastic.

(b)   At $p = 6$, we know the demand is inelastic. A 10% increase in p would result in

$$\left(\frac{2p^2}{192 - p^2}\right) \cdot 10\% = \left(\frac{2 \cdot 36}{192 - 36}\right) \cdot 10\% \approx 4.6\% \text{ increase in demand}$$

(c)   At $p = 10$, we know the demand is elastic. A 10% increase in p would result in

$$\left(\frac{2p^2}{192 - p^2}\right) \cdot 10\% = \left(\frac{2 \cdot 100}{192 - 100}\right) \cdot 10\%$$
$$\approx 21.7\% \text{ decrease in demand} \qquad \blacksquare$$

Practice Problem 1 With your graphing calculator draw the graph of

$$y = \frac{2p^2}{192 - p^2}$$

to check the preceding discussion. Use the range $0 \le p \le 20$ and $-10 \le y \le 10$.

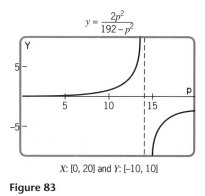

$$y = \frac{2p^2}{192 - p^2}$$

X: [0, 20] and Y: [−10, 10]

Figure 83

ANSWER   The graph in Figure 83 shows that $E(p)$ is inelastic when $0 < p < 8$, and $E(p)$ is elastic for $8 < p < \sqrt{192}$. The graph below the axis where $x > \sqrt{192}$ is not in the domain of p. The graph appears to agree with Example 52.

EXAMPLE 53   Returning to Example 51, where $D = f(p) = 1600 - 80p$ for $0 < p < 20$, we noted that

$$E(p) = \frac{80p}{1600 - 80p} = \frac{p}{20 - p}$$

Let's discover where $E(p) < 1$ and $E(p) > 1$.

SOLUTION   The inequality $E(p) < 1$ means that

$$\frac{p}{20 - p} < 1$$

$$p < 20 - p \qquad 20 - p \text{ is given as positive.}$$

$$2p < 20$$

$$p < 10$$

For $0 \leq p < 10$, $E(p) < 1$ and the demand is inelastic.
  For $E(p) > 1$,

$$\frac{p}{20 - p} > 1$$

$$p > 20 - p \qquad 20 - p \text{ is given as positive.}$$

$$2p > 20$$

$$p > 10$$

For $10 \leq p < 20$, $E(p) > 1$ and the demand is elastic.

Revenue Function

Now let's examine where a revenue function is increasing and where it is decreasing. Using the preceding example,

$$R = p \cdot D \qquad \text{Price} \cdot \text{Demand}$$
$$= pf(p)$$
$$= p(1600 - 80p)$$
$$= 1600p - 80p^2$$

$$\frac{dR}{dp} = 1600 - 160p$$

$$\text{If} \quad 1600 - 160p = 0 \qquad \text{Setting } \frac{dR}{dp} = 0$$

$$\text{then} \quad p = 10$$

For $0 < p < 10$, dR/dp is positive and the function is increasing; therefore, we have increasing revenue. For $10 < p < 20$, dR/dp is negative and the function is decreasing; therefore we have decreasing revenue.

  For the preceding example, we note that revenue is increasing precisely when demand is inelastic and revenue is decreasing when demand is elastic. This is further emphasized by Figure 84.

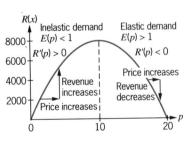

Figure 84

Practice Problem 2 $E(p) = \dfrac{p^2}{80 - p^2}$. Is the demand elastic or inelastic at

(a) $p = 4$?
(b) $p = 8$?

ANSWER

(a) At $p = 4$, $E(p) = 0.25$, so demand is inelastic.
(b) At $p = 8$, $E(p) = 4$, so demand is elastic.

SUMMARY

In business it is very important to understand the relationship between revenue, cost, profit, and the changes in these as prices change.

1. When demand is elastic at price p [i.e., $E(p) > 1$], an increase in the price per unit will cause a decrease in revenue. However, a decrease in the price per unit will result in an increase in revenue.
2. When demand is inelastic at price p [i.e., $E(p) < 1$], then increasing the price per unit results in an increase in revenue. Likewise, a decrease in the price per unit causes the total revenue to decrease.
3. If the demand has unit elasticity at price p [i.e., $E(p) = 1$], an increase in the price per unit will result in very little change in the revenue.

Exercise Set 3.7

1. Consider the function $y = x^2$.
   (a) Find the relative change in x as x changes from 2 to 4.
   (b) Find the relative change in y.
   (c) Find the average elasticity.
2. Consider the function $y = x^3$.
   (a) Find the relative change in x as x changes from 1 to 3.
   (b) Find the relative change in y.
   (c) Find the average elasticity.

Applications (Business and Economics)

Find the elasticity (point elasticity) of the given function $y = f(x)$ at the designated point. Tell whether the function is elastic, inelastic, or has unit elasticity.

3. $y = x^2 + 2$ at $x = 1$
4. $y = x(x + 1)$ at $x = 2$

5. $y = \dfrac{1}{x}$ at $x = 3$

6. $y = 3x^2 + 2$ at $x = 3$
7. $y = 3x(x^2 + 1)$ at $x = 1$

8. $y = \dfrac{1}{x^2}$ at $x = 2$

9. *Elasticity of Demand.* The following demand equation is given:

$$D(p) = 200(40 - p), \qquad 0 \le p \le 40$$

   (a) Find the elasticity of demand, $E(p)$.
   (b) What is the elasticity when $p = \$10$?
   (c) Classify part (b) as elastic, inelastic, or as having unit elasticity.
   (d) What will be the approximate change in demand if p is increased by 5% when $p = \$10$?
   (e) What is the elasticity when $p = \$30$?
   (f) Classify part (e) as elastic, inelastic, or as having unit elasticity.

(g) What will be the approximate change in demand if p is increased by 5% when $p = \$30$?

10. **Elasticity of Demand.** The following demand equation is given:

$$D(p) = 20(10 - p), \qquad 0 \le p \le 10$$

(a) Find the elasticity of demand, $E(p)$.
(b) What is the elasticity when $p = \$3$?
(c) Classify part (b) as elastic, inelastic, or as having unit elasticity.
(d) What will be the approximate change in demand if p is decreased by 10% when $p = \$3$?
(e) What is the elasticity when $p = \$8$?
(f) Classify part (e) as elastic, inelastic, or as having unit elasticity.
(g) What will be the approximate change in demand if p is decreased by 5% when $p = \$8$?

For the following demand equations, determine if demand is elastic, inelastic, or has unit elasticity for the given values of p.

11. $D(p) = 100 - p^2$ for $p = 8$
12. $D(p) = 100 - p^2 - p$ for $p = 6$
13. $D(p) = 1000 - p^2$ for $p = 10$
14. $D(p) = 1000 - p^2 - 4p$ for $p = 10$

For each of the following demand equations, determine intervals of p that produce a demand that is elastic and the intervals that produce a demand that is inelastic.

15. $D(p) = 200(40 - p), 0 \le p \le 40$
16. $D(p) = 20(10 - p), 0 \le p \le 10$
17. $D(p) = 20(100 - p)^2, 0 \le p \le 100$
18. $D(p) = 10(64 - 2p)^2, 0 \le p \le 32$

19. **Elasticity of Cost.** Suppose that the cost of producing x units weekly is

$$C(x) = \frac{1}{2}x^2 + 6x + 200$$

Find the elasticity of cost when $x = 4$, $x = 10$, and $x = 20$. Would you classify the elasticity of cost as being elastic, inelastic, or as having unit elasticity at these three points? An increase of 10% in production at these points produces what change in cost?

For each of the following demand equations, find the revenue equation, sketch the graph of the revenue equation, label where the revenue graph is increasing and where it is decreasing, and find the regions of elastic and inelastic demand and mark these on the graph of the revenue equation.

20. $D(p) = 40(20 - p), 0 \le p \le 20$
21. $D(p) = 100(100 - p), 0 \le p \le 100$
22. $D(p) = 40(20 - p)^2, 0 \le p \le 20$
23. $D(p) = 100(100 - p)^2, 0 \le p \le 100$

Review Exercises

Find the critical points, intervals where the graph is increasing and where decreasing, relative maxima and minima, points of inflection, intervals where the graph is concave upward and concave downward, horizontal and vertical asymptotes if they exist, and sketch the graph of the following functions.

24. $y = x + x^{-1}$

25. $y = \frac{1}{3}(x^4 - 4x^3)$

26. $y = 2x\sqrt{x - 1}$

27. $y = 2x(x - 2)^2$

Chapter Review

Review the following concepts to ensure that you understand and can use them.

Important Terms

Inflection points	Concavity
Curve sketching	First derivative test
Relative maxima	Second derivative test

Relative minima	Average elasticity
Absolute maxima	Point elasticity
Absolute minima	Elastic
Asymptote	Inelastic
Critical point	Unit elasticity
Increasing and decreasing function	Elasticity of demand

Make sure you understand the following concepts and formulas.

$f'(c) = 0$ or $f'(c)$ is undefined when c is a critical value

$f'(x) > 0$; graph is increasing

$f'(x) < 0$; graph is decreasing

$f''(x) > 0$; graph concave upward

$f''(x) < 0$; graph concave downward

$\lim_{x \to \infty} f(x) = c$; then $y = c$ is a possible horizontal asymptote

$\lim_{x \to c} f(x) = \infty$; then $x = c$ is a possible vertical asymptote

$$\text{Average elasticity} = \frac{x}{y}\left[\frac{f(x + \Delta x) - f(x)}{\Delta x}\right]$$

$$E(x) = \frac{xf'(x)}{f(x)}$$

$$E(p) = \frac{-pf'(p)}{f(p)}$$

Chapter Test

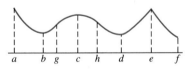

Use the figure to answer questions 1–4. Give answers as the largest possible intervals if possible (otherwise as points).

1. From the figure determine over (a, f).
   (a) Where is the curve increasing?
   (b) Where is the curve decreasing?
   (c) For what value or values of x are there relative maxima?

2. From the figure determine over (a, f).
   (a) For what value(s) are there relative minima?
   (b) On $a \le x \le h$, where is there an absolute minimum?
   (c) On $a \le x \le h$, where is there an absolute maximum?

3. From the figure determine over (a, f).
   (a) Where are the points of inflection?
   (b) Intervals where the second derivative is positive?
   (c) Intervals where the second derivative is negative?

4. From the figure determine over (a, f).
   (a) Where is the curve concave upward?
   (b) Where is the curve concave downward?
   (c) Where is the first derivative undefined?

5. Given $y = -2x^2 + 4x$.
   (a) Where is the curve increasing and where is it decreasing?
   (b) Find all points of relative extrema.
   (c) Sketch the curve.

6. For $y = x^4 - x^3$ use both your graphing calculator and calculus techniques to draw the graphs of y, y', and y'' to answer each question.
   (a) Where is the curve concave upward? Concave downward?
   (b) Where are the points of inflection?
   (c) Find all points of relative extrema.

7. For $y = x + x^{-1}$ use both your graphing calculator and calculus techniques to draw the graphs of y, y', and y'' to answer each question.
   (a) Where is the curve increasing? Decreasing?
   (b) Where is the curve concave upward? Concave downward?
   (c) Find points of relative extrema.
   (d) Find a vertical asymptote.
   (e) Sketch the curve.

8. An electronics store sells 5000 radios each year. It costs $10 to store a radio for a year. To reorder there is a fixed fee of $10 plus $6 for each radio on inventory. How many times a year should the store reorder radios in order to minimize costs, and what should the size of the orders be?

9. When 200 cars are sold per month at an automobile dealership, the profit per car is $400. For each sale above 200 cars, the profit decreases by $1 per car. How many cars should be sold for maximum profit?

10. For the demand equation $D(p) = -p + 6$, is demand elastic or inelastic at $p = 2$? Interpret this result for a price increase of 5%.

CHAPTER 4

Additional Derivative Topics

In this chapter, we introduce procedures for differentiation of exponential functions and logarithmic functions that are useful in the discussion of applications of calculus. These functions serve as models for population growth, growth of investments, depreciation of capital goods, decay of radioactive material, the rate of learning, the spread of epidemics, and compound interest. You may need to review exponents and exponential functions in Chapter 1.

Sometimes functions cannot be given with one variable expressed explicitly as a function of the other. When this occurs, a procedure for finding a derivative, called implicit differentiation, is most useful. Similarly, sometimes two variables are functions of a third variable. In this chapter, we investigate the relationship between the derivatives of the two variables in terms of the derivative of the third variable.

4.1 FINDING DERIVATIVES OF EXPONENTIAL FUNCTIONS

OVERVIEW This is an important section because exponential functions are used extensively in applications. Growth and decay models are based on exponentials. Exponential functions are used by economists to study the growth rate of the money supply. They are needed by businesspeople to study the rate of change in sales, by biologists to study the rate of growth of organisms, and by social scientists to study the rate of population growth. In this section we

- Find the derivative of exponential functions
- Use the chain rule with exponential functions
- Study graphing techniques
- Introduce applications of exponential functions

We introduced exponential functions in Chapter 1. Most of our work with exponential functions will involve the base e.

$$e = \lim_{n \to \infty} \left[1 + \frac{1}{n} \right]^n$$

if this limit exists. This concept of the constant e, defined as a limit, is in agreement with our use of e in Chapter 1.

To get a feel for the size of e, we use the Table feature of our graphing calculator to approximate e. We see that e does appear to exist and is approximately 2.7182818 (see Table 1).

In calculus, e is an excellent choice for the base of an exponential function because of the simplicity of the derivative. If $y = e^x$, then $dy/dx = e^x$. This fact will be proved after we apply this formula in several examples. First consider the more general case in which the exponent of e is a function of x. If $y = e^{u(x)}$, we apply the chain rule to obtain

$$\frac{dy}{dx} = \frac{dy}{du} \cdot \frac{du}{dx}$$

$$= e^u u'(x) \qquad \frac{d}{du} e^u = e^u$$

These ideas are summarized as follows.

TABLE 1

n	$\left(1 + \dfrac{1}{n} \right)^n$
100	2.7048138
500	2.7155685
1,000	2.7169239
10,000	2.7181459
100,000	2.7182682
1,000,000	2.7182805
100,000,000	2.7182818

DIFFERENTIATION FORMULAS FOR EXPONENTIAL FUNCTIONS

$$\frac{d}{dx}(e^x) = e^x$$

If u is a differentiable function of x, then

$$\frac{d}{dx}[e^{u(x)}] = e^{u(x)} \frac{du}{dx} = u'(x)e^{u(x)}$$

The following examples illustrate the use of these formulas.

EXAMPLE 1   If $y = e^{3x}$, find $\dfrac{dy}{dx}$.

SOLUTION   Let $u(x) = 3x$. Then $u'(x) = 3$, and

$$\frac{dy}{dx} = e^{u(x)}u'(x)$$
$$= e^{3x} \cdot 3$$
$$= 3e^{3x}$$

EXAMPLE 2   If $y = e^{3x^2}$, find $\dfrac{dy}{dx}$.

SOLUTION   Let $u(x) = 3x^2$.

$$\frac{dy}{dx} = e^u u'(x)$$
$$= e^{3x^2} \cdot 6x \qquad u = 3x^2, \text{ so } u' = 6x.$$
$$= 6xe^{3x^2}$$

EXAMPLE 3   If $y = e^{\sqrt{x}}$, find $\dfrac{dy}{dx}$.

SOLUTION   Let $u(x) = \sqrt{x}$.

$$\frac{dy}{dx} = e^u u'$$
$$= e^{\sqrt{x}}\frac{1}{2\sqrt{x}} \qquad u = \sqrt{x}, \text{ so } u' = \frac{1}{2\sqrt{x}}.$$
$$= \frac{1}{2\sqrt{x}}e^{\sqrt{x}}$$

EXAMPLE 4   If $y = e^{3x^2 + x - 2}$, find $\dfrac{dy}{dx}$.

SOLUTION   Let $u(x) = 3x^2 + x - 2$.

$$\frac{dy}{dx} = e^u u'$$
$$= e^{3x^2 + x - 2}(6x + 1) \qquad u = 3x^2 + x - 2, \text{ so } u' = 6x + 1.$$
$$= (6x + 1)e^{3x^2 + x - 2}$$

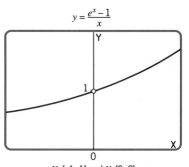

$y = \dfrac{e^x - 1}{x}$

0

X: [–1, 1] and Y: [0, 2]

Figure 1

In the process of developing the formula for the derivative of e^x, we will need $\lim\limits_{h \to 0} \dfrac{e^h - 1}{h}$. To get this limit we investigate with our calculator what happens to the function $y = (e^h - 1)/h$ as h approaches 0 from both the left and the right. To find this limit, consider the graph of $y = (e^x - 1)/x$ in Figure 1. This function is not defined at $x = 0$, but it seems that

$$\lim_{x \to 0} \frac{e^x - 1}{x} = 1$$

as we use the TRACE key to look at x values that are very close to 0.

Using the Table feature of our calculator, Table 2 presents a picture of $(e^x - 1)/x$ approaching 1 as x approaches 0 from both the left and right.

TABLE 2

x	0.01	0.001	0.0001	$\to 0 \leftarrow$	-0.0001	-0.001	-0.01
$(e^x - 1)/x$	1.005	1.0005	1.00005	$\to 1 \leftarrow$	0.99995	0.9995	0.995

Since

$$\lim_{h \to 0} \frac{e^h - 1}{h} \quad \text{is the same as} \quad \lim_{x \to 0} \frac{e^x - 1}{x}$$

we use the fact that

$$\lim_{h \to 0} \frac{e^h - 1}{h} = 1$$

in the following derivation of the derivative of e^x. We use the four-step procedure from Chapter 8.

$$f(x) = e^x$$

$$f(x + h) = e^{x+h} = e^x \cdot e^h$$

$$f(x + h) - f(x) = e^x \cdot e^h - e^x$$

$$\frac{f(x + h) - f(x)}{h} = \frac{e^x \cdot e^h - e^x}{h} \qquad h \neq 0$$

$$\frac{dy}{dx} = \lim_{h \to 0} \frac{f(x + h) - f(x)}{h} = \lim_{h \to 0} \frac{e^x(e^h - 1)}{h}$$

$$= e^x \lim_{h \to 0} \frac{e^h - 1}{h}$$

$$= e^x \cdot 1 = e^x$$

Consequently, if $y = e^x$, then $dy/dx = e^x$.

EXAMPLE 5   If $y = e^x$, find $\dfrac{d^2y}{dx^2}$.

SOLUTION   Since $\dfrac{dy}{dx} = e^x$, then $\dfrac{d^2y}{dx^2} = e^x$ and, in general,

$$\frac{d^n y}{dx^n} = e^x$$

EXAMPLE 6   If $y = xe^x$, find $\dfrac{dy}{dx}$.

SOLUTION   Notice that y is a product of two functions, $f(x) = x$ and $g(x) = e^x$. Therefore, by the derivative formula for a product we have

$$\frac{dy}{dx} = x \cdot \frac{d}{dx}(e^x) + \frac{d}{dx}(x) \cdot e^x \qquad FS' + F'S$$
$$= xe^x + 1 \cdot e^x$$
$$= xe^x + e^x$$
$$= e^x(x + 1)$$

Practice Problem 1   Given $f(x) = 3xe^{x^2}$, find $f'(x)$.

ANSWER $f'(x) = 6x^2e^{x^2} + 3e^{x^2} = e^{x^2}(6x^2 + 3)$

EXAMPLE 7   Find y' where $y = \sqrt[3]{e^{2x} - 3}$.

SOLUTION

$$y = (e^{2x} - 3)^{1/3}$$

$$y' = \frac{1}{3}(e^{2x} - 3)^{-2/3} \cdot \frac{d}{dx}(e^{2x} - 3) \qquad \frac{d}{dx}(u^n) = nu^{n-1}\frac{du}{dx}$$

$$= \frac{1}{3}(e^{2x} - 3)^{-2/3} \cdot e^{2x} \cdot (2) \qquad \frac{d}{dx}(e^{2x} - 3) = e^{2x}(2)$$

$$= \frac{2e^{2x}}{3(e^{2x} - 3)^{2/3}}$$

How can we tell whether the graphs obtained from data represent mathematical models for growth or decay? Since many growth and decay models can be represented by graphs of exponential functions, a knowledge of the shape and characteristics of graphs of exponential functions helps us to answer this question. We now investigate characteristics of the graph of $y = e^x$.

The derivative of $y = e^x$ is $y' = e^x$, which is positive for all finite values of x. Thus, the graph of $y = e^x$ is always increasing. Since y'' is also e^x, the graph is concave upward for all x. When $x = 0$, $y = e^0 = 1$ (the y-intercept). A calculator

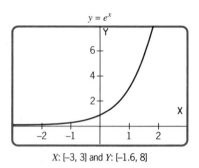

$y = e^x$

X: [-3, 3] and Y: [-1.6, 8]

Figure 2

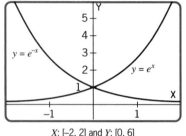

$y = e^{-x}$ $y = e^x$

X: [-2, 2] and Y: [0, 6]

Figure 3

graph of $y = e^x$ is shown in Figure 2. By shifting the viewing window to the left, y seems to approach 0 as x gets smaller and smaller. In fact, $\lim\limits_{x \to -\infty} e^x = 0$ and the x-axis is an asymptote of the graph.

Practice Problem 2 With your graphing calculator, draw the graph of both $y = e^x$ and $y = e^{-x}$. Do these graphs seem to be mirror images of each other about some line?

ANSWER The graph is shown in Figure 3. The graphs appear to be mirror images about the y-axis.

Compounding Continuously

Recall that if interest is compounded continuously, then the amount of a loan, P, at the end of t years is

$$A = Pe^{rt}$$

where r is the rate expressed in terms of a year.

EXAMPLE 8 Suppose that a loan is for $1000 and the interest rate is 8% compounded continuously. Find the rate of change of the loan amount with respect to time.

SOLUTION

$$A = 1000e^{0.08t} \quad \text{and} \quad \frac{dA}{dt} = 1000 \frac{d}{dt}[e^{0.08t}]$$

Now

$$\frac{d}{dt}(e^u) = e^u \frac{du}{dt}$$

and we have

$$\frac{dA}{dt} = 1000e^{0.08t}(0.08) \qquad u = 0.08t, \text{ so } \frac{du}{dt} = 0.08$$
$$= 80e^{0.08t}$$

Therefore, $80e^{0.08t}$ is the rate of change of the loan amount at time t. ∎

Growth equations and decay equations (dependent variable decreasing) are important in business, economics, biology, and the social sciences. At this time we discuss three of the many growth equations, often named for the people who first used them.

The **logistic curve** (discussed in Exercise 46) is defined by

$$y = \frac{A}{1 + Be^{-kt}}$$

and is a suitable model for defining natural phenomena where there is at first rapid growth and then a slow down of growth because of overcrowding, scarcity of food, and other factors.

One family of curves that has been used to describe both growth and deterioration is called the **Gompertz curve**, whose equation is of the form

$$N = Ca^{kt}$$

where N is the possible number of individuals at a given time t. For example, a business executive may predict the number of employees of his company by the Gompertz curve $N = 1000(0.5)^{0.6t}$, where t represents the number of years after starting the new business.

In Exercise 47 we discuss the characteristics of the Von Bertalanffy curve, which is useful in biology. A discussion of this and other exponential models is found in *Mathematical Models and Applications* by D. Maki and M. Thompson (Englewood Cliffs, N.J.: Prentice Hall, 1973), pp. 312–317.

EXAMPLE 9   The following equation is an example of a Gompertz curve.

$$N(t) = 1000e^{-0.05t}$$

where t is measured in days. Find the rate of decay (or decrease) after 4 days.

SOLUTION   The rate of decay is the derivative of the decay function with respect to time or $N'(t)$.

$$N'(t) = (-0.05)(1000)e^{-0.05t} \qquad \frac{d}{dt}(-0.05t) = -0.05$$

$$N'(4) = -50e^{-0.05(4)} = -50e^{-0.2}$$
$$= -40.94$$

The population is decreasing at a rate of approximately 41 per day at the end of 4 days. In fact, the population is always decreasing, but the rate of decrease approaches zero for large t. ■

Calculator Note

If all that is needed is the rate of decay after 4 days, finding the numerical value of a derivative on a calculator saves a great deal of time. On page 511 of Chapter 8 we learned to obtain the numerical value of a derivative using nDeriv ($1000\,e^{-0.05x}$, x, 4), which equals -40.9365.

SUMMARY

In this section we introduced two important derivative formulas:

1.　If $y = e^x$, then $\dfrac{dy}{dx} = e^x$.

2.　If $y = e^{u(x)}$, then $\dfrac{dy}{dx} = e^{u(x)}u'(x)$.

These derivatives are important when working with growth and decay models.

Exercise Set 4.1

Find dy/dx for the following functions.

1. $y = e^{4x}$
2. $y = 4e^{3x}$
3. $y = 5 + e^{2x}$
4. $y = 3x^2 - 2e^{4x}$
5. $y = 4x - e^{-3x}$
6. $y = 8x^2 + e^{-5x}$
7. $y = 4x - e^{3x}$
8. $y = e^{3x} + e^{4x}$
9. $y = e^{x^2}$
10. $y = 3 \cdot e^{x+4}$
11. $y = \sqrt{e^x + 4}$
12. $y = \sqrt{e^{4x+1}}$
13. $y = (3 + e^x)(2 - e^{-x})$
14. $y = \dfrac{3 + e^x}{3 - e^{-x}}$
15. $y = 3x^2e^{4x}$
16. $y = (3 + 5x)e^{2x}$
17. $y = e^{3x} - e^x$
18. $y = e^{3x} - 2e^{4x}$
19. $y = \sqrt{e^{2x} + e^x}$
20. $y = x^2e^x - 5xe^{2x} + 2e^x$

Find f"(x) for each of the following.

21. $y = xe^{x-1}$
22. $y = xe^{x^2}$
23. $y = e^{x^2+x}$
24. $y = (2x + 5)e^{-3x}$
25. Find the equation of the tangent line to the graph of $y = e^{3x-2}$ at the point $(\frac{2}{3}, 1)$.
26. Find the equation of the tangent line to the graph of $y = e^{x^2}$ at the point $(0, 1)$.

Find dy/dx for the following functions.

27. $y = (e^{2x} + 5)e^x$
28. $y = \dfrac{e^{4x} - 1}{e^x - 5}$
29. $y = xe^{2x} + e^{2x^2}$
30. $y = e^{3x} + e^{x^2} - 5$
31. $y = e^{3x} + e^{x^2}$
32. $y = \dfrac{e^{5x}}{e^{3x} + 2}$

Applications (Business and Economics)

33. **Salvage Value.**　If the salvage value of an airplane after t years is

$$V(t) = 400{,}000e^{-0.1t}$$

what is the rate of depreciation, dV/dt, after　(a) 1 year,　(b) 3 years, and　(c) 10 years?

34. **Marginal Revenue Function.**　If the price demand and revenue functions for x units of an item are, respectively,

$$p(x) = 100e^{-0.06x} \quad \text{and}$$
$$R(x) = xp(x) = 100xe^{-0.06x}$$

find the marginal revenue function. What number of units gives the maximum revenue?

35. **Sales Decay.**　Sales at an automobile dealership were excellent when the 1995 cars were introduced, but then leveled off with time according to the model

$$S(t) = 160 - 80e^{-0.2t}$$

where t represents time in weeks. Find the rate of change of sales at the end of　(a) 4 weeks,　(b) 10 weeks, and　(c) 20 weeks.

36. **Compounding Continuously.**　If $P = \$1000$ and interest is compounded continuously at a rate of 10%, find the rate of change of A with respect to time　(a) at the end of 2 years and　(b) at the end of 10 years.

Applications (Social and Life Sciences)

Sometimes exponential graphs are classified as unlimited growth, limited growth, logistic growth, and exponential decay. In Exercises 37–44, use your cal-

culator to compare the graphs of the functions. First use $0 < t < 20$ and $0 < y < 500$. Change your window if you do not feel you are getting a useful section of each graph.

37. Unlimited growth: $y = 500e^{0.15t}$
38. Unlimited growth: $y = 1000e^{0.06t}$
39. Limited growth: $y = 500(1 - e^{-0.15t})$
40. Limited growth: $y = 1000(1 - e^{-0.06t})$
41. Logistic growth: $y = \dfrac{500}{1 + 200e^{-0.15t}}$
42. Logistic growth: $y = \dfrac{1000}{1 + 400e^{-0.06t}}$
43. Exponential decay: $y = 500e^{-0.15t}$
44. Exponential decay: $y = 1000e^{-0.06t}$
45. **Learning Function.**   A learning function is given by

$$N(t) = 100(-e^{-0.03t})$$

Find the rate of learning, $N'(t)$. What is the rate (a) after 10 hours and   (b) after 20 hours?

46. **Logistic Growth Curve.**   The equation of a logistic growth curve of bacteria is of the form

$$w(t) = \frac{1000}{1 + 50e^{-0.2t}}$$

where t is measured in days and $w(t)$ is the weight after t days. Find the rate of change of weight w with respect to time t.

47. **Von Bertalanffy Curve.**   The number of bacteria in a culture after t hours can be given by

$$N(t) = 1000(1 - .4e^{-0.2t})^3$$

Find the rate of change of $N(t)$ with respect to time t in hours.

4.2 LOGARITHMIC FUNCTIONS AND THEIR PROPERTIES

OVERVIEW   Approximately 400 years ago John Napier discovered an ingenious way to multiply by adding and to divide by subtracting. He did this by developing **logarithms**. In 1614 Napier published his discovery in a paper entitled *A Description of the Marvelous Rule of Logarithms*. For many years logarithms were used to facilitate the multiplication and division of large numbers. Today, with modern calculators we no longer need logarithms for computation. However, logarithmic functions are important because they are used to express relationships in such areas as advertising, archeology, and biochemistry. In business, some cost functions and demand functions are expressed in terms of logarithms. The well-known Richter scale for earthquakes is expressed in terms of logarithms. As you will see, logarithmic functions are closely related to exponential functions. In this section we

- Define a logarithm in terms of exponents
- State the properties of logarithms
- Examine the characteristics of logarithmic functions
- Draw graphs of logarithmic functions
- Solve exponential equations

In Chapter 1 we determined that one bacteria can divide into two new ones. If the split occurs each minute, how long will it take for 1 bacteria to become 256 bacteria? To solve this problem, we observe the pattern that occurs. After 1 minute we have 2 or 2^1 bacteria; after 2 minutes we have 4 or 2^2 bacteria; and after

3 minutes we have 8 or 2^3 bacteria. In general, after y minutes we have 2^y bacteria. For our problem we have $2^y = 256$. Thus it takes $y = 8$ minutes for 1 bacteria to become 256 bacteria since $2^8 = 256$.

The exponential equation $2^8 = 256$ is often rewritten in logarithmic form as

$$\log_2 256 = 8$$

The symbol $\log_2 256$ is read the "logarithm, base 2, of 256." Another example of an exponential equation and its logarithmic equivalent is $10^3 = 1000$ and $\log_{10} 1000 = 3$. In general, if $x = b^y$, we say that y is the log of x to the base b and denote this by

$$y = \log_b x$$

That is, in $x = b^y$ the exponent y can be written in terms of another function called a logarithm. Remember that logarithms are exponents.

LOGARITHM

If x and b are positive numbers and b is not equal to 1, then the **logarithm** of x to the base b is equal to y, which is written as

$$\log_b x = y, \qquad x \geq 0$$

if and only if $x = b^y$.

TABLE 3

Logarithmic Form	Exponential Form
$\log_{10} 100 = 2$	$10^2 = 100$
$\log_{10} 0.01 = -2$	$10^{-2} = 0.01$
$\log_2 0.25 = -2$	$2^{-2} = 0.25$
$\log_{10} 1 = 0$	$10^0 = 1$
$\log_b 1 = 0$	$b^0 = 1$
$\log_b b = 1$	$b^1 = b$

We read $\log_b x$ as "the logarithm of x to the base b." The function defined is called the **logarithmic function**. Table 3 illustrates a number of logarithmic expressions and their exponential equivalents.

EXAMPLE 10

(a) Write $81 = 3^4$ in logarithmic form.
(b) Write $\log_2 \frac{1}{32} = -5$ in exponential form.
(c) If $3 = \log_2 x$, find x.

SOLUTION

(a) $\log_3 81 = 4$ (b) $2^{-5} = \frac{1}{32}$ (c) $x = 2^3 = 8$

Practice Problem 1 Find $\log_3 \frac{1}{81}$.

ANSWER -4

EXAMPLE 11 Sketch the graph of $y = \log_2 x$.

SOLUTION We know that $\log_2 8 = 3$ because $2^3 = 8$, so the point $x = 8$, $y = 3$ is on the graph. In a similar manner, $(1, 0)$, $(2, 1)$, $(\frac{1}{2}, -1)$, $(\frac{1}{8}, -3)$ are points on the graph. The graph of $y = \log_2 x$ is shown in Figure 4.

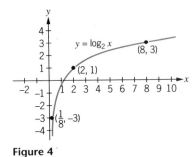

Figure 4

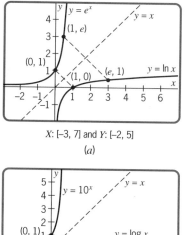

X: [–3, 7] and Y: [–2, 5]

(a)

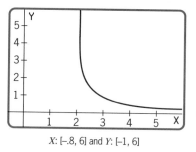

X: [–4, 10] and Y: [–3, 6]

(b)

Figure 5

X: [–.8, 6] and Y: [–1, 6]

Figure 6

In the preceding example, notice that the domain of $y = \log_2 x$ is the set of *positive* real numbers and the range is the set of all real numbers such that

$$y = \log_2 x \quad \begin{cases} < 0, & \text{for } 0 < x < 1 \\ = 0, & \text{for } x = 1 \\ > 0, & \text{for } x > 1 \end{cases}$$

The x-intercept is at $(1, 0)$. The graph approaches the y-axis as an asymptote.

As we have noted, the base of a logarithm may be any positive number except 1. However, most logarithms use the base 10 (called **common logarithms**) or the base e (called **natural logarithms**). On your calculator you will see a $\boxed{\text{LOG}}$ key representing common logarithms and an $\boxed{\text{LN}}$ key representing natural logarithms.

With your graphing calculator draw the graphs of $y = e^x$ and $y = \ln x$ on the same coordinate system, as shown in Figure 5(a). Notice that the graphs of $y = e^x$ and $y = \ln x$ are reflections of each other about the line $y = x$. Likewise, draw the graphs of $y = 10^x$ and $y = \log x$ on the same coordinate system, as shown in Figure 5(b). Notice that $y = 10^x$ and $y = \log x$ are also reflections of each other about $y = x$.

The pairs of functions in Figure 5 have special names; $y = e^x$ and $y = \ln x$ are called inverse functions. In general, we say that $y = f(x)$ and $y = g(x)$ are **inverse functions** if, whenever the pair (a, b) satisfies $y = f(x)$, the pair (b, a) satisfies $y = g(x)$. Furthermore, because the values of the x- and y-coordinates are interchanged for inverse functions, their graphs are reflections of each other about the line $y = x$.

Practice Problem 2 Use a calculator to identify the domain of

$$y = \ln \frac{|x|}{x - 2}$$

ANSWER As shown in Figure 6, the domain is $x > 2$.

Logarithmic functions have several useful properties that follow directly from their definitions and from the properties of exponents.

PROPERTIES OF LOGARITHMS

For $b > 0$, $b \neq 1$, and n a real number, if M and N are positive numbers, then the following properties are true.

1. $\log_b(M \cdot N) = \log_b M + \log_b N$
2. $\log_b \dfrac{M}{N} = \log_b M - \log_b N$
3. $\log_b M^N = N \log_b M$
4. $\log_b 1 = 0$
5. $\log_b b = 1$
6. $\log_b M = \log_b N$ if and only if $M = N$

These properties are stated in words to assist you in remembering them.

1.   The logarithm of a product is the sum of the logarithms of the factors.
2.   The logarithm of a quotient is the difference of the logarithm of the numerator and the logarithm of the denominator.
3.   The logarithm of a number to a power is the power times the logarithm of the number.
4.   The logarithm of 1 to any base is 0.
5.   The logarithm of a number which equals the base is one.
6.   If the logarithms of two numbers to the same base are equal, then the numbers are equal.

We can demonstrate the first three properties by comparing the graphs in Figure 7. In part (a) we have the graph of $y = \ln 6x$. Note that this is exactly the same graph as $y = \ln 6 + \ln x$ in part (b); that is, $\ln 6x = \ln 6 + \ln x$. The graph of $y = \ln (x/6)$ in part (c) is the same as the graph of $y = \ln x - \ln 6$ in part (d); that is, $\ln (x/6) = \ln x - \ln 6$. In part ($e$) the graph of $y = \ln x^3$ is the same as the graph of $y = 3 \ln x$ in part (f); that is, $\ln x^3 = 3 \ln x$.

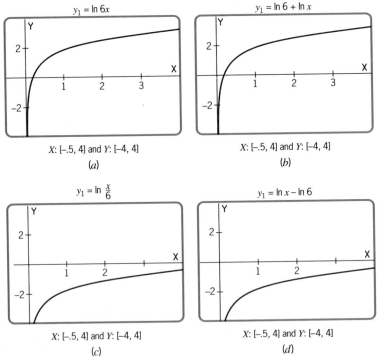

$y_1 = \ln 6x$

X: [−.5, 4] and Y: [−4, 4]

(a)

$y_1 = \ln 6 + \ln x$

X: [−.5, 4] and Y: [−4, 4]

(b)

$y_1 = \ln \frac{x}{6}$

X: [−.5, 4] and Y: [−4, 4]

(c)

$y_1 = \ln x - \ln 6$

X: [−.5, 4] and Y: [−4, 4]

(d)

Figure 7

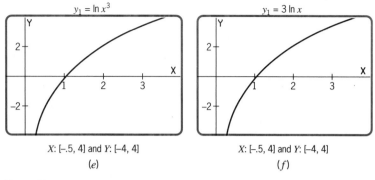

X: [-.5, 4] and Y: [-4, 4]   X: [-.5, 4] and Y: [-4, 4]
(e)   (f)

Figure 7

EXAMPLE 12 Using $\ln 2 = 0.693147$ and $\ln 3 = 1.098612$, find $\ln 6$ using the properties of logarithms.

SOLUTION

$$\ln 6 = \ln (2 \cdot 3) = \ln 2 + \ln 3 \qquad \text{Property 1}$$

$$\ln 6 = 0.693147 + 1.098612 = 1.791759$$

EXAMPLE 13 Find $\ln e$.

SOLUTION $\ln e = 1$ by property 5.

Calculator Note

To find ln e use $\boxed{\text{LN}}\ \boxed{e}\ \boxed{\wedge}\ 1$.

EXAMPLE 14 Find $\ln 8$ using the properties of logarithms. (See Example 12.)

SOLUTION

$$\ln 8 = \ln 2^3 = 3 \ln 2 \qquad \text{Property 3}$$

$$\ln 8 = 3(0.693147) = 2.079441$$

EXAMPLE 15 Solve for x in $3^x = 4$ given that $\log 2 = 0.301030$ and $\log 3 = 0.477121$. (Verify these values with your calculator.)

SOLUTION Take the logarithm of both sides to obtain

$$\log 3^x = \log 4$$
$$= \log 2^2$$
$$x \log 3 = 2 \log 2 \qquad \text{Property 3}$$
$$x = \frac{2 \log 2}{\log 3}$$
$$= \frac{0.602060}{0.477121}$$
$$= 1.261860$$

EXAMPLE 16 Express

$$\log \left(\frac{x^2 y^{1/2}}{z^3} \right)$$

as a sum or difference of logarithms without exponents.

SOLUTION

$$\log \left(\frac{x^2 y^{1/2}}{z^3} \right) = \log x^2 + \log y^{1/2} - \log z^3 \qquad \text{Properties 1 and 2}$$
$$= 2 \log x + \frac{1}{2} \log y - 3 \log z \qquad \text{Property 3}$$

EXAMPLE 17 Solve for x in $\log_5 x + \log_5(x - 4) = 1$.

SOLUTION

$$\log_5 x + \log_5(x - 4) = 1$$
$$\log_5 x(x - 4) = 1 \qquad \log M + \log N = \log MN$$
$$x(x - 4) = 5^1 \qquad \text{Definition of a logarithm}$$
$$x^2 - 4x - 5 = 0$$
$$(x - 5)(x + 1) = 0$$
$$x = 5 \quad \text{or} \quad x = -1$$

It would seem that $x = 5$ and $x = -1$ are solutions of the given equation, but a check of these values in the original equation shows that only $x = 5$ is a solution. Why is $x = -1$ not a solution?

Practice Problem 3 Use your calculator to find the value of $\log \frac{120}{62}$.

ANSWER 0.286790

Logarithms are very useful in solving for exponents in exponential equations. For example, we can use logarithms to compute the length of time necessary for an investment to double using the formula $A = P(1 + i)^n$.

EXAMPLE 18 How long will it take for an investment of $10,000 to double if a bank pays 8% interest compounded annually?

SOLUTION If P dollars are invested at $i\%$ compounded yearly, the investment will accumulate to A dollars in n years by the formula

$$A = P(1 + i)^n$$

If $P = \$10,000$ is to double, then $A = 2P = \$20,000$, and i is 0.08. The problem is to find n.

$$A = P(1 + i)^n$$
$$20{,}000 = 10{,}000(1 + 0.08)^n \qquad \text{Substitution}$$
$$2 = (1.08)^n$$
$$\log 2 = \log(1.08)^n \qquad \text{Solve by taking log or ln of both sides.}$$
$$= n \log 1.08 \qquad \text{Property 3}$$
$$n = \frac{\log 2}{\log 1.08}$$
$$= 9.0065$$

Thus, in approximately 9 years the investment will have doubled. ■

NOTE: Compare the answer in the preceding example with the Rule of 72 often used in the business world. This rule gives the doubling time of an investment at $i\%$ as being approximately $72/i$, where i is not changed to a decimal. For Example 18 the doubling time would be estimated to be $72/8 = 9$ years.

SUMMARY

In this section we defined a logarithm in terms of exponents and learned how to use logarithmic functions and properties of logarithms. We also solved logarithmic equations. All of these will be important as we learn to take derivatives of logarithmic functions in the next section.

Exercise Set 4.2

1. Write the following equations in logarithmic notation.
   (a) $3^4 = 81$ (b) $2^7 = 128$
   (c) $3^{-2} = \dfrac{1}{9}$

2. Write the following equations in logarithmic notation.
   (a) $5^4 = 625$ (b) $8^{1/3} = 2$
   (c) $16^{1/4} = 2$

3. Write the following in exponential notation.
   (a) $\log_7 49 = 2$ (b) $\log_6 36 = 2$
   (c) $\log_3 \dfrac{1}{9} = -2$

4. Write the following in exponential notation.
   (a) $\log_3 \dfrac{1}{27} = -3$ (b) $\log_{10} 1 = 0$
   (c) $\log_{10} 1000 = 3$

5. Find the following logarithms.
   (a) $\log_2 16$ (b) $\log_2 \dfrac{1}{8}$

   (c) $\log_{1/2} \dfrac{1}{8}$ (d) $\log_3 27$

6. Find the following logarithms.
   (a) $\log_{10} 10$ (b) $\log_{10} 1$
   (c) $\log_{10} 0.1$ (d) $\log_{10} 0.01$

7. Find the base of the following logarithms.
   (a) $\log_b 2 = \dfrac{1}{2}$ (b) $\log_b 2 = -1$

8. Find the base of the following logarithms.
   (a) $\log_b \dfrac{1}{8} = -3$ (b) $\log_b 100 = 2$

9. Determine the solution for x by inspection or by first
   writing the equation in exponential form.
   (a) $\log_3 9 = x$ (b) $\log_{10} x = 3$

10. Determine the solution for x by inspection or by first
    writing the equation in exponential form.
    (a) $\log_{10} x = -3$ (b) $\log_3 27 = x$

*Express each of the following as a sum or difference of
logarithms without exponents.*

11. $\ln \dfrac{x^2 z^4}{y}$

12. $\ln \dfrac{x^2 y^{-2}}{z^3}$

13. $\log_b \sqrt[3]{\dfrac{x^2}{y}}$

14. $\log_b \sqrt{\dfrac{x}{y^3}}$

15. $\ln \sqrt{\dfrac{x^2 y}{z}}$

16. $\ln \sqrt{\dfrac{x}{z^3 y}}$

*Use your calculator to find the value of each of the
following.*

17. $\log \sqrt[3]{165}$

18. $\log \sqrt{16^3 \cdot 71}$

19. $\log \dfrac{\sqrt{131}}{\sqrt[3]{9}}$

20. $\ln \sqrt[3]{7}$

21. $\ln \sqrt{18^3 \cdot 27}$

22. $\ln \sqrt[5]{\dfrac{161}{5}}$

Solve the following equations for x.

23. $\log_6 1 = x$ 24. $\log_4 x = 1$

25. $\log x = 2 \log 3 - 3 \log 2$

26. $\log \sqrt{x} = 2 \log 4 - \dfrac{1}{2} \log 9$

27. $\ln x^2 = 2$

28. $\ln x^2 + \ln 4 = \ln x + \ln 8$

29. $\ln e^{-0.01x} = 4$ 30. $\ln e^{x^2} = 25$

31. $\log x + \log(x + 1) = \log 2$

32. $\ln(x + 2) - \ln 3 = \ln 4x$

*Use your calculator to solve the following equations.
Round your answers to four decimal places.*

33. $8^x = 61$ 34. $14^y = 0.1$

35. $74^x = 16$ 36. $e^{2x} = 6$

37. $5e^{-3x} = 8$ 38. $e^{-x/2} = 0.4$

Applications (Business and Economics)

39. **Doubling of Sales.**   A corporation has formulated
    the model

    $$y = 1,000,000e^{0.06t}$$

    to predict sales growth, where y is annual sales in
    dollars in year t when t starts at 0 in 1990. In what
    year will the annual sales reach \$2,000,000?

40. **Compound Interest.**   Brooke has placed \$10,000 in
    a savings and loan that pays 9% interest compounded
    annually. When will she have \$25,000 in her ac-
    count? [**Hint:** $A = P(1 + i)^n$.]

41. **Compounded Continuously**.   How long will it take to double an investment if a bank pays 8% interest compounded continuously?

42. **Cost Function**.   The cost function for manufacturing a certain commodity is given as

$$C(x) = 10,000 + 600 \ln (2x^2 + 1)$$

where x is the number of items to be manufactured.
   (a) What is the fixed cost (the cost when $x = 0$)?
   (b) What is the cost of manufacturing 10,000 items?
   (c) What is the cost of manufacturing 1000 items?

Applications (Social and Life Sciences)

43. **Earthquake**.   A Richter scale measurement of the intensity of an earthquake is given as

$$RS = \log \frac{I}{I_0}$$

where I_0 is the intensity used for comparison.
   (a) Find RS if I is 1,000,000 times I_0.
   (b) The 1983 earthquake measured 7.7 on the Richter scale. Find the intensity in terms of I_0.

Review Exercises

Find dy/dx for the following functions.

44. $y = e^{x^2}$
45. $y = e^{5x^3}$
46. $y = x^2 e^{2x^2}$

4.3   DERIVATIVES OF LOGARITHMIC FUNCTIONS

OVERVIEW   Many functions in business and economics are expressed in terms of logarithms. In order to obtain marginal functions, such as marginal revenue and marginal cost, we need to know the derivative of a logarithmic function. In this section we use formulas for dy/dx when $y = \ln x$ and $y = \ln u(x)$. The derivation of dy/dx when $y = \ln x$ using our four-step procedure is rather cumbersome. So we postpone this derivation until Section 10.5 where the derivation is simple using implicit differentiation. In this section we study

   • Formulas for the derivative of logarithmic functions
   • The chain rule
   • Graphing techniques

The derivative formula for $y = \ln x$ is very simple.

$$\text{If } y = \ln x, \text{ then } \frac{dy}{dx} = \frac{1}{x}.$$

With your calculator, on some range $x > 0$, sketch the graph of $y = $ nDeriv $(\ln x, x, x)$ and then sketch $y = 1/x$ and note that the two graphs are identical for $x > 0$. We postpone the derivation of this derivative formula until Section 10.5, as implicit derivatives facilitate an easy derivation. The derivative formula for $y = \ln x$, along with the properties of logarithms, enable us to differentiate a wide variety of functions.

EXAMPLE 19   Find $\dfrac{dy}{dx}$ if $y = \ln x^3$.

SOLUTION

$$y = \ln x^3$$
$$= 3 \ln x \qquad \ln M^N = N \ln M$$

$$\frac{dy}{dx} = 3 \cdot \frac{1}{x} = \frac{3}{x}$$

For $y = \ln u$, where u is a function of x, we use the chain rule to find dy/dx, where $u(x) > 0$.

$$y = \ln u(x)$$

$$\frac{dy}{dx} = \frac{d}{du} \ln u \cdot \frac{du}{dx}$$

$$= \frac{1}{u} \cdot \frac{du}{dx} \qquad \frac{d}{du} \ln u = \frac{1}{u}$$

Now let's rework Example 19 using this formula.

$$y = \ln x^3 \qquad \text{Let } u = x^3.$$

$$\frac{dy}{dx} = \frac{1}{u} \cdot \frac{du}{dx}$$

$$= \frac{1}{x^3} \cdot (3x^2) \qquad \frac{du}{dx} = 3x^2$$

$$= \frac{3}{x}$$

We state these formulas as follows.

DERIVATIVES OF LOGARITHMIC FUNCTIONS

1. If $f(x) = \ln x$, then

$$f'(x) = \frac{1}{x}$$

for all $x > 0$.

2. If $f(x) = \ln u(x)$, where $u(x) > 0$ and has a derivative with respect to x, then

$$f'(x) = \frac{u'(x)}{u(x)}$$

EXAMPLE 20   If $f(x) = \ln(2x + 5)$, find $f'(x)$.

SOLUTION   Let $u(x) = 2x + 5$. Then $u'(x) = 2$, and

$$f'(x) = \frac{u'(x)}{u(x)} = \frac{2}{2x + 5}$$

EXAMPLE 21   If $f(x) = \ln(3x^2 + x - 2)$, find $f'(x)$.

SOLUTION   Let $u(x) = 3x^2 + x - 2$. Then $u'(x) = 6x + 1$, and

$$f'(x) = \frac{u'(x)}{u(x)} = \frac{6x + 1}{3x^2 + x - 2}$$

EXAMPLE 22   If $f(x) = \ln(x^{3/2} - 2x^{5/2})$, find $f'(x)$.

SOLUTION

$$f'(x) = \frac{u'(x)}{u(x)} = \frac{u'(x)}{x^{3/2} - 2x^{5/2}}$$

If $u(x) = x^{3/2} - 2x^{5/2}$, then $u'(x) = \frac{3}{2}x^{1/2} - 5x^{3/2}$. Therefore,

$$
\begin{aligned}
f'(x) &= \frac{\frac{3}{2}x^{1/2} - 5x^{3/2}}{x^{3/2} - 2x^{5/2}} \\[6pt]
&= \frac{x^{1/2}[\frac{3}{2} - 5x]}{x^{1/2}[x - 2x^2]} \qquad \text{\small Factor } x^{1/2} \text{ from the numerator and denominator.}\\[6pt]
&= \frac{\frac{3}{2} - 5x}{x - 2x^2} \qquad\qquad \text{\small Divide out } x^{1/2}.\\[6pt]
&= \frac{3 - 10x}{2x - 4x^2}
\end{aligned}
$$

EXAMPLE 23   If $f(x) = x \ln(x + 1)$, find $f'(x)$.

SOLUTION

$$
\begin{aligned}
f'(x) &= x \frac{d}{dx} \ln(x + 1) + \frac{d}{dx}(x) \ln(x + 1) \qquad \text{\small Product formula}\\[6pt]
&= x \cdot \frac{1}{x + 1} + 1 \cdot \ln(x + 1)\\[6pt]
&= \frac{x + (x + 1)\ln(x + 1)}{x + 1}
\end{aligned}
$$

Practice Problem 1   Given $f(x) = x^2 \ln(x^2 + 2)$, find $f'(x)$.

ANSWER $f'(x) = 2x \ln(x^2 + 2) + \dfrac{2x^3}{x^2 + 2}$

Whenever the equation is given, such as the revenue function in Example 24, calculus procedures can be used to maximize profit.

EXAMPLE 24 A company can produce men's ties at a cost of $4 per tie. The revenue equation is

$$R(x) = 10x - 0.8x \ln x$$

where x is the number of ties manufactured each week. Find the weekly production that will maximize profit.

SOLUTION

$$P(x) = R(x) - C(x)$$
$$= 10x - 0.8x \ln x - 4x = 6x - 0.8x \ln x \qquad C(x) = 4x$$

Then we have

$$P'(x) = 6 - \frac{0.8x}{x} - 0.8 \ln x \qquad \text{Derivative of a product}$$

$$0 = 5.2 - 0.8 \ln x \qquad \text{Set } P'(x) = 0.$$

$$\ln x = \frac{5.2}{0.8} = 6.5$$

$$x = e^{6.5} \approx 665 \qquad \text{Definition of } \ln x$$

Since $P''(x) = -0.8/x$, which is negative for $x > 0$, there is a relative maximum at $x = 665$. Therefore, a production of 665 ties per week would produce a maximum profit. ■

Sometimes it is helpful to apply the properties of logarithms before taking derivatives. This procedure saves time when working problems such as Example 25.

EXAMPLE 25 If $f(x) = \ln[(x + 2)^2(x^3 + 4)]$, find $f'(x)$.

SOLUTION

$$f(x) = \ln[(x + 2)^2(x^3 + 4)]$$
$$= \ln(x + 2)^2 + \ln(x^3 + 4) \qquad \ln MN = \ln M + \ln N$$
$$= 2 \ln(x + 2) + \ln(x^3 + 4) \qquad \ln M^N = N \ln M$$

$$f'(x) = \frac{2}{x + 2} + \frac{3x^2}{x^3 + 4} \qquad \frac{d}{dx}(x^3 + 4) = 3x^2$$

■

SUMMARY

In this section we learned to use two formulas involving logarithms.

1.   If $y = \ln x$, then

$$y' = \frac{1}{x} \quad \text{for } x > 0$$

2.   If $y = \ln u$, where u is a function of x, then

$$y' = \frac{u'}{u} \quad \text{for } u > 0$$

These formulas can be combined with the rules of differentiation in Chapter 8 to differentiate a wide variety of functions.

Exercise Set 4.3

Find dy/dx for the following functions and simplify.

1. $y = 3 \ln x$
2. $y = \ln x^2$
3. $y = \ln(2x + 3)$
4. $y = \ln(3x + 10)$
5. $y = 4 \ln(3x + 2)$
6. $y = 7 \ln(2x + 5)$
7. $y = \ln(3x^2 + 2x + 5)^2$
8. $y = 3 \ln(2x^2 + x + 3)^2$
9. $y = 3x^2 + 2x + \ln(2x + 5)$
10. $y = 2x \ln(3x + 7)$
11. $y = 3x^2 \ln(3x + 2)$
12. $y = (2x^2 + 1) \ln 5x^3$
13. $f(x) = (\ln x)^3$
14. $f(x) = (\ln 2x^2)^2$
15. $y = 10x^2 \ln(x^2 + 1)^3$
16. $y = 2x^3 \ln(3x^2 + x)$
17. $y = \ln[3\sqrt{x}(2x + 1)^3]$
18. $y = \ln[2x^3\sqrt{3x + 2}]$
19. $y = [\ln(3x^2 - 2x + 5)]^2$
20. $y = \ln\left[\dfrac{3x^2 + 4x + 5}{(2x + 3)^4}\right]$

Applications (Business and Economics)

21. **Maximum Profit.**   If the revenue function of an industry is given as

$$R(x) = 400 \ln(2x - 10)$$

and the cost function is $C(x) = 40x$ for the production of x items, find the number of items that must be produced to have a maximum profit.

22. **Demand Equation.**   The market research of a company indicates that the demand equation is $p = D(x) = 20 - 4 \ln x$, where p is the price and x is the number of units in demand. Find dp/dx.

23. **Cost Function.**   The cost function for selling a product is given as

$$C(x) = 100 + 60 \ln(2x^2 - 2x + 1)$$

where x is the number of units of 10,000 pounds sold. Find the number of pounds that should be sold to keep cost at a minimum. First, approximate by using your graphing calculator.

24. **Business Advertising.**   The number of responses (automobiles sold) to x thousands of dollars of advertising by an automobile dealership seems to follow the model

$$N(x) = 200x - 2000 \ln x.$$

Find the amount of advertising that would produce the maximum number of responses. First, approximate using your graphing calculator.

Applications (Social and Life Sciences)

25. **Earthquakes.** Earthquakes are reported in units R on the Richter scale with R being defined as

$$R = \frac{0.67}{\ln 10} \ln(E) - 7.9$$

where E is the energy in ergs released by the earthquake. Find an expression for the rate of change of R with respect to E.

26. **Archaeology.** The formula

$$\ln(P) = \frac{1}{2} \ln(A)$$

relates the population of a site to the area of the site. Find the rate of change of the population with respect to the area.

27. **pH Factor.** Recall that the pH of a solution is defined to be

$$pH = \frac{\ln\left(\dfrac{1}{H^+}\right)}{\ln 10}$$

where H^+ represents the concentration of hydrogen ions per liter. Compute the derivative of pH with respect to H^+, and show that pH is decreasing for $H^+ > 0$.

Review Exercises

28. Write as a sum:

$$y = \ln\left(\frac{\sqrt{x} \cdot y^{-1/3}}{z^{-1/3}}\right)$$

29. Write $\log_5 x = 3$ as an exponential expression and then find x.

For the following functions, find dy/dx.

30. $y = e^{x-2}$

31. $y = xe^{-3/x}$

32. $y = (4 + 2x^2)(3 - e^{-2x})$

33. $y = \dfrac{4 - 2e^{2x}}{3 + e^{-3x}}$

4.4 GRAPHING EXPONENTIAL AND LOGARITHMIC FUNCTIONS (Optional)

OVERVIEW   An easy way to graph exponential and logarithmic functions is to use a calculator with graphing capabilities. This will be our first step in this section. However, the main objective of this section is to apply the calculus concepts of Chapters 8 and 9 to graphing exponential and logarithmic functions. Our objective is to show how calculus can give more specific information about graphs than that obtained from calculator graphs. In this section we

- Discuss the characteristics of a graph using a graphing calculator
- Discuss the characteristics of a graph using calculus

Our first step in working problems in this section is to draw a given curve with a calculator. Then we discuss intuitively what seem to be the characteristics of the curve.

EXAMPLE 26   With a calculator, draw the graphs of y, y', and y'' for the function $y = f(x) = xe^x$ and discuss each of the following.

(a)   Intercepts of $f(x)$
(b)   Points where maxima or minima occur for $f(x)$
(c)   Where the graph of $f(x)$ is increasing or decreasing

(d)   What happens to $f(x)$ when $x \to \infty$ and when $x \to -\infty$
(e)   Changes in concavity of $f(x)$

SOLUTION   The graphs of y, y', and y'' in Figure 8 suggest the following characteristics of $y = xe^x$.

(a)   Both the x- and y-intercepts of $f(x)$ are at $(0, 0)$, as shown in Figure 8(a).
(b)   A relative minimum occurs when $x = -1$, since we can see in Figure 8(b) that y' changes from negative to positive at $x = -1$.
(c)   The graph of $f(x)$ is increasing on $(-1, \infty)$ and decreasing on $(-\infty, -1)$.
(d)   As $x \to \infty$, $y \to \infty$; as $x \to -\infty$, $y \to 0$ as a horizontal asymptote.
(e)   The second derivative is 0 at $x = -2$ in Figure 8(c). At this point there is a point of inflection or a change in concavity. The curve is concave upward for $x > -2$ and concave downward for $x < -2$. ∎

Of course, the solutions in the preceding example could have been obtained to greater accuracy; however, the procedure we are following at this time is to obtain approximate answers quickly using a graphing calculator and then to obtain exact answers using calculus procedures.

EXAMPLE 27   For the curve $y = xe^x$, use calculus techniques to

(a)   find the intercepts.
(b)   locate points of relative extrema.
(c)   discuss where the graph is increasing or decreasing.
(d)   discuss concavity.
(e)   determine what happens when $x \to \infty$ and when $x \to -\infty$.
(f)   sketch the graph of $y = xe^x$.

SOLUTION

(a)   When $x = 0$, $y = 0$, so the curve goes through the origin.
(b)   First we find y' to be $y' = xe^x + e^x$. To find the critical points, find where $y' = 0$.

$$y' = e^x(x + 1) = 0$$

Since e^x cannot be 0, $x = -1$ is the only solution and the only critical value. Next we have $y(-1) = -1 \cdot e^{-1} = -1/e$. So, $(-1, -1/e)$ is a critical point and we apply the second derivative test to determine whether this critical point is an extremum.

$$y'' = e^x(1 + x) + e^x \qquad \text{Product formula}$$
$$= e^x(2 + x)$$

$$y''(-1) = e^{-1}[2 + (-1)]$$
$$= e^{-1} \cdot 1$$

$$= \frac{1}{e} > 0$$

Therefore, there is a relative minimum at $(-1, -1/e)$.

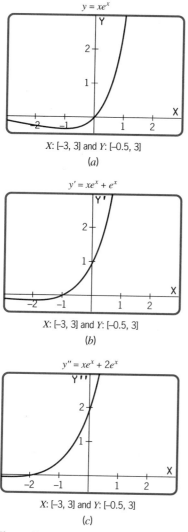

$y = xe^x$

X: [-3, 3] and Y: [-0.5, 3]

(a)

$y' = xe^x + e^x$

X: [-3, 3] and Y: [-0.5, 3]

(b)

$y'' = xe^x + 2e^x$

X: [-3, 3] and Y: [-0.5, 3]

(c)

Figure 8

(c) To determine where the function is increasing and where it is decreasing, we examine the first derivative $y' = e^x(1 + x)$. The slope changes at $x = -1$. Looking at the calculator graph of y', the derivative seems to be negative for $x < -1$ and positive for $x > -1$, or since e^x is always positive, $e^x(1 + x)$ is negative when $x < -1$ and positive when $x > -1$. Therefore, the function is decreasing for x on $(-\infty, -1)$ and increasing for x on $(-1, \infty)$.

(d) To determine concavity, we find the second derivative and determine where is it zero.

$$y'' = e^x(2 + x)$$

Set y'' equal to zero and solve for x.

$$e^x(2 + x) = 0$$

$$2 + x = 0 \qquad e^x \text{ cannot be zero.}$$

$$x = -2$$

Since $y'' = 0$ at $x = -2$, concavity can change only at $x = -2$. For $(-\infty, -2)$, e^x is always positive and $e^x(2 + x) < 0$; the curve is concave downward. For $(-2, \infty)$, $e^x(2 + x) > 0$ and the curve is concave upward.

(e) We can use a calculator to verify that as x gets larger, y gets larger, or as $x \to \infty$, $y \to \infty$ and as $x \to -\infty$, $y \to 0$.

(f) To sketch the graph shown in Figure 9, we use the preceding information for $y = xe^x$:

The y-intercept is $(0, 0)$.
A minimum point exists at $(-1, -1/e)$.
It is decreasing over $(-\infty, -1)$ and increasing over $(-1, \infty)$.
It is concave downward over $(-\infty, -2)$ and upward over $(-2, \infty)$.
$\lim_{x \to -\infty} y = 0$ and $\lim_{x \to \infty} y \to \infty$. ■

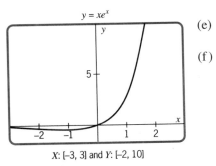

$y = xe^x$

X: [-3, 3] and Y: [-2, 10]

Figure 9

In a similar manner, we discuss the graph of a function involving logarithms by drawing the graph on a calculator and then obtaining characteristics of the graph intuitively.

EXAMPLE 28 With your graphing calculator draw the graph of $y = x \ln x$, the graph of the slope function $y' = \ln x + 1$, and the graph of $y'' = 1/x$. Then discuss each of the following intuitively.

(a) Intercepts of y
(b) Domain of y
(c) What happens to y as $x \to 0$ and as $x \to \infty$

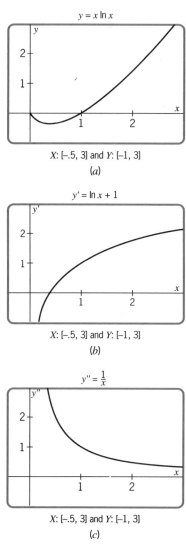

$y = x \ln x$

X: [-.5, 3] and Y: [-1, 3]

(a)

$y' = \ln x + 1$

X: [-.5, 3] and Y: [-1, 3]

(b)

$y'' = \frac{1}{x}$

X: [-.5, 3] and Y: [-1, 3]

(c)

Figure 10

(d)   Points of relative extrema of y

(e)   Where the graph of y is increasing and decreasing

(f)   Where the graph of y is concave up and concave down

SOLUTION   A study of the graphs of y, y' and y'' in Figure 10 suggest the following.

(a)   From Figure 10(a) we see that an x-intercept is at $x = 1$. The graph is approaching $(0, 0)$; however, $(0, 0)$ may not be an intercept.

(b)   The domain is $x > 0$.

(c)   As $x \rightarrow 0$, the graph seems to approach $y = 0$. As $x \rightarrow \infty$, y seems to approach ∞.

(d)   There is a relative minimum at x slightly less than $x = \frac{1}{2}$ because the slope is 0 at this point; see Figures 10(a) and (b).

(e)   The graph in Figure 10(a) is decreasing for x between 0 and some x slightly less than $x = \frac{1}{2}$ (also $y' < 0$ in (b)) and increasing from there on ($y' > 0$).

(f)   The graph is concave up for all x in the domain [$y'' > 0$ for all x in Figure 10(c)].

EXAMPLE 29   For $y = x \ln x$, use calculus to discuss each of the following.

(a)   Intercepts

(b)   Domain

(c)   What happens as $x \rightarrow 0$, and as $x \rightarrow \infty$

(d)   Points of relative extrema

(e)   Where the function is decreasing and where it is increasing

(f)   Where the graph is concave up and where it is concave down

(g)   Sketch the graph.

SOLUTION

(a)   When $x = 1$, $\ln 1 = 0$, so an x-intercept is at $x = 1$. There is no y-intercept because $\ln 0$ is undefined.

(b)   Since $\ln x$ is defined only for $x > 0$, the domain is $x > 0$.

(c)   By substituting values closer and closer to $x = 0$, it seems that as $x \rightarrow 0$, $y \rightarrow 0$. (We cannot be certain at this time.) As $x \rightarrow \infty$, we see from the following table that y seems to get very large or $y \rightarrow \infty$.

x	100	1000	10,000
y	461	6908	92,103

(d)   We determine the critical points by finding the first derivative and setting it equal to 0.

$$y = x \ln x$$

$$y' = x \left(\frac{1}{x} \right) + \ln x \qquad \text{Product formula}$$

$$= 1 + \ln x$$

Set the first derivative to zero and solve.

$$\ln x + 1 = 0$$

$$\ln x = -1$$

$$x = e^{-1} \qquad \text{Definition of } \ln x$$

$$\approx 0.368$$

When $x \approx 0.368$, $y = 0.368 \ln 0.368 \approx -0.368$. Thus, a critical point occurs at approximately $(0.368, -0.368)$. We use the second derivative to determine whether we have a relative maximum or minimum.

$$y'' = \frac{1}{x}$$

Since $1/x$ is always positive for $x > 0$, a relative minimum occurs at approximately $(0.368, -0.368)$.

(e) Since $y' = 1 + \ln x$ and $1 + \ln x$ is 0 only for x at approximately 0.368, we can easily verify that y' is negative for $x < 0.368$ and positive for $x > 0.368$. Therefore, the function is decreasing on $(0, 0.368)$ and increasing on $(0.368, \infty)$.

(f) The second derivative, $y'' = 1/x$, is positive for all $x > 0$. Therefore, the graph is concave up for all x in the domain.

(g) Using these characteristics, we can sketch the graph shown in Figure 11. ∎

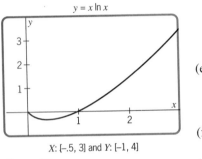

$y = x \ln x$

$X: [-.5, 3]$ and $Y: [-1, 4]$

Figure 11

SUMMARY

In this section, we used the calculus concepts of Chapters 8 and 9 along with the graphing calculator to investigate graphs of exponential and logarithmic functions.

Exercise Set 4.4

With a graphing calculator, graph the following functions.

1. $y = e^{x^2}$
2. $y = e^{-(4x+2)}$
3. $y = \ln(2x^2 + 1)$
4. $y = \ln(3x - 4)$

With a graphing calculator, graph the given functions and their derivatives and then find each of the following intuitively.

(a) *Intercepts*
(b) *Domain*
(c) *Where the function is increasing and where it is decreasing*
(d) *Points where there are relative extrema*
(e) *Where the graph is concave up and where it is concave down*
(f) *What happens when $x \to \infty$ and when $x \to -\infty$*

5. $y = e^{x-4}$
6. $y = e^{-(3x+2)}$
7. $y = \ln(x - 4)$
8. $y = \ln(x^2 + 2)$

For each of the following use algebra and calculus techniques to

(a) *find the intercepts.*
(b) *find where the function is increasing or decreasing and locate points where there are relative extrema.*

(c) *find where the curve is concave up and where it is concave down.*

(d) *discuss what happens to y as $x \to \infty$ and as $x \to -\infty$.*

(e) *sketch the graph.*

9. $y = e^{x-4}$. (Check Exercise 5.)

10. $y = e^{-(3x+2)}$. (Check Exercise 6.)

11. $y = \ln(x - 4)$. (Check Exercise 7.)

12. $y = \ln(x^2 + 2)$. (Check Exercise 8.)

13. $y = xe^{-x}$

14. $y = \dfrac{e^x}{x}$

15. $y = x^2 \ln x$

16. $y = e^x \ln x$

17. Use a graphing calculator to find the points at which there are relative maxima or minima for $y = e^{-|x^2 - 1|}$.

Applications (Business and Economics)

18. **Sales Decay.** Automobile sales at an automobile dealership were excellent when the 1995 cars were introduced but then leveled off with time according to the model

$$S(t) = 160 - 80e^{-0.2t}$$

where t represents time in weeks. Sketch the graph of this function and verify with a graphing calculator.

19. **Compounding Continuously.** Sketch the graph of the rate of change of A with respect to time where $P = \$1000$ and $i = 10\%$. (Recall that $A = Pe^{rt}$.)

20. **Marginal Revenue Function.** The price–demand and revenue functions for x units of an item are, respectively,

$$p(x) = 100e^{-0.06x} \quad \text{and}$$
$$R(x) = xp(x) = 100xe^{-0.06x}$$

Find the marginal revenue function. On the same coordinate system, sketch the graphs of $R(x)$ and $R'(x)$. Discuss the relationship between these two graphs.

Applications (Social and Life Sciences)

21. **Standard Normal Curve.** The equation of a standard normal curve is given by

$$y = \frac{1}{\sqrt{2\pi}} e^{-x^2/2}$$

(a) Find the points of relative extrema and the points of inflection.

(b) Sketch the curve.

22. **Logistic Growth Curve.** The equation of a logistic growth curve of bacteria is of the form

$$w(t) = \frac{1000}{1 + 50e^{-0.2t}}$$

where t is measured in days and $w(t)$ is the weight after t days.

(a) Discuss where the curve is increasing and decreasing.

(b) Discuss where the curve is concave up and concave down.

(c) What happens when $x \to \infty$? (Use your calculator.)

(d) Sketch the curve.

23. **Von Bertalanffy Curve.** The number of bacteria in a culture after t hours can be given by

$$N = 1000(1 - .4e^{-0.2t})^3$$

(a) Discuss where the curve is increasing and decreasing.

(b) Discuss where the curve is concave up and concave down.

(c) What happens when $x \to \infty$? (Use your calculator.)

(d) Sketch the curve.

24. **Learning Function.** A learning function is given by

$$N(t) = 100(1 - e^{-0.03t})$$

Find the rate of learning, $N'(t)$. On the same coordinate system, sketch the graphs of $N(t)$ and $N'(t)$. Discuss the relationship between these two graphs.

Review Exercises

Find dy/dx for each function.

25. $y = xe^{3x^2}$

26. $y = \ln(1 + x^2)$

27. $y = x^2 \ln x$

28. $y = e^{1/x^3}$

29. $y = \ln \left[\dfrac{1 + x^2}{x} \right]$

30. $y = 4x^3 e^{2x}$

4.5 IMPLICIT DIFFERENTIATION

OVERVIEW   In Chapter 8 the functions for which we found derivatives were given in the form

$$y = f(x)$$

which are called **explicit functions**. For example, $y = -4x^3 + 3$ defines explicitly a function $f(x)$ with x as the independent variable and y as the dependent variable. This equation can be written as

$$f(x, y) = y + 4x^3 - 3 = 0$$

The y in this equation is the same y as in $y = -4x^3 + 3$. The equation $y + 4x^3 - 3 = 0$ gives **implicitly** y as a function of x. To find dy/dx when y is an implicit function of x is called **implicit differentiation**. In this section we

- Do implicit differentiation
- Develop a formula for y' where $y = \ln x$
- Find derivatives of $y = a^x$, where $a > 0$

To introduce the procedures for this differentiation, we find dy/dx by differentiating both sides of the equation with respect to the independent variable x.

$$y + 4x^3 - 3 = 0$$

$$\frac{d}{dx}(y + 4x^3 - 3) = \frac{d}{dx}(0)$$

By using the fact that the derivative of a sum is the sum of the derivatives, we have

$$\frac{d}{dx}(y) + \frac{d}{dx}(4x^3) + \frac{d}{dx}(-3) = \frac{d}{dx}(0)$$

$$\frac{d}{dx}(y) + 12x^2 + 0 = 0$$

Since y is a function of x, we write

$$\frac{d}{dx}(y) \quad \text{as} \quad \frac{dy}{dx} \text{ or } y'$$

Solving

$$\frac{dy}{dx} + 12x^2 = 0$$

for dy/dx gives

$$\frac{dy}{dx} = -12x^2$$

Note that this is the same result that is obtained by rewriting the equation as an explicit function and then differentiating.

$$y = -4x^3 + 3$$

$$\frac{dy}{dx} = -12x^2$$

Of greater importance is finding dy/dx for a function that cannot be expressed explicitly as a function of the independent variable. Sometimes it may be difficult or even impossible to express a relation as an explicit function, and yet the relation may still be such that it defines y as a function of x. Such a function is called an **implicit function**. The derivative of an implicit function can be found, if it exists, as in the preceding example, by differentiating both sides of the equation with respect to the independent variable. To illustrate implicit differentiation we consider

$$y^3 + y^2 + x^2 = 0$$

where y is an implicit function of x. Note that when we differentiate each term with respect to x we encounter

$$\frac{d}{dx}(y^3) \quad \text{and} \quad \frac{d}{dx}(y^2)$$

Recall the formula in Chapter 8 for the derivative with respect to x of a function of x raised to the nth power. We can write this derivative as

$$\frac{d}{dx}u^n = nu^{n-1}\frac{du}{dx}$$

where u is a function of x. If we consider y as a function of x, then

$$\frac{d}{dx}(y^n) = ny^{n-1}\frac{dy}{dx}$$

Therefore,

$$\frac{d}{dx}(y^2) = 2y\frac{dy}{dx} = 2yy'$$

$$\frac{d}{dx}(y^3) = 3y^2\frac{dy}{dx} = 3y^2y'$$

$$\frac{d}{dx}(y^4) = 4y^3\frac{dy}{dx} = 4y^3y'$$

Returning to our example, we differentiate both sides of the equation $y^3 + y^2 + x^2 = 0$ with respect to x.

$$\frac{d}{dx}(y^3 + y^2 + x^2) = \frac{d}{dx}(0)$$

$$\frac{d}{dx}(y^3) + \frac{d}{dx}(y^2) + \frac{d}{dx}(x^2) = 0 \qquad \text{Derivative of a sum}$$

$$3y^2y' + 2yy' + 2x = 0 \qquad \frac{d}{dx}(y^2) = 2y \cdot y'; \frac{d}{dx}(y^3) = 3y^2y'$$

$$(3y^2 + 2y)y' = -2x \qquad \text{Factor.}$$

$$y' = \frac{-2x}{3y^2 + 2y} \qquad \text{Solve for } y'.$$

EXAMPLE 30 Find y' for the function defined implicitly by

$$y^2 - x^2 - 3y = 0$$

SOLUTION Differentiate both sides with respect to x to obtain

$$\frac{d}{dx}(y^2 - x^2 - 3y) = \frac{d}{dx}(0)$$

$$\frac{d}{dx}(y^2) + \frac{d}{dx}(-x^2) + \frac{d}{dx}(-3y) = 0$$

$$2yy' - 2x - 3y' = 0 \qquad \frac{d}{dx}(y^2) = 2yy'$$

$$(2y - 3)y' = 2x$$

$$y' = \frac{2x}{2y - 3} \qquad \text{Solve for } y'.$$

■

Calculator Note

As support for implicit differentiation, let's explore what happens when we separate $x^2 + y^2 = 25$ into two explicit functions by solving for y.

$$y = \pm\sqrt{25 - x^2}$$

We draw $y_1 = \sqrt{25 - x^2}$ and then $y_2 = -\sqrt{25 - x^2}$ on the same window and note what happens. Do you get the complete curve for $x^2 + y^2 = 25$? Using your calculator, find the slope of each graph when $x = 3$ and when $x = -3$. Do you get a slope of $-\frac{3}{4}$ at $(3, 4)$ and a slope of $\frac{3}{4}$ at $(-3, 4)$ on $y_1 = \sqrt{25 - x^2}$? Do you get a slope of $\frac{3}{4}$ at $(3, -4)$ and a slope of $-\frac{3}{4}$ at $(-3, -4)$ on $y_2 = -\sqrt{25 - x^2}$? The corresponding tangent lines are shown in Figure 12.

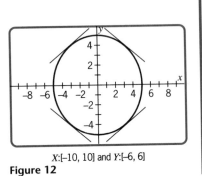

X:[-10, 10] and Y:[-6, 6]

Figure 12

Now we find the derivative of $x^2 + y^2 = 25$ implicitly.

$$\frac{d}{dx}(x^2) + \frac{d}{dx}(y^2) = \frac{d}{dx}(25)$$

$$2x + 2yy' = 0$$

$$y' = -\frac{x}{y}$$

Thus the slope of the graph at $(3, 4)$ is $-\frac{3}{4}$; at $(-3, 4)$ is $\frac{3}{4}$; at $(3, -4)$ is $\frac{3}{4}$; and at $(-3, -4)$ is $-\frac{3}{4}$. Note these slopes are in agreement with our calculator answers for explicit functions.

Practice Problem 1 Given $y^3 + x^2 = y - 1$, find dy/dx.

ANSWER $\dfrac{dy}{dx} = \dfrac{-2x}{3y^2 - 1}$

COMMON ERROR   For equations involving a product such as x^2y, students often forget that the product rule must be used when taking the derivative.

$$\frac{d}{dx}(x^2y) \neq 2xy' \qquad \frac{d}{dx}(x^2y) = 2xy + x^2y'$$

EXAMPLE 31   If $x^2y = x^3 + 4y$, find $\dfrac{dy}{dx}$.

SOLUTION   We first take the derivative with respect to x of each term to obtain

$$\frac{d}{dx}(x^2y) = \frac{d}{dx}(x^3) + \frac{d}{dx}(4y)$$

Now, as mentioned above, x^2y is a product and we must use the product formula.

$$x^2\frac{d}{dx}(y) + 2xy = 3x^2 + 4\frac{d}{dx}(y) \qquad \text{Derivative of a product}$$

$$x^2\frac{dy}{dx} - 4\frac{dy}{dx} = 3x^2 - 2xy$$

$$\frac{dy}{dx}(x^2 - 4) = 3x^2 - 2xy \qquad \text{Factor.}$$

$$\frac{dy}{dx} = \frac{3x^2 - 2xy}{x^2 - 4} \qquad \text{Divide by } x^2 - 4 \neq 0. \qquad \blacksquare$$

Practice Problem 2 Given $xy + y^2 = 4x$, find y'.

ANSWER $y' = \dfrac{4 - y}{x + 2y}$

EXAMPLE 32   Find dy/dx and the slope of the function

$$y^3 + y + 3x^2 + 2x + 1 = 0 \quad \text{at} \quad (-1, -1)$$

SOLUTION   Differentiating both sides with respect to x gives

$$\frac{d}{dx}(y^3 + y + 3x^2 + 2x + 1) = \frac{d}{dx}(0)$$

$$3y^2 \frac{dy}{dx} + \frac{dy}{dx} + 6x + 2 = 0 \qquad \frac{d}{dx}y^3 = 3y^2 \cdot \frac{dy}{dx}$$

$$3y^2 \frac{dy}{dx} + \frac{dy}{dx} = -6x - 2 \qquad \text{Subtraction}$$

$$(3y^2 + 1)\frac{dy}{dx} = -6x - 2 \qquad \text{Factoring}$$

$$\frac{dy}{dx} = \frac{-6x - 2}{3y^2 + 1} \qquad \text{Division}$$

By substituting $(-1, -1)$, we have

$$\left.\frac{dy}{dx}\right|_{(-1, -1)} = \frac{-6(-1) - 2}{3(-1)^2 + 1} = \frac{6 - 2}{3 + 1} = \frac{4}{4} = 1$$

Note that

$$\left.\frac{dy}{dx}\right|_{(-1, -1)}$$

means to evaluate dy/dx at $x = -1$, $y = -1$. The slope of the function is 1 at $(-1, -1)$.

EXAMPLE 33   Find y'' for $xy + y^2 = 4$.

SOLUTION   First we find y' by implicit differentiation.

$$\frac{d}{dx}(xy) + \frac{d}{dx}(y^2) = \frac{d}{dx}(4)$$

$$xy' + y + 2yy' = 0 \qquad \text{Note the product.}$$

$$(x + 2y)y' = -y$$

$$y' = \frac{-y}{x + 2y} \qquad \text{Solve for } y'.$$

Now

$$\frac{d}{dx}(y') = y''$$

but the derivative of the right side of the equation involves a quotient.

$$y'' = \frac{\dfrac{d}{dx}(-y) \cdot (x + 2y) - \dfrac{d}{dx}(x + 2y) \cdot (-y)}{(x + 2y)^2}$$

$$= \frac{-xy' - 2yy' + y + 2yy'}{(x + 2y)^2}$$

$$= \frac{-xy' + y}{(x + 2y)^2}$$

$$= \frac{-x\left(\dfrac{-y}{x + 2y}\right) + y}{(x + 2y)^2} \qquad\qquad y' = \frac{-y}{x + 2y}$$

$$= \frac{xy + (xy + 2y^2)}{(x + 2y)^3}$$

$$= \frac{2(xy + y^2)}{(x + 2y)^3}$$

In Section 10.3 we introduced a formula for the derivative of $y = \ln x$ without proof that the formula was valid. We now use implicit differentiation to prove that if $y = \ln x$, then

$$\frac{dy}{dx} = \frac{1}{x}$$

First, we write $y = \ln x$ as $e^y = x$. Then we take the derivative of both sides of the equation with respect to x.

$$\frac{d}{dx}(e^y) = \frac{d}{dx}(x)$$

$$e^y \frac{dy}{dx} = 1 \qquad\qquad d[e^{u(x)}] = e^{u(x)}\frac{dy}{dx}$$

$$\frac{dy}{dx} = \frac{1}{e^y} = \frac{1}{x} \qquad\qquad \text{Divide by } e^y.$$

In a like manner, we use implicit differentiation to develop the formula for the derivative of

$$y = a^x$$

$$\ln y = \ln a^x \qquad\qquad \text{Take ln of both sides.}$$

$$= x \ln a \qquad\qquad \ln M^N = N \ln M$$

Then we differentiate both sides with respect to x.

$$\frac{1}{y} \cdot \frac{dy}{dx} = \ln a \qquad\qquad \frac{d}{dx}(\ln y) = \frac{1}{y}\frac{dy}{dx}$$

$$\frac{dy}{dx} = y \ln a \qquad\qquad \text{Multiply both sides by } y.$$

$$\text{or}\quad \frac{dy}{dx} = a^x \ln a \qquad\qquad \text{Substitute } a^x \text{ for } y.$$

Note that if $a = e$, then $\ln e = 1$, and the differentiation formula becomes

$$\frac{dy}{dx} = e^x \qquad\qquad \text{When } y = e^x$$

Out of all the possible choices for bases of logarithmic and exponential functions, the simplest derivative formulas occur when the base is e.

Implicit differentiation of logarithmic functions gives a procedure for simplifying the derivative of products and quotients.

EXAMPLE 34   Find dy/dx if

$$y = x^3 \left(\frac{x^2 + 1}{\sqrt{x - 1}} \right)$$

SOLUTION   Before taking the derivative we take the natural logarithm of both sides of the equation. Then we find dy/dx implicitly.

$$y = x^3 \left(\frac{x^2 + 1}{\sqrt{x - 1}} \right)$$

$$\ln y = \ln \left[x^3 \left(\frac{x^2 + 1}{\sqrt{x - 1}} \right) \right]$$

$$= \ln x^3 + \ln(x^2 + 1) - \ln \sqrt{x - 1}$$

$$= 3 \ln x + \ln(x^2 + 1) - \frac{1}{2} \ln(x - 1)$$

$$\frac{d}{dx} \ln y = \frac{d}{dx}(3 \ln x) + \frac{d}{dx}[\ln(x^2 + 1)] - \frac{d}{dx}\left[\frac{1}{2} \ln(x - 1) \right]$$

$$\frac{y'}{y} = \frac{3}{x} + \frac{2x}{x^2 + 1} - \frac{\frac{1}{2}}{x - 1}$$

$$y' = y \left[\frac{3}{x} + \frac{2x}{x^2 + 1} - \frac{\frac{1}{2}}{x - 1} \right] \qquad\qquad \text{Solve for } y'$$

$$= x^3 \left(\frac{x^2 + 1}{\sqrt{x - 1}} \right) \left[\frac{3}{x} + \frac{2x}{x^2 + 1} - \frac{\frac{1}{2}}{x - 1} \right]$$

SUMMARY

In this section we introduced the following procedure for differentiating implicitly.

1. Differentiate both sides of the equation with respect to the independent variable.
2. If y is the dependent variable, use the chain rule for terms that involve y to a power.
3. Solve for the derivative, y', in terms of the other variables.

Exercise Set 4.5

Find dy/dx by implicit differentiation. Solve for the explicit function $y = f(x)$, and then differentiate to check your results.

1. $y - 4x^2 + 3x = 0$
2. $x - y + 3x^3 = 0$
3. $x + 2y + 3x^2 = 0$
4. $3x - 2y + 5x^3 = 0$
5. $4x^2 - 3x + 7y = 2$
6. $9y - 3x + 4x^3 - 2 = 0$
7. $x^2 - 3x + 2 + 5y = 0$
8. $x^2 + 4x - 3y = 7$
9. $x^2 - 3x^3 + x - 2y = 0$
10. $2x^2 - 4x - 7y = 5$

Find dy/dx by implicit differentiation.

11. $3xy - 4x^2 = 0$
12. $4xy + 5x^2 = 0$
13. $2x^2y - 3x + 5 = 0$
14. $3x^2y + 4x^2 - 2 = 0$
15. $y + 3xy - 4 = 0$
16. $y + 2xy - 3x^3 = 0$
17. $y + 2x^2y + 3x = 0$
18. $y + 3x^2y - 4x = 0$
19. $y - 3x^2y - 2x = 0$
20. $y - 4x^2y + 5x = 0$

Find dy/dx by implicit differentiation.

21. $y^2 + y - 3x = 0$
22. $y^3 + y + 4x = 0$
23. $3y^2 - y + 4x^2 = 0$
24. $4y^2 - 2y + 3x = 0$
25. $y^2 + xy - 4x = 5$
26. $y^2 + 3xy - x^3 = 4$
27. $3xy^2 - 2y - 3 = 0$
28. $4x^2y^2 - 3y + 6 = 0$
29. $2xy^2 - 3xy + x^3 = 3$
30. $3xy^2 - 2xy + 4x^2 = 0$

Find the equation(s) of the tangent(s) to the graphs at the indicated value of x.

31. $3xy - x - 2 = 0$ at $x = 1$

32. $y^2 + 2y - x = 0$ at $x = 0$
33. $y + 3xy - 7 = 0$ at $x = 2$
34. $y^2 - 2xy - 8 = 0$ at $x = -1$
35. $x^3y + 3x^2 + 4 = 0$ at $x = -2$

Find dy/dx and the value of the slope of the graph at the indicated point.

36. $(1 + y)^3 + y = 2x + 7$ at $(1, 1)$
37. $(y - 2)^2 + x = y$ at $(2, 3)$
38. $(x + y)^2 + 3y = -3$ at $(1, -1)$
39. $(2x + y)^2 + 2x = -1$ at $(-1, 3)$
40. $(x + 2y)^2 - x^2 + 8 = 0$ at $(3, -1)$
41. $xy = 6$ at $(3, 2)$

Find dy/dx for the following exponential functions.

42. $y = 4^{x^2}$
43. $y = 10^{3x+4}$

Find the derivatives of the following functions by first taking the ln of both sides.

44. $y = \dfrac{x^3(x^2 + 1)}{x + 3}$
45. $y = \dfrac{\sqrt{x + 1}}{x^2}$
46. $y = \dfrac{(3x + 2)^{1/3}}{(2x - 4)^2}$
47. $y = \dfrac{3x^2(2x + 3)^{1/3}}{(4x^3 - 3x + 1)^2}$

Find y″ for the following and leave the answer in terms of y′.

48. $xy + x^2 + y^2 = 4$
49. $x^3 + y^3 = xy$

50. Separate $3x^2 + 4y^2 = 16$ into two explicit functions and draw both with a common window. Two of the following points are on one graph and the other two are on the second graph: $(2, 1)$, $(-2, 1)$, $(2, -1)$, and $(-2, -1)$. With your calculator, find slopes of

the graphs at the appropriate points. Draw the tangent lines. Then find the derivative implicitly and find the slopes of the tangent lines at the points. Is there agreement?

51. Follow the instructions in Exercise 50 for

$$5x^2 - 4y^2 = 4$$

at $(2, 2)$, $(-2, 2)$, $(2, -2)$, and $(-2, -2)$.

52. **Exam.** What is the slope of the line tangent to the curve $y^3 - x^2y + 6 = 0$ at the point $(1, -2)$?

(a) $\dfrac{-2}{5}$ (b) $\dfrac{-4}{11}$ (c) $\dfrac{4}{11}$

(d) $\dfrac{11}{4}$ (e) 8

53. **Exam.** $\dfrac{d}{dx}(2^{-x^3}) = ?$

(a) $-3x^2 \ln 2$ (b) $-6x^{-2x^3}$
(c) $-3x^2 \cdot 2^{-x^3} \ln 2$ (d) $6x^{2-x^3}$
(e) $6x^{2-x^3} \ln 2$

Applications (Business and Economics)

54. **Instantaneous Change in Sales.** Suppose that a company's sales S, in hundreds of thousands of dollars, is related to the amount x, in thousands of dollars spent on training, by

$$Sx + S = 900 + 40x$$

(a) Find dS/dx, the instantaneous rate of change of S.
(b) Find

$$\left. \frac{dS}{dx} \right|_{(1,470)}$$

55. **Instantaneous Demand.** If x is the number of

items that can be sold at a price of p dollars in the demand equation

$$x^3 + p^3 = 1200$$

find dp/dx.

56. Rework Exercise 55 for the demand equation

$$p^3 - 3p^2 - x + 300 = 0$$

Applications (Social and Life Sciences)

57. **Pollution.** Suppose that pollution P, in parts per million, x yards away from the source, is given by

$$P + 2xP + x^2P = 600$$

Find the instantaneous rate of pollution 10 yards away.

58. **Learning Function.** A learning function L is given in terms of t hours as

$$L^2 - 256t = 0$$

(a) Find the instantaneous rate of learning.
(b) Find the instantaneous rate of learning at the end of 1 hour.
(c) Find the instantaneous rate of learning at the end of 9 hours.

Review Exercises

59. If $y = \ln(x^2 + 3)^2 \sqrt{5x + 1}$, find dy/dx.
60. Using logarithms, find dy/dx for

$$y = \frac{\sqrt[3]{3x + 2}}{\sqrt{5x^2 + 1}}$$

4.6 DERIVATIVES AND RELATED RATES (Optional)

OVERVIEW   The rate of increase of the radius of a balloon is related to the rate of pumping helium into the balloon. If the bottom of a ladder is pulled from the wall, the rate of descent of the top is related to the rate at which the bottom is pulled away. Production changes with time; that is, production is a function of time. Cost, revenue, and profit are all functions of time. The rate of change of cost with respect to time is related to the rate of change of production with respect to time. In this section we

- Define related rates
- Outline a procedure for finding related rates
- Apply related rates to application problems

Suppose that y is a function of x, say $y = f(x)$, and x and y vary with time t. That is, both x and y are functions of time t. The chain rule gives

$$\frac{dy}{dt} = \frac{dy}{dx} \cdot \frac{dx}{dt}$$

Thus the rate of change of y with respect to time is related to the rate of change of x with respect to t. That is, if $y = f(x)$ and both y and x are functions of t, by the chain rule we have

$$\frac{dy}{dt} = \frac{d}{dx} [f(x)] \cdot \frac{dx}{dt}$$

Now let's use the chain rule to solve problems.

EXAMPLE 35   If $y = 3x^2$, find dy/dt when $dx/dt = \frac{1}{2}$ and $x = 2$.

SOLUTION

$$\frac{dy}{dt} = 6x \frac{dx}{dt} \qquad\qquad \frac{d}{dt}(3x^2) = \frac{d}{dx}(3x^2)\frac{dx}{dt}$$

$$= 6 \cdot 2 \cdot \frac{1}{2} = 6 \qquad x = 2; \frac{dx}{dt} = \frac{1}{2}$$

■

Often, two or more variables may be differentiable functions of another variable, such as time, and yet the explicit function may not be given. Suppose that x and y are related by the equation $x^2 + y^2 = 36$ and that both x and y are functions of t. If we differentiate both sides implicitly with respect to t, we obtain

$$\frac{d}{dt}(x^2 + y^2) = \frac{d}{dt}(36)$$

$$2x \frac{dx}{dt} + 2y \frac{dy}{dt} = 0 \qquad\qquad x \text{ and } y \text{ are functions of } t.$$

Practice Problem 1 Given $y = x^2 + 2x$, find dy/dt in terms of dx/dt where x and y are functions of t.

ANSWER $\dfrac{dy}{dt} = (2x + 2)\dfrac{dx}{dt}$

EXAMPLE 36   Given $x^2 + y^2 = 169$ and $dy/dt = 2$, find dx/dt at (5, 12), where x and y are functions of t.

SOLUTION The equation relating the variables is $x^2 + y^2 = 169$. The rate dy/dt is known, and the rate dx/dt is to be found. Differentiate both sides of the given equation with respect to t.

$$\frac{d}{dt}(x^2 + y^2) = \frac{d}{dt}(169) \qquad \text{Implicit differentiation}$$

$$2x\frac{dx}{dt} + 2y\frac{dy}{dt} = 0$$

$$2(5)\frac{dx}{dt} + 2(12)(2) = 0 \qquad\qquad x = 5, y = 12, \frac{dy}{dt} = 2$$

$$10\frac{dx}{dt} + 48 = 0$$

$$\frac{dx}{dt} = -\frac{48}{10} = -\frac{24}{5} \qquad\qquad\qquad ■$$

Practice Problem 2 Given $xy + y^2 = 3$, find dy/dt at $(2, 1)$ where $dx/dt = 5$.

ANSWER $\dfrac{dy}{dt} = -\dfrac{5}{4}$

The procedure for solving application problems involving related rates is outlined as follows.

Procedure for Solving Related Rates Problems

1. First read the problem and draw a picture if possible.
2. Carefully identify all the variables involved. This step is one of the most important steps in the procedure and is the one that is most often neglected or only partly done.
3. Relate the variables by some equation that holds generally and not just at some particular time. A sketch will often help you find this equation.
4. List all the rates involved in the problem, those that are known and those that are to be found.
5. If the variables are functions of time, differentiate, with respect to time, both sides of the equation relating the variables.
6. Substitute the values of the known rates and the known variables at the given time, and solve for the unknown rate.

EXAMPLE 37 Suppose that the radius of a circular oil slick is increasing at the rate of 12 yards per hour. Find the rate at which the area is increasing when $r = 40$ yards.

SOLUTION   Let r be the radius and A be the area. We know that the area of a circle is $A = \pi r^2$. We also know that $dr/dt = 12$ yd/hr and dA/dt is unknown. Then

$$\frac{dA}{dt} = \frac{d}{dt}(\pi r^2) = \frac{d}{dr}(\pi r^2)\frac{dr}{dt} \qquad \text{Chain rule}$$

$$= 2\pi r \cdot \frac{dr}{dt}$$

$$= 2\pi(40 \text{ yd}) \cdot (12 \text{ yd/hr}) \qquad r = 40; \frac{dr}{dt} = 12$$

$$= 960\pi \text{ yd}^2/\text{hr}$$

■

EXAMPLE 38   A toy manufacturer has found that cost, revenue, and profit functions can be expressed as functions of production. If x is the number of toys produced in a week and C, R, and P represent cost, revenue, and profit, respectively, then

$$C(x) = 6000 + 2x, \qquad R(x) = 20x - \frac{x^2}{2000}, \qquad P(x) = R(x) - C(x)$$

Suppose that production is increasing at the rate of 200 toys per week from a production level of 1000 toys. Find the rate of increase in   (a) cost,   (b) revenue, and   (c) profit.

SOLUTION   Since production is changing with respect to time, production must be a function of time. Hence, cost, revenue, and profit are all functions of time. To find the rate of increase, we differentiate each with respect to time.

(a)   Cost:

$$C(x) = 6000 + 2x$$

$$\frac{dC}{dt} = \frac{d}{dt}(6000) + \frac{d}{dt}(2x) \qquad \text{Implicit differentiation}$$

$$= 0 + 2\frac{dx}{dt}$$

$$\frac{dC}{dt} = 2(200) = 400 \qquad \frac{dx}{dt} = 200$$

Cost is increasing at a rate of \$400 per week.

(b)   Revenue:

$$R = 20x - \frac{x^2}{2000}$$

$$\frac{dR}{dt} = \frac{d}{dt}(20x) - \frac{d}{dt}\left(\frac{x^2}{2000}\right) \qquad \text{Implicit differentiation}$$

$$= 20\frac{dx}{dt} - \left(\frac{x}{1000}\right)\frac{dx}{dt} \qquad \text{Chain rule}$$

$$= 20(200) - \frac{1000}{1000}(200) \qquad x = 1000;\ \frac{dx}{dt} = 200$$

$$= 3800$$

Revenue is increasing at a rate of $3800 per week.

(c)   Profit:

$$P = R - C$$

$$\frac{dP}{dt} = \frac{dR}{dt} - \frac{dC}{dt}$$

$$= 3800 - 400$$

$$= 3400$$

Profit is increasing at a rate of $3400 per week.

EXAMPLE 39   The amount of time A, in minutes, required to perform an operation on an assembly line is given as a function of the number of trials x by the equation

$$A = 9 + 9x^{-1/2}$$

Find dA/dt if $dx/dt = 4$ when $x = 25$.

SOLUTION

$$A = 9 + 9x^{-1/2}$$

$$\frac{dA}{dt} = \frac{d}{dt}(9) + \frac{d}{dt}(9x^{-1/2}) \qquad \text{Implicit differentiation}$$

$$= 0 + 9\frac{d}{dt}(x^{-1/2})$$

$$= 9\left(\frac{-1}{2}\right)x^{-3/2}\frac{dx}{dt} \qquad \text{Chain rule}$$

$$= \left(\frac{-9}{2}\right) \cdot (25)^{-3/2} \cdot (4) \qquad x = 25,\ \frac{dx}{dt} = 4$$

$$= \frac{-18}{125}$$

SUMMARY

The chain rule is very important in this section. If $y = f(x)$ and both x and y are functions of t, then

$$\frac{dy}{dt} = \frac{d}{dx} f(x) \cdot \frac{dx}{dt}$$

If x and y are functions of t, then

$$\frac{d}{dt}(x^2) = 2x \cdot \frac{dx}{dt} \quad \text{and} \quad \frac{d}{dt}(y^3) = 3y^2 \cdot \frac{dy}{dt}$$

Exercise Set 4.6

For the following equations assume that x and y are functions of t. Find dy/dt given that dx/dt = 3 when x = 2.

1. $x + y = 3$
2. $2x + y = 5$
3. $3x - 2y = 4$
4. $4x - 3y = 2$
5. $y - 2\sqrt{x} = 0$
6. $x^2 - 2y = 0$
7. $y - 3\sqrt{x} = 0$
8. $y - 2\sqrt[3]{x} = 0$
9. $y - x^2 = 4$
10. $y - 2x^2 + x = 0$
11. $xy = 4$
12. $3xy = 1$
13. $x + xy = 6$
14. $2x - xy = 4$
15. $y + xy = 5$
16. $y - 2xy = 3$
17. $y - xy + x^2 = 5$
18. $y + 2xy - x^2 = 3$
19. $y + x^2y - 3x = 2$
20. $y - x^2y - 4x = 4$

21. A particle travels along the curve $y = \sqrt{x}$ so that $dx/dt = 4$ centimeters per second. How fast is y changing when (a) $x = 2$ and (b) $x = 9$?

22. The radius of a circle is increasing at the rate of 2 centimeters per second. At what rate is the area of the circle increasing when the radius is (a) 10 centimeters and (b) 15 centimeters?

23. The length of a rectangle is 4 times the width. If the length is increasing at the rate of 8 centimeters per second, how fast is the area changing when the width is (a) 1 centimeter and (b) 2 centimeters?

24. The edges of a cube are increasing at a rate of 2 centimeters per second. At what rate is the volume changing when the edge is 3 centimeters?

25. At what rate is the surface area of the cube in Exercise 24 increasing when the edge is 3 centimeters?

Applications (Business and Economics)

26. **Cost, Revenue, and Profit Rates.** Suppose that the revenue from the sale of x stereos is given by $R(x) = 500x - x^2$ and the cost is given by $C(x) = 2000 + 30x$. If the company is selling six stereos per day, then when 50 stereos are sold find each of the following.
    (a) Rate of change in revenue.
    (b) Rate of change in cost.
    (c) Rate of change in profit.

27. **Demand Rates.** Suppose that a demand equation is $p^2 + p + 3x = 39$. Find dp/dt if $dx/dt = 3$ when $p = 2$.

Applications (Social and Life Sciences)

28. **Pollution.** Oil is leaking from a tanker and has formed a circular oil slick 3 centimeters thick. To estimate the rate of leakage, the radius was measured and found to be 100 meters and increasing at the rate of 5 centimeters per minute. Assume that the depth is constant.
    (a) Use these results to find the rate of leakage.
    (b) After finding the rate of leakage, assume that the rate of leakage is constant and find how fast the radius is changing 5 hours after the leakage began.

29. **Learning Function.** A learning function is given as required time R after x tries,

$$R(x) = 7 + 7x^{-1/2}$$

Find dR/dt if $dx/dt = 5$ when $x = 49$.

30. **Medicine.** The cross section area of a tumor is given by

$$A = \pi r^2$$

The radius is increasing at a rate of 0.4 millimeters per day. At a radius of 20 millimeters, how fast is the area of the tumor increasing? If

$$V = \frac{4}{3}\pi r^3$$

how fast is the volume increasing per day?

Review Exercises

Find dy/dx at $x = 1$, $y = 3$.

31. $xy = 3$
32. $x^2 + 3xy = 10$
33. $2x^2 + y^2 = 11$

Find dy/dx.

34. $x^2y + y^2 = 4$
35. $2xy^2 + y = 7$
36. $4x^3 + 5xy + 7y^3 = 2$
37. $ye^x + x = y$
38. $ye^{x^2} + x^4 = 4$
39. $x \ln(x^2 + 2) = xy$
40. $xy - \ln x = 4$

Chapter Review

Review the following concepts to ensure that you understand and can use them.

Logarithmic Functions

Graphs $\log_b b = 1$ and $\log_b 1 = 0$

$\log x$ $\log_b b^x = x$

$\ln x$

Implicit derivatives Related rates

Make sure you understand the following relations.

$\log_b x = y$ if and only if $b^y = x$ $\log_b M^N = N \log_b M$

$\log_b M \cdot N = \log_b M + \log_b N$ $\log_b x = \log_b y$ if and only if $x = y$

$\log_b \dfrac{M}{N} = \log_b M - \log_b N$

Can you use each of the following formulas?

$$\frac{d}{dx}(e^u) = e^u \frac{du}{dx}, \quad u > 0 \qquad\qquad \frac{d}{dx}(\ln u) = \frac{1}{u} \cdot \frac{du}{dx}, \quad u > 0$$

Chapter Test

1. If $3xy - 4x^2 = 0$, find dy/dx.
2. If $3xy - 2x = 4$, find dy/dt if $dx/dt = 4$ when $x = 1$.
3. If $w = 4e^{z^2 + 5}$, find dw/dz.

4. If $r = 6 \ln(s^2 + 4s)$, find dr/ds.

5. If $y = x \ln x^2$, find dy/dx.

6. (a) With your graphing calculator, find an approximation to the maximum value of the function

$$y = \frac{\ln x}{x}$$

(b) Using calculus, find the exact value of the maximum value of

$$y = \frac{\ln x}{x}$$

7. If $y = \ln(1 + e^x)$, find y'.

8. (a) Using your graphing calculator, for $y = xe^{-x}$ find each of the following and round to two decimal places if necessary.
   (i) Find the intervals over which y is increasing or where it is decreasing.
   (ii) Find the intervals where y is concave up and where it is concave down.
   (iii) Sketch the graph.
   (b) Repeat part (a) using calculus techniques.

9. If $y = x^2 \ln x + xe^y$, find y'.

10. The amount of money in a savings account is $A = 10,000e^{0.08t}$, where t is the time measured in years. What is the rate of increase of the account at the end of 5 years?

11. If $x^2 e^y + \ln y = \ln(x + 1)$, find y'.

12. Make a calculator sketch of $y = x^2 e^x$.

13. For Problem 12, find the relative maxima and minima.

14. For Problem 12, discuss where the curve is increasing or decreasing.

15. For Problem 12, discuss concavity.

CHAPTER 5

Integral Calculus

A s indicated earlier, calculus is divided into two parts, differential calculus, discussed in the three preceding chapters, and integral calculus, which we introduce in this chapter. We considered the derivative of a function as a rate of change and found many ways to apply that concept. In integral calculus the rate of change is known and our problem is to find the function. Again, many interesting applications arise from this type of problem. For example:

A businessperson knows the rate of change of profit in terms of items produced and would like to know the profit for a given production.

A biologist introduces a bactericide into a culture and knows the rate of change of the number of bacteria present. The biologist would like to know the number of bacteria present at a given time.

This chapter begins with a discussion of antiderivative problems in which information about the derivative of a function is given and the function is unknown. The set of antiderivatives is called the indefinite integral of a function. Another concept, the definite integral of a function, seems quite different from the indefinite integral; however, a remarkable theorem, the fundamental theorem of calculus, will show how the two are intimately related.

5.1   THE ANTIDERIVATIVE (INDEFINITE INTEGRAL)

OVERVIEW   Many of the operations we have studied so far have inverse operations. For example, subtraction is the inverse operation for addition, division is the inverse operation for multiplication, and taking the square root is the inverse operation for squaring ($x \geq 0$). Inverse functions exist for many functions. Thus, it seems natural to consider an inverse operation for differentiation. If the derivative of a function is known, can the function be found? We answer this question by introducing the concept of **antiderivative**. In this section we learn

- The meaning of antiderivative
- Ways to find antiderivatives
- Relationships between indefinite integrals and antiderivatives
- Notation for indefinite integrals

We learned in the preceding chapters that the derivative of x^2 is $2x$. Now let's perform an inverse operation of $f(x) = 2x$, called **antidifferentiation**. Instead of going forward to find the derivative

$$x^2 \rightarrow 2x \qquad \text{Differentiation}$$

we will work backward to find a function whose derivative is $2x$.

$$x^2 \leftarrow 2x \qquad \text{Antidifferentiation}$$

That is, an antiderivative of $2x$ is x^2.

ANTIDERIVATIVE

A function $F(x)$ is an **antiderivative** of $f(x)$ on the interval (a, b) if

$$\frac{d}{dx} F(x) = f(x)$$

for all x in (a, b).

$F(x) = x^2 + 4$
$F(x) = x^2$
$F(x) = x^2 - 2$

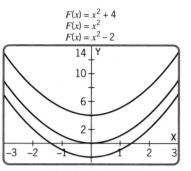

X: [-3, 3] and Y: [-3, 14]
A few antiderivatives of $f(x) = 2x$

Figure 1

$y = x^3 - 4x + 4$
$y = x^3 - 4x$
$y = x^3 - 4x - 3$

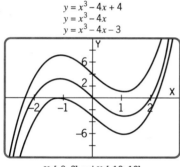

X: [-3, 3] and Y: [-10, 10]

Figure 2

In the definition above, we could have also used the notation $D_x[F(x)] = F'(x) = f(x)$.

Did you note the word *an* in the preceding definition? This word implies that there is more than one antiderivative of a function. As we shall see, there is a set of functions that are all antiderivatives of a given function.

EXAMPLE 1   Let's continue to look at the antiderivatives of $2x$. Note that

$$\frac{d}{dx}x^2 = 2x, \qquad \frac{d}{dx}(x^2 + 4) = 2x, \qquad \frac{d}{dx}(x^2 - 2) = 2x$$

Thus, x^2, $x^2 + 4$, and $x^2 - 2$ are all antiderivatives of $2x$. We can see that the only difference in each function is the constant. This is demonstrated in Figure 1. Since the only difference in the antiderivatives is the constant, we can use the notation $F(x) = x^2 + C$, where C is any constant, as an antiderivative of x, to indicate the entire set of antiderivatives of $f(x) = 2x$.   ■

Practice Problem 1 With a graphing calculator, draw the graphs of $y = x^3 - 4x$, $y = x^3 - 4x + 4$, and $y = x^3 - 4x - 3$. Then take the derivative of each function. The functions represented by the three graphs are antiderivatives of what function?

ANSWER   The graphs are shown in Figure 2. $y' = 3x^2 - 4$. The graphs are antiderivatives of $3x^2 - 4$.

In general, if $F(x)$ is an antiderivative of $f(x)$, then each of the functions $F(x) + C$ is also an antiderivative, where C is any real number. It can be shown that there are no other antiderivatives, so that this is the entire set. This property can be stated as follows.

PROPERTY OF ANTIDERIVATIVES

If the derivatives of two functions are equal, then the functions differ at most by a constant.

The preceding property suggests that we need a notation or a symbol to represent all antiderivatives of $f(x)$. Such a symbol is called the **indefinite integral** and is denoted by

$$\int f(x)\, dx$$

where $\int$ is called the **integral sign**, $f(x)$ is the **integrand**, and dx is the **differential**.

The importance of dx will be discussed in Section 5.3. At this time we consider dx simply as a part of the indefinite integral, which can be used to designate the independent variable.

INDEFINITE INTEGRAL

Let $F(x)$ be an antiderivative of the function $f(x)$. The **indefinite integral** of $f(x)$ is defined to be

$$\int f(x)\, dx = F(x) + C$$

The $\int$ and dx indicate that antidifferentiation is to be performed with respect to x on the function $f(x)$, and C is called the constant of integration. Hence, $F'(x) = f(x)$.

Before we introduce the rules of integration, let's look at several examples and use our knowledge of derivatives to find some antiderivatives.

EXAMPLE 2 Find the indefinite integral $\int 5x^4\, dx$.

SOLUTION Since the derivative of x^5 is $5x^4$, x^5 is an antiderivative of $5x^4$; hence, $x^5 + C$ is the indefinite integral of $5x^4$.

$$\int 5x^4\, dx = x^5 + C \quad \text{because} \quad D_x[x^5 + C] = 5x^4$$

EXAMPLE 3 Find $\int 2x^3\, dx$.

SOLUTION The derivative of x^4 is $4x^3$ and not $2x^3$. Since $2x^3$ is one-half of $4x^3$, we should try $\frac{1}{2}x^4$ as an antiderivative of $2x^3$. The derivative of $\frac{1}{2}x^4$ is $2x^3$; therefore, $\frac{1}{2}x^4 + C$ is the indefinite integral of $2x^3$.

$$\int 2x^3\, dx = \frac{1}{2}x^4 + C \quad \text{because} \quad D_x\left[\frac{x^4}{2} + C\right] = 2x^3$$

EXAMPLE 4 Find the indefinite integral $\int 3\, dx$.

SOLUTION The derivative of $3x$ is 3. Thus $3x$ is an antiderivative of 3. The indefinite integral of 3 is $3x + C$.

$$\int 3 \, dx = 3x + C \quad \text{because} \quad D_x[3x + C] = 3$$

The preceding examples should suggest the following indefinite integral formulas.

INDEFINITE INTEGRAL FORMULAS

For k and C constants:

1. $\displaystyle\int k \, dx = kx + C$

2. $\displaystyle\int x^n \, dx = \frac{x^{n+1}}{n+1} + C, \qquad n \neq -1$ Power rule for antiderivatives

To verify the power rule for antiderivatives, all we need to do is take the derivative of the right side and show that this is the integrand. Since

$$\frac{d\left(\dfrac{x^{n+1}}{n+1}\right)}{dx} = \left(\frac{1}{n+1}\right)\frac{d(x^{n+1})}{dx} \qquad \frac{1}{n+1} \text{ is a constant}$$

$$= \left(\frac{1}{n+1}\right)[(n+1)x^{n+1-1}]$$

$$= x^n$$

then

$$\int x^n \, dx = \frac{x^{n+1}}{n+1} + C, \qquad n \neq -1$$

A special formula for $n = -1$ is discussed in the next section. The present formula states that to find the indefinite integral of x^n with respect to x, you must increase the exponent of x by 1 and divide by the new exponent. (You may have noticed that this is the reverse of the power rule for differentiation. When we take the derivative of a term x^n, we *subtract* 1 from n and *multiply*. When we integrate, we *add* 1 to n and *divide*.)

EXAMPLE 5

$$\int x^4 \, dx = \frac{x^5}{5} + C \qquad n = 4; n + 1 = 5$$

EXAMPLE 6

$$\int x^{2/3} \, dx = \frac{x^{(2/3)+1}}{\frac{5}{3}} + C = \frac{3}{5}x^{5/3} + C \qquad n = \frac{2}{3}; n + 1 = \frac{5}{3}$$

EXAMPLE 7   Find $\int \dfrac{1}{x^2}\,dx$.

SOLUTION   First, we rewrite the function so that we can apply the power rule.

$$\int \frac{1}{x^2}\,dx = \int x^{-2}\,dx \qquad \text{Now apply the power rule.}$$

$$= \frac{x^{-2+1}}{-1} + C \qquad n = -2;\ n+1 = -1$$

$$= \frac{x^{-1}}{-1} + C = -\frac{1}{x} + C \qquad \blacksquare$$

The properties of derivatives allow us to develop and use the following properties for antiderivatives.

PROPERTIES OF INDEFINITE INTEGRALS

If both f and g have antiderivatives and k is any constant, then

1.   Constant rule: $\displaystyle \int kf(x)\,dx = k \int f(x)\,dx$

2.   Sum rule: $\displaystyle \int [f(x) + g(x)]\,dx = \int f(x)\,dx + \int g(x)\,dx$

3.   Difference rule: $\displaystyle \int [f(x) - g(x)]\,dx = \int f(x)\,dx - \int g(x)\,dx$

EXAMPLE 8   Find $\int 10x^3\,dx$.

SOLUTION

$$\int 10x^3\,dx = 10 \int x^3\,dx = 10\left(\frac{x^4}{4}\right) + C = \frac{5x^4}{2} + C \qquad \text{Property 1} \quad \blacksquare$$

EXAMPLE 9   Find $\int 30x^5\,dx$.

SOLUTION

$$\int 30x^5\,dx = 30 \int x^5\,dx = 30\left(\frac{x^6}{6}\right) + C = 5x^6 + C \qquad \text{Property 1} \quad \blacksquare$$

NOTE: It is important to remember that property 1 states that a *constant* can be factored from the integrand; a *variable* cannot be factored in this manner.

COMMON ERROR   Students sometimes try to treat a variable like a constant.

Correct	Incorrect

$$\int 5\sqrt{x}\, dx = 5 \int \sqrt{x}\, dx \qquad \int x\sqrt{x}\, dx = x \int \sqrt{x}\, dx$$

EXAMPLE 10

$$\int (x^3 + x^{1/2})\, dx = \int x^3\, dx + \int x^{1/2}\, dx = \frac{x^4}{4} + \frac{2}{3} x^{3/2} + C \qquad \text{Property 2}$$

EXAMPLE 11

$$\int (x^3 + x^2 + 2x + 3)\, dx = \int x^3\, dx + \int (x^2 + 2x + 3)\, dx \qquad \text{Property 2}$$

$$= \int x^3\, dx + \int x^2\, dx + \int (2x + 3)\, dx \qquad \text{Property 2}$$

$$= \int x^3\, dx + \int x^2\, dx + \int 2x\, dx + \int 3\, dx \qquad \text{Property 2}$$

$$= \frac{x^4}{4} + \frac{x^3}{3} + x^2 + 3x + C$$

The C's for the four integrands are combined into one C, since the constant is arbitrary.

This example shows that by repeated application of the sum formula, the indefinite integral of a function that is the sum of a finite number of functions can be obtained by taking the sum of the indefinite integrals of the functions.

EXAMPLE 12

$$\int (3x^4 - 2x^3 + x - 3)\, dx = \int 3x^4\, dx + \int (-2)x^3\, dx + \int x\, dx + \int (-3)\, dx$$

$$= 3 \int x^4\, dx - 2 \int x^3\, dx + \int x\, dx + \int (-3)\, dx$$

$$= \frac{3x^5}{5} - \frac{x^4}{2} + \frac{x^2}{2} - 3x + C$$

Practice Problem 2 Find $\int (3x^2 + 4x)\, dx$.

ANSWER $x^3 + 2x^2 + C$

Some algebraic manipulation or simplification will often be necessary before the integral can be found. This is demonstrated in the following examples.

EXAMPLE 13   Find $\int \dfrac{x^3 + 2x^2 + 3}{x^2}\, dx.$

SOLUTION   Before the power rule or the other rules can be used, the expression must first be simplified. We rewrite the expression so that each term can be reduced.

$$\int \frac{x^3 + 2x^2 + 3}{x^2}\, dx = \int \frac{x^3}{x^2}\, dx + \int \frac{2x^2}{x^2}\, dx + \int \frac{3}{x^2}\, dx$$

$$= \int x\, dx + \int 2\, dx + \int 3x^{-2}\, dx$$

$$= \frac{x^2}{2} + 2x + \frac{3x^{-1}}{-1} + C \qquad\qquad \text{Only one constant is needed.}$$

$$= \frac{x^2}{2} + 2x - \frac{3}{x} + C$$

Practice Problem 3   Find $\int \dfrac{x^5 - 2}{x^2}\, dx.$

ANSWER $\dfrac{x^4}{4} + \dfrac{2}{x} + C$

EXAMPLE 14   Find each of the following.

(a) $\int x\,(x^2 + 2x + 1)\, dx$

(b) $\int (x + 1)(2x - 1)\, dx$

(c) $\int \sqrt{x}\,(x^2 + 3)\, dx$

SOLUTION   Each integral is multiplied before integrating.

(a) $\displaystyle\int x(x^2 + 2x + 1)\, dx = \int (x^3 + 2x^2 + x)\, dx = \frac{x^4}{4} + \frac{2x^3}{3} + \frac{x^2}{2} + C$

(b) $\displaystyle\int (x + 1)(2x - 1)\, dx = \int (2x^2 + x - 1)\, dx = \frac{2x^3}{3} + \frac{x^2}{2} - x + C$

(c) $\displaystyle\int \sqrt{x}\,(x^2 + 3)\, dx = \int x^{1/2}(x^2 + 3)\, dx = \int (x^{5/2} + 3x^{1/2})\, dx$

$$= \frac{x^{7/2}}{\frac{7}{2}} + \frac{3x^{3/2}}{\frac{3}{2}} + C = \frac{2}{7}x^{7/2} + 2x^{3/2} + C$$

Practice Problem 4 Find each of the following.

(a) $\displaystyle\int x^2(x^3 - 2x)\,dx$ (b) $\displaystyle\int (x + 2)(x - 1)\,dx$

(c) $\displaystyle\int \sqrt{x}(3 - 2x)\,dx$

ANSWER

(a) $\dfrac{x^6}{6} - \dfrac{x^4}{2} + C$

(b) $\dfrac{x^3}{3} + \dfrac{x^2}{2} - 2x + C$

(c) $2x^{3/2} - \dfrac{4}{5}x^{5/2} + C$

COMMON ERROR

Do not try to integrate a product as you would a sum. In the same way that the derivative is not the product of the derivatives, the integral of a product is not the product of the integrals.

Correct

$\displaystyle\int x(x^2 + x)\,dx = \int (x^3 + x^2)\,dx$

Incorrect

$\displaystyle\int x(x^2 + x)\,dx = \int x\,dx \cdot \int (x^2 + x)\,dx$

Finding a Particular Antiderivative

We have observed that there is more than one, in fact there are an infinite number of antiderivatives associated with a function. By requiring that the graph of an antiderivative go through a given point or by requiring the antiderivative $F(x)$ to be equal to some value for a fixed value of x, we can evaluate the constant of integration and attain a particular antiderivative.

EXAMPLE 15   Find the particular function $F(x)$ whose derivative is

$$F'(x) = f(x) = x^2 + 3x + 2$$

and $F(0) = 4$.

SOLUTION   First, we find the set of antiderivatives.

$$F(x) = \int (x^2 + 3x + 2)\,dx$$

$$= \int x^2\,dx + \int 3x\,dx + \int 2\,dx \qquad \text{Property 2}$$

$$= \frac{x^3}{3} + \frac{3x^2}{2} + 2x + C$$

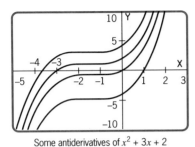

Some antiderivatives of $x^2 + 3x + 2$

(a)

$f(x) = \frac{x^3}{3} + \frac{3x^2}{2} + 2x + 4$

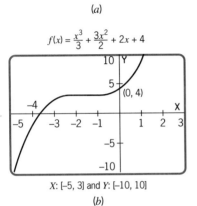

$X: [-5, 3]$ and $Y: [-10, 10]$

(b)

Figure 3

Looking at the graphs in Figure 3(a), we see some of the antiderivatives. However, we can see that only one of the graphs contains the point (0, 4). Let's now find that particular function.

$$F(0) = \frac{0^3}{3} + \frac{3(0)^2}{2} + 2(0) + C = 4 \qquad \text{Substitute } x = 0; F(0) = 4$$

$$C = 4 \qquad \text{Solve for } C.$$

The specific function sought is

$$f(x) = \frac{x^3}{3} + \frac{3x^2}{2} + 2x + 4$$

and is graphed in Figure 3(b).

An **initial condition** is often used to find the constant of integration. In Example 15, the initial condition given was $F(0) = 4$.

EXAMPLE 16 Given $f'(x) = 8x - 6$ and $f(1) = 6$, find $f(4)$.

SOLUTION The function $f(1) = 6$ is the initial condition and is used to find the constant of integration. Since $f'(x) = 8x - 6$, we know that

$$f(x) = \int (8x - 6) \, dx = 4x^2 - 6x + C$$

We were given that $f(1) = 6$; therefore, $f(1) = 4(1)^2 - 6 \cdot 1 + C = 6$, so we have that $C = 8$. Thus,

$$f(x) = 4x^2 - 6x + 8 \quad \text{and} \quad f(4) = 4(4)^2 - 6 \cdot 4 + 8 = 48$$

The graph is shown in Figure 4.

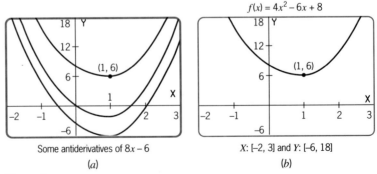

Some antiderivatives of $8x - 6$

(a)

$f(x) = 4x^2 - 6x + 8$

$X: [-2, 3]$ and $Y: [-6, 18]$

(b)

Figure 4

Practice Problem 5 Given $F'(x) = 2x + 3$ and $F(2) = 4$. Find the particular function $F(x)$.

ANSWER $F(x) = x^2 + 3x - 6$

COMMON ERROR

Students often make careless mistakes when working with negative exponents. They subtract instead of add.

Correct

$$\int x^{-4}\, dx = \frac{x^{-3}}{-3} + C = -\frac{1}{3x^3} + C$$

Incorrect

$$\int x^{-4}\, dx = \frac{x^{-5}}{-5} + C = -\frac{1}{5x^5} + C$$

SUMMARY

You must understand the meaning of antiderivative to be successful in the remainder of this chapter. An antiderivative is a function that is found from its derivative. We use the process of antidifferentiation to accomplish this. Before using the constant rule or the general power rule, it is sometimes necessary to rewrite the integrand so that the rule can be applied. You can always check your work by differentiating the antiderivative to make certain your answer is identical to the integrand. Checking your work with differentiation also reinforces your differentiation skills. When you find a particular antiderivative, you are finding only one of a set of antiderivatives all sharing the integrand as their derivative.

Exercise Set 5.1

Find each indefinite integral and check by differentiation.

1. $\int 5\, dx$

2. $\int 3\, dx$

3. $\int \sqrt{3}\, dx$

4. $\int \pi\, dx$

5. $\int -2x\, dx$

6. $\int 7x\, dx$

7. $\int \sqrt{2x}\, dx$

8. $\int \pi x\, dx$

9. $\int 6x^2\, dx$

10. $\int 9x^2\, dx$

11. $\int 3x^3\, dx$

12. $\int 5x^3\, dx$

Find an expression for all the antiderivatives of each of the following derivatives.

13. $\dfrac{dy}{dx} = \dfrac{6}{x^2}$

14. $\dfrac{dy}{du} = 4u^3$

15. $\dfrac{dy}{du} = 16u$

16. $\dfrac{dy}{dt} = 4$

17. $\dfrac{dy}{dt} = 4\sqrt{t}$

18. $\dfrac{dy}{du} = \dfrac{3}{u^2} + 4u$

Find each indefinite integral and check by differentia-
tion.

19. $\int x^{1/2}\, dx$

20. $\int x^{-1/2}\, dx$

21. $\int x^{-3}\, dx$

22. $\int 2x^{1/3}\, dx$

23. $\int 3x^{-3}\, dx$

24. $\int x^{2/3}\, dx$

25. $\int x^{-2}\, dx$

26. $\int 3x^{-2}\, dx$

27. $\int \left(\frac{1}{x^2} - 2x \right) dx$

28. $\int \left(6u^2 - \frac{7}{u^2} \right) du$

29. $\int \left(\frac{4}{\sqrt{u}} - \sqrt{u} \right) du$

30. $\int x^2 \left(x + 2x^2 - \frac{3}{x} \right) dx$

31. $\int \frac{1}{x^3} \left(x^4 + x^3 + \frac{4}{x} \right) dx$

32. $\int (x^2 - 1)(x - 3)\, dx$

Find the particular antiderivative that satisfies each
condition. Use your graphing calculator to graph the
function.

33. $f'(x) = 3x + 2, \quad f(0) = 10$

34. $f'(u) = 3u - 2, \quad f(1) = 7$

35. $\dfrac{dy}{du} = \dfrac{10}{u^2}, \quad y(5) = 2$

36. $\dfrac{dy}{dx} = \dfrac{10}{\sqrt{x}}, \quad y(4) = 2$

37. $C'(x) = 0.2x + 4x^2, \quad C(0) = 10$

38. $C'(x) = x + \dfrac{1}{x^2}, \quad C(4) = 8$

Find each indefinite integral.

39. $\int \sqrt{x}\, (x^2 + 2x - 1)\, dx$

40. $\int \sqrt[3]{x}\, (x^2 - 3x + 2)\, dx$

41. $\int \dfrac{4x^5 - 2x^4 + 3x^2 - 1}{x^2}\, dx$

42. $\int \dfrac{5x^4 - 2x^3 + 2x - 4}{x^3}\, dx$

43. $\int (4x^{1/2} - 3)\, dx$

44. $\int \dfrac{x^3 + 7}{\sqrt{x}}\, dx$

45. $\int \dfrac{x^4 - 7x^3 + x^2 - 3}{x^2}\, dx$

46. $\int \dfrac{x^2 - x}{x^{4/3}}\, dx$

47. $\int \dfrac{2x^3 - 3x}{\sqrt[3]{x}}\, dx$

48. $\int \dfrac{3x^2 - 5}{\sqrt[4]{x}}\, dx$

49. $\int \dfrac{x^2 - 2x + 3}{\sqrt[3]{x}}\, dx$

50. $\int \left(\dfrac{x^{4.3} - x^{2.1}}{x^2} \right) dx$

51. $\int \left(\dfrac{-3}{\sqrt[5]{x^2}} + \dfrac{2}{3\sqrt{x}} \right) dx$

52. $\int \left(\dfrac{2}{\sqrt[4]{x^3}} - \dfrac{1}{2\sqrt{x}} \right) dx$

Applications (Business and Economics)

53. ***Sales Function.*** The rate of sales of an item is

$$\dfrac{dS}{dt} = 8t + 6$$

Find the sales function and the number of sales at $t = 2$ if $S = 0$ when $t = 0$.

54. ***Cost Function.*** The marginal cost for producing x items is given by

$$C'(x) = -0.002x^2 - 0.6x + 100$$

Find the cost function if $C(0) = 60$.

55. ***Profit Function.*** The marginal profit for producing x items is given by

$$P'(x) = 500 - 4x$$

Find the profit function if $P = 0$ when $x = 0$.

56. ***Revenue Function.*** A marginal revenue function is given by

$$R'(x) = 400 - 0.8x$$

Find the revenue function if $R(0) = 0$, and find the revenue for a sale of 300 items.

Applications (Social and Life Sciences)

57. **Bacteria.** After introducing a bactericide into a culture, a biologist gives the rate of change of the number of bacteria present as

$$\frac{dN}{dt} = 60 - 12t$$

If $N(0) = 1200$, find $N(t)$, $N(5)$, and $N(8)$. When will the number of bacteria be 0?

58. **Flu Epidemic.** A city has a flu epidemic. The health department gives the rate of change of the number of people with the flu to be

$$\frac{dN}{dt} = 500t + 10$$

where t is the number of days after the start of the epidemic. If $N(0) = 600$, find a function $N(t)$ for the number of people with the flu in terms of t. Use this function to find the number of people with the flu 20 days after the epidemic began.

59. **Learning Rate.** A learning rate is given by

$$\frac{dL}{dt} = 0.06t - 0.0006t^2$$

where L is the number of words learned and t is time in minutes. Find $L(30)$ and $L(40)$ if $L = 0$ when $t = 0$.

60. **Population.** The change in the population of a certain area is estimated in terms of time t years as

$$\frac{dP}{dt} = 600 + 500\sqrt{t}, \qquad 0 \le t \le 5$$

If the current population is 8000, what will the population be in 4 years?

5.2   INTEGRATION FORMULAS AND MARGINAL ANALYSIS

OVERVIEW   The integration formulas introduced in this section are important. The functions that we study in this section are of the form $y = e^{kx}$ and $y = \ln |x|$. Such functions are useful in studying the growth rate of money supply, the rate of decline or the rate of increase in sales, the rate of growth of organisms in biology, and human population growth. In this section we

- Introduce new integration formulas
- Take a new look at marginal analysis

In the preceding chapter we learned that

$$\frac{d(e^{kx})}{dx} = ke^{kx}$$

The antiderivatives of the expressions on the left side should equal the functions on the right side; that is,

$$\int ke^{kx}\, dx = e^{kx} + C_1$$

$$k \int e^{kx}\, dx = e^{kx} + C_1 \qquad \text{Property 1}$$

$$\int e^{kx}\, dx = \frac{e^{kx}}{k} + C \qquad C = \frac{C_1}{k}$$

When $k = 1$, this formula becomes

$$\int e^x \, dx = e^x + C$$

Also recall that for $x > 0$,

$$\frac{d}{dx}(\ln |x|) = \frac{d(\ln x)}{dx} = \frac{1}{x}$$

If $x < 0$, then $|x| = -x$ or for this case

$$\frac{d(\ln |x|)}{dx} = \frac{d(\ln (-x))}{dx}$$

$$= \frac{1}{-x} \frac{d(-x)}{dx} \qquad \text{Chain rule: } D_x \ln |u| = \frac{1}{u} D_x u$$

$$= \frac{1}{-x} \cdot (-1) = \frac{1}{x} \qquad \frac{d(-x)}{dx} = -1$$

So $\dfrac{d}{dx} \ln |x| = \dfrac{1}{x}$

Therefore $\displaystyle\int \frac{1}{x} \, dx = \ln |x| + C$

These indefinite integrals are stated as follows.

INTEGRATION FORMULAS

1. $\displaystyle\int e^x \, dx = e^x + C$

2. $\displaystyle\int e^{kx} \, dx = \frac{e^{kx}}{k} + C = \frac{1}{k} e^{kx} + C$

3. $\displaystyle\int \frac{dx}{x} = \ln |x| + C$

**COMMON
ERROR**

Correct

$$\int \frac{dx}{x} = \ln |x| + C$$

Incorrect

$$\int \frac{dx}{x} = \int x^{-1} \, dx = \frac{x^{-1+1}}{-1+1} + C = \frac{x^0}{0} + C$$

Division by 0 is undefined.

The following examples utilize these integration formulas along with the properties and formulas from Section 5.1.

EXAMPLE 17   Find $\int 5e^x \, dx$.

SOLUTION

$$\int 5e^x \, dx = 5 \int e^x \, dx = 5e^x + C \qquad \text{Formula 1}$$

EXAMPLE 18   Find $\int 3x^{-1} \, dx$.

SOLUTION

$$\int 3x^{-1} \, dx = 3 \int \frac{dx}{x} = 3 \ln |x| + C \qquad \text{Formula 3}$$

EXAMPLE 19   Find $\int e^{3x} \, dx$.

SOLUTION

$$\int e^{3x} \, dx = \frac{e^{3x}}{3} + C \qquad k = 3 \text{ in Formula 2}$$

EXAMPLE 20   Find $\int \left(\frac{1}{x} - e^{2x} \right) dx$.

SOLUTION

$$\int \left(\frac{1}{x} - e^{2x} \right) dx = \int \frac{dx}{x} - \int e^{2x} \, dx \qquad \text{Property 3 and Formula 2}$$

$$= \ln |x| - \frac{e^{2x}}{2} + C$$

Now let's look at some examples that involve using multiple formulas and properties.

EXAMPLE 21   Find $\int \left(e^{3x} - \frac{1}{x^2} + \frac{3}{x} \right) dx$.

SOLUTION

$$\int \left(e^{3x} - \frac{1}{x^2} + \frac{3}{x} \right) dx = \int e^{3x} \, dx - \int \frac{1}{x^2} \, dx + 3 \int \frac{dx}{x} \qquad \text{Properties 2 and 3}$$

Now we have

$$\int e^{3x} \, dx = \frac{e^{3x}}{3} + C_1 \qquad\qquad \text{Formula 2}$$

$$\int \frac{dx}{x^2} = \int x^{-2} \, dx = \frac{x^{-1}}{-1} + C_2 = \frac{-1}{x} + C_2 \qquad \int x^n \, dx$$

$$3 \int \frac{dx}{x} = 3 \ln |x| + C_3 \qquad\qquad \text{Formula 3}$$

Substituting for these three integrals and combining the three constants into one constant called C yields

$$\int \left(e^{3x} - \frac{1}{x^2} + \frac{3}{x} \right) dx = \frac{e^{3x}}{3} + \frac{1}{x} + 3 \ln |x| + C$$

EXAMPLE 22   Find $\displaystyle\int \frac{dx}{e^{4x}}$.

SOLUTION

$$\int \frac{dx}{e^{4x}} = \int e^{-4x} \, dx = \frac{-1}{4} e^{-4x} + C = \frac{-1}{4e^{4x}} + C \qquad \text{Formula 2}$$

Practice Problem 1 Find each indefinite integral.

(a) $\displaystyle\int \frac{dx}{e^{3x}}$

(b) $\displaystyle\int \left(e^{2x} - \frac{1}{e^x} - \frac{3}{x} \right) dx$

ANSWER

(a) $\dfrac{-1}{3e^{3x}} + C$

(b) $\dfrac{1}{2} e^{2x} + \dfrac{1}{e^x} - 3 \ln |x| + C$

EXAMPLE 23   Find $\displaystyle\int \left(\frac{5xe^{3x} - 4 + x^2}{x} \right) dx$.

SOLUTION   This expression must be rewritten and then simplified.

$$\int \left(\frac{5xe^{3x} - 4 + x^2}{x} \right) dx = \int \left(5e^{3x} - \frac{4}{x} + x \right) dx \qquad \text{Simplify each term.}$$

$$= \int 5e^{3x} \, dx - \int \frac{4}{x} \, dx + \int x \, dx \qquad \text{Property 2}$$

$$= 5 \int e^{3x} \, dx - 4 \int \frac{dx}{x} + \int x \, dx \qquad \text{Property 1}$$

$$= \frac{5e^{3x}}{3} - 4 \ln |x| + \frac{x^2}{2} + C$$

Practice Problem 2 Find each indefinite integral.

(a) $\displaystyle\int \frac{x^2 - 2x + 1}{x} \, dx$

(b) $\displaystyle\int \left(e^{-2x} - \frac{1}{x} \right) dx$

ANSWER

(a) $\dfrac{x^2}{2} - 2x + \ln |x| + C$

(b) $\dfrac{-1}{2e^{2x}} - \ln |x| + C$

A New Look at Marginal Analysis

In the preceding chapters we have seen that cost, average cost, revenue, profit, and productivity functions (sometimes called *total functions*) have derivatives that we call marginal functions. Now that we understand indefinite integrals, we can work backward from marginal functions to the total functions. In practice, economists often obtain marginal functions from empirical data and then obtain total functions by integration.

To illustrate, assume that a corporation has the information given in Table 1. The company finds the marginal productivity function to be

$$MPD = \frac{d(PD)}{dx} = 5x - x^2$$

where x is the number of units ($\$100{,}000$ each) of investment capital. (Verify the values in Table 1 by substituting the inputs in the expression for marginal productivity.) **Marginal productivity**, *MPD*, is the rate at which productivity changes (increases or decreases) with changes in capital (money available). From marginal productivity we can obtain productivity, *PD*, as a function of capital available.

TABLE 1

Input	Marginal Productivity
1	4
2	6
3	6
4	4
5	0
6	−6

EXAMPLE 24   Let the marginal productivity of a corporation be expressed as $5x - x^2$, where x is investment capital in hundreds of thousands of dollars. If productivity is zero when investment capital is zero, express productivity as a function of investment capital.

SOLUTION   Since

$$\frac{d}{dx}[PD(x)] = 5x - x^2$$

$$PD(x) = \int (5x - x^2)\, dx \qquad \text{Indefinite integral}$$

$$= \frac{5x^2}{2} - \frac{x^3}{3} + C$$

$$PD(0) = \frac{5(0)^2}{2} - \frac{0^3}{3} + C = 0 \qquad \text{Set } x = 0.$$

$$C = 0$$

Therefore,

$$PD(x) = \frac{5x^2}{2} - \frac{x^3}{3}$$

EXAMPLE 25

(a)   Find the total revenue function $R(x)$ if the marginal revenue is

$$R'(x) = 500 - 0.6x$$

(b)   Find the total revenue for a sale of 800 items.

SOLUTION

(a)   Since the marginal revenue function is the derivative of the total revenue function, the total revenue function is an antiderivative of the marginal revenue function.

$$R'(x) = 500 - 0.6x$$

$$\int R'(x)\, dx = \int (500 - 0.6x)\, dx \qquad \text{Integrate both sides of the equation.}$$

$$R(x) = \int (500 - 0.6x)\, dx \qquad \int R'(x) = R(x)$$

$$= 500x - 0.3x^2 + C$$

Now we find the constant C by setting $R = 0$ when $x = 0$. (The total revenue is 0 when the number of items sold is 0.)

$$0 = 500(0) - 0.3(0)^2 + C \qquad \text{Substitute } x = 0.$$

$$0 = C \qquad \text{Solve for } C.$$

Hence, the total revenue function is

$$R(x) = 500x - 0.3x^2$$

(b) Setting $x = 800$ gives

$$R = 500(800) - 0.3(800)^2$$
$$= \$208,000$$

EXAMPLE 26

(a) Find the total profit function if the marginal profit function is

$$P'(x) = 600 - 4x$$

where x is the number of items produced. Assume that there is a loss of $100 when no items are produced.

(b) Find the maximum profit.

SOLUTION

(a) Since the marginal profit function is the derivative of the total profit function, the total profit function is an antiderivative of the marginal profit function.

$$P'(x) = 600 - 4x \qquad \text{Marginal profit function}$$

$$P(x) = \int (600 - 4x)\, dx \qquad \text{Integrate both sides.}$$
$$= 600x - 2x^2 + C$$

Substituting $P(x) = -100$ when $x = 0$ gives

$$-100 = 600(0) - 2(0)^2 + C \qquad \text{Solve for } C.$$
$$= C$$

Hence,

$$P(x) = 600x - 2x^2 - 100 \qquad \text{Total profit function}$$

(b)   To find the maximum profit, first set $P'(x)$ equal to 0, and find the critical value(s).

$$P'(x) = 600 - 4x$$
$$0 = 600 - 4x$$
$$x = 150 \qquad \text{Critical value}$$

Since $P''(x) = -4 < 0$, the graph is concave downward and the absolute maximum occurs at $x = 150$. The absolute maximum value of the total profit function is then

$$P(150) = 600(150) - 2(150)^2 - 100 \qquad \text{Substitute } x = 150 \text{ in } P(x).$$
$$= \$44{,}900$$

The graph is shown in Figure 5. From the graph we see that the maximum profit occurs when $x = 150$. Use a range of $0 \le x \le 200$ with the increments set at 50 and a range of $0 \le y \le 50{,}000$ with the increments set at 5000.

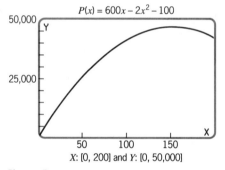

Figure 5

SUMMARY

Finding the antiderivatives of functions involving the natural exponent and those problems where the natural logarithm is the antiderivative are fairly straightforward. With some practice, these should not be difficult for you. Be sure to be careful when working with negative exponents.

Exercise Set 5.2

Find the following indefinite integrals.

1. $\int 6x^{-1} \, dx$

2. $\int 8x^{-1} \, dx$

3. $\int 4e^x \, dx$

4. $\int 10e^x \, dx$

5. $\int (6e^x + 2x) \, dx$

6. $\int (7e^x + 3x^2 - 5) \, dx$

7. $\int (2x^{-1} + e^x) \, dx$

8. $\int (5x^{-1} + x^2) \, dx$

9. $\int \left(\frac{3}{x}\right) \, dx$

10. $\int e^{-3t} \, dt$

11. $\int (e^{2t} - t^{-2}) \, dt$

12. $\int (x^{-1} + e^{-x}) \, dx$

13. $\int \left(\frac{2}{x} - e^{-10x}\right) \, dx$

14. $\int (-3x^{-2} + 4x^{-1}) \, dx$

15. $\int 3e^{-0.02t} \, dt$

16. $\int \frac{e^{-x} - 3x^{-1}}{6} \, dx$

17. $\int \frac{e^{-2x} - 2x^{-1}}{4} \, dx$

18. $\int (e^{-x} - e^{-3x}) \, dx$

19. $\int \frac{(xe^x - 2x)e^x}{x} \, dx$

20. $\int \frac{(xe^{2x} - 3x)e^x}{2x} \, dx$

21. $\int \frac{x^2 + 1}{x} \, dx$

22. $\int \frac{3x^2 + x + 2}{x} \, dx$

23. $\int (e^{-0.1x} + 2x^{-1} + e^{-0.3x}) \, dx$

24. $\int (e^{-0.2x} - e^{-0.1x}) \, dx$

Find the particular antiderivative of each derivative that satisfies the given condition.

25. $f'(x) = e^{2x} + 1, \quad f(0) = 6$

26. $f'(x) = e^{-x}, \quad f(0) = 6$

27. $f'(x) = \dfrac{3 + x}{x}, \quad f(1) = 4$

28. $f'(x) = \dfrac{3x^2 + 4}{x^2}, \quad f(1) = 2$

Applications (Business and Economics)

29. **Total Revenue.** A marginal revenue function is given by

$$R'(x) = 800 - 0.4x$$

Find the revenue function and the revenue for a sale of 1000 items. What is the maximum revenue? (**Hint:** $R = 0$ when $x = 0$.)

30. **Profit Function.** ABC Company has determined its marginal profit function to be

$$MP = 200 - 5x$$

If ABC Company loses $50 when no items are produced, find the company's total profit function. What is ABC Company's profit when 30 items are produced? What number of items should ABC Company produce in order to have maximum profit? What is the maximum profit?

31. **Price–Demand Equation.** The marginal price dp/dx at x units demand per month for a car is given by

$$\frac{dp}{dx} = -300e^{-0.05x}$$

Find the price–demand equation if at a price of $10,000 each, the demand is 10 cars per month.

32. **Cost Function.** The marginal average cost for producing x items is given by

$$\overline{C}'(x) = \frac{-600}{x^2}$$

where $\overline{C}(x)$ is the average cost in dollars. If $\overline{C}(50) = 20$, find the average cost function.

33. **Revenue.** Find the revenue function given the marginal revenue

$$R'(x) = 4000 - 5x$$

and knowing that $R(0) = 0$.

34. **Cost.** Find the cost for 100 units given the marginal cost

$$C'(x) = 3000 - 4x$$

when $C(0) = \$5000$.

35. **Profit.** Use the marginal functions in Exercises 33 and 34 to answer the following.
    (a) Write an expression for marginal profit.
    (b) If $P(0) = -\$5000$, find the profit for 100 units.
    (c) Find the profit when $x = 10$.

Applications (Social and Life Sciences)

36. **Population.** The rate of change of a population of bacteria in terms of time t in hours is given in thousands by

$$\frac{dP}{dt} = 60 - 0.06t$$

If $P(0) = 100$, find $P(t)$ and $P(60)$.

Review Exercises

Find the following indefinite integrals.

37. $\displaystyle\int 5x^{1/2}\, dx$

38. $\displaystyle\int \left(\frac{2}{x^2} - 3x\right) dx$

39. $\displaystyle\int \left(\frac{1}{\sqrt{x}} - x^{3/2}\right) dx$

5.3　DIFFERENTIALS AND INTEGRATION

OVERVIEW　When we studied differentiation, we did not treat dy/dx as the quotient of two separate quantities. In this section we give meaning to both dy and dx and study how they are related to integral calculus. We begin this section with a review of increments. Then we define differentials and show how the two are related. In this section we study

- Increments
- Differentials
- Integration

Increments

Recall from previous work that for two points $P(x_1, y_1)$ and $Q(x_2, y_2)$ in Figure 6 that

$$\Delta x = x_2 - x_1 \qquad \text{Increment: change in } x$$

$$\Delta y = y_2 - y_1 \qquad \text{Increment: change in } y$$

We are sometimes interested in how y is affected by a small change in x. For example, let

$$y = f(x) = x^2$$

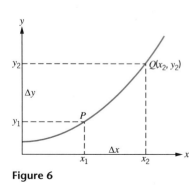

Figure 6

What is the change in y as x changes from 2 to 2.1? We have

$$\Delta x = 2.1 - 2 = 0.1 \qquad \text{Change in } x$$

$$\begin{aligned}\Delta y &= f(2.1) - f(2) \\ &= (2.1)^2 - (2)^2 \\ &= 4.41 - 4 = 0.41 \qquad \text{Corresponding change in } y\end{aligned}$$

Differentials

An understanding of increments leads to the definition of differentials. We have defined dy/dx to be one symbol. However, by themselves the symbols dy and dx may be given distinct meanings.

DIFFERENTIALS

Let $y = f(x)$ be a differentiable function of x. Then,

(a) dx, the differential of the independent variable x, is an arbitrary increment of x.

$$dx = \Delta x$$

(b) dy, the differential of the dependent variable y, is a function of x and dx given by

$$dy = f'(x)\, dx$$

Now let's use these definitions to find differentials.

EXAMPLE 27

(a)   If $u = g(x) = 2x^3$, find du.
(b)   If $y = h(x) = x^{-1}$, find dy.
(c)   If $w = f(t) = \sqrt{t^2 + 1}$, find dw.

SOLUTION

(a)
$$u = g(x) = 2x^3$$

$$du = g'(x)\, dx \qquad \text{Definition of differential}$$

$$= 6x^2\, dx \qquad \frac{d(2x^3)}{dx} = 6x^2$$

(b)
$$y = h(x) = x^{-1}$$

$$dy = h'(x)\, dx \qquad \text{Definition of differential}$$

$$= -x^{-2}\, dx \qquad \frac{d(x^{-1})}{dx} = -x^{-2}$$

$$= \frac{-1}{x^2}\, dx$$

(c)
$$w = f(t) = \sqrt{t^2 + 1}$$

$$dw = f'(t)\, dt \qquad \text{Definition of differential}$$

$$= \frac{1}{2}(t^2 + 1)^{-1/2}(2t)\, dt \qquad \frac{d(t^2 + 1)^{1/2}}{dt} = \frac{1}{2}(t^2 + 1)^{-1/2}(2t)$$

$$= \frac{t}{\sqrt{t^2 + 1}}\, dt$$

Practice Problem 1 If $u = f(x) = x^2 + 3x$, find du.

ANSWER $du = (2x + 3)\, dx$

EXAMPLE 28

(a) If $u = g(x) = e^{-3x}$, find du.
(b) If $y = h(x) = \ln(1 + 2x)$, find dy.
(c) If $w = f(t) = e^{3t^2}$, find dw.

SOLUTION

(a)
$$u = g(x) = e^{-3x}$$

$$du = g'(x)\, dx \qquad \frac{d}{dx}(-3x) = -3$$

$$= -3e^{-3x}\, dx$$

(b)
$$y = h(x) = \ln(1 + 2x)$$

$$dy = h'(x)\, dx \qquad D_u(\ln u) = \frac{1}{u}u'$$

$$= \frac{1}{1 + 2x} \cdot 2\, dx \qquad \frac{d}{dx}(1 + 2x) = 2$$

$$= \frac{2}{1 + 2x}\, dx$$

(c)
$$w = f(t) = e^{3t^2}$$

$$dw = f'(t)\, dt \qquad D_u e^u = e^u u'$$

$$= e^{3t^2}(6t)\, dt \qquad \frac{d}{dy}(3t^2) = 6t$$

$$= (6t)e^{3t^2}\, dt$$

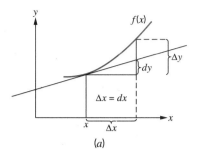

(a)

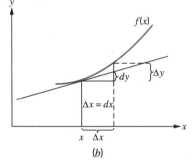

(b)

$m = f'(x) = dy/dx$, which now can be considered as either a derivative or a quotient of differentials.

Figure 7

Practice Problem 2 If $w = f(z) = e^{-4z^2}$, find dw.

ANSWER $dw = -8ze^{-4z^2}\, dz$

We can represent dy and dx pictorially as in Figure 7. Let x change by Δx. The corresponding change in the function is Δy. Now let $dx = \Delta x$. The differential dy is the change in the tangent line to the curve at x. Note that $dy \neq \Delta y$. However, for small values of dx, $dy \approx \Delta y$.

Integration

The integration formulas presented in the preceding two sections were expressed in terms of the variable x; however, they could have been expressed in terms of any variable. For instructional purposes we repeat the formulas in terms of a variable u, where we consider u as a function of x.

INTEGRATION FORMULAS

1. $\displaystyle\int u^n\, du = \frac{u^{n+1}}{n+1} + C, \quad n \neq -1$

2. $\displaystyle\int e^u\, du = e^u + C$

3. $\displaystyle\int \frac{du}{u} = \ln|u| + C$

Integration is not as straightforward as differentiation. When we differentiate $y = (2x^2 + 3)^4$ by the chain rule, we obtain

$$\frac{dy}{dx} = 4(2x^2 + 3)^3(4x) = (16x)(2x^2 + 3)^3$$

So, if we are to integrate $y = 16x(2x^2 + 3)^3$, we must think about using the chain rule in reverse. Suppose that we wish to integrate

$$\int 2(2x + 3)^2\, dx$$

Note that a function of x, namely $(2x + 3)$, is raised to the second power. If we call this function u, so that we have $u = 2x + 3$, then u is raised to the second power. To use integration formula 1, we must have an expression corresponding to du. Now, if $u = 2x + 3$, then

$$du = u'\, dx \qquad \text{Definition of the differential}$$

$$du = 2\, dx$$

Thus,

$$\int 2(2x + 3)^2 \, dx = \int (2x + 3)^2 \, 2 \, dx$$

would be in the form $\int u^2 \, du$ if $(2x + 3)$ were replaced by u and $2 \, dx$ were replaced by du, so that the expression can now be written as

$$\int 2(2x + 3)^2 \, dx = \int \overbrace{(2x + 3)^2}^{u^2} \overbrace{2 \, dx}^{du} = \int u^2 \, du = \frac{u^3}{3} + C = \frac{(2x + 3)^3}{3} + C$$

Now, suppose that the integral above had been $\int (2x + 3)^2 \, dx$, with $du = 2 \, dx$ not within the integrand. An adjustment can be made to solve a problem like this, as we see in Example 29.

EXAMPLE 29   Find $\displaystyle\int (3x + 5)^4 \, dx$.

SOLUTION   To use $\int u^n \, du$ you must consider $3x + 5$ as u, that is,

$$u = 3x + 5$$

$$\text{so}\quad du = 3 \, dx \qquad \frac{d(3x + 5)}{dx} = 3$$

$$\text{Therefore,}\quad dx = \frac{du}{3}$$

We now proceed to find the antiderivative.

$$\int (3x + 5)^4 \, dx = \int \overbrace{(3x + 5)^4}^{u^4} \cdot \overbrace{dx}^{\frac{du}{3}}$$

$$= \int u^4 \frac{du}{3} \qquad\qquad \text{Substituting } u = 3x + 5 \text{ and } dx = \frac{du}{3}$$

$$= \frac{1}{3} \int u^4 \, du \qquad\qquad \text{Constant rule}$$

$$= \frac{1}{3} \cdot \frac{u^5}{5} + C$$

$$= \frac{(3x + 5)^5}{15} + C$$

Practice Problem 3 Find $\int (5x - 1)^3\, dx.$

ANSWER $\dfrac{(5x - 1)^4}{20} + C$

EXAMPLE 30 Find $\int e^{3x}\, dx.$

SOLUTION We already have a formula for this integral. However, let's consider the integral as $\int e^u\, du.$

$$\text{Let} \quad u = 3x$$

$$\text{Then} \quad du = 3 \cdot dx$$

$$dx = \frac{du}{3}$$

$$\text{Therefore,} \quad \int e^{3x}\, dx = \int e^u\, \frac{du}{3}$$

$$= \frac{1}{3} \int e^u\, du$$

$$= \frac{1}{3} e^u + C$$

$$= \frac{1}{3} e^{3x} + C$$

This is exactly the same answer you would get using the formula on page 695. ∎

EXAMPLE 31 Find $\int \dfrac{dx}{6x + 7}.$

SOLUTION The integral seems to be of the form

$$\int \frac{du}{u}$$

If this is true, then

$$u = 6x + 7$$

$$du = 6 \cdot dx \qquad \frac{d(6x + 7)}{dx} = 6$$

$$dx = \frac{du}{6}$$

To put the integral in the form

$$\int \frac{du}{u}$$

we substitute and get

$$\int \frac{dx}{6x + 7} = \int \frac{1}{6x + 7}\, dx$$
$$= \int \frac{1}{u} \frac{du}{6}$$
$$= \frac{1}{6} \int \frac{du}{u}$$
$$= \frac{1}{6} \ln |u| + C$$
$$= \frac{1}{6} \ln |6x + 7| + C$$

Practice Problem 4 Find $\displaystyle\int \frac{dx}{7x - 3}$.

ANSWER $\dfrac{1}{7} \ln |7x - 3| + C$

SUMMARY

Much of what has been done in this section is used and amplified in the next section, when we begin to integrate using a technique called substitution. Therefore, make certain that you have a good grasp of the integration formulas and how to use them before proceeding to the next section.

Exercise Set 5.3

In Exercises 1–6, find dy and Δy if $x = 2$, $\Delta x = 0.1$, and dx = 0.1.

1. $y = 2x + 3$
2. $y = 3 - 2x$
3. $y = x^2 - 2$
4. $y = 3 - x^2$
5. $y = x + \dfrac{1}{x}$
6. $y = \dfrac{1}{x - 1}$

In Exercises 7–12, find dy and Δy if $x = 2$, $\Delta x = 0.01$ and dx = 0.01. How do the differences in Δy and dy compare to the differences found in Exercises 1–6?

7. $y = 2x + 3$
8. $y = 3 - 2x$
9. $y = x^2 - 2$
10. $y = 3 - x^2$
11. $y = x + \dfrac{1}{x}$
12. $y = \dfrac{1}{x - 1}$

In Exercises 13–26, find the differentials specified.

13. $u = t^2 + 4t + 1;$ du

14. $w = \sqrt{x};$ dw

15. $u = e^{0.3x};$ du

16. $w = e^{3x^2};$ dw

17. $v = \dfrac{1}{x - 1};$ dv

18. $u = \sqrt{7x^2 + 2};$ du

19. $p = \ln(1 + t^2);$ dp

20. $u = \ln(1 + e^{2x});$ du

21. $u = e^{0.3x^2 + 6};$ du

22. $w = e^{3x^2 + 4x};$ dw

23. $v = \dfrac{1}{\sqrt{x^2 - 1}};$ dv

24. $u = \sqrt{7x^2 + 2};$ du

25. $p = \ln\sqrt{1 + t^2};$ dp

26. $u = \ln\sqrt{1 + e^{2x}};$ du

Find the following indefinite integrals.

27. $\displaystyle\int (x - 3)^2 \, dx$

28. $\displaystyle\int (x + 4)^2 \, dx$

29. $\displaystyle\int (2x - 3)^2 \, dx$

30. $\displaystyle\int (2x + 4)^2 \, dx$

31. $\displaystyle\int (3x + 2)^2 \, dx$

32. $\displaystyle\int e^{3x} \, dx$

33. $\displaystyle\int e^{5x} \, dx$

34. $\displaystyle\int \dfrac{dx}{2x + 1}$

35. $\displaystyle\int \dfrac{dx}{3x + 2}$

Review Exercises

Find the following indefinite integrals.

36. $\displaystyle\int 6x^{-2} \, dx$

37. $\displaystyle\int (3x^2 + 5x - 7) \, dx$

38. $\displaystyle\int (2x^3 - 3x + 5) \, dx$

5.4 INTEGRATION BY SUBSTITUTION

OVERVIEW   In this section we work with integration by substitution. We will not be as concerned with building up the differential as we did in Section 11.3; instead, we proceed with the method of substitution. The work in this section is similar to that in the last section but with much more varied functions. In this section we

- Learn the process of formal substitution
- Learn to use algebraic simplification in integration

The key step in using the method of substitution with the integration formulas listed in the preceding section is recognizing which function of x to set equal to u. This will involve some practice, but you will soon find that it is not too difficult to recognize the avenue of substitution. Remember, so far we have only three key formulas of integration. These are on page 272 in Section 5.3.

When you are trying to find an integral, you must "think backwards" because you are looking at the result of differentiation. Therefore, you must decide which formula is a "match" for the particular integration problem you are working with and then decide on the proper substitution.

EXAMPLE 32 Find $\int (2x + 3)^{1/2}\, dx$.

SOLUTION This integration problem is of the form $\int u^n\, du$, so we use formula 1. If we let $u = 2x + 3$, then

$$du = \frac{d}{dx}(2x + 3) \cdot dx$$
$$= 2 \cdot dx$$
$$2\, dx = du$$

Solving for dx yields

$$dx = \frac{1}{2}\, du$$

Now we substitute into the original problem and integrate as follows.

$$\int (2x + 3)^{1/2}\, dx = \int u^{1/2} \cdot \frac{1}{2}\, du \qquad \text{Substitute } u = 2x + 3 \text{ and } dx = \tfrac{1}{2}\,du.$$

$$= \frac{1}{2} \int u^{1/2}\, du$$

$$= \frac{\frac{1}{2}u^{3/2}}{\frac{3}{2}} + C$$

$$= \frac{1}{3} u^{3/2} + C$$

$$= \frac{1}{3}(2x + 3)^{3/2} + C \qquad \text{Substitute } u = 2x + 3.$$

Practice Problem 1 Find $\int (3x + 2)^{1/3}\, dx$ by substituting $u = 3x + 2$.

ANSWER $\dfrac{(3x + 2)^{4/3}}{4} + C$

EXAMPLE 33 Find $\int \sqrt{5 - x}\, dx$.

SOLUTION Since $\sqrt{5 - x} = (5 - x)^{1/2}$, we use formula 1.

$$\int (5 - x)^{1/2}\, dx = \int u^{1/2}(-\, du) = -\int u^{1/2}\, du \qquad \begin{array}{l}\text{Substitution: } u = 5 - x; \text{ so} \\ du = -\,dx \text{ and } dx = -\,du.\end{array}$$

$$= \frac{-u^{3/2}}{\frac{3}{2}} + C = -\frac{2}{3}u^{3/2} + C$$

$$= -\frac{2}{3}(5 - x)^{3/2} + C \qquad \text{Substitute } u = 5 - x.$$

Practice Problem 2 Find $\int \sqrt[3]{x - 3}\, dx$.

ANSWER $\frac{3}{4}(x - 3)^{4/3} + C$

In the previous examples, there was really only one substitution possible. However, some problems may be a product or quotient of functions and the proper substitution must be carefully considered. Several problems of this nature are demonstrated in the next examples.

EXAMPLE 34 Find $\int 2x^2 \sqrt{3 + 5x^3}\, dx$.

SOLUTION   Here we must decide which part to use for substitution. Since $\sqrt{3 + 5x^3} = (3 + 5x^3)^{1/2}$ is of the form u^n and is the more complicated term, we use formula 1 and let $u = 3 + 5x^3$.

$$\int 2x^2 \sqrt{3 + 5x^3}\, dx = 2 \int \sqrt{3 + 5x^3}\, (x^2)\, dx$$

$$= 2 \int \sqrt{u}\, \frac{du}{15}$$

Substitution: $u = 3 + 5x^3$; so $du = 15x^2\, dx$ and $du/15 = x^2\, dx$.

$$= \frac{2}{15} \int u^{1/2}\, du$$

$$= \frac{2}{15}\, \frac{u^{3/2}}{\frac{3}{2}} + C$$

$$= \frac{4}{45}(3 + 5x^3)^{3/2} + C$$

Substitute $u = 3 + 5x^3$. ∎

Practice Problem 3 Find $\int 6x^2 \sqrt{x^3 + 5}\, dx$.

ANSWER $\frac{4}{3}(x^3 + 5)^{3/2} + C$

EXAMPLE 35 Find $\int x^2 e^{x^3}\, dx$.

SOLUTION

$$\int x^2 e^{x^3}\, dx = \int e^{x^3} x^2\, dx$$

Substitution: $u = x^3$; so $du = 3x^2\, dx$ and $x^2\, dx = \dfrac{du}{3}$.

$$= \int e^u \frac{du}{3}$$

$$= \frac{1}{3} \int e^u \, du$$

$$= \frac{e^u}{3} + C$$

$$= \frac{e^{x^3}}{3} + C \qquad \text{Substitute } u = x^3.$$

Practice Problem 4 Find $\int xe^{-4x^2} \, dx$.

ANSWER $\dfrac{-1}{8} e^{-4x^2} + C$

EXAMPLE 36 Find $\int \dfrac{x}{1 + x^2} \, dx$.

SOLUTION

$$\int \frac{x}{1 + x^2} \, dx = \int \frac{x \, dx}{1 + x^2} \qquad \begin{array}{l} \text{Substitution: } u = 1 + x^2; \text{ so} \\ du = 2x \, dx \text{ and } x \, dx = \dfrac{du}{2}. \end{array}$$

$$= \int \frac{du/2}{u}$$

$$= \frac{1}{2} \int \frac{du}{u}$$

$$= \frac{1}{2} \ln |u| + C$$

$$= \frac{1}{2} \ln |1 + x^2| + C \qquad \text{Substitution: } u = 1 + x^2.$$

$$= \frac{1}{2} \ln (1 + x^2) + C \qquad \text{Since } x^2 + 1 > 0$$

Practice Problem 5 Find $\int \dfrac{(\ln x)^2}{x} \, dx$.

ANSWER $\dfrac{(\ln x)^3}{3} + C$

EXAMPLE 37 Find $\int \dfrac{1}{\sqrt{x}} e^{\sqrt{x}} \, dx$.

SOLUTION   The integral seems to be of the form $\int e^u \, du$. If this is true, then $u = \sqrt{x}$ or $x = u^2$.

$$\int \frac{1}{\sqrt{x}} e^{\sqrt{x}} \, dx = \int e^{\sqrt{x}} \frac{dx}{\sqrt{x}}$$

Substitution: $x = u^2$; so $u = \sqrt{x}$

and $du = \dfrac{dx}{2\sqrt{x}} \Rightarrow 2 \, du = \dfrac{dx}{\sqrt{x}}$.

$$= \int e^u \, 2 \, du$$

$$= 2 \int e^u \, du$$

$$= 2e^u + C$$

$$= 2e^{\sqrt{x}} + C$$

Substitute $u = \sqrt{x}$. ∎

Practice Problem 6 Find $\displaystyle\int \frac{1}{x^2} e^{1/x} \, dx$.

ANSWER $-e^{1/x} + C$

Sometimes an integration problem will not seemingly match any of the three formulas. When this occurs, you must make an initial substitution to simplify the problem and then use other techniques as demonstrated in the following examples to complete the process of integration. Note that there may be times when more than one substitution is possible, and if the substitution you choose does not seem to be working, go back and try another substitution. There will, of course, be many functions for which substitution simply will not work and other methods must be used.

EXAMPLE 38   Find $\displaystyle\int \frac{x}{x+1} \, dx$.

SOLUTION   There is no clear-cut substitution in this problem. Therefore, let's try an initial substitution of $u = x + 1$. If $u = x + 1$, then $x = u - 1$ and $du = dx$. Now let's substitute these into the problem.

$$\int \frac{x}{x+1} \, dx = \int \frac{u-1}{u} \, du$$

Substitution: $u = x + 1$, $x = u - 1$, and $du = dx$

$$= \int \left(\frac{u}{u} - \frac{1}{u} \right) du$$

Algebraic simplification

$$= \int \left(1 - \frac{1}{u} \right) du$$

$$= u - \ln|u| + C$$

$$= (x + 1) - \ln|x + 1| + C$$

Substitute $u = x + 1$.

$$= x - \ln|x + 1| + C$$

The 1 is unnecessary, since C is arbitrary. ∎

Practice Problem 7 Find $\int \dfrac{x}{2 - x}\, dx$.

ANSWER $-x - 2 \ln |2 - x| + C$

EXAMPLE 39   Find $\int x\sqrt{x - 3}\, dx$.

SOLUTION   At first it appears that we should substitute with $u = \sqrt{x - 3}$, but this would not help us with this problem because $u' = \frac{1}{2}(x - 3)^{-1/2}$ is not part of the expression to be integrated.

For our substitution we use $u = x - 3$ so that we have $x = u + 3$ and $du = dx$.

$$\int x\sqrt{x - 3}\, dx = \int x(x - 3)^{1/2}\, dx$$

Substitution: $u = x - 3$; therefore, $du = dx$ and $x = u + 3$.

$$= \int (u + 3)u^{1/2}\, du$$

$$= \int (u^{3/2} + 3u^{1/2})\, du$$

Simplify algebraically.

$$= \frac{u^{5/2}}{\frac{5}{2}} + 3\,\frac{u^{3/2}}{\frac{3}{2}} + C$$

$$= \frac{2}{5}(x - 3)^{5/2} + 2(x - 3)^{3/2} + C$$

Substitute $u = x - 3$.

Practice Problem 8 Find $\int \dfrac{x}{\sqrt{x + 4}}\, dx$.

ANSWER $\frac{2}{3}(x + 4)^{3/2} - 8(x + 4)^{1/2} + C$

NOTE: You may have noticed a pattern to the work we have done. When using substitution, try to let the denominator be u, or the term that is under a radical, or the term raised to the highest power. These are guidelines that usually lead to a successful substitution.

EXAMPLE 40   Find $\int \dfrac{x\, dx}{\sqrt{4 - x^2}}$.

SOLUTION   We let the term under the radical in the denominator be u.

$$\int \frac{x\, dx}{\sqrt{4 - x^2}} = \int \frac{\dfrac{-1}{2}\, du}{\sqrt{u}}$$

Substitution: $u = 4 - x^2$ and $du = -2x\, dx$, $\dfrac{-1}{2}\, du = x\, dx$

$$= \frac{-1}{2} \int \frac{du}{u^{1/2}} = \frac{-1}{2} \int u^{-1/2}du$$

$$= -\frac{1}{2}(2u^{1/2}) + C = -u^{1/2} + C$$

$$= -\sqrt{4 - x^2} + C$$

Practice Problem 9 Find $\displaystyle\int \frac{(x + 1)\, dx}{\sqrt[3]{3x^2 + 6x + 5}}$.

ANSWER $\frac{1}{4}(3x^2 + 6x + 5)^{2/3} + C$

COMMON ERROR   When using substitution in finding an integral, do not forget to substitute back. The variable in your answer should be the same as in the original problem.

SUMMARY

It is important to realize that it is not always necessary or advantageous to use substitution. In the two integrals listed below, it would be better to multiply the first one out and then use the simple integration techniques. In the second integral, substitution would be a good choice.

$$\int x(3x^2 + 4)^2\, dx \qquad \text{Multiply first before integrating.}$$

$$\int x(3x^2 + 4)^3\, dx \qquad \text{Use substitution.}$$

It is also important to be able to distinguish between problems requiring you to use the formula involving $\ln x$ and the formula involving a term such as u^{-n}, as shown below.

$$\int \frac{1}{3x + 4}\, dx \qquad \text{requires} \qquad \int \frac{1}{u}\, du = \ln|u| + C \qquad \text{Why?}$$

$$\int \frac{1}{(3x + 4)^2}\, dx \quad \text{requires} \quad \int u^n\, du = \frac{u^{n+1}}{n + 1} + C, \quad n \neq -1 \qquad \text{Why?}$$

When integrating using the technique of substitution, you may realize that certain substitutions are obvious while others are not. When a substitution is not obvious, some algebraic simplification will usually be necessary after the initial substitution. Since these problems involve more steps and are somewhat more involved than those in previous sections, it is wise to check your solutions by differentiating your answer.

Exercise Set 5.4

Find the following indefinite integrals.

1. $\int 6\sqrt{x + 3}\, dx$

2. $\int 5\sqrt{x - 4}\, dx$

3. $\int 6\sqrt{4 - 3x}\, dx$

4. $\int 5\sqrt{2x - 3}\, dx$

5. $\int 2xe^{3x^2}\, dx$

6. $\int \frac{x\, dx}{1 - x^2}$

7. $\int 3xe^{-5x^2}\, dx$

8. $\int \frac{-3x\, dx}{2x^2 + 3}$

9. $\int \sqrt[3]{3 + 7x}\, dx$

10. $\int \sqrt[3]{x + 3}\, dx$

11. $\int x(2x^2 - 7)^{3/4}\, dx$

12. $\int x\sqrt{x^2 + 1}\, dx$

13. $\int x\sqrt{x^2 + 4}\, dx$

14. $\int 2x(3x^2 - 4)^{2/3}\, dx$

15. $\int x^2\sqrt{1 + x^3}\, dx$

16. $\int x^2\sqrt[3]{1 + x^3}\, dx$

17. $\int \frac{x + 1}{x - 1}\, dx$

18. $\int \frac{2x}{\sqrt{x + 2}}\, dx$

19. $\int \frac{-2x}{3 - x}\, dx$

20. $\int \frac{4x}{x + 2}\, dx$

21. $\int x(x + 1)^4\, dx$

22. $\int 2x(x - 3)^3\, dx$

23. $\int 2x\sqrt{x - 5}\, dx$

24. $\int x\sqrt[3]{x + 3}\, dx$

25. $\int \frac{e^{2x}}{1 + e^{2x}}\, dx$

26. $\int 3x(e^{x^2})^4\, dx$

27. $\int \frac{x}{1 - 3x}\, dx$

28. $\int (x + 2)\sqrt{2 - x}\, dx$

29. $\int (3 - x)\sqrt[3]{x + 1}\, dx$

30. $\int \frac{x}{(x - 3)^4}\, dx$

31. $\int \frac{e^{\sqrt{x - 1}}}{\sqrt{x - 1}}\, dx$

32. $\int \frac{1/\sqrt{x}}{\sqrt{x} - 1}\, dx$

33. $\int \frac{(\sqrt{x} + 3)^5}{\sqrt{x}}\, dx$

34. $\int \frac{(2x + 1)}{x^2 + x}\, dx$

35. Why does

$$\int \frac{1}{3x + 4}\, dx$$

require the use of the formula

$$\int \frac{1}{u}\, du$$

while

$$\int \frac{1}{(3x + 4)^2}\, dx$$

requires the use of the formula

$$\int u^n du?$$

Applications (Business and Economics)

36. **Revenue.** If the marginal revenue for the sale of x units per week is

$$R'(x) = x(x^2 + 1)^2 \quad \text{and} \quad R(0) = 0$$

find $R(10)$.

37. **Cost.** If the marginal cost of x items is

$$C'(x) = 40 + 1000e^{-2x}$$

with $C(0) = 1000$, find $C(10)$.

38. **Profit.** Use Exercises 36 and 37 to find an expression for marginal profit. Then with $P(0) = -1000$, find $P(10)$. Is $P(10) = R(10) - C(10)$?

39. **Revenue Function.** Suppose that a marginal revenue function is given by

$$R'(x) = \frac{x}{\sqrt{x^2 + 16}}$$

Find the revenue function. Find the revenue for a production of three items. [**Hint:** $R(0) = 0$.]

Applications (Social and Life Sciences)

40. ***Pollution.*** Suppose that the radius R, in feet, of an oil slick is increasing at the rate of

$$\frac{dR}{dt} = 80(t + 16)^{-1/2}$$

where t is time in minutes. If $R = 0$ when $t = 0$, find the radius of the slick after 20 minutes.

41. ***Learning Function.*** Suppose that the rate of change of a learning function is given as a function of time t in hours as

$$N'(t) = -1.6e^{0.02t}$$

If $N(0) = 80$, find the learning function as a function of time.

Review Exercises

Find the following indefinite integrals.

42. $\displaystyle\int (x + 3)^{-2}\, dx$ 43. $\displaystyle\int e^{-4x}\, dx$

44. $\displaystyle\int e^{-2x}\, dx$ 45. $\displaystyle\int \frac{dx}{x + 1}$

5.5 DEFINITE INTEGRALS

OVERVIEW   In the previous sections, we have been concerned with finding the indefinite integral (antiderivative) $F(x)$ of some function $f(x)$ denoted by

$$F(x) = \int f(x)\, dx$$

In this section, we are interested in finding a function that gives the total change over an interval in an antiderivative of a function. For example, if $f(x) = 2x$, then $F(x) = x^2 + C$. The total change in F over the interval $[1, 4]$ will be

$$F(4) - F(1) = [4^2 + C] - [1^2 + C] = 15$$

This total change over an interval in the antiderivative is called the **definite integral** and is one of the important concepts of calculus. A definite integral can be used to evaluate the total depreciation of a machine over a period of time or the total amount of money that a machine will generate over several years. Our approach in this section is intuitive and informal. In Section 5.7 we consider the definite integral as a Riemann integral defined as the limit of a sum. This more formal approach is needed to develop many of the applications of the definite integral. In this section we study

- Definite integrals
- Properties of definite integrals
- Use of substitution in definite integrals
- Applications

We illustrate the usefulness of the definite integral with the following example.

EXAMPLE 41 The marginal cost in producing a certain model of a radio at an electronics corporation is

$$\frac{dC}{dx} = 6 - 0.04x$$

where x is the number of radios produced in a day. If the number produced in a day changes from 50 to 100 radios, what is the change in cost?

SOLUTION Recall that $C(x)$ is the cost function. We seek

$$C(100) - C(50)$$

Using the indefinite integral, we find

$$C(x) = \int (6 - 0.04x)\, dx$$
$$= 6x - 0.02x^2 + C \qquad \text{Antiderivative of } 0.04x = 0.02x^2.$$

Therefore,

$$C(100) - C(50) = [6(100) - 0.02(100)^2 + C] - [6(50) - 0.02(50)^2 + C]$$
$$= 150$$

We see that the constant C in each term is eliminated by the subtraction process and does not play a part in the computation. ∎

Notice in the preceding example that the change in $C(x)$ was computed using an antiderivative of $C'(x)$. Since the antiderivative is symbolized by

$$C(x) = \int C'(x)\, dx$$

we can indicate the change in $C(x)$ by the following notation:

$$\text{Change in } C(x) = C(600) - C(400) = \int_{400}^{600} C'(x)\, dx$$

This integral form is called the definite integral.

DEFINITE INTEGRAL

The **definite integral** of a nonnegative, continuous function $f(x)$ over the interval from $x = a$ to $x = b$ is the net change in an antiderivative of $f(x)$ over the interval. This fact is symbolized by

$$\int_a^b f(x)\, dx = F(x) \Big|_a^b = F(b) - F(a)$$

where $F'(x) = f(x)$, for all x in $[a, b]$.

In Example 41 we noted that the constant of integration subtracted out. Since the definite integral was defined for any antiderivative, it is customary to take the simplest antiderivative where $C = 0$. In the symbolic form

$$\int_a^b f(x)\, dx$$

b is called the **upper limit** of integration, a is the **lower limit** of integration, and $f(x)$ is called the **integrand**.

Even though many graphing calculators can evaluate definite integrals, it is important for you to understand the definition and what is being found when a definite integral is used. A definite integral calculates the change in a function over an interval.

The following examples are done using the traditional calculus methods. Before you use a graphing calculator exclusively to evaluate definite integrals, you should be able to do them as shown in the examples.

EXAMPLE 42   Evaluate $\displaystyle\int_2^6 x^2\, dx$.

SOLUTION

$$\int_2^6 x^2\, dx = \frac{x^3}{3}\Big|_2^6 = \frac{6^3}{3} - \frac{2^3}{3} = 69\tfrac{1}{3} \qquad \text{Simplest antiderivative of } x^2 \text{ is } \frac{x^3}{3}. \qquad \blacksquare$$

EXAMPLE 43   Evaluate $\displaystyle\int_0^3 e^{2x}\, dx$.

SOLUTION

$$\int_0^3 e^{2x}\, dx = \frac{e^{2x}}{2}\Big|_0^3 = \frac{e^6}{2} - \frac{e^0}{2}$$

$$= \frac{e^6 - 1}{2} \approx 201.21 \qquad \text{Antiderivative of } e^{2x} = \frac{e^{2x}}{2}. \qquad \blacksquare$$

Practice Problem 1 Evaluate $\displaystyle\int_1^2 (x - x^{-1})\, dx$.

ANSWER $\frac{3}{2} - \ln 2 \approx 0.807$

Practice Problem 2 Evaluate $\displaystyle\int_1^2 e^{3x}\, dx$.

ANSWER $\dfrac{e^6 - e^3}{3} \approx 127.78$

NOTE: We should take time now to note the distinction between the definite integral and the indefinite integral. The **definite integral is a real number** (a value), whereas the **indefinite integral is a set of functions** [all of the antiderivatives of $f(x)$].

The following is a list of properties of definite integrals. Properties 1 and 2 parallel properties of indefinite integrals. Properties 3, 4, and 5 follow from the definition of the integral given in this section. Even when using a graphing calculator, it is important to know these properties.

DEFINITE INTEGRAL PROPERTIES

Assume that $f(x)$ and $g(x)$ are continuous functions on the indicated intervals and k is a constant. Then

1. $\displaystyle\int_a^b kf(x)\,dx = k\int_a^b f(x)\,dx$

2. $\displaystyle\int_a^b [f(x) \pm g(x)]\,dx = \int_a^b f(x)\,dx \pm \int_a^b g(x)\,dx$

3. $\displaystyle\int_a^b f(x)\,dx = -\int_b^a f(x)\,dx$

4. $\displaystyle\int_a^a f(x)\,dx = 0$

5. $\displaystyle\int_a^b f(x)\,dx = \int_a^c f(x)\,dx + \int_c^b f(x)\,dx, \quad a \le c \le b$

Calculator Note

Now let's look at how to evaluate a definite integral with your graphing calculator. On the MATH menu you will find a key for evaluating the integral. It may be fnInt(or something similar to that notation. When evaluating the definite integral, you enter the function, the variable, and the limits of integration. For example, to find the integral from Practice Problem 1,

$$\int_1^2 (x - x^{-1})\,dx$$

your entry and answer would look like

$$\text{fnInt}(x - x \wedge -1, x, 1, 2) = .8068528194$$

If your calculator has a slightly different notation, it should work the same way.

We now give some examples to illustrate properties of definite integrals. We evaluate definite integrals using both the properties and then with the graphing calculator to check our work.

EXAMPLE 44

1. $\displaystyle\int_a^b kf(x)\,dx = k\int_a^b f(x)\,dx$

$$\int_2^3 6x\,dx = 6\int_2^3 x\,dx = 6\left(\frac{x^2}{2}\right)\Big|_2^3$$

$$= 6\left(\frac{3^2}{2}\right) - 6\left(\frac{2^2}{2}\right) \qquad F(3) - F(2)$$

$$= 6\left(\frac{9}{2} - \frac{4}{2}\right) = 15$$

> ✓ Calculator Check: fnInt(6x, x, 2, 3) = 15

2. $\displaystyle\int_a^b [f(x) \pm g(x)]\,dx = \int_a^b f(x)\,dx \pm \int_a^b g(x)\,dx$

$$\int_2^3 (6x + 4)\,dx = \int_2^3 6x\,dx + \int_2^3 4\,dx$$

$$= \left(\frac{6x^2}{2}\right)\Big|_2^3 + (4x)\Big|_2^3$$

$$= \left[\frac{6\cdot 3^2}{2} - \frac{6\cdot 2^2}{2}\right] \qquad F(3) - F(2) \text{ for each part}$$

$$+ [4\cdot 3 - 4\cdot 2]$$

$$= 15 + 4 = 19$$

> ✓ Calculator Check: fnInt(6x + 4, x, 2, 3) = 19

3. $\displaystyle\int_a^b f(x)\,dx = -\int_b^a f(x)\,dx$

$$\int_2^3 6x\,dx = -\int_3^2 6x\,dx$$

$$3x^2\Big|_2^3 = -3x^2\Big|_3^2$$

$$27 - 12 = -(12 - 27)$$

$$15 = -(-15)$$

✓ Calculator Check: $(-)$fnInt$(6x, x, 3, 2) = 15$

4. $\displaystyle\int_a^a f(x)\, dx = 0$

$$\int_2^2 6x\, dx = 3x^2 \Big|_2^2 = 12 - 12 = 0$$

✓ Calculator Check: fnInt$(6x, x, 2, 2) = 0$

5. $\displaystyle\int_a^b f(x)\, dx = \int_a^c f(x)\, dx + \int_c^b f(x)\, dx$

$$\int_1^3 6x\, dx = \int_1^2 6x\, dx + \int_2^3 6x\, dx \qquad C = 2 \text{ is selected arbitrarily.}$$

$$3x^2 \Big|_1^3 = 3x^2 \Big|_1^2 + 3x^2 \Big|_2^3$$

$$27 - 3 = 12 - 3 + 27 - 12$$

$$24 = 9 + 15$$

✓ Calculator Check: fnInt$(6x, x, 1, 2)$ + fnInt$(6x, x, 2, 3) = 24$. Note: Try this same problem with your calculator using another value for C.

Practice Problem 3 Evaluate each integral using the properties and then check your answer with your graphing calculator.

(a) $\displaystyle\int_1^4 5x^2\, dx$

(b) $\displaystyle\int_2^4 (3x^2 - 2x)\, dx$

(c) $\displaystyle\int_1^3 \left(\frac{4}{x^2} - 2\right) dx$

ANSWER

(a) 105   (b) 44   (c) $-4/3$

Sometimes algebraic simplification is used to find the definite integral.

EXAMPLE 45   Evaluate $\displaystyle\int_4^9 \dfrac{x+1}{\sqrt{x}}\, dx.$

SOLUTION

$$\int_4^9 \frac{x+1}{\sqrt{x}}\, dx = \int_4^9 \left(\frac{x}{\sqrt{x}} + \frac{1}{\sqrt{x}}\right) dx = \int_4^9 (x^{1/2} + x^{-1/2})\, dx$$

$$= \left(\frac{x^{3/2}}{\frac{3}{2}} + \frac{x^{1/2}}{\frac{1}{2}}\right)\Bigg|_4^9 = \left(\frac{2}{3}x^{3/2} + 2x^{1/2}\right)\Bigg|_4^9$$

$$= \left[\frac{2}{3}\cdot 9^{3/2} + 2\cdot 9^{1/2}\right] - \left[\frac{2}{3}\cdot 4^{3/2} + 2\cdot 4^{1/2}\right]$$

$$= (18 + 6) - \left(\frac{16}{3} + 4\right) = 14\tfrac{2}{3}$$

✔ Calculator Check: fnInt$((x+1)/\sqrt{x}, x, 4, 9) = 14.66666667$

Practice Problem 4   Evaluate $\displaystyle\int_0^4 \sqrt{x}\,(3 + x)\, dx.$

ANSWER $\dfrac{144}{5} = 28\tfrac{4}{5}.$

Integration by Substitution

A definite integral in which substitution is required, can be evaluated using the properties of integration with two different methods. Method 1 involves actually changing the limits of integration by finding corresponding values for the new variable (u). Method 2 involves obtaining the antiderivative in terms of the original variable before evaluating the definite integral. We demonstrate both methods in Example 46. Generally, Method 1 results in less complicated calculations. As before, after we find the integral, we will check our answer with a calculator.

EXAMPLE 46   Evaluate $\displaystyle\int_1^5 \sqrt{2x-1}\, dx.$

SOLUTION

> *Method 1*   This method involves substitution and changing the limits of integration to correspond to the new variable. For the integral

$$\int_1^5 \sqrt{2x-1}\, dx$$

we use the following substitutions:

$$u = 2x - 1 \quad \text{gives} \quad du = 2\,dx, \quad \text{which yields} \quad dx = \frac{1}{2}\,du$$

We now change the limits of integration to match the new variable of integration, u. When $x = 1$, $u = 2(1) - 1 = 1$; and when $x = 5$, $u = 2(5) - 1 = 9$.

$$
\begin{aligned}
\int_1^5 \sqrt{2x - 1}\,dx &= \int_1^9 u^{1/2}\,\frac{1}{2}\,du \qquad \text{Substitution}\\[6pt]
&= \frac{1}{2}\int_1^9 u^{1/2}\,du\\[6pt]
&= \left.\frac{u^{3/2}}{3}\right|_1^9 \qquad\qquad \text{Integration}\\[6pt]
&= \frac{9^{3/2}}{3} - \frac{1^{3/2}}{3}\\[6pt]
&= \frac{26}{3}
\end{aligned}
$$

> ✓ Calculator Check: fnInt($\sqrt{(2x - 1)}$, x, 1, 5) = 8.666666667 and (1/2)fnInt ($x \wedge (1/2)$, x, 1, 9) = 8.666666667.

Method 2   For this method, we find the antiderivative in terms of x (the original variable). We use the same substitutions as above, but notice that we do not place the limits of integration on the indefinite integral.

$$
\begin{aligned}
\int \sqrt{2x - 1}\,dx &= \int u^{1/2}\,\frac{1}{2}\,du \qquad\qquad\qquad \text{Substitutions}\\[6pt]
&= \frac{u^{3/2}}{3} = \frac{(2x - 1)^{3/2}}{3} \qquad\qquad \text{Substitute back.}\\[10pt]
\int_1^5 \sqrt{2x - 1}\,dx &= \left.\frac{(2x - 1)^{3/2}}{3}\right|_1^5\\[6pt]
&= \frac{(2 \cdot 5 - 1)^{3/2}}{3} - \frac{(2 \cdot 1 - 1)^{3/2}}{3} \qquad F(5) - F(1)\\[6pt]
&= \frac{27}{3} - \frac{1}{3} = \frac{26}{3}
\end{aligned}
$$

From this point on, we will show only Method 1 when evaluating definite integrals requiring substitution. Method 1 reduces the amount of actual calculation that has to be done because you deal with simpler terms.

Calculator Note

We will no longer show the Calculator Check, but you should continue to check all work with your calculator.

EXAMPLE 47   Evaluate $\displaystyle\int_0^2 x(2x^2 + 1)^3 \, dx$.

SOLUTION

$$\int_0^2 x(2x^2 + 1)^3 \, dx = \int_1^9 u^3 \frac{1}{4} \, du$$

Substituting $u = 2x^2 + 1$ yields $du = 4x \, dx$ and $du/4 = x \, dx$.

$$= \frac{1}{4} \int_1^9 u^3 \, du$$

Therefore, when $x = 0$, $u = 1$; and when $x = 2$, $u = 9$.

$$= \frac{u^4}{16} \bigg|_1^9$$

$$= \frac{9^4}{16} - \frac{1^4}{16} = \frac{6560}{16} = 410$$

Practice Problem 5 Evaluate $\displaystyle\int_0^2 x^2(3x^3 + 2)^2 \, dx$.

ANSWER $650\frac{2}{3}$

COMMON ERROR   When evaluating a definite integral by using substitution and the methods of changing the limits, students often go back to the original terms to evaluate the limit. Once the limits have been changed, you no longer use the variable x. You use only the variable u.

Practice Problem 6 Evaluate $\displaystyle\int_0^3 \frac{dx}{(3x + 1)^2}$.

ANSWER $\dfrac{3}{10}$

Integration is used in many important applications in many fields. In applied problems we may know the rate of change of the function, but we may be interested in computing the actual change in the function as the value of the variable changes. Let's look at some examples here, and then we will expand the topic in Section 5.8.

EXAMPLE 48   Suppose that a company's marginal cost, marginal revenue, and marginal profit are given in thousands of dollars in terms of the number x of units produced as

$$C'(x) = 1$$
$$R'(x) = 12 - 2x \quad \text{for } 0 \leq x < 12$$
$$P'(x) = R'(x) - C'(x)$$

If production changes from 3 units to 6 units, find the change in (a) cost, (b) revenue, and (c) profit.

SOLUTION

(a)   The change in cost is

$$\int_3^6 C'(x)\, dx = \int_3^6 1\, dx = x \Big|_3^6 = 6 - 3 = 3$$

Therefore, the change in cost by the increase in production from 3 units to 6 units is $3000.

(b)   The change in revenue is

$$\int_3^6 R'(x)\, dx = \int_3^6 (12 - 2x)\, dx = (12x - x^2) \Big|_3^6$$
$$= (72 - 36) - (36 - 9) = 9$$

Therefore, revenue increases by $9000 with the change in production.

(c)   The change in profit is

$$\int_3^6 P'(x) = \int_3^6 [R'(x) - C'(x)]\, dx$$
$$= \int_3^6 R'(x)\, dx - \int_3^6 C'(x)\, dx = 9 - 3 = 6$$

Therefore, profit increases by $6000 by the change in production.

Note that we are not calculating the total amount of the cost, revenue, and profit functions. We are calculating the change in the amount of the function resulting from a change in production.   ■

EXAMPLE 49   The marginal revenue that a manufacturer receives is given by $R'(x) = 2 - 0.02x + 0.003x^2$ dollars per unit sold. How much additional revenue is received if sales increase from 100 to 200 units?

SOLUTION

$$\int_{100}^{200} (2 - 0.02x + 0.003x^2)\, dx = 2x - 0.01x^2 + 0.001x^3 \Big|_{100}^{200}$$

$$= [2 \cdot 200 - 0.01 \cdot 200^2 + 0.001 \cdot 200^3]$$
$$- [2 \cdot 100 - 0.01 \cdot 100^2 + 0.001 \cdot 100^3]$$
$$= 8000 - 1100$$
$$= 6900$$

Therefore, the revenue will increase by a total of $6900 if production is increased from 100 to 200 units.

SUMMARY

Make certain that you understand the difference between an indefinite integral (antiderivative) and a definite integral. When you are determining an indefinite integral (antiderivative), you are actually finding an entire set of functions. When you evaluate a definite integral, you are calculating a change in the antiderivatives as the variable changes from one value to another. The constant of integration does not play a role in the definite integral, since it subtracts out during the process of evaluation. Therefore, we always choose the simplest antiderivative, with $C = 0$, when evaluating a definite integral. Remember, if you use a graphing calculator to evaluate a definite integral, you should first make certain that you know how to do the problem using the properties of integration.

Exercise Set 5.5

Evaluate the following definite integrals using the properties of integration.

1. $\int_1^3 dx$

2. $\int_3^5 e^2\, dx$

3. $\int_2^3 7\, dx$

4. $\int_1^6 \frac{4}{x}\, dx$

5. $\int_1^3 3x\, dx$

6. $\int_2^5 4x\, dx$

7. $\int_2^5 (2x + 3)\, dx$

8. $\int_1^3 (4x - 1)\, dx$

9. $\int_0^4 e^{2t}\, dt$

10. $\int_0^2 (x^2 + x)\, dx$

11. $\int_2^5 (x^2 - x + 1)\, dx$

12. $\int_0^1 e^{-t/3}\, dt$

13. $\int_1^3 (x - 3)^2\, dx$

14. $\int_0^2 (x - 3)^2\, dx$

15. $\int_3^6 \sqrt[4]{x - 3}\, dx$

16. $\int_3^5 \sqrt{2x - 3}\, dx$

17. $\int_{-2}^2 x^{2/3}\, dx$

18. $\int_{-1}^2 3x^{2/3}\, dx$

Evaluate using the properties of integration and then use a graphing calculator to check your work.

19. $\int_3^6 2x^{-1}\, dx$

20. $\int_1^2 \frac{dt}{4t + 6}$

21. $\int_0^4 \frac{dt}{e^{4t}}$

22. $\int_1^3 xe^{2x^2}\, dx$

23. $\int_0^1 x^2 e^{x^3}\, dx$

24. $\int_0^4 (1 + xe^{-x^2})\, dx$

25. $\int_0^1 x\sqrt{x^2 + 1}\, dx$

26. $\int_0^1 x\sqrt[3]{x^2 + 2}\, dx$

27. $\int_2^4 \frac{x + 1}{x - 1}\, dx$

28. $\int_4^9 \frac{x + 4}{\sqrt{x}}\, dx$

29. $\int_0^2 3x(x + 1)^4\, dx$

30. $\int_0^2 \sqrt{x}(x + 1)\, dx$

If $\int_2^5 f(x)\, dx = 6$ and $\int_2^5 g(x) = -4$ use the properties of definite integrals to find the following.

31. $\int_2^5 [f(x) - g(x)]\, dx$

32. $\int_5^2 [2f(x) + g(x)]\, dx$

33. $\int_2^5 [4f(x) - 3g(x)]\, dx$

If $\int_1^3 f(x)\, dx = 10$ and $\int_3^7 f(x) = 8$ use the properties of definite integrals to find the following.

34. $\int_1^7 f(x)\, dx$

35. $\int_7^1 f(x)\, dx$

36. $\int_1^7 f(x)\, dx - \int_3^7 f(x)\, dx$

Applications (Business and Economics)

37. *Assembly Function*. From assembling several units of a product, the production manager obtained the following rate of assembly function:

$$f(x) = 120x^{-1/2}$$

where $f(x)$ represents the rate of labor hours required to assemble x units. The company plans to bid on a new order of 12 additional units. Help the manager estimate the total labor requirements for assembling the additional 12 units, if the current production level is 50 units.

38. *Manufacturing Costs*. A corporation has determined that the marginal cost per week of manufacturing electric shavers is given by

$$C'(x) = 40 - 0.2x + \frac{x^2}{10}$$

where x is the number of units manufactured per week. What will be the total cost change if the company decides to increase weekly production from 400 to 450 razors per week?

39. *Revenue*. A corporation determines a marginal revenue function associated with selling x shavers to be

$$R'(x) = 80 - 0.1x$$

Find the additional revenue generated by increasing sales from 200 to 300 shavers.

40. *Cost*. The daily marginal cost function associated with producing x large-screen televisions is given by

$$C'(x) = 0.0009x^2 - 0.004x + 4$$

where $C'(x)$ is measured in dollars per unit and x denotes the number of units produced. The daily fixed cost of production is $250.00. Find the total cost involved in the production of the first 1000 units.

41. *Cost, Revenue, Profit*. A company's marginal cost, revenue, and profit equations (in thousands of dollars per day) where x is the number of units produced per day are

$$C'(x) = 1$$
$$R'(x) = 10 - 2x \quad \text{for } 0 \le x \le 10$$
$$P'(x) = R'(x) - C'(x)$$

Find the change in cost, revenue, and profit in going from a production level of 3 units per day to 5 units per day.

42. *Cost, Revenue, Profit*. The marginal cost and marginal revenue functions (in dollars per item) of a manufacturer of small computers are found to be

$$C'(x) = 300 - 0.2x \quad \text{and} \quad R'(x) = 400 + 0.01x$$

where x is the level of production. Assume that each computer manufactured is sold.
(a) Find the change in revenue received when the level of production increases from 20 to 40 computers per week.
(b) Find the revenue received from the manufacture of 50 computers.

(c) If the fixed cost is $200, find the cost function and the cost of producing 50 computers.

(d) Find the profit received from the manufacture and sale of 50 computers.

Applications (Social and Life Sciences)

43. **Temperature of a Patient.**   The rate of change of a patient's temperature 1 hour after x milligrams of a drug is administered is

$$\frac{dT}{dx} = 0.3x - \frac{x^2}{4}, \qquad 0 \le x \le 4$$

What total change of temperature occurs when the dosage changes from (a) 0 to 3 milligrams and (b) 1 to 4 milligrams?

44. **Poiseville's Law.**   Suppose that V represents the total flow in cubic units per second of blood through an artery whose radius is R, and suppose that r is the distance of a particle of blood from the center of the artery. Assume that

$$\frac{dV}{dr} = 2\pi C(Rr - r^2)$$

where C is a constant that depends on the units used.

(a) Compute

$$V = \int_{R}^{1.1R} 2\pi C(Rr - r^2)\, dr$$

to find the increase in blood flow when the radius of the artery is increased by 10%.

(b) Find the increase in blood flow when the radius of the artery is increased by 20%.

45. **Learning Function.**   A person's learning rate of new words per minute is given by

$$\frac{dL}{dt} = 30t^{-1/2}, \qquad 1 \le t \le 16$$

Find the number of words learned (i.e., find L) from $t = 4$ to $t = 9$ minutes.

Review Exercises

Find the following indefinite integrals.

46. $\int \sqrt{3x - 7}\, dx$

47. $\int \sqrt[3]{2x + 3}\, dx$

48. $\int \frac{x + 1}{x - 2}\, dx$

49. $\int x\sqrt{x - 1}\, dx$

5.6   THE DEFINITE INTEGRAL AND AREA

OVERVIEW   The definite integral can be used to find the area under a curve [see Figure 8(a)]. In this section, we illustrate how the indefinite integral, the definite integral, and the area under a curve are related. As one application, we note that the area between the marginal revenue curve and marginal cost curve can be interpreted as profit over an interval [Figure 8(b)]. In this section we consider

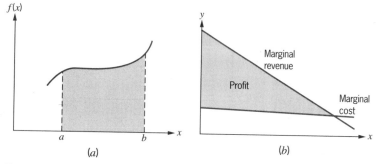

(a)   (b)

Figure 8

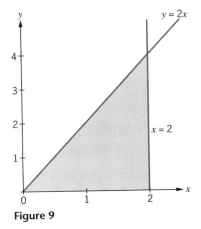

Figure 9

- The area under a curve
- Integrands with negative values
- Area between curves

To introduce the concept of the area under a curve, consider the following example which we will work in two ways.

EXAMPLE 50   Find the area of the shaded region in Figure 9 that is bounded by the line $y = 2x$, the vertical line $x = 2$, and the x-axis.

SOLUTION   The shaded area is a right triangle with base $= 2$ and altitude $= 4$, so the area is

$$\text{Area of triangle} = \frac{1}{2} \cdot 2 \cdot 4 = 4$$

Now suppose that we find the definite integral of the function $y = f(x) = 2x$ as x changes from 0 to 2. We have

$$\int_0^2 2x \, dx = x^2 \Big|_0^2 = 2^2 - 0^2 = 4$$

The definite integral of the bounding function over $[0, 2]$ gives the area under the curve from $x = 0$ to $x = 2$.

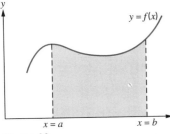

Figure 10

Practice Problem 1   Find the area of the region that is bounded by the line $y = 4x$, the vertical line $x = 3$, and the x-axis. Use the area of the triangle formula and then evaluate the definite integral

$$\int_0^3 4x \, dx$$

ANSWER   18

You should always draw the graphs involved in an area problem. Generally, the area under a curve cannot be found with the area formula of a geometric figure. For example, the shaded area of Figure 10 is given by

$$\int_a^b f(x) \, dx$$

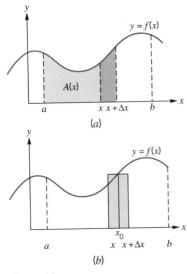

Figure 11

Let's investigate now why the definite integral gives the exact area under a curve. In Figure 11(a), let $A(x)$ be the area under the curve from a to x. We show first that $A(x)$ is an antiderivative of $f(x)$, for $A'(x) = f(x)$. Now, $A(x + \Delta x)$ is the area under the curve from a to $x + \Delta x$ in Figure 11(a).

In Figure 11(a), $A(x + \Delta x) - A(x)$ is the darker shaded area. This area can be approximated by a rectangle in Figure 11(b). The width of the rectangle is $(x + \Delta x) - x = \Delta x$ and the height of the rectangle where $x \leq x_0 \leq (x + \Delta x)$ is $f(x_0)$. Thus, the area of the rectangle is approximately $f(x_0) \Delta x$. Hence,

$$A(x + \Delta x) - A(x) \approx f(x_0) \Delta x$$

This approximation of the area of a rectangle improves as Δx becomes smaller and smaller because the area of the rectangle being calculated will get closer and closer to the actual area. Dividing both sides by Δx gives

$$\frac{A(x + \Delta x) - A(x)}{\Delta x} \approx f(x_0) \frac{\Delta x}{\Delta x} \qquad \Delta x \neq 0$$

$$\lim_{\Delta x \to 0} \frac{A(x + \Delta x) - A(x)}{\Delta x} = \lim_{\Delta x \to 0} f(x_0) \qquad \frac{\Delta x}{\Delta x} = 1$$

$$A'(x) = \lim_{\Delta x \to 0} f(x_0) \qquad \text{Definition of a derivative}$$

$$A'(x) = \lim_{\Delta x \to 0} f(x_0) = f(x) \qquad f(x) \text{ is continuous and } x \leq x_0 \leq (x + \Delta x).$$

Thus, $A(x)$ is an antiderivative of $f(x)$. Hence,

$$A(b) - A(a) = \int_a^b f(x)\, dx \qquad \text{Definition of definite integral}$$

Since $A(b) - A(a)$ represents the area under a curve from $x = a$ to $x = b$,

$$\int_a^b f(x)\, dx$$

is the area under the curve defined by $y = f(x)$ from $x = a$ to $x = b$.

AREA UNDER A CURVE

The area bounded by the continuous function $f(x) \geq 0$ on $a \leq x \leq b$, the x-axis, and the lines $x = a$ and $x = b$ is given by

$$A = \int_a^b f(x)\, dx$$

Later, this area will be expressed as a limit of a sum leading to the fundamental theorem of integral calculus.

EXAMPLE 51   Find the area of the region bounded by $f(x) = 3x^2$, the x-axis, and the lines $x = 2$ and $x = 4$. This region is illustrated in Figure 12.

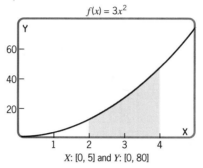

$f(x) = 3x^2$

X: [0, 5] and Y: [0, 80]

Figure 12

SOLUTION   This area is

$$\int_a^b f(x)\,dx = \int_2^4 3x^2\,dx = 3\int_2^4 x^2\,dx$$

$$= 3\left(\frac{x^3}{3}\right)\Bigg|_2^4 = x^3\,\Bigg|_2^4$$

$$= 4^3 - 2^3 = 64 - 8 = 56$$

Practice Problem 2   Find the area under $y = 2x^2 - x + 1$ from $x = 1$ to $x = 4$. First look at the area on your graphing calculator and make an estimate for the area.

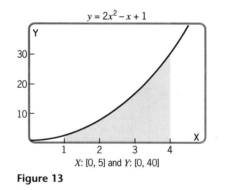

$y = 2x^2 - x + 1$

X: [0, 5] and Y: [0, 40]

Figure 13

ANSWER   The answer is 37.5. The graph is shown in Figure 13.

The definite integral for area was given for $f(x) \geq 0$. Now let's look at what happens if the graph of $f(x)$ falls below the x-axis between $x = a$ and $x = b$

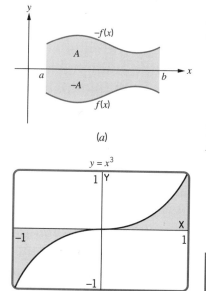

(a)

y = x³

X: [–1, 1] and Y: [–1, 1]

(b)

Figure 14

[Figure 14(a)]. The definite integral is the total change of f. If f is below the x-axis, the change is negative, the integral is negative, and the "area" is negative. Therefore, to calculate the area we use $-f(x)$ in our formula and we have

$$\int_a^b f(x)\, dx = -\int_a^b -f(x)\, dx = -A$$

To get the area bounded by the x-axis, where $f(x) \le 0$ on $[a, b]$ and $x = a$ and $x = b$, we take the absolute value of the definite integral:

$$\left| \int_a^b f(x)\, dx \right|$$

Calculator Note

Your graphing calculator will shade an area under a curve and then state the value of $\int f(x)\, dx$ for the function being graphed with the established lower and upper bounds. However, the calculator function that gives you a value for a definite integral is not calculating area. A definite integral can be negative or zero, but area must be positive. It is important to know when to use the absolute value of a function in calculating the definite integral, especially when you are using a graphing calculator. For example, have your graphing calculator draw x^3 over X:$[-1, 1]$ and Y:$[-1, 1]$ and then use $\int f(x)\, dx$ by using $x = -1$ and $x = 1$ as the lower and upper bounds. The calculator will show

$$\int_{-1}^1 x^3\, dx = 0$$

but this is certainly not the area shaded in Figure 14(b). This is one example of why you must look at the graph and use the properties of integrals to calculate the area correctly using a graphing calculator. To do this, on the home screen enter 2nd ABS MATH (select 9) (fnInt($x \wedge 3, x, -1, 0$) + fnInt($x \wedge 3, x, 0, 1$) = 0.5. Also note that when we are working the examples, we will often indicate the answer as a fraction, but your graphing calculator will give answers in decimal form unless you have it convert to fraction form using MATH (select ▶Frac).

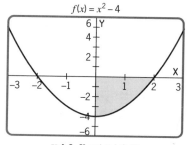

f(x) = x² – 4

X: [–3, 3] and Y: [–6, 6]

Figure 15

EXAMPLE 52 Find the area bounded by the x-axis and $f(x) = x^2 - 4$ from $x = 0$ to $x = 2$. The graph of the region bounded by this function, the x-axis, and the lines $x = 0$ to $x = 2$ is shaded in Figure 15.

SOLUTION   The bounded area is given by the absolute value of the definite integral:

$$\int_a^b f(x)\,dx = \int_0^2 (x^2 - 4)\,dx = \left(\frac{x^3}{3} - 4x\right)\Bigg|_0^2$$

$$= \left[\frac{2^3}{3} - 4(2)\right] - \left[\frac{0^3}{3} - 4(0)\right]$$

$$= \frac{8}{3} - 8 = -\frac{16}{3} \approx -5.3$$

Thus, the area of this region is $\left|-\frac{16}{3}\right| = \frac{16}{3}$.

Calculator Note

When you have your calculator shade the area under a function to calculate $\int f(x)\,dx$, you may not be able to set the lower and upper boundaries exactly where you want by using $\boxed{\text{TRACE}}$. You can get around this problem by setting the Xmin and Xmax to the values you want as boundaries for the integration. Now the proper area can be shaded. Be careful, however, because the integral being calculated by the calculator is the definite integral and not the area. You must use absolute value at the proper places to calculate the area. For instance, in Example 52 you could go to the home screen and enter $\boxed{\text{2nd}}$ $\boxed{\text{ABS}}$ $\boxed{\text{MATH}}$ (select 9)fnInt $(x^2 - 4, x, 0, 2)$ $\boxed{\text{MATH}}$ (select ▶Frac) $\boxed{\text{ENTER}}$ and get the result 16/3.

Practice Problem 3   Find the area bounded by the x-axis and $f(x) = x^2 - 4$ from $x = -2$ to $x = 2$.

ANSWER $\dfrac{32}{3} \approx 10.67$

EXAMPLE 53   Find the area of the region bounded by $f(x) = x^2 - 4$ and the x-axis from $x = 2$ to $x = 4$. This area is shaded in Figure 16(a).

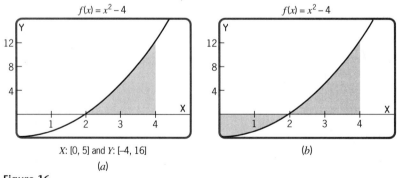

X: [0, 5] and Y: [-4, 16]

(a)

(b)

Figure 16

SOLUTION   Notice that since this area lies above the x-axis, we do not need to use the absolute value to evaluate the area.

$$\int_{2}^{4} (x^2 - 4)\, dx = \left(\frac{x^3}{3} - 4x \right)\Bigg|_{2}^{4}$$
$$= \left[\frac{4^3}{3} - 4(4) \right] - \left[\frac{2^3}{3} - 4(2) \right]$$
$$= \left(\frac{64}{3} - 16 \right) - \left(\frac{8}{3} - 8 \right)$$
$$= \frac{16}{3} + \frac{16}{3} = \frac{32}{3}$$

EXAMPLE 54   Find the area of the region bounded by the y-axis, the x-axis, the graph of $f(x) = x^2 - 4$, and the line $x = 4$. This region is shaded in Figure 16(b).

SOLUTION   Since part of the shaded region is below the x-axis and part is above, the area of the total region must be found by finding the area of the region below the x-axis and the area of the region above the x-axis and adding their absolute values. The area of the region below the x-axis was found in Example 52 to be $\left| -\frac{16}{3} \right|$. The area of the region above the x-axis was found in Example 53 to be $\frac{32}{3}$. Therefore, the total area is

$$\left| -\frac{16}{3} \right| + \left| \frac{32}{3} \right| = \frac{16}{3} + \frac{32}{3} = \frac{48}{3} = 16$$

NOTE: Notice that the answer in Example 54 could not have been found by evaluating the definite integral

$$\int_{0}^{4} (x^2 - 4)\, dx = \frac{16}{3}$$

since the definite integral combines the values $-\frac{16}{3} + \frac{32}{3} = \frac{16}{3}$. With your graphing calculator, you can use absolute value and determine the correct answer.

COMMON ERROR   When calculating areas that lie above and below the x-axis, students often neglect to separate the integrals and to use the absolute value. You should always draw the function being integrated with your graphing calculator so that you can avoid this error.

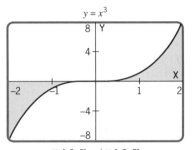

$y = x^3$

X: [-2, 2] and Y: [-8, 8]

Figure 17

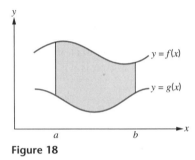

Figure 18

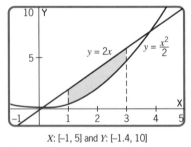

X: [-1, 5] and Y: [-1.4, 10]

Figure 19

Practice Problem 4 On your calculator, shade the area bounded by the graph of $f(x) = x^3$ and the x-axis from $x = -2$ to $x = 2$ and use this graph to help you evaluate the area.

ANSWER The graph is shown in Figure 17. The area = 8.

Up until this point, we have only considered areas that were bounded by a function and the x-axis from one value of x to another. Now let's consider the area bounded by $f(x)$ and $g(x)$, with $f(x) \geq g(x)$, between $x = a$ and $x = b$. Using definite integrals, we determine this area by finding the area under the curve $f(x)$ and subtracting the area under $g(x)$. This situation can be seen in Figure 18. The difference in total change of $f(x)$ and $g(x)$ is the area between the two curves.

$$\text{Area of shaded region} = \text{Area under } f(x) - \text{Area under } g(x)$$

$$= \int_a^b f(x)\, dx - \int_a^b g(x)\, dx$$

$$= \int_a^b [f(x) - g(x)]\, dx \qquad \text{Property of definite integrals}$$

AREA BETWEEN TWO CURVES

If $y = f(x)$ lies above $y = g(x)$, where $f(x) \geq g(x)$, for $a \leq x \leq b$, the area of the region bounded by $f(x)$ and $g(x)$ from $x = a$ to $x = b$ is

$$\int_a^b [f(x) - g(x)]\, dx$$

EXAMPLE 55 Find the area between the curves $y = 2x$ and $y = x^2/2$ from $x = 1$ to $x = 3$ (see Figure 19).

SOLUTION The graph shows the $2x \geq x^2/2$ over [1, 3]; therefore, by the preceding property, the area is given by

$$A = \int_1^3 \left(2x - \frac{x^2}{2}\right) dx = x^2 - \frac{x^3}{6} \Big|_1^3 \qquad \text{Antiderivative}$$

$$= \left(3^2 - \frac{3^3}{6}\right) - \left(1^2 - \frac{1^3}{6}\right) \qquad F(3) - F(1)$$

$$= \left(9 - \frac{9}{2}\right) - \left(1 - \frac{1}{6}\right)$$

$$= 3\tfrac{2}{3}$$

Calculator Note

How would you calculate the problem in Example 55 using your calculator? It is necessary to look at the graphs to determine which function is above the other over the designated interval. After that is done, you can use the fnInt function by subtracting the lower function from the one above.

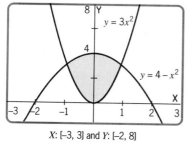

X: [-3, 3] and Y: [-2, 8]

Figure 20

Practice Problem 5 Using your calculator, find the area of the region bounded by the functions in Example 55 from $x = 0$ to $x = 4$.

ANSWER $\dfrac{16}{3}$. Enter fnInt($2x$, x, 0, 4) $-$ fnInt($x^2/2$, x, 0, 4).

EXAMPLE 56   Find the area of the region bounded by $y = 4 - x^2$ and $y = 3x^2$. This region is illustrated in Figure 20.

SOLUTION   Note that these graphs intersect at the points $(1, 3)$ and $(-1, 3)$. You can locate these points with your calculator using the intersect function. However, if the points of intersection given by the calculator are decimal approximations, you can find the exact points of intersection algebraically by setting the two functions equal and solving the resulting equation as we have done below. (Your calculator will convert some decimals to fraction form, but there may be a restriction on how many places it will allow in the denominator.)

$$3x^2 = 4 - x^2$$
$$4x^2 = 4$$
$$x^2 = 1$$
$$x = 1 \quad \text{or} \quad x = -1$$

Using the preceding property, the area is given by

$$A = \int_{-1}^{1} [(4 - x^2) - 3x^2]\, dx = \int_{-1}^{1} (4 - 4x^2)\, dx$$

$$= \left(4x - \frac{4}{3}x^3 \right) \bigg|_{-1}^{1} \qquad \text{Antiderivative}$$

$$= \left[4 \cdot 1 - \frac{4}{3} \cdot 1 \right] - \left[4 \cdot (-1) - \frac{4}{3}(-1) \right] \qquad F(1) - F(-1)$$

$$= \frac{16}{3}$$

Calculator Note

It is not necessary to simplify expressions before entering them in the calculator. Thus, an expression such as $4 - x^2 - 3x^2$ is allowed. However, if the second function has more than one term, you will need to use parentheses or distribute the negative sign.

Practice Problem 6 Use a calculator to draw the graphs of $y = x^2 - 2$ and $y = x + 4$. Then find the points of intersection and the area bounded by the two curves.

ANSWER   As shown in Figure 21, the points of intersection are at $x = -2$ and $x = 3$. Area $= 125/6$.

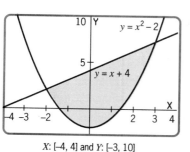

X: [-4, 4] and Y: [-3, 10]

Figure 21

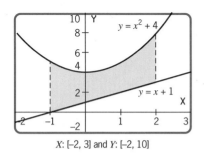

X: [-2, 3] and Y: [-2, 10]

Figure 22

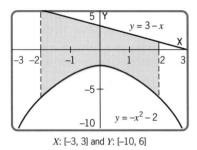

X: [-3, 3] and Y: [-10, 6]

Figure 23

EXAMPLE 57   Find the area of the region bounded by the functions $y = x^2 + 4$ and $y = x + 1$ from $x = -1$ to $x = 2$.

SOLUTION   Before we calculate this area, let's look at the graphs in Figure 22 so that we can accurately set up the definite integral. Since the function $y = x^2 + 4$ lies above $y = x + 1$, we set up and evaluate the integral as follows:

$$A = \int_{-1}^{2} [(x^2 + 4) - (x + 1)]\, dx = \int_{-1}^{2} (x^2 - x + 3)\, dx$$

$$= \frac{x^3}{3} - \frac{x^2}{2} + 3x \, \Big|_{-1}^{2} \qquad \text{Antiderivative}$$

$$= \left[\frac{2^3}{3} - \frac{2^2}{2} + 3(2)\right] - \left[\frac{(-1)^3}{3} - \frac{(-1)^2}{2} + 3(-1)\right] \qquad F(2) - F(-1)$$

$$= 10\tfrac{1}{2}$$

Practice Problem 7   Use your calculator to draw the graphs of $y = -x^2 - 2$ and $y = 3 - x$. Find the area bounded by these functions from $x = -2$ to $x = 2$.

ANSWER   The graphs are shown in Figure 23. Area = 76/3

EXAMPLE 58   Find the area between the curves $y = x^2$ and $y = 2x$ from $x = 1$ to $x = 3$.

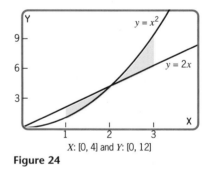

X: [0, 4] and Y: [0, 12]

Figure 24

SOLUTION   From the graph in Figure 24, we can see that the functions intersect and that neither function lies above the other over the entire interval $1 \le x \le 3$. Therefore, we must find the point of intersection within the given interval and use that point to find the area. You can use your calculator to find the point of intersection or you can find the point of intersection algebraically by setting the two functions equal and solving the resulting equation.

$$x^2 = 2x$$

$$x^2 - 2x = x(x - 2) = 0$$

$$x = 0 \quad \text{or} \quad x = 2$$

The point of intersection within the desired interval is at $x = 2$. Notice that in Figure 24, $y = 2x$ lies above $y = x^2$ over the interval $1 \le x \le 2$, and $y = x^2$ lies above $y = 2x$ over the interval $2 \le x \le 3$. Therefore, the area of the region between the two curves is

$$A = \int_1^2 [2x - x^2] \, dx + \int_2^3 [x^2 - 2x] \, dx$$

$$= \left[x^2 - \frac{x^3}{3} \Big|_1^2 \right] + \left[\frac{x^3}{3} - x^2 \Big|_2^3 \right] \qquad \text{Antiderivative}$$

$$= \left[2^2 - \frac{2^3}{3} \right] - \left[1^2 - \frac{1^3}{3} \right]$$

$$+ \left[\frac{3^3}{3} - 3^2 \right] - \left[\frac{2^3}{3} - 2^2 \right] \qquad F(b) - F(a) \text{ for each}$$

$$= 4 - \frac{8}{3} - \frac{2}{3} + \frac{4}{3}$$

$$= 2$$

■

EXAMPLE 59   Find the area bounded by the curves $y = x^3$ and $y = x^2 + 2x$.

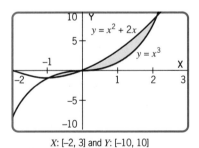

X: [-2, 3] and Y: [-10, 10]

Figure 25

SOLUTION   From the graphs shown in Figure 25, we see that the curves intersect at three points. We find the points of intersection algebraically by setting the two functions equal and solving the resulting equation. (They can be found on your calculator using the intersect function.)

$$x^3 = x^2 + 2x$$

$$x^3 - x^2 - 2x = 0$$

$$x(x^2 - x - 2) = x(x - 2)(x + 1) = 0$$

$$x = -1 \quad \text{or} \quad x = 0 \quad \text{or} \quad x = 2$$

Therefore, we set up and solve the integral using the interval $-1 \le x \le 0$ where $y = x^3$ is the top function and the interval $0 \le x \le 2$ where $y = x^2 + 2x$ is the top function.

$$A = \int_{-1}^{0} [x^3 - (x^2 + 2x)]\, dx + \int_{0}^{2} [(x^2 + 2x) - x^3]\, dx$$

$$= \left[\frac{x^4}{4} - \frac{x^3}{3} - x^2 \;\Big|_{-1}^{0} \right] + \left[-\frac{x^4}{4} + \frac{x^3}{3} + x^2 \;\Big|_{0}^{2} \right] \qquad \text{Antiderivatives}$$

$$= \left[\left(\frac{0^4}{4} - \frac{0^3}{3} - 0^2 \right) - \left(\frac{(-1)^4}{4} - \frac{(-1)^3}{3} - (-1)^2 \right) \right]$$

$$+ \left[\left(-\frac{2^4}{4} + \frac{2^3}{3} + 2^2 \right) - \left(-\frac{0^4}{4} + \frac{0^3}{3} + 0^2 \right) \right]$$

$$= \frac{37}{12}$$

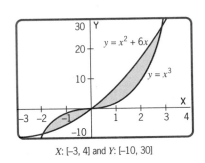

X: [-3, 4] and Y: [-10, 30]

Figure 26

Practice Problem 8 Use your calculator to draw the graphs of $y = x^3$ and $y = x^2 + 6x$ and find the area bounded by the two curves.

ANSWER   The graphs are shown in Figure 26. Area = 253/12

SUMMARY

It is important to draw the graphs of functions when evaluating the area bounded by their curves. There are times when the points of intersection are obvious or when the limits of integration are given. However, there will also be problems for which you must find the point or points of intersection algebraically or use your calculator to find the intersection points. Consequently, you will need a graph to assist you in setting up the definite integrals with proper limits of integration.

Exercise Set 5.6

Find the area of the region bounded by the following curves using techniques of integration. Use your graphing calculator to draw the curves and check your work.

1. $f(x) = 5$, $f(x) = 0$, $x = 2$, $x = 5$
2. $f(x) = x$, $f(x) = 0$, $x = 1$, $x = 3$
3. $f(x) = 2x$, $f(x) = 0$, $x = 2$, $x = 4$
4. $f(x) = x^2$, the x-axis on $1 \le x \le 3$
5. $f(x) = x^3$, the x-axis on $0 \le x \le 2$
6. $f(x) = x^2 - 4$, $f(x) = 0$, $x = -3$, $x = 4$
7. $f(x) = x^2 - x$, $f(x) = 0$, $x = 1$, $x = 3$
8. $f(x) = \sqrt{x}$, the x-axis on $1 \le x \le 4$
9. $f(x) = 1 + \sqrt{x}$, the x-axis on $1 \le x \le 9$

10. $f(x) = \dfrac{1}{\sqrt{x}}$, the x-axis on $1 \le x \le 4$
11. $f(x) = e^{2x}$, the x-axis on $0 \le x \le 2$
12. $f(x) = \dfrac{1}{x}$, the x-axis on $1 \le x \le 2$
13. $f(x) = \dfrac{1}{2x + 1}$, the x-axis on $\dfrac{1}{2} \le x \le 1$
14. $f(x) = 3x$, $f(x) = 1$, $x = 1$, $x = 4$
15. $f(x) = 7x$, $f(x) = 2$, $x = 2$, $x = 3$
16. $f(x) = (x - 2)^2$, the x-axis, $x = 0$, $x = 5$
17. $y = x^2 + 2x + 2$, $y = 2x + 3$
18. $f(x) = x^2$, $f(x) = x$
19. $y = \sqrt{x}$, $x = 9$
20. $y = x^2$, $y = \sqrt{x}$

Find the area of the region bounded by the following curves using either techniques of integration, your graphing calculator, or both methods.

21. $y = 16 - x^2$, $y = 6x$

22. $y = x^2 - 4$, the x-axis, $x = -1$, $x = 2$

23. $y = x^3$, the x-axis, $x = -2$, $x = 1$

24. $y = 2xe^{-x^2}$, the x-axis, $x = 0$, $x = 2$

25. $y = x^2 - 4x$, $y = 2$, $x = 0$, $x = 2$

26. $y = x^2 - 1$, $y = 0$, $x = -1$, $x = 3$

27. $y = 2x + 2$, $y = x^2 + 2$

28. $y = x + 2$, $y = x^2 + x - 7$

29. $y = 3x + 3$, $y = x^2 + 2x + 1$

30. $y = x^2$, $y = x^3 - 6x$

31. $y = x^3 - 3x^2 + 3x$, $y = x^2$

32. $f(x) = 2x^2$, $f(x) = x^4 - 2x^2$

33. $y = -x^2 + 4x + 3$, $y = x^2 - 4x + 3$

34. $y = x^2 - x$, $y = 2x + 4$

35. $y = 2x^2$, $y = x^3 - 8x$

36. $y = x + 4$, $y = -x^2 - 2$, $x = -2$, $x = 4$

37. $y = x - 4$, $y = -x^2 + 3$, $x = -2$, $x = 2$

38. $y = x^2 + 3$, $y = -x^2 - 3$, $x = -1$, $x = 2$

39. $y = 2x^2 + 1$, $y = 2 - x^2$, $x = -5$, $x = 5$

Review Exercises

Find the following indefinite integrals.

40. $\displaystyle\int xe^{-x^2}\, dx$

41. $\displaystyle\int x^2 e^{-x^3}\, dx$

Find the value of the following definite integrals using your graphing calculator.

42. $\displaystyle\int_0^3 (9x - x^3)^2 dx$

43. $\displaystyle\int_2^3 \frac{2x\, dx}{(1 + x^2)^2}$

5.7   RIEMANN SUMS AND THE DEFINITE INTEGRAL (Optional)

OVERVIEW   In the preceding section, we used the definite integral

$$\int_a^b f(x)\, dx$$

to find the area between the x-axis and $y = f(x) > 0$ from $x = a$ to $x = b$. In this section, we look at an alternative approach to the definition of a definite integral. This procedure was developed by George Bernhard Riemann, a nineteenth century mathematician. In this section we

- Approximate the area under a curve using the sum of the areas of approximating rectangles
- Define the definite integral as the limit of such sums as the number of rectangles increases without bound
- Use the definition of the definite integral as an infinite sum to give meaning to such applications as the average value of a function over a closed interval

First, we consider how to approximate the area under a curve by dividing it into subregions that are *almost* rectangles. In Figure 27(a), [a, b] has been divided into four subintervals of equal width, Δx.

$$\Delta x = \frac{b - a}{4}$$

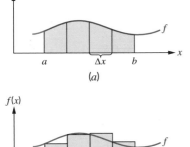

(a)

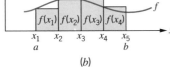

(b)

Figure 27

Note the four regions that are *almost* rectangles.

In Figure 27(b), we approximate the area under the curve with rectangles. We select the left edge of each subinterval (we could have selected the right edge or any point in the subinterval) and use $f(x_1), f(x_2), f(x_3)$, and $f(x_4)$ as the heights of four rectangles. The sum of the areas of these four rectangles approximates the area under the curve. Thus, the area under the curve is approximately

$$f(x_1)\, \Delta x + f(x_2)\, \Delta x + f(x_3)\, \Delta x + f(x_4)\, \Delta x$$

We can simplify this by using summation notation

$$\sum_{i=1}^{4} f(x_i)\, \Delta x = f(x_1)\, \Delta x + f(x_2)\, \Delta x + f(x_3)\, \Delta x + f(x_4)\, \Delta x$$

For notational purposes we will use the following definitions:

1. Let $s_{\min}$ be defined as the sum of the areas of the rectangles formed using the minimum value of $f(x)$ for x on the base of each rectangle.
2. Let $s_{\max}$ be defined as the sum of the areas of the rectangles formed using the maximum value of $f(x)$ for x on the base of each rectangle.

EXAMPLE 60   Approximate the area bounded by $y = x^2$, the x-axis, $x = 1$, and $x = 3$ using eight intervals.

SOLUTION   The graph of $y = x^2$ is shown in Figure 28. The width of each rectangle is Δx and

$$\Delta x = \frac{3-1}{8} = \frac{1}{4} = 0.25$$

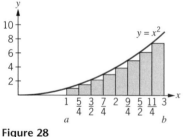

Figure 28

so the sides of rectangles occur at

$$x = 1, \frac{5}{4}, \frac{6}{4}, \frac{7}{4}, 2, \frac{9}{4}, \frac{10}{4}, \frac{11}{4}, 3$$

To obtain $s_{\min}$ we use the minimum height of the function, which occurs at the left endpoint in each interval, to find the area of each rectangle. The area of the first rectangle is $f(1)$ times the width Δx; the area of the second rectangle is $f(\frac{5}{4})$ times the width Δx. Thus,

$$s_{min} = f(1)\, \Delta x + f\left(\frac{5}{4}\right) \Delta x + f\left(\frac{6}{4}\right) \Delta x + f\left(\frac{7}{4}\right) \Delta x$$

$$+ f(2)\, \Delta x + f\left(\frac{9}{4}\right) \Delta x + f\left(\frac{10}{4}\right) \Delta x + f\left(\frac{11}{4}\right) \Delta x$$

$$= 1\left(\frac{1}{4}\right) + \frac{25}{16}\left(\frac{1}{4}\right) + \frac{9}{4}\left(\frac{1}{4}\right) + \frac{49}{16}\left(\frac{1}{4}\right)$$

$$+ 4\left(\frac{1}{4}\right) + \frac{81}{16}\left(\frac{1}{4}\right) + \frac{25}{4}\left(\frac{1}{4}\right) + \frac{121}{16}\left(\frac{1}{4}\right)$$

$$= \left(\frac{16 + 25 + 36 + 49 + 64 + 81 + 100 + 121}{16}\right)\left(\frac{1}{4}\right)$$

$$\approx 7.69$$

$\Delta x = \frac{1}{4}$

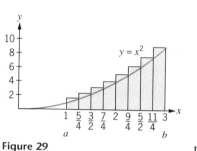

Figure 29

The graph of $y = x^2$ and associated rectangles using the maximum heights of the rectangles (here the right endpoints of each interval) is shown in Figure 29. We have

$$s_{max} = f\left(\frac{5}{4}\right) \Delta x + f\left(\frac{6}{4}\right) \Delta x + f\left(\frac{7}{4}\right) \Delta x + f(2)\, \Delta x$$

$$+ f\left(\frac{9}{4}\right) \Delta x + f\left(\frac{10}{4}\right) \Delta x + f\left(\frac{11}{4}\right) \Delta x + f(3)\, \Delta x$$

$$= \frac{25}{16}\left(\frac{1}{4}\right) + \frac{9}{4}\left(\frac{1}{4}\right) + \frac{49}{16}\left(\frac{1}{4}\right) + 4\left(\frac{1}{4}\right)$$

$$+ \frac{81}{16}\left(\frac{1}{4}\right) + \frac{25}{4}\left(\frac{1}{4}\right) + \frac{121}{16}\left(\frac{1}{4}\right) + 9\left(\frac{1}{4}\right)$$

$$= \left(\frac{25 + 36 + 49 + 64 + 81 + 100 + 121 + 144}{16}\right)\left(\frac{1}{4}\right)$$

$$\approx 9.69$$

The actual area is a number between s_{min} and s_{max}, so we know that

$$7.69 \le A \le 9.69$$

From the preceding discussion, two facts become evident about the approximation of the area under a curve using rectangles.

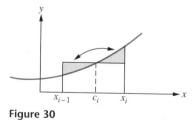

Figure 30

1. If a point is taken to get a height of each rectangle so that there is about as much area of the rectangle above the curve as there is missing below the curve, each rectangle gives a better approximation for the area under the curve (see Figure 30).

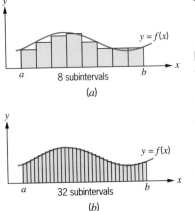

(a)

(b)

Figure 31

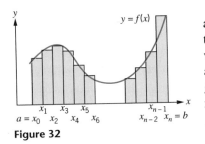

Figure 32

2. As the number of rectangles increases or as Δx gets smaller and smaller, the sum of the areas of the rectangles gets closer and closer to the area under the curve (see Figure 31).

Practice Problem 1 On your graphing calculator, draw the graph of $y = x^2$, using $2 \leq x \leq 3$ and $4 \leq y \leq 9$.

(a) What characteristic of the curve do you see on this portion of the graph?
(b) If the endpoints for s_{min} are at $x = 2, 2.2, 2.4, 2.6, 2.8$ and the endpoints for s_{max} are at $x = 2.2, 2.4, 2.6, 2.8, 3$, calculate bounds for the area using rectangles of width 0.2 and interpret this result.

ANSWER

(a) The curve appears to be almost linear.
(b) $s_{min} = 5.84$ and $s_{max} = 6.84$. The actual area is somewhere between these values.

Now let's consider an area A bounded by $y = f(x)$, the x-axis, $x = a$, and $x = b$ as shown in Figure 32. This time let's divide the interval $[a, b]$ into n subintervals that may or may not be of equal length. Let $\Delta x_1 = x_1 - x_0$ and let c_1 be a point where $x_0 \leq c_1 \leq x_1$ such that $f(c_1) \Delta x_1$ gives a fairly good approximation for the area under the curve from x_0 to x_1. Likewise, $f(c_2) \Delta x_2$, where $\Delta x_2 = x_2 - x_1$, gives a good approximation for the area from x_1 to x_2. After forming n such rectangles, the area under the curve can be expressed as

$$A \approx f(c_1) \Delta x_1 + f(c_2) \Delta x_2 + \cdots + f(c_n) \Delta x_n$$

where $x_{i-1} \leq c_i \leq x_i$. Of course, as n gets larger (with the understanding that as n increases the largest Δx_i decreases), this expression gets closer and closer to the exact area under the curve. This discussion leads to the following definition.

THE DEFINITE INTEGRAL USING RIEMANN SUMS

Assume that $y = f(x)$ is a function defined on $a \leq x \leq b$ and that

1. $a = x_0 \leq x_1 \leq \cdots \leq x_{n-1} \leq x_n = b$
2. $\Delta x_i = x_i - x_{i-1}$ for $i = 1, 2, \ldots n$
3. $\Delta x_i \rightarrow 0$ as $n \rightarrow \infty$
4. $x_{i-1} \leq c_i \leq x_i$ for $i = 1, 2, \ldots n$

If

$$\lim_{n \rightarrow \infty} [f(c_1) \Delta x_1 + f(c_2) \Delta x_2 + \cdots + f(c_n) \Delta x_n]$$

exists, then this limit is called the **definite integral** of $f(x)$ from a to b. Symbolically, this is written as

$$\lim_{n \rightarrow \infty} \sum_{i=1}^{n} f(c_i) \Delta x_i = \int_a^b f(x) \, dx$$

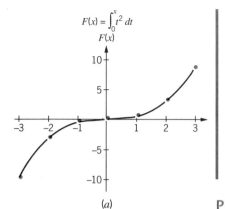

$$F(x) = \int_0^x t^2 \, dt$$

(a)

$$y = \frac{x^3}{3}$$

X: [-3, 3] and Y: [-10, 10]

(b)

Figure 33

Calculator Note

You should be aware that

$$F(x) = \int_0^x t^2 \, dt$$

is really the antiderivative of x^2, namely $x^3/3$. On your calculator, evaluate $F(x)$ at several points. Then, on paper, plot the points and connect them to make a graph as in Figure 33(a). Compare your graph with the graph of $y = x^3/3$ in Figure 33(b). Do these two graphs appear to be the same? We have plotted the points $F(-3) = -9$, $F(-2) = \frac{-8}{3}$, $F(-1) = \frac{-1}{3}$, $F(0) = 0$, $F(1) = \frac{1}{3}$, $F(2) = \frac{8}{3}$, $F(3) = 9$.

Practice Problem 2 Use the technique described in the preceding Calculator Note to plot the graph of

$$F(x) = \int_0^x t^3 \, dt$$

Then plot the graph of $y = x^4/4$. Are they the same graph?

ANSWER　After plotting $F(x)$ for $x = -2, -1, 0, 1, 2$ to obtain the graph in Figure 34(a) and comparing this to the calculator graph of $y = x^4/4$ in Figure 34(b), the graphs appear to be the same.

Recall from the preceding section that if $F(x)$ is an antiderivative of $f(x)$, then

$$F(b) - F(a) = \int_a^b f(x) \, dx$$

Note that for $f(x) \geq 0$, $F(b) - F(a)$ represents the area under a curve from $x = a$ to $x = b$. Since

$$\lim_{n \to \infty} [f(c_1) \, \Delta x_1 + f(c_2) \, \Delta x_2 + \cdots + f(c_n) \, \Delta x_n]$$

also represents the area under a curve from $x = a$ to $x = b$, these results can be combined to give the **fundamental theorem of integral calculus**.

FUNDAMENTAL THEOREM OF INTEGRAL CALCULUS

If $F(x)$ is any antiderivative of $f(x)$ over $a \leq x \leq b$, then

$$\lim_{n \to \infty} [f(c_1) \, \Delta x_1 + f(c_2) \, \Delta x_2 + \cdots + f(c_n) \, \Delta x_n] = \int_a^b f(x) \, dx$$

$$= F(b) - F(a)$$

where $x_{i-1} \leq c_i \leq x_i$ for $i = 1$ to n.

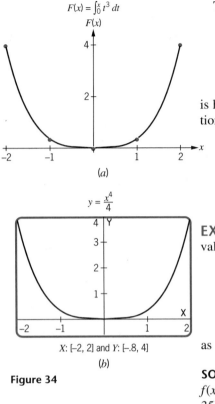

$F(x) = \int_0^x t^3\, dt$

(a)

$y = \dfrac{x^4}{4}$

X: [–2, 2] and Y: [–.8, 4]

(b)

Figure 34

This theorem provides a way to evaluate a definite integral. The notation

$$\int_a^b f(x)\, dx = F(x) \Big|_a^b = F(b) - F(a)$$

is helpful. Also, notice that it is not necessary to include the constant of integration since

$$\int_a^b f(x)\, dx = [F(b) + C] - [F(a) + C] = F(b) - F(a)$$

EXAMPLE 61   Consider $f(x)$ on [1, 3] and let $\Delta x = (3 - 1)/n$. Find the value of

$$\lim_{n \to \infty} \sum_{i=1}^{n} x_i^3\, \Delta x$$

as a definite integral.

SOLUTION   As $n \to \infty$, $\Delta x = (3 - 1)/n \to 0$. Furthermore, x_i^3 is the value of $f(x)$ for a value of x on the ith interval. Therefore, $f(x)$ could be x^3 (see Figure 35). Thus, we have

$$\lim_{n \to \infty} \sum_{i=1}^{n} x_i^3\, \Delta x = \int_1^3 x^3\, dx = \frac{x^4}{4} \Big|_1^3$$

$$= \frac{3^4}{4} - \frac{1^4}{4}$$

$$= \frac{80}{4} = 20 \qquad \blacksquare$$

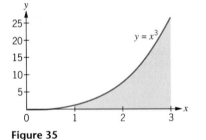

Figure 35

Practice Problem 3   Consider $f(x) = x^3$ on [2, 4] and let $\Delta x = (4 - 2)/n$. Find the value of

$$\lim_{n \to \infty} \sum_{i=1}^{n} x_i^3\, \Delta x$$

as a definite integral. To reinforce your work visually, draw $f(x) = x^3$ on your calculator using $2 \le x \le 4$ and $0 \le y \le 64$.

ANSWER   60. The graph is shown in Figure 36.

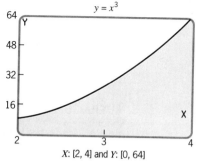

X: [2, 4] and Y: [0, 64]

Figure 36

Average Value

To illustrate one of the many ways that the Riemann sum definition of a definite integral can be used in application problems, let's consider the problem of finding the average value for a function over a desired interval. Let $y = f(x)$ and let $\bar{y}$ denote the average value of $y_1, y_2, \ldots, y_n$. That is,

$$\bar{y} = \frac{y_1 + y_2 + y_3 + \cdots + y_n}{n} = \frac{1}{n}[f(x_1) + f(x_2) + \cdots + f(x_n)]$$

This expression for $\bar{y}$ gives the average value of the function for n values. To find the average value of a continuous function over a closed interval $[a, b]$, it seems natural to generalize this expression for $\bar{y}$ by taking the limit of the expression as n increases without bound.

$$\bar{y} = \lim_{n \to \infty} \frac{1}{n}[f(x_1) + f(x_2) + \cdots + f(x_n)], \qquad y_i = f(x_i), \quad i = 1, 2, \ldots, n$$

To use a definite integral to evaluate this limit, we multiply the numerator and denominator by $b - a$ and then use summation and limit properties to put the expression in the following form:

$$\bar{y} = \lim_{n \to \infty} \left(\frac{1}{n}\right) \left(\frac{b - a}{b - a}\right) [f(x_1) + f(x_2) + \cdots + f(x_n)]$$

$$= \frac{1}{b - a} \lim_{n \to \infty} [f(x_1) + f(x_2) + \cdots + f(x_n)] \left(\frac{b - a}{n}\right)$$

Now let $\Delta x = (b - a)/n$ and substitute this value in the preceding equation.

$$\bar{y} = \frac{1}{b-a} \lim_{n \to \infty} [f(x_1)\,\Delta x + f(x_2)\,\Delta x + \cdots + f(x_n)\,\Delta x]$$

$$= \frac{1}{b-a} \lim_{n \to \infty} \sum_{i=1}^{n} f(x_i)\,\Delta x$$

$$= \frac{1}{b-a} \int_a^b f(x)\,dx$$

AVERAGE VALUE

If $y = f(x)$ is a continuous function over $a \le x \le b$, then the **average value** $\bar{y}$ of the function over $a \le x \le b$ is

$$\bar{y} = \frac{1}{b-a} \int_a^b f(x)\,dx$$

EXAMPLE 62   The average cost of a certain auto part is expected to increase over the next 2 years. Find the average cost of the part over the next 2 years if $C(x) = 8 + x^2$.

SOLUTION

$$\bar{y} = \frac{1}{b-a} \int_a^b f(x)\,dx = \frac{1}{2-0} \int_0^2 (8 + x^2)\,dx$$

$$= \frac{1}{2}\left[8x + \frac{x^3}{3} \right]\Bigg|_0^2 = \frac{1}{2}\left[8(2) + \frac{2^3}{3} \right] - \frac{1}{2}\left[8(0) + \frac{0^3}{3} \right]$$

$$= \frac{1}{2}\left(16 + \frac{8}{3} \right) - \frac{1}{2}(0 + 0) = \frac{1}{2}\left(\frac{48}{3} + \frac{8}{3} \right) - 0$$

$$= \frac{1}{2}\left(\frac{56}{3} \right) = \frac{28}{3}$$

As shown in Figure 37, the average value is 28/3 over the interval [0, 2].   ■

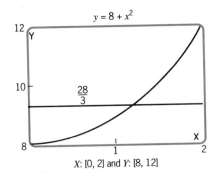

$$X: [0, 2] \text{ and } Y: [8, 12]$$

Figure 37

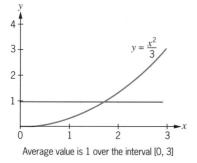

Average value is 1 over the interval [0, 3]

Figure 38

Practice Problem 4 Find the average of $f(x) = x^2/3$ from $x = 0$ to $x = 3$. Draw the graph of f on $[0, 3]$ and a horizontal line at the average value.

ANSWER $\dfrac{1}{3 - 0} \displaystyle\int_0^3 \dfrac{x^2}{3}\, dx = 1$. The graph is shown in Figure 38.

EXAMPLE 63 Suppose that the temperature, in Celsius degrees, t hours after midnight at a certain station, is given by the function

$$C(t) = t^3 - 3t + 15, \qquad 0 \le t \le 3$$

(a) What was the average temperature over this period?
(b) What was the average temperature over $1 \le t \le 3$?

SOLUTION

(a)
$$\overline{C} = \frac{1}{b - a} \int_a^b f(x)\, dx = \frac{1}{3 - 0} \int_0^3 (t^3 - 3t + 15)\, dt$$

$$= \frac{1}{3}\left[\frac{t^4}{4} - \frac{3t^2}{2} + 15t\right]\Big|_0^3 = \frac{1}{3}\left[\frac{3^4}{4} - \frac{3}{2}(3)^2 + 15(3)\right]$$

$$= \frac{1}{3}\left(\frac{81}{4} - \frac{27}{2} + 45\right) = \frac{1}{3}\left(\frac{27}{4} + \frac{180}{4}\right) = \frac{1}{3}\left(\frac{207}{4}\right)$$

$$= \frac{69}{4} = 17.25°C$$

(b)
$$\overline{C} = \frac{1}{b - a} \int_a^b f(x)\, dx = \frac{1}{3 - 1} \int_1^3 (t^3 - 3t + 15)\, dt$$

$$= \frac{1}{2}\left[\frac{t^4}{4} - \frac{3t^2}{2} + 15t\right]\Big|_1^3$$

$$= \frac{1}{2}\left[\frac{3^4}{4} - \frac{3}{2}(3)^2 + 15(3)\right] - \frac{1}{2}\left[\frac{1^4}{4} - \frac{3}{2}(1)^2 + 15(1)\right]$$

$$= \frac{1}{2}\left(\frac{81}{4} - \frac{27}{2} + 45\right) - \frac{1}{2}\left(\frac{1}{4} - \frac{3}{2} + 15\right)$$

$$= \frac{207}{8} - \frac{1}{2}\left(\frac{1}{4} - \frac{6}{4} + \frac{60}{4}\right) = \frac{207}{8} - \frac{1}{2}\left(\frac{55}{4}\right)$$

$$= \frac{207}{8} - \frac{55}{8} = \frac{152}{8} = 19°C$$

Calculator Note

There are functions, such as

$$f(x) = e^{-x^2} \quad \text{and} \quad f(x) = \frac{\log(1 + 2x)}{x}$$

that do not have antiderivatives. However, we can obtain values of definite integrals and graphical antiderivatives.

SUMMARY

Evaluating definite integrals using Riemann sums is more rigorous than what was done in Section 5.6. Even though it is not necessary to calculate a Riemann sum in order to evaluate a definite integral, you should work through several problems to reinforce your understanding of the definite integral. Also, being able to find the average value of a function is useful in many application areas such as the average cost of equipment.

Exercise Set 5.7

1. Use Riemann sums to approximate

$$\int_{7}^{37} f(x)\, dx$$

where $f(x)$ is defined by the following table. (**Hint**: $\Delta x = (37 - 7)/5 = 6$.)

x	10	16	22	28	34
$f(x)$	14	20	26	32	38

2. Use Riemann sums to approximate

$$\int_{1}^{11} f(x)\, dx$$

where $f(x)$ is defined by the following table. (**Hint**: $\Delta x = (11 - 1)/5 = 2$.)

x	2	4	6	8	10
$f(x)$	2	8	18	32	50

3. Find an approximation of the area of the region bounded above by $f(x)$ and below by the x-axis. Use the rectangles as constructed in the figure. (**Hint**: The curve goes through the middle of the top of each rectangle.)

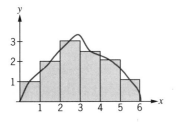

4. Given $f(x) = x^2$, compute the Riemann sum of f over the interval $[0, 1]$ by partitioning the interval into five subintervals of the same length.

5. Repeat Exercise 4 using 10 subintervals of the same length.

Find the average value of each of the following functions over the indicated interval. Draw each function with your graphing calculator and try to estimate the average value before you actually evaluate it.

6. $f(x) = 5, \quad 1 \le x \le 4$
7. $f(x) = x, \quad 2 \le x \le 4$
8. $f(x) = 10 - 2x, \quad 3 \le x \le 4$
9. $f(t) = 3t^2, \quad 0 \le t \le 4$
10. $c(t) = t^2 - t + 3, \quad 0 \le t \le 2$
11. $f(x) = 3x^2 - 2x + 1, \quad 1 \le x \le 3$

Find an approximation for the areas of the following regions using the midpoint of each interval as the point at which to evaluate $f(x)$.

12. $y = 3x + 4$, $[1, 5]$, $n = 4$
13. $y = \dfrac{x^2}{4}$, $[1, 4]$, $n = 3$
14. $y = x^2 - 1$, $[1, 7]$, $n = 6$
15. $y = e^x$, $[0, 4]$, $n = 8$
16. $y = e^x$, $[0, 2]$, $n = 4$

For $F(x)$, calculate at least five points with your graphing calculator, plot the points on paper, and draw the graph of F. Then have your graphing calculator draw the graph of $f(x)$ to verify your work.

17. $F(x) = \displaystyle\int_{0}^{x} \sqrt{t}\, dt, \quad f(x) = \dfrac{2}{3} x^{3/2}$

18. $F(x) = \int_0^x e^{2t}\, dt, \quad f(x) = \dfrac{1}{2} e^{2x}$

19. $F(x) = \int_0^x (t^3 - 2t)\, dt, \quad f(x) = \dfrac{x^4}{4} - x^2$

20. $F(x) = \int_0^x \dfrac{1}{t}\, dt, \quad f(x) = \ln x \quad (x > 0)$

21. **Exam.**   Which of the following is the average value of $1/x$ for $e \le x \le e^e$?

   (a) $\dfrac{2(e^{e-1} - 1)}{e^e}$

   (b) $2(e^{e-1} - 1)$

   (c) $\dfrac{e - 1}{e^e - e}$

   (d) $\dfrac{e - 1}{2}$

   (e) $e - 1$

Applications (Business and Economics)

22. **Cost of Production.**   A corporation finds the marginal cost of a product at various production levels to be

Production (units)	20	24	28	32	36
Marginal Cost ($)	300	310	320	300	240

   Approximate the cost of increasing production from 18 to 38 units (use midpoints to obtain intervals). [**Hint**: $\Delta x = (38 - 18)/5 = 4$.]

23. **Increase in Production.**   Suppose that production at a new company is increasing monthly at a rate of

$$p'(t) = 100e^{0.5t}$$

   How much has production increased from the sixth to the twelfth month? Approximate using the midpoint of six intervals.

Applications (Social and Life Sciences)

24. **Concentration of a Drug.**   The amount of a certain drug in the body of a patient after t hours is

$$A(t) = 6e^{-0.4t}$$

   Find the average amount of the drug present over a 3-day period after the drug has been administered.

Review Exercises

Find each indefinite integral.

25. $\displaystyle\int \dfrac{2x^2 + x - 3}{\sqrt{x}}\, dx$

26. $\displaystyle\int \sqrt{x}(x - 1)^2\, dx$

27. $\displaystyle\int \dfrac{e^x}{e^x + 1}\, dx$

28. $\displaystyle\int 3x^2 - 2e^{4x}\, dx$

29. $\displaystyle\int \dfrac{5x}{\sqrt{x + 2}}\, dx$

30. $\displaystyle\int \dfrac{3x - 1}{x + 1}\, dx$

Evaluate.

31. $\displaystyle\int_0^3 \dfrac{x + 1}{x - 4}\, dx$

32. $\displaystyle\int_2^3 \dfrac{2x}{x^2 + 1}\, dx$

Find the area bounded by the following curves.

33. $f(x) = x^2, \quad f(x) = 0, \quad 4 \le x \le 7$

34. $f(x) = -x^2, \quad f(x) = x$

5.8   APPLICATIONS OF THE DEFINITE INTEGRAL

OVERVIEW   The number of application problems that can be solved using integral calculus is inexhaustible. Many interesting applications in economics involve what we will call consumers' surplus and producers' surplus. Other interesting problems involve continuous money flow. We also consider problems that determine who gets what proportion of the income in a company or in a country. To solve such problems we learn to compute a coefficient of inequality. In this section, we use definite integrals to find

   • Consumers' surplus
   • Producers' surplus

- The amount of a continuous money flow
- The coefficient of inequality

Figure 39(a) shows a demand curve for a product. The equation $p = D(x)$ indicates the price p per unit at which consumers will demand (or purchase) x units. Of course, the higher the price the smaller the demand, and the lower the price the greater the demand. Figure 39(b) shows a supply curve for a given product. The equation $p = S(x)$ indicates the price p per unit at which the manufacturer will supply x units. The greater the price the more units the manufacturer will supply. In Figure 39(a) at price p_1, customers demand x_1 units. Likewise, at price p_1, manufacturers supply x_1 units.

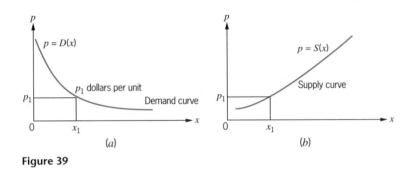

Figure 39

As we can see in Figure 39, some consumers are willing to pay more than the price of p_1 dollars per unit for the product. The consumers who would have paid higher prices have saved money. However, when an industry sets the market value at p_1, it loses the revenue of those who would have paid a higher price. The revenue received at price p_1 is $R = p_1 x_1$ (the area of the rectangle in Figure 40). The income that could have been received is the area under the curve from $x = 0$ to $x = x_1$. The loss of income is the shaded area. Since the loss by the industry is a gain by the consumer, the shaded area is called the **consumers' surplus**, CS, at market price p_1. Since this shaded area is the area between two curves, $p = D(x)$ and $p = p_1$, we have

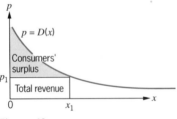

Figure 40

$$CS = \int_0^{x_1} [D(x) - p_1]\, dx$$

CONSUMERS' SURPLUS

If (x_1, p_1) is a point on the price–demand curve for $p = D(x)$, then the **consumers' surplus** at p_1 is

$$CS = \int_0^{x_1} [D(x) - p_1]\, dx$$

EXAMPLE 64   Find the consumer's surplus at a price level of $5 for the demand function

$$p = D(x) = 10 - \frac{5}{6}x$$

SOLUTION   Find the demand for given price, $p_1 = 5$.

$$5 = 10 - \frac{5}{6}x_1 \qquad \text{Replace } p_1 \text{ with 5.}$$

$$-30 = -5x_1 \qquad \text{Multiply by 6.}$$

$$x_1 = 6$$

Copy the calculator graph and shade in the area of consumers' surplus (see Figure 41).

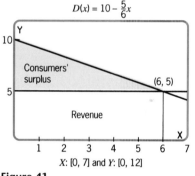

$D(x) = 10 - \frac{5}{6}x$

X: [0, 7] and Y: [0, 12]

Figure 41

Find the consumers' surplus (the shaded region in Figure 41).

$$CS = \int_0^6 \left[10 - \frac{5}{6}x - 5 \right] dx \qquad \text{Substitute in formula.}$$

$$= 5 \int_0^6 \left(1 - \frac{x}{6} \right) dx$$

$$= 5 \left(x - \frac{x^2}{12} \right) \Big|_0^6 = 5(6 - 3) = 15 \qquad \text{Definite integral}$$

EXAMPLE 65   Solve Example 64 at price level $4.

SOLUTION   When $p_1 = \$4$, x_1 is obtained from

$$4 = 10 - \frac{5}{6}x_1 \qquad \text{Replace } p_1 \text{ with 4.}$$

$$24 - 60 = -5x_1 \qquad \text{Multiply by 6.}$$

$$-36 = -5x_1$$

$$x_1 = \frac{36}{5} \qquad \text{Solve for } x_1.$$

So the consumer's surplus is

$$CS = \int_0^{36/5} \left[10 - \frac{5}{6}x - 4 \right] dx \qquad \text{Substitute in formula.}$$

$$= \int_0^{36/5} \left[6 - \frac{5}{6}x \right] dx$$

$$= 6x - \frac{5}{12}x^2 \Big|_0^{36/5} \qquad \text{Definite integral}$$

$$= 6 \cdot \frac{36}{5} - \frac{5}{12} \left(\frac{36}{5} \right)^2$$

$$= \frac{108}{5} = 21.6$$

We see that for a \$1 decrease in price, the consumers' surplus increases from 15 to 21.6, an increase of 44%. ∎

For a given market price, say p_1, some producers are willing to sell a product for less than this price. These producers actually gain by selling at the given market price. In Figure 42, the rectangle gives the total revenue at fixed price p_1, and the area under the curve gives the revenue with the price varying with the price–supply curve. The additional money that the producers gain from the higher price is called the **producers' surplus**, PS (the dark shaded region in Figure 42), and can be expressed in terms of a definite integral, as we did with consumers' surplus. This integral gives the area between the two curves $p = S(x)$ and $p = p_1$ from $x = 0$ to $x = x_1$.

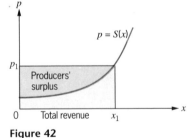

Figure 42

PRODUCERS' SURPLUS

If (x_1, p_1) is a point on the price–supply curve representing $p = S(x)$, then the **producers' surplus** at price p_1 is

$$PS = \int_0^{x_1} [p_1 - S(x)] \, dx$$

EXAMPLE 66   Find the producers' surplus at a price level of $5 for the price–supply equation

$$p = S(x) = \frac{x}{2} + 2$$

SOLUTION   Find the supply x_1 when the price is $5.

$$5 = \frac{x_1}{2} + 2 \qquad P_1 = 5$$

$$x_1 = 6 \qquad \text{Solve for } x_1.$$

Copy the calculator graph and shade in the area of producers' surplus (see Figure 43).

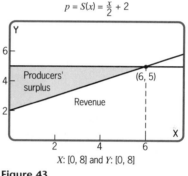

$$p = S(x) = \frac{x}{2} + 2$$

Producers' surplus

(6, 5)

Revenue

X: [0, 8] and Y: [0, 8]

Figure 43

Now find the producers' surplus.

$$PS = \int_0^6 \left[5 - \left(\frac{x}{2} + 2 \right) \right] dx \qquad \text{Substitute in formula.}$$

$$= 3x - \frac{x^2}{4} \Big|_0^6 \qquad \text{Definite integral}$$

$$= 18 - 9 = 9 \qquad \blacksquare$$

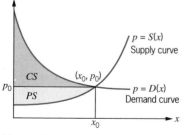

$p = S(x)$
Supply curve

$p = D(x)$
Demand curve

Figure 44

Typical demand and supply curves are shown in Figure 44. The supply curve is usually increasing and the demand curve decreasing. In general, there is a tendency for supply and demand to balance. Let (x_0, p_0) be the point where these two curves intersect. This point where the supply is equal to the demand is called the **equilibrium point**, and p_0 is called the **equilibrium price**. The regions representing consumers' surplus and producers' surplus are shown in the Figure 44.

EXAMPLE 67   Sketch the calculator graphs of the price–demand function

$$p = D(x) = 10 - \frac{5x}{6}$$

and the price–supply function $p = S(x) = \dfrac{x}{2} + 2$ on the same coordinate system. Locate graphically and find algebraically the equilibrium point. Shade with lighter shading the region representing the producers' surplus and with darker shading the region representing the consumers' surplus.

SOLUTION

$$D(x) = S(x) \qquad \text{Solve for } x_0.$$

$$10 - \frac{5x_0}{6} = \frac{x_0}{2} + 2 \qquad \text{Multiply by 6.}$$

$$60 - 5x_0 = 3x_0 + 12$$

$$48 = 8x_0$$

$$x_0 = 6$$

$$p_0 = \frac{x_0}{2} + 2 = \frac{6}{2} + 2 = 5 \qquad \text{Find } p_0.$$

So (6, 5) is the equilibrium point. The shaded regions for the producers' surplus and consumers' surplus are shown in Figure 45. The actual computations of consumers' surplus and producers' surplus are found in Examples 64 and 66.

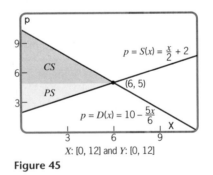

X: [0, 12] and Y: [0, 12]

Figure 45

Practice Problem 1 With your graphing calculator, graph on the same coordinate system $y = 8 - x$ (demand function) and $y = x^2 + 2$ (supply function) and locate the point of intersection, y_0. Then graph $y = y_0$. Can you locate the region of consumers' surplus and producers' surplus? Calculate *CS* and *PS*.

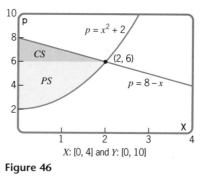

Figure 46

ANSWER   The graph is shown in Figure 46. The equilibrium point is $(2, 6)$: $CS = 2$; and $PS = \frac{16}{3}$.

Continuous Money Flow

A viable business will continually receive income from its activities. Most of the money a business receives is reinvested. For our purposes, we will assume that the income is received continuously over a year and is immediately reinvested at a certain rate of interest compounded continuously. This can be thought of as a continuous money flow.

If we select an initial time to be $t = 0$, then at a later time t, we define the rate of the money flowing in to be

$$\text{Rate of flow} = f(t)$$

This money is immediately invested at rate r and compounded continuously $[f(t)e^{rt}]$. Now, if $A(t)$ is the total amount of money at time t (t can be any unit of time: day, month, year), then

$$\frac{dA}{dt} = f(t)e^{rt} \quad \text{and} \quad A(T) - A(0) = \int_0^T f(t)e^{rt}\, dt$$

AMOUNT OF A CONTINUOUS MONEY FLOW

If $f(t)$ is the rate of flow of money into an investment, then the accumulated value of the amount of continuous money flow over time T at an interest rate r compounded continuously is given by

$$A = \int_0^T f(t)e^{rt}\, dt$$

EXAMPLE 68   An amount of $5000 a year is being invested at 10% compounded continuously for 12 years. Find the amount of the continuous money flow.

SOLUTION   For this problem $f(t)$ is a constant function, $f(t) = \$5000$, $r = 0.10$, and $T = 12$ years.

$$\int_0^{12} 5000e^{0.10t}\, dt = \left. \frac{5000e^{0.10t}}{0.10} \right|_0^{12} \qquad \text{Antiderivative of } e^{kx} \text{ is } \frac{e^{kx}}{k}.$$

$$= \frac{5000e^{(0.10)(12)}}{0.10} - \frac{5000e^{(0.10)(0)}}{0.10}$$

$$= \frac{5000e^{1.2}}{0.10} - \frac{5000}{0.10} = 116{,}005.85$$

The investment will amount to $116,005.85 in 12 years.

Practice Problem 2   An amount of $4000 a year is being invested at 4.5% compounded continuously for 10 years. Find the amount of the continuous money flow.

ANSWER $50,516.64

If $5000 is invested at 10% compounded continuously for 5 years, the amount will be $8243.61. We say that $5000 is the present value of $8243.61 in 5 years. Present value is the principal that will grow to the given sum at a specified future date at a constant rate of interest.

Now if $A = Pe^{rt}$ is the formula for the amount where a fixed P dollars has been invested at an interest rate of r compounded continuously, then the present value P is

$$P = Ae^{-rt} \qquad \text{Divide by } e^{-rt}.$$

Suppose that the sum of money A is to be received continuously (in a series of frequent payments). That is, let $f(t)$ be the rate of flow of money from the fund. The accumulation of all the present values can be obtained from the following integral.

PRESENT VALUE OF A CONTINUOUS MONEY FLOW

If $f(t)$ is the rate of flow of a continuous money flow, then the present value from now until some time T in the future is

$$\int_0^T f(t)e^{-rt}\, dt$$

where r is the interest rate compounded continuously.

EXAMPLE 69   Find the present value of an investment over a 10-year period if there is a continuous money flow of $1000 yearly and interest is at 8% compounded continuously.

SOLUTION   The function $f(t)$ is a constant function equal to $1000 yearly.

$$\int_0^{10} 1000e^{-0.08t}\, dt = \left.\frac{1000e^{-0.08t}}{-0.08}\right|_0^{10} \qquad \text{Antiderivative of } e^{kx} \text{ is } \frac{e^{kx}}{k}.$$

$$= \frac{-1000e^{-0.8} + 1000}{0.08}$$

$$= 6883.39$$

The present value of this money flow is $6883.39.   ■

Practice Problem 3 Find the present value of an investment over a 6-year period if there is a continuous money flow of $3000 yearly and interest is at 3.5% compounded continuously.

ANSWER $16,235.64

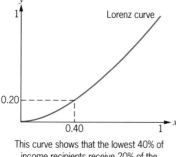

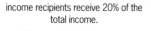

This curve shows that the lowest 40% of income recipients receive 20% of the total income.

Figure 47

Coefficient of Inequality

A **Lorenz curve**, named for the American statistician, M. D. Lorenz, is used by economists to study the distribution of income. It shows what percent of income is received by x percent of the population of a country or state, or perhaps the employees of a corporation. In Figure 47, x represents the cumulative proportion of income recipients, ranked from those with lowest income to those with highest income. The point $x = 0.40$ indicates that 40% of the people have incomes below this point. The function $f(x)$ represents the proportion of total income received. The Lorenz curve in Figure 47 indicates that the lowest 40% of income recipients receive about 20% of the total income. The function has domain [0, 1], range [0, 1], $f(0) = 0$, $f(1) = 1$, and $f(x) \leq x$ for all x in [0, 1].

EXAMPLE 70   The income distribution of a developing country is described by

$$f(x) = \frac{9}{10}x^2 + \frac{1}{10}x$$

(a)   Use your graphing calculator to draw the Lorenz curve for the given function. Use X:[0, 1] with Xscl=.2 and Y:[0, 1] with Yscl = .2.

(b)   The lowest 20% of income recipients receive what percent of the total income?

(c)   The top 10% of income recipients receive what percent of the total income?

SOLUTION

(a)   The graph is shown in Figure 48.

(b) $f(0.20) = \frac{9}{10}(0.20)^2 + \frac{1}{10}(0.20) = 0.056$. The lowest 20% of income recipients receive 5.6% of the total income.

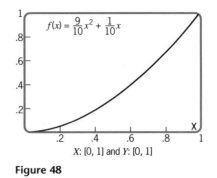

$$f(x) = \frac{9}{10}x^2 + \frac{1}{10}x$$

X: [0, 1] and Y: [0, 1]

Figure 48

(c) $f(0.90) = 0.819$, so the bottom 90% of income recipients receive 81.9% of the total income, leaving $1 - 0.819 = 0.181$ or 18.1% of the total income for the top 10%.

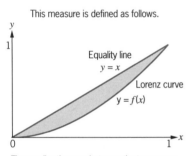

This measure is defined as follows.

The smaller the area between the two curves, the more equitable the distribution of income.

Figure 49

One of the most interesting applications of the area between two curves when one of the curves is a Lorenz curve is the **Gini index** or **coefficient of inequality**. The function $y = f(x)$ represents **equality of income** if the lowest 25% of income recipients receive 25% of the total income, the lowest 50% of income recipients receive 50% of the total income, and so on. The area between the equality line and a Lorenz curve (the shaded region in Figure 49), gives a measure of inequality. This measure is defined as follows.

GINI INDEX OR COEFFICIENT OF INEQUALITY

If $y = f(x)$ is the equation of a Lorenz curve then

$$2 \int_0^1 [x - f(x)] \, dx$$

is called the **Gini index** or the **coefficient of inequality**.

EXAMPLE 71   Find the coefficient of inequality for the Lorenz curve defined in Example 70.

SOLUTION

$$\text{Coefficient of inequality} = 2 \int_0^1 [x - f(x)] \, dx$$

$$= 2 \int_0^1 \left[x - \frac{9}{10}x^2 - \frac{1}{10}x \right] dx \qquad \text{Substitute for } f(x).$$

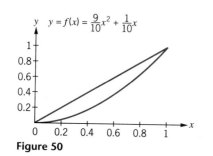

Figure 50

$$= 2 \int_0^1 \left(\frac{9}{10} x - \frac{9}{10} x^2 \right) dx$$

$$= 2 \left[\frac{9}{20} x^2 - \frac{9}{30} x^3 \right] \Big|_0^1 \qquad \text{Definite integral}$$

$$= 2 \left[\frac{9}{20} - \frac{9}{30} \right] = \frac{9}{30} = \frac{3}{10}$$

See Figure 50.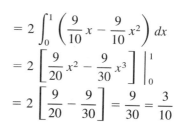

The coefficient of inequality can be used to compare income distributions for various countries, or to check to see if there is a change in the income distribution in a given country. For example, a few years ago the coefficient of inequality for the United States was 0.26 while that for Sweden was 0.18 and that for Brazil was 0.34. From these values we can see that among these three countries income was more equitably distributed in Sweden than in the United States and was the least equitably distributed in Brazil.

According to the *1993 World Almanac,* the gap between rich and poor did not widen in 1991, but the long-term trend pointed to an increasing inequality of income. The richest 20% of all households got 46.5% of all household income in 1991, up from 43.5% in 1971 and 44.4% in 1981. The poorest 20% got 3.8% of all income in 1991, contrasted with 4.1% in 1971 and 1981. (Note that most government assistance to the poor is not included and does affect the income inequity.)

Income distribution can be applied to situations other than nations and states.

Practice Problem 4 The salary or pay distribution of a corporation is

$$f(x) = 0.95x^2 + 0.05x$$

Compute the coefficient of inequality.

ANSWER $\quad 2 \int_0^1 (x - 0.95x^2 - 0.05x) \, dx \approx 0.32$

SUMMARY

We have looked at only a few of the applications of the definite integral. Many are beyond the scope of this book but are used throughout economics and business. You should investigate this topic by doing some research in the areas of economics and business.

Exercise Set 5.8

Applications (Business and Economics)

1. **Coefficient of Inequality.** Find the coefficient of inequality for each of the Lorenz curves.
   (a) $f(x) = x^2$ (b) $f(x) = x^3$

2. **Coefficient of Inequality.** Find the coefficient of inequality for each of the Lorenz curves.
   (a) $f(x) = x^{1.5}$ (b) $f(x) = x^{5/2}$

3. **Continuous Money Flow.** Find the total income produced in 6 years by the continuous money flow if the rate of flow is $f(t) = 2000$ per year and $r = 8\%$ compounded continuously.

4. **Continuous Money Flow.** Find the total income in 10 years by a continuous money flow with a rate of $f(t) = e^{0.04t}$ and $r = 10\%$.

5. **Continuous Money Flow.** Find the total income in 8 years by a continuous money flow with a rate of $f(t) = e^{0.06t}$ and $r = 10\%$.

6. **Continuous Money Flow.** Find the present value at 8% compounded continuously for 6 years for a continuous money flow where $f(t) = 2000$.

7. **Continuous Money Flow.** Find the present value at 8% compounded continuously for 6 years for a continuous money flow where $f(t) = e^{-0.02t}$.

8. **Continuous Money Flow.** Find the present value at 10% compounded continuously for 8 years for a continuous money flow $f(t) = e^{-0.06t}$.

For each pair of equations: (a) *Use your graphing calculator to graph the pair on the same screen,* (b) *find the equilibrium point,* (c) *add the graph of $y = p$ (equilibrium price) and* (d) *find the consumers' surplus and the producers' surplus.*

9. $D(x) = -2x + 6, \quad S(x) = x + 2$

10. $D(x) = -3x + 4, \quad S(x) = 2x + 1$

11. $D(x) = -2x + 16, \quad S(x) = x^2 + 7$

12. $D(x) = -3x + 4, \quad S(x) = x^2$

13. $D(x) = 9 - x, \quad S(x) = x + 3$

14. $D(x) = 8 - 2x, \quad S(x) = x + 2$

15. $D(x) = 6 - x, \quad S(x) = \frac{1}{4}x + 1$

16. $D(x) = 7 - x, \quad S(x) = \frac{1}{4}x + 2$

17. $D(x) = 8 - \frac{1}{4}x^2, \quad S(x) = \frac{1}{8}x^2 + 2$

18. $D(x) = 9 - \frac{1}{12}x^2, \quad S(x) = \frac{1}{12}x^2 + 3$

19. **Lorenz Curve.** The wages of a given industry are described by a Lorenz curve

$$f(x) = 0.8x^2 + 0.2x$$

   (a) What percent of the total wages of the industry are earned by workers classified as the lowest-paid 20% of the work force of the industry?
   (b) What percent of the total wages of the industry are earned by the workers classified as the lowest-paid 80% of the work force of the industry?
   (c) Find the coefficient of inequality for the Lorenz curve.

20. **Coefficient of Inequality.** The Lorenz curve for a laborer in a plant is $f(x) = \frac{11}{12}x^2 + \frac{1}{12}x$ and that of a supervisor is $f(x) = \frac{8}{11}x^4 + \frac{3}{11}x$.
   (a) Compute the coefficient of inequality for each Lorenz curve.
   (b) Which has a more equitable income distribution?

21. **Consumers' Surplus.** The demand for electric shavers in x hundred units per week in terms of the wholesale price p in dollars is given as

$$p = D(x) = \sqrt{400 - 6x}$$

   Determine the consumers' surplus if the wholesale price is set at $16.

22. **Producers' Surplus.** The supply function for the electric shavers in Exercise 21 is

$$p = S(x) = \sqrt{40 + 2x}$$

   Determine the producers' surplus if the wholesale price is set at $10.

23. **Price Equilibrium**
   (a) Using the demand function from Exercise 21 and the supply function from Exercise 22, find the point of equilibrium.
   (b) Find the consumers' surplus at this point.

(c) Find the producers' surplus at this point.

(d) Show parts (a), (b), and (c) on a graph.

24. **Continuous Money Flow.** Find the amount of a continuous money flow if $1000 is being invested each year at 10% compounded continuously for 12 years.

25. **Continuous Money Flow.** The Formans want to establish a fund for the education of their children. If they deposit $5000 yearly for 10 years in a savings account that is paying 8% compounded continuously, find how much is available for the college education of their children. What is the present value of this fund?

26. **Lorenz Curve.** The income distribution in the United States in 1986 was approximated to be the Lorenz curve $f(x) = x^{2.2}$. Find the coefficient of inequality.

27. **Lorenz Curve.** The income distribution of a certain country is given by

$$f(x) = \frac{19}{20} x^2 + \frac{1}{20} x$$

(a) Sketch the Lorenz curve for this function.

(b) Compute $f(0.60)$ and interpret your answer.

(c) Compute the coefficient of inequality.

28. **Coefficient of Inequality.** The Lorenz curve for the distribution of U.S. income in 1929 was $f(x) = x^{3.2}$; find the coefficient of inequality.

29. **Coefficient of Inequality.** An approximation for the income distribution in Argentina in 1986 is given by

$$f(x) = 0.4x - 2.2x^2 + 2.8x^3$$

Compute the coefficient of inequality.

30. **Consumer Demand.** For a certain product, the consumer demand is $p = D(x) = 600 - 10x$ and market price is $100.

(a) How many units will be sold at this price?

(b) What is the total cost of these units?

(c) Find the total value of these units (area under the curve) to the consumer.

(d) Find the consumers' surplus.

(e) Is (d) = (c) − (b)?

31. **Consumer Demand.** For a certain product, the consumer demand is given by $p = D(x) = 356.25 - 0.01x^2$ and the market price is $200.

(a) How many units will be sold at this price?

(b) What is the total cost of these units?

(c) Find the total value of these units (area under the curve) to the consumer.

(d) Find the consumers' surplus.

(e) Is (d) = (c) − (b)?

Review Exercises

Find the values of the following definite integrals.

32. $\displaystyle \int_0^7 \sqrt[3]{x + 1} \, dx$

33. $\displaystyle \int_0^3 \sqrt{4 - x} \, dx$

Chapter Review

Make certain that you understand the following important terms.

Antiderivative	Indefinite integral
Integration constant	Integral sign
Integrand	Definite integral
Lower limit of integration	Upper limit of integration
Area under a curve	Area between curves
Integration by substitution	Consumers' surplus
Continuous money flow	Producers' surplus
Coefficient of inequality	Lorenz curve

Be sure that you are proficient in working with the following formulas.

$$\int x^n \, dx = \frac{x^{n+1}}{n+1} + C, \quad n \neq -1$$

$$\int e^{kx} \, dx = \frac{e^{kx}}{k} + C$$

$$\int kf'(x) \, dx = k \int f'(x) \, dx = kf(x) + C$$

$$\int [f(x) + g(x)] \, dx = \int f(x) \, dx + \int g(x) \, dx$$

$$\int \frac{1}{x} \, dx = \ln |x| + C \qquad \int_a^b f(x) \, dx = F(b) - F(a)$$

Chapter Test

Find the following indefinite integrals.

1. $\displaystyle \int \left(x^3 + 4x^2 + \frac{1}{2} \right) dx$

2. $\displaystyle \int \frac{dx}{(x+1)^{1/3}}$

3. $\displaystyle \int \left(\frac{-4x^2 - 7x + 6}{x^4} \right) dx$

4. $\displaystyle \int x(1 - x^2)^{1/2} \, dx$

5. $\displaystyle \int 2(4x + 1)^{-1} \, dx$

6. $\displaystyle \int (20e^{4x} + 4x^{-3}) \, dx$

7. $\displaystyle \int 12e^{3x^2} x \, dx$

8. $\displaystyle \int \frac{2x + 1}{x - 1} \, dx$

9. $\displaystyle \int x\sqrt{x^2 - 3} \, dx$

10. $\displaystyle \int x(x + 3)^4 \, dx$

Find the value of the following definite integrals using both the properties of integrals and your graphing calculator.

11. $\displaystyle \int_1^4 (6x^2 - 4x + 3) \, dx$

12. $\displaystyle \int_{-2}^0 (4 - 6x)^{1/2} \, dx$

13. Find the differential du if $u = \sqrt{3x^2 - 1}$.

14. Find the area bounded by the curves $y = x^2 - 1$ and $y = 0$. Use your graphing calculator to draw the graph.

15. The Goad Coal Company's marginal cost function is $C'(x) = 12x + 30x^{1/2}$. Find the cost of increasing production from 25 tons to 40 tons.

16. Find the producers' surplus at a price level of $17 for the price–supply equation

$$p = S(x) = 1 + \frac{1}{100} x^2$$

17. If a Lorenz curve is given by $f(x) = 0.9x^2 + 0.1x$, find the coefficient of inequality.

CHAPTER

6

Techniques of Integration and Numerical Integration

In the preceding chapter, we defined the definite integral $\int_a^b f(x)\,dx$ to be the total change in any derivative of $f(x)$ from $x = a$ to $x = b$. We developed or applied this idea to areas under curves. In fact, our discussion of topics relating to the fundamental theorem of integral calculus was associated with the area under a curve. This theorem, however, is important in a multitude of applications. The definition of a definite integral, as the limit of a sum, allows us to apply the fundamental theorem of integral calculus to various application problems.

Sometimes it is difficult or even impossible to find an antiderivative for a given integrand. We introduce two procedures to approximate the value of a definite integral when it cannot be evaluated by any other method. Also in this chapter, we introduce three new techniques for integration. A technique known as integration by parts assists in finding an antiderivative when the function can be considered as a product. We expand our notion of the definite integral to encompass cases in which the interval of integration is infinite. We include in this chapter a discussion of the use of a table of integrals. Infinite (improper) integrals and integration by parts enable us to compute the capital value of a continuous money flow where interest is compounded continuously and where time increases without bound. Improper integrals enable us to study statistical distributions and probability density functions.

6.1 INTEGRATION BY PARTS

OVERVIEW   Recall that the derivative of a product is not the product of the derivatives. Hence, you would expect that the integral of a product is not the product of the integrals. We introduce in this section a very powerful method of integration called **integration by parts**. This method of integration is based on the formula for the derivative of a product. This method is particularly useful in finding integrals involving the product of an exponential (or logarithmic) function and an algebraic function. The need for such a procedure is quite evident. Suppose that we desire to find

$$\int xe^x \, dx$$

Prior to this section we have had no procedure for integrating such a product. In this section we

- Find indefinite integrals using integration by parts
- Evaluate definite integrals using integration by parts

To derive a formula for the antiderivative of a product, recall that

$$\frac{d(uv)}{dx} = u \cdot \frac{dv}{dx} + v \frac{du}{dx}$$ Where u and v are functions of x. Now rewrite the formula.

$$\frac{d(uv)}{dx} = uv' + vu'$$ $v' = \frac{dv}{dx}$ and $u' = \frac{du}{dx}$

$$uv + C = \int (uv' + vu') \, dx$$ Take antiderivatives of both sides of this equation.

Hence $uv + C = \int uv' \, dx + \int vu' \, dx$ Integral of a sum

or $\int uv' \, dx = uv - \int vu' \, dx + C$

INTEGRATION-BY-PARTS FORMULA

If $u(x)$ and $v(x)$ are functions whose derivatives exist, then

$$\int uv' \, dx = uv - \int vu' \, dx + C$$

or

$$\int u \, dv = uv - \int v \, du + C$$

The second part of the preceding formula can be obtained by substituting

$$dv = v' \, dx \quad \text{and} \quad du = u' \, dx \qquad dv = \frac{dv}{dx} \cdot dx$$

When an antiderivative is found by using the formula just derived, the process is called **integration by parts**. The factors u and dv should be chosen so that $\int v \, du$ is simpler to evaluate than the original problem.

In Example 1, we return to the problem that we stated at the beginning of the section.

EXAMPLE 1   Find $\int xe^x \, dx$.

SOLUTION   We use the following substitutions:

$$x = u \quad \text{and} \quad e^x \, dx = dv$$

We can see that $dx = du$ and we now find v by integrating:

$$dv = e^x \, dx$$

$$v = \int e^x \, dx$$

$$v = e^x \qquad \qquad \text{Constant is added later.}$$

Therefore,

$$\int \overset{u}{x} \, \overset{dv}{\overbrace{e^x \, dx}} = \overset{u}{x} \, \overset{v}{e^x} - \int \overset{v}{e^x} \, \overset{du}{\overbrace{dx}} + C = xe^x - e^x + C$$

For some functions, it does not matter which factor is chosen for u or which is selected for dv in $\int u\ dv$; however, for many functions the choice of u and dv is critical in the integration process. For example, suppose that in Example 1 we had chosen $u = e^x$ and $dv = x\ dx$. We would then have $du = e^x\ dx$ and $v = \int x\ dx = x^2/2$. Then by using the formula we would have

$$\int e^x x\ dx = \frac{x^2}{2} \cdot e^x - \int \frac{x^2}{2} e^x\ dx$$

which is more complicated than the original integral. Thus, our first choice for u and dv was the appropriate choice in order to perform the integration. Undoubtedly you are asking, How do I know which factor to call u and which to call dv? To a large extent, it is a matter of some trial and error and practice. You should, however, work toward making the second integral, $\int v\ du$, one which you can readily integrate. Also, you must be able to find v, given dv. After we have done some examples and you have had the opportunity to see some problems solved, we will give some helpful hints for selecting the proper substitutions.

NOTE: In Example 2, we could use substitution as we did in Section 5.4. However, we integrate by parts to illustrate that there are times when you have a choice of techniques of integration to use in a particular problem.

EXAMPLE 2　Find $\int x\sqrt{1 + x}\ dx$.

SOLUTION　We use the following substitutions because dv can be integrated to find v.

$$u = x \quad \text{and} \quad dv = \sqrt{1 + x}\ dx$$

From these substitutions we obtain

$$du = dx \quad \text{and} \quad v = \int dv = \int (1 + x)^{1/2}\ dx = \frac{2}{3}(1 + x)^{3/2}$$

Substitution in the integration-by-parts formula

$$\int u\ dv = uv - \int v\ du$$

yields

$$\int \overbrace{x}^{u} \overbrace{\sqrt{1+x}\ dx}^{dv} = \overbrace{x}^{u} \overbrace{\frac{2}{3}(1+x)^{3/2}}^{v} - \int \overbrace{\frac{2}{3}(1+x)^{3/2}}^{v} \overbrace{dx}^{du} + C$$

Since

$$\int (1+x)^{3/2}\ dx = \frac{2}{5}(1+x)^{5/2}$$

we now have

$$\int x\sqrt{1+x}\ dx = \frac{2}{3}x(1+x)^{3/2} - \frac{4}{15}(1+x)^{5/2} + C$$

Calculator Note

Use your graphing calculator to find several values for

$$F(x) = \int_{0}^{x} t\sqrt{t+1}\ dt$$

and then plot these points on a graph and draw a curve through the points. Then have your calculator draw the graph of $y(x) = \frac{2}{3}x(1+x)^{3/2} - \frac{4}{15}(1+x)^{5/2}$. How do these graphs compare? Are you surprised that they differ only by the constant of integration, C?

Practice Problem 1   Find $\int xe^{2x}\ dx$.

ANSWER $\dfrac{xe^{2x}}{2} - \dfrac{e^{2x}}{4} + C$

EXAMPLE 3   Find $\int \ln x\ dx$.

SOLUTION   By just looking at this integrand, it would appear that there is only one factor. However, let's use $(\ln x)$ as one factor and (dx) as the other factor because $\ln x$ can be differentiated and dx can be integrated. Since we cannot integrate $(\ln x)$, we choose the following substitutions and work as follows:

$$u = \ln x \qquad dv = dx$$

$$du = \frac{1}{x}\ dx \qquad v = \int dv = \int dx$$

$$v = x$$

$$\int \overbrace{(\ln x)}^{u} \overbrace{(1\ dx)}^{dv} = \overbrace{(\ln x)}^{u} \overbrace{(x)}^{v} - \int \overbrace{x \frac{1}{x} dx}^{v\ du} + C$$

$$= x \ln x - \int dx + C$$

$$= x \ln x - x + C$$

Repeated Integration by Parts

Sometimes it is necessary to repeat the integration by parts in order to integrate a function. When this occurs, you usually select for u a type of function that will become a constant after several differentiations. Let's take a look at this situation in the next example.

EXAMPLE 4 Find $\int x^2 e^x\ dx.$

SOLUTION Since x^2 will become a constant after two differentiations, we select our substitutions and work as follows:

$$u = x^2 \qquad dv = e^x\ dx$$

$$du = 2x\ dx \qquad v = \int dv = \int e^x\ dx$$

$$v = e^x$$

$$\int \overbrace{(x^2)}^{u} \overbrace{(e^x\ dx)}^{dv} = \overbrace{(x^2)}^{u} \overbrace{(e^x)}^{v} - \int \overbrace{(e^x)}^{v} \overbrace{(2x\ dx)}^{du} + C$$

$$= x^2 e^x - 2 \int x e^x\ dx + C$$

Note that $\int x^2 e^x\ dx$ is now expressed in terms of $\int x e^x\ dx$, which we found in Example 1. Therefore, using the substitution $\int x e^x\ dx = x e^x - e^x$ from that example, we now have

$$\int x^2 e^x\ dx = x^2 e^x - 2[x e^x - e^x] + C = x^2 e^x - 2x e^x + 2e^x + C$$

We now list some hints for using the integration-by-parts formula.

Hints for Integration by Parts

1. You must be able to integrate dv. It will usually be the most complicated part that fits an integration formula.
2. An application of the formula should produce an integral that is easier to integrate.
3. For $\int x^p e^{kx}\, dx$, $p > 0$, let $u = x^p$ and $dv = e^{kx}\, dx$.
4. For $\int x^p (\ln x)^q\, dx$, $p \geq 0$, let $u = (\ln x)^q$ and $dv = x^p\, dx$.

EXAMPLE 5   Find $\int x \ln x\, dx$.

SOLUTION   Using hint 4, we let

$$u = \ln x \qquad dv = x\, dx$$

$$du = \frac{1}{x}\, dx \qquad v = \int dv = \int x\, dx = \frac{x^2}{2}$$

$$\int \overbrace{(\ln x)}^{u}\, \overbrace{(x\, dx)}^{dv} = \overbrace{(\ln x)}^{u} \overbrace{\left(\frac{x^2}{2}\right)}^{v} - \int \overbrace{\left(\frac{x^2}{2}\right)}^{v} \overbrace{\left(\frac{1}{x}\, dx\right)}^{du} + C$$

$$= \frac{x^2 \ln x}{2} - \frac{1}{2} \int x\, dx + C$$

$$= \frac{x^2 \ln x}{2} - \frac{1}{2} \cdot \frac{x^2}{2} + C$$

$$= \frac{x^2 \ln x}{2} - \frac{x^2}{4} + C$$

EXAMPLE 6   Evaluate $\int \frac{xe^x}{(x + 1)^2}\, dx$.

SOLUTION   We use the following substitutions:

$$u = xe^x \qquad\qquad dv = \frac{1}{(x + 1)^2}$$

$$du = (xe^x + e^x)\, dx \qquad\qquad v = \int dv = \int \frac{1}{(1 + x)^2}\, dx$$

$$= e^x(1 + x)\, dx \qquad\qquad v = \frac{-1}{x + 1}$$

Therefore, we have

$$\int \frac{xe^x}{(x+1)^2}\,dx = \overset{u}{\overbrace{(xe^x)}} \overset{v}{\overbrace{\left(\frac{-1}{x+1}\right)}} - \int \overset{v}{\overbrace{\left(\frac{-1}{x+1}\right)}} \overset{du}{\overbrace{[e^x(x+1)\,dx]}}$$

$$= \frac{-xe^x}{x+1} + \int e^x\,dx$$

$$= \frac{-xe^x}{x+1} + e^x + C$$

$$= \frac{e^x}{x+1} + C$$

We can evaluate definite integrals as well as indefinite integrals using the integration-by-parts formula.

DEFINITE INTEGRAL: INTEGRATION BY PARTS

$$\int_a^b u\,dv = uv\,\Big|_a^b - \int_a^b v\,du$$

where the limits of integration are for x since both u and v are functions of x.

In Chapter 5 we stated that definite integrals can be found on most graphing calculators. However, we will continue to show the techniques involved in the calculation of the definite integral and check with a graphing calculator.

EXAMPLE 7   Find

$$\int_1^e \ln x\,dx$$

using integration by parts and check with numerical integration on your graphing calculator.

SOLUTION   From Example 3 we know that

$$\int \ln x\,dx = x \ln x - x + C$$

Therefore, we now have

$$\int_1^e \ln x\,dx = (x \ln x - x)\,\Big|_1^e = (e \ln e - e) - (1 \ln 1 - 1)$$

$$= (e - e) - (0 - 1) = 1$$

> ✓ Calculator Check: fnInt(ln x, x, 1, e) = 1

We know that definite integrals are associated with the area under a curve. Draw the graph of $f(x) = \ln x$ with your calculator and shade the area from $x = 1$ to $x = e$. (Use e^1 on your calculator to get e.) See Figure 1.

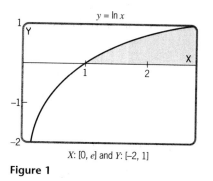

X: [0, e] and Y: [−2, 1]

Figure 1

Practice Problem 2 Find

$$\int_1^e 2x \ln x \, dx$$

using integration by parts and check with numerical integration on your graphing calculator.

ANSWER $\displaystyle\int_1^e 2x \ln x \, dx = \frac{e^2}{2} + \frac{1}{2} \approx 4.2$

> ✓ Calculator Check: fnInt($2x$ ln x, x, 1, $e \wedge 1$) ≈ 4.2

EXAMPLE 8   Find $\displaystyle\int_0^4 x^2\sqrt{1 + 2x} \, dx$ and check.

SOLUTION   Let

$$u = x^2 \qquad\qquad dv = (1 + 2x)^{1/2} \, dx$$

$$du = 2x \, dx \qquad v = \int dv = \int (1 + 2x)^{1/2} \, dx = \frac{(1 + 2x)^{3/2}}{3}$$

We now substitute into the formula to get

$$\int_0^4 x^2\sqrt{1 + 2x} \, dx = x^2\,\frac{(1 + 2x)^{3/2}}{3}\,\bigg|_0^4 - \int_0^4 \frac{(1 + 2x)^{3/2}}{3}\,2x \, dx$$

Before we proceed, we must find

$$\int \frac{(1 + 2x)^{3/2}}{3} 2x \, dx$$

Let

$$u = 2x \qquad\qquad dv = \frac{(1 + 2x)^{3/2}}{3} \, dx$$

$$du = 2 \, dx \qquad\qquad v = \int dv = \frac{(1 + 2x)^{5/2}}{15}$$

We can now complete the problem.

$$\int_0^4 x^2 \sqrt{1 + 2x} \, dx = \frac{x^2(1 + 2x)^{3/2}}{3} \Big|_0^4 - \frac{2x(1 + 2x)^{5/2}}{15} \Big|_0^4$$

$$+ \int_0^4 \frac{(1 + 2x)^{5/2}}{15} \cdot 2 \, dx$$

$$= \left[\frac{x^2(1 + 2x)^{3/2}}{3} - \frac{2x(1 + 2x)^{5/2}}{15} + \frac{2(1 + 2x)^{7/2}}{105} \right] \Big|_0^4$$

$$= \left[\frac{16(9)^{3/2}}{3} - \frac{8(9)^{5/2}}{15} + \frac{2(9)^{7/2}}{105} \right] - \left[\frac{2}{105} \right]$$

$$= 144 - \frac{648}{5} + \frac{1458}{35} - \frac{2}{105}$$

$$= \frac{5884}{105} \approx 56.04$$

✓ Calculator Check: fnInt($x \wedge 2 \sqrt{(1 + 2x)}$, x, 0, 4) ≈ 56.04

Draw the graph of $y = x^2 \sqrt{1 + 2x}$ with your graphing calculator and shade the area being found (see Figure 2).

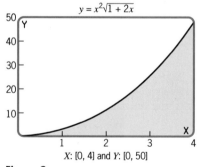

$$y = x^2\sqrt{1 + 2x}$$

X: [0, 4] and Y: [0, 50]

Figure 2

EXAMPLE 9 The rate at which natural gas will be produced from a well t years after production begins is estimated to be

$$P(t) = 100te^{-0.1t}$$

thousand cubic feet per year. How many cubic feet will be produced in the first 2 years that the well is in production? Draw the graph of the production function with your calculator and shade the area being calculated.

SOLUTION

$$\int_0^2 100te^{-0.1t}\, dt = 100 \int_0^2 te^{-0.1t}\, dt$$

Let

$$u = t \qquad dv = e^{-0.1t}\, dt$$

$$du = dt \qquad v = \int dv = \int e^{-0.1t}\, dt$$

$$v = -\frac{1}{0.1} e^{-0.1t} = -10e^{-0.1t}$$

Thus

$$100 \int_0^2 te^{-0.1t}\, dt = 100 \left[\left. (-10te^{-0.1t}) \right|_0^2 + 10 \int_0^2 e^{-0.1t}\, dt \right]$$

$$= 100[-10te^{-0.1t} - 100e^{-0.1t}] \Big|_0^2$$

$$= -1000te^{-0.1t} - 10{,}000e^{-0.1t} \Big|_0^2$$

$$= [(-1000)(2)e^{-0.2} - 10{,}000e^{-0.2}] - (0 - 10{,}000e^0)$$

$$= 175.23$$

The graph of the production function is shown in Figure 3.

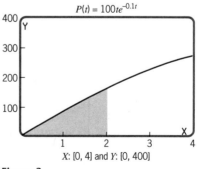

$$P(t) = 100te^{-0.1t}$$

X: [0, 4] and Y: [0, 400]

Figure 3

> ✓ Calculator Check: fnInt($100xe^{-.1x}$, x, 0, 2) ≈ 175.23

Therefore, 175.23 · 1000 = 175,230 cubic feet will be produced in the first 2 years.   ∎

SUMMARY

Knowing *how* to use an integration technique is important, but you must also know *when* to use a particular technique. When do you use substitution, or a formula, or integration by parts? You must practice *recognizing* which technique to use as much as how to use the technique. Just a slight change in the integrand calls for a different technique.

Integrand	Technique
$\int x \ln x \, dx$	Integration by parts
$\int \dfrac{\ln x}{x} \, dx$	Substitution and power rule
$\int \dfrac{1}{x \ln x} \, dx$	Substitution and ln rule

Exercise Set 6.1

Find the following indefinite integrals.

1. $\int 3xe^x \, dx$

2. $\int xe^{3x} \, dx$

3. $\int 2xe^{-x} \, dx$

4. $\int xe^{-2x} \, dx$

5. $\int (x-1)e^x \, dx$

6. $\int (x+1)e^x \, dx$

7. $\int (2x-1)e^{3x} \, dx$

8. $\int (3x+2)e^{4x} \, dx$

9. $\int (3x+1)e^{-2x} \, dx$

10. $\int (4x-3)e^{-3x} \, dx$

11. $\int 5x \ln x \, dx$

12. $\int 3x^2 \ln x \, dx$

13. $\int x^2 \ln x \, dx$

14. $\int 3x^3 \ln x \, dx$

15. $\int x^3 \ln x \, dx$

16. $\int 2x^3 \ln x \, dx$

17. $\int 3x\sqrt{1+x} \, dx$

18. $\int 3x(1+x)^{-1/2} \, dx$

19. $\int (3x-1)(x+2)^{-1/2} \, dx$

20. $\int (xe^{2x} - \ln x) \, dx$

21. $\int \dfrac{x}{(x+4)^5} \, dx$

22. $\int \dfrac{x}{\sqrt{2+3x}} \, dx$

23. $\int \dfrac{1}{x(\ln x)^3} \, dx$

24. $\int \dfrac{(\ln x)^2}{x} \, dx$

25. $\int \dfrac{xe^{2x}}{(2x+1)^2} \, dx$

26. $\int \dfrac{x^3 e^{x^2}}{(x^2+1)^2} \, dx$

Evaluate the definite integrals using both the integration-by-parts formula and your graphing calculator.

27. $\int_1^3 xe^{4x}\, dx$

28. $\int_0^4 xe^{-2x}\, dx$

29. $\int_{1/2}^{e/2} \ln{(2x)}\, dx$

30. $\int_0^2 \dfrac{x^2}{e^x}\, dx$

31. $\int_0^1 \ln{(1+2x)}\, dx$

32. $\int_1^e x^4 \ln x\, dx$

33. $\int_0^1 x^2 e^x\, dx$

Locate with your graphing calculator the areas bounded by the following and use calculus techniques to find the areas. Check your answers with your graphing calculator.

34. $y = xe^x$, $y = 0$, $x = 1$, and $x = 2$

35. $y = \ln x$, $y = 0$, $x = 1$, and $x = 2$

36. $y = x\sqrt{2x+1}$, $y = 0$, $x = 0$, and $x = 4$

37. $y = \dfrac{x}{(x+2)^2}$, $y = 0$, $x = 0$, and $x = 1$

The following exercises may require the use of the integration-by-parts formula more than once, or they may be integrable by previous methods. Check your answers with your graphing calculator.

38. $\int x^2 e^{3x}\, dx$

39. $\int xe^{3x^2}\, dx$

40. $\int \dfrac{3 \ln x}{x}\, dx$

41. $\int \dfrac{2 \ln x}{x^2}\, dx$

42. $\int x^3 e^{x^2}\, dx$

Use your graphing calculator to approximate the integral to four decimal places.

43. $\int_0^1 e^{x^3}\, dx$

44. $\int_0^4 \dfrac{1}{\sqrt{x^3+2}}\, dx$

45. $\int_{-1}^1 \sqrt{x^6+1}$

46. **Exam**

$$\int_{-1}^0 x\sqrt{x+1}\, dx =$$

(a) $-\dfrac{2}{3}$

(b) $-\dfrac{4}{15}$

(c) $-\dfrac{1}{30}$

(d) 0

(e) $\dfrac{1}{3}$

Applications (Business and Economics)

47. **Total Revenue.**   Find the total revenue for a product over the next 5 years if the demand is given by

$$D(t) = 1000(1 - e^{-t})$$

and the price is given by

$$P(t) = 1.04t$$

[**Hint**: $R(t) = P(t)D(t)$]

48. **Marginal Profit.**   A company gives its marginal profit function as

$$MP = 3x - xe^{-x}$$

If $P(0) = 0$, find $P(x)$.

49. **Production.**   The instantaneous rate of production of a company is given by

$$\frac{dP}{dt} = 20te^{-0.2t}$$

Find the total production for a year, if t is measured in months.

50. **Continuous Cash Flow.**   Find the present value at 10% compounded continuously for 6 years for a continuous cash flow of

$$f(t) = 10{,}000 - 400t$$

51. **Total Cost.**   The marginal cost for producing x electric shavers at a manufacturing company is

$$C'(x) = 0.040x \ln x$$

What is the total cost of producing 100 shavers?

Applications (Social and Life Sciences)

52. **Population**. Suppose that the population P, in thousands, of a town is approximated by

$$P = 30 + 5t - 4te^{-0.1t}$$

where t is time in years after 1980. Find the average population from $t = 1$ to $t = 5$.

53. **Medicine**. The rate, dA/dt, of assimilation of a drug after t minutes is

$$\frac{dA}{dt} = te^{-0.3t}$$

Find the total amount assimilated after 8 minutes.

6.2 MORE TECHNIQUES FOR INTEGRATION

OVERVIEW In Section 11.4, we used the method of substitution to integrate when a direct application of a formula was not possible. In this section we

- Look at the process of substitution again
- Examine graphically what happens in a substitution problem when dealing with the definite integrals and the area under a curve

You should have a list of the basic integration formulas with you as a quick reference when you are using the technique of substitution.

Integration Formulas	
$\displaystyle\int k\,dx = kx + C$	Constant
$\displaystyle\int x^n\,dx = \frac{x^{n+1}}{n+1} + C, \quad n \neq -1$	Simple power rule
$\displaystyle\int u^n\,du = \frac{u^{n+1}}{n+1} + C, \quad n \neq -1$	General power rule
$\displaystyle\int e^{kx}\,dx = \frac{e^{kx}}{k} + C$	Exponential rule
$\displaystyle\int e^u\,du = e^u + C$	General exponential rule
$\displaystyle\int \frac{1}{x}\,dx = \ln|x| + C$	Log rule
$\displaystyle\int \frac{1}{u}\,du = \ln|u| + C$	General log rule
$\displaystyle\int u\,dv = uv - \int v\,du + C$	Integration by parts

When using substitution for indefinite integrals, you can generally follow these guidelines.

Guidelines for Substitution

1. Let u be some function of x. (It will usually be an expression that is raised to a power, an expression under a radical, or an expression in the denominator.)
2. Solve for x and dx in terms of u and du.
3. Substitute into the integral and determine which basic formula to use. You may find it necessary to use more than one substitution.
4. After integrating, you must substitute back and rewrite the antiderivative in terms of x.

EXAMPLE 10   Find $\displaystyle\int \frac{x}{(x + 1)^4}\, dx$.

SOLUTION   Let u equal the expression in the denominator that is raised to the fourth power: $u = x + 1$ (guideline 1). Therefore, $du = dx$ and $x = u - 1$ (guidelines 2 and 3). We are now ready to substitute. We integrate using u and then substitute back for u in terms of x.

$$
\begin{aligned}
\int \frac{x}{(x + 1)^4}\, dx &= \int \frac{u - 1}{u^4}\, du \\
&= \int \left(\frac{u}{u^4} - \frac{1}{u^4} \right) du \\
&= \int (u^{-3} - u^{-4})\, du \\
&= \frac{u^{-2}}{-2} - \frac{u^{-3}}{-3} + C \qquad \text{Substitute back.} \\
&= \frac{-1}{2(x + 1)^2} + \frac{1}{3(x + 1)^3} + C
\end{aligned}
$$

■

EXAMPLE 11   Find $\displaystyle\int \frac{e^{4x}}{1 + e^{4x}}\, dx$.

SOLUTION   Let $u = 1 + e^{4x}$. From this we obtain $du = 4e^{4x}\, dx$, which gives us $\frac{1}{4}\, du = e^{4x}\, dx$. The key here is to recognize that e^{4x} is part of the expression for du. Now we substitute, integrate, and substitute back in terms of x.

$$
\begin{aligned}
\int \frac{e^{4x}}{1 + e^{4x}}\, dx &= \int \left(\frac{1}{1 + e^{4x}} \right) (e^{4x}\, dx) \\
&= \int \left(\frac{1}{u} \right) \left(\frac{1}{4}\, du \right) \\
&= \frac{1}{4} \int \frac{1}{u}\, du
\end{aligned}
$$

$$= \frac{1}{4} \ln |u| + C$$

$$= \frac{1}{4} \ln (1 + e^{4x}) + C \qquad \text{Absolute value is not needed here,} \\ \text{since } 1 + e^{4x} > 0 \text{ for all } x.$$

Calculator Note

With your calculator, compare the values of

$$y_1 = \int_0^x \frac{e^{4t}}{1 + e^{4t}} \, dt \quad \text{and} \quad y_2 = \frac{1}{4} \ln (1 + e^{4x})$$

for $x = -1$, $x = 0$, $x = 1$, $x = 2$. Do you see that the values are the same *constant* apart? How would the two graphs compare?

Practice Problem 1

(a) Find $\displaystyle\int \frac{2x}{(x-1)^5} \, dx$.   (b) Find $\displaystyle\int \frac{e^{-2x}}{1 - e^{-2x}} \, dx$.

ANSWER

(a) $\displaystyle\frac{1 - 4x}{6(x-1)^4} + C$ (b) $\displaystyle\frac{\ln |1 - e^{-2x}|}{2} + C$

COMMON ERROR   When working substitution problems, students often forget to substitute back to the original variable. You must substitute back to the original variable in an indefinite integral.

In some integration problems, there can be more than one way to perform a substitution. There are times when the specific substitution will make little difference, but in others, one way can be much more efficient than the other. With experience and practice, you should be able to determine which substitution will be the most efficient.

EXAMPLE 12   Find $\displaystyle\int \frac{3}{\sqrt{x} - 2} \, dx$.

SOLUTION   None of the basic formulas can be used here. Let's try the substitution

$$u = \sqrt{x} - 2$$

In order to find dx in terms of u and du, we must first solve for x.

$$\sqrt{x} = u + 2$$
$$x = (u + 2)^2 = u^2 + 4u + 4 \qquad \text{Square both sides.}$$
$$dx = (2u + 4)\, du \qquad \text{Differentiate with respect to } u.$$

Now we can substitute into the original integral, perform the integration, and substitute back to x.

$$\int \frac{3}{\sqrt{x} - 2}\, dx = \int \left(\frac{3}{u}\right)(2u + 4)\, du$$

$$= \int \frac{6u + 12}{u}\, du$$

$$= \int \left(6 + \frac{12}{u}\right) du$$

$$= 6u + 12 \ln|u| + C$$
$$= 6(\sqrt{x} - 2) + 12 \ln|\sqrt{x} - 2| + C$$
$$= 6\sqrt{x} + 12 \ln|\sqrt{x} - 2| + C \qquad -12 \text{ is part of the constant, } C.$$

Practice Problem 2 Find $\displaystyle\int \frac{4}{2 - \sqrt{x}}\, dx$.

ANSWER $-8\sqrt{x} - 16 \ln|2 - \sqrt{x}| + C$

Substitution and Definite Integrals

In Section 5.6 we saw that definite integrals are related to the area under a curve. In Section 5.5 we looked at definite integrals and substitution. We now combine these two and see graphically what happens when substitution is used and the limits of integration are changed in a definite integral.

Calculator Note

On page 301 we demonstrated two methods for evaluating a definite integral using substitution. In Method 2, we kept the original limits of integration and obtained the integration in terms of the original variable. In Method 1, we changed the limits of integration to the new variable. We now give a calculator verification that these procedures give the same answers. Then we use Method 1 for the examples in this section. For example, using substitution and changing the limits in the second integral, we have

$$\int_1^4 \frac{dx}{2\sqrt{x}\,(1 + \sqrt{x})^2} = \int_2^3 \frac{du}{u^2} \qquad \text{Using } u = 1 + \sqrt{x} \text{ and } du = \frac{dx}{2\sqrt{x}}$$

With a calculator, we get an answer of $\frac{1}{6}$ for both integrals.

EXAMPLE 13 Find the area bounded by

$$y = \frac{x}{\sqrt{x+1}}$$

and the x-axis from $x = 3$ to $x = 8$.

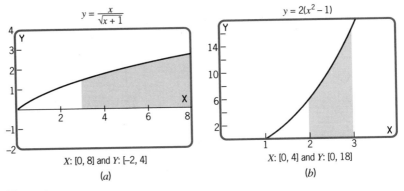

X: [0, 8] and Y: [−2, 4]

(a)

X: [0, 4] and Y: [0, 18]

(b)

Figure 4

SOLUTION   In Figure 4(a) we see the area to be calculated in the shaded portion of the graph. We know that

$$A = \int_3^8 \frac{x}{\sqrt{x+1}} \, dx$$

To evaluate the integral we let

$$u = \sqrt{x+1}$$

$$u^2 = x + 1 \qquad \text{Square both sides and solve for } x.$$

$$x = u^2 - 1$$

$$dx = 2u \, du \qquad \text{Differentiate with respect to } u.$$

We must now change the limits of integration to match the new variable of integration, u.

Change the lower limit: $x = 3$ gives $u = \sqrt{3+1} = 2$
Change the upper limit: $x = 8$ gives $u = \sqrt{8+1} = 3$

We are now ready to integrate.

$$\int_3^8 \frac{x}{\sqrt{x+1}}\, dx = \int_2^3 \frac{u^2-1}{u}\, 2u\, du$$

$$= 2\int_2^3 (u^2-1)\, du$$

$$= 2\left(\frac{u^3}{3} - u\right)\Big|_2^3$$

$$= 2\left[\left(\frac{3^3}{3} - 3\right) - \left(\frac{2^3}{3} - 2\right)\right]$$

$$= \frac{32}{3}$$

Now, we know that

$$\int_3^8 \frac{x}{\sqrt{x+1}}\, dx = \int_2^3 2(u^2-1)\, du = \frac{32}{3}$$

Geometrically, this means that the two regions have the same area [see Figure 4(b)]. ■

Practice Problem 3 Find $\int_0^5 \frac{x}{(x+5)^2}\, dx$.

ANSWER $\ln 2 - \dfrac{1}{2} \approx 0.193$

EXAMPLE 14   A company has fixed costs of \$150. It is found that the marginal cost is given by $C'(x) = x\sqrt{x^2 + 9}$. Find the company's total cost function.

SOLUTION   Fixed costs of \$150 means that $C(0) = 150$. Therefore,

$$C(x) = \int C'(x)\, dx = \int x\sqrt{x^2 + 9}\, dx$$

$$= \frac{1}{2}\int u^{1/2}\, du \qquad \begin{array}{l}\text{Let } u = x^2 + 9 \text{ so that } du = 2x\, dx \\ \text{and } \tfrac{1}{2}\, du = x\, dx.\end{array}$$

$$= \frac{\tfrac{1}{2}u^{3/2}}{\tfrac{3}{2}} + C$$

$$= \frac{u^{3/2}}{3} + C$$

$$= \frac{(x^2 + 9)^{3/2}}{3} + C$$

$$C(0) = 150 \qquad \text{We must now find } C.$$

$$\frac{9^{3/2}}{3} + C = 150$$

$$C = 150 - 9 = 141 \qquad \text{The constant is 141.}$$

Thus, we now have

$$C(x) = \frac{(x^2 + 9)^{3/2}}{3} + 141$$

■

SUMMARY

You may come across a problem where, after substitution, the lower limit is greater than the upper limit of integration. Do not rearrange the limits. It is not unusual for this to occur and it is not incorrect. Also, if you encounter a problem where one substitution does not work or seems to be too involved, try a different substitution.

Exercise Set 6.2

Find each indefinite integral. (Some may require integration by parts.)

1. $\displaystyle\int \frac{12x + 2}{3x^2 + x}\, dx$

2. $\displaystyle\int \frac{6x^2 + 2}{x^3 + x}\, dx$

3. $\displaystyle\int \frac{x^2}{x + 2}\, dx$

4. $\displaystyle\int \frac{2x}{x + 2}\, dx$

5. $\displaystyle\int e^{8x}\, dx$

6. $\displaystyle\int \frac{e^x}{1 + e^x}\, dx$

7. $\displaystyle\int \frac{x}{(x + 3)^3}\, dx$

8. $\displaystyle\int \frac{x^2}{(x + 1)^4}\, dx$

9. $\displaystyle\int \frac{1}{6 + \sqrt{2x}}\, dx$

10. $\displaystyle\int \frac{2\sqrt{x} + 3}{x}\, dx$

11. $\displaystyle\int \frac{1}{1 + \sqrt{x}}\, dx$

12. $\displaystyle\int \frac{1 - \sqrt{x}}{1 + \sqrt{x}}\, dx$

13. $\displaystyle\int x^2(x^3 + 1)^8\, dx$

14. $\displaystyle\int (4x - 1)(4x^2 - 2x + 1)^{1/3}\, dx$

15. $\displaystyle\int \frac{x - 1}{x^2 - 2x + 5}\, dx$

16. $\displaystyle\int \frac{x}{x^2 + 1}\, dx$

17. $\displaystyle\int \left(x + \frac{1}{2}\right) e^{x^2 + x + 1}\, dx$

18. $\displaystyle\int \frac{e^{-x} - 1}{(e^{-x} + x)^2}\, dx$

19. $\displaystyle\int \frac{\sqrt{\ln x}}{x}\, dx$

20. $\displaystyle\int \frac{(\ln x)^5}{x}\, dx$

21. $\displaystyle\int \frac{x}{e^x}\, dx$

22. $\displaystyle\int x \ln(x + 1)\, dx$

Evaluate each definite integral using both the properties of integration and your graphing calculator. (Some may require integration by parts.)

23. $\displaystyle\int_1^4 x\sqrt{3x^2 + 1}\, dx$

24. $\displaystyle\int_0^6 \frac{1}{\sqrt{2x + 4}}\, dx$

25. $\displaystyle\int_2^5 \frac{x}{\sqrt{x - 1}}\, dx$

26. $\displaystyle\int_1^3 \frac{x}{x^2 + 1}\, dx$

27. $\displaystyle\int_0^5 \frac{4x}{x^2 + 9}\, dx$

28. $\displaystyle\int_0^4 \frac{1}{x + 1}\, dx$

29. $\int_0^2 x(4 - x^2)^{1/2}\, dx$

30. $\int_0^4 \sqrt{3x + 4}\, dx$

31. $\int_1^{e^2} \frac{4(\ln x)^3}{x}\, dx$

32. $\int_2^4 \frac{1 + 3x^2}{5(x + x^3)}\, dx$

33. $\int_0^6 \frac{x}{\sqrt{5x + 6}}\, dx$

34. $\int_1^3 (x + 1)(\sqrt{3 - x})\, dx$

35. $\int_0^1 \frac{xe^x}{(x + 1)^2}\, dx$

36. $\int_2^4 \frac{\ln x}{x^2}\, dx$

Applications (Business and Economics)

37. **Demand.** A marginal demand function is defined to be

$$D'(x) = \frac{-2000x}{\sqrt{25 - x^2}}$$

Find the demand function, $D(x)$, if $D(3) = 13,000$.

38. **Profit.** A distributing company has found that the marginal profit function for their company is

$$P'(x) = \frac{9000 - 3000x}{(x^2 - 6x + 10)^2}$$

If $P(3) = 1500$, find $P(x)$.

39. **Production.** A toy manufacturing company finds that the marginal production t hours after a shift begins is

$$p'(t) = \frac{t}{\sqrt{t^2 + 1}}, \qquad 0 \le t \le 8$$

What is the total volume of production during an 8-hour shift?

40. **Revenue.** If a marginal revenue function is given as

$$R'(x) = \frac{20x}{x^2 + 1}$$

what would be the total change in revenue if x is increased from 3 to 7?

Review Exercises

Integrate using integration by parts.

41. $\int 4x^2 \ln x\, dx$

42. $\int 3xe^x\, dx$

6.3   IMPROPER INTEGRALS AND STATISTICAL DISTRIBUTIONS

OVERVIEW   Some applications of integral calculus in business, in statistics, and the social and life sciences involve finding areas that extend infinitely to the right or left or extend infinitely in both directions, as shown by the shaded regions in Figure 5.

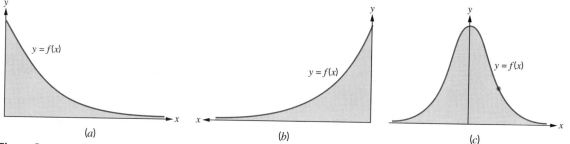

Figure 5

In this section, we show how to determine if the area of such a region exists and how to find the area if it does exist. To do this we must define improper integrals. In this section we

- Determine improper integrals
- Determine whether an improper integral does or does not exist
- Use improper integrals to compute capital values

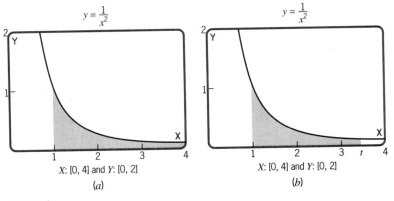

$y = \dfrac{1}{x^2}$

X: [0, 4] and Y: [0, 2]

(a)

$y = \dfrac{1}{x^2}$

X: [0, 4] and Y: [0, 2]

(b)

Figure 6

We introduce improper integrals by finding the area between $y = 1/x^2$ and the x-axis to the right of $x = 1$ [see Figure 6(a)]. Notice that this region is not bounded. Since we have not yet considered how to find the area of such an unbounded region, we consider the following problem. Find the area under this curve from $x = 1$ to t [see Figure 6(b)].

$$A = \int_1^t \frac{1}{x^2}\, dx = -\frac{1}{x}\bigg|_1^t \qquad \text{Antiderivative of } \frac{1}{x^2}.$$

$$= -\frac{1}{t} - \frac{-1}{1}$$

$$= 1 - \frac{1}{t}$$

Now as t becomes large, $1/t$ becomes very small and actually approaches 0. [Remember from Chapter 9 that $\lim\limits_{x \to \infty} (1/x) = 0$.] We can see that the area gets closer and closer to 1. You can verify the above conclusion by substituting values for t and finding A.

$t = 1000$ $\qquad A = 0.999$

$t = 100{,}000$ $\qquad A = 0.99999$

$t = 1{,}000{,}000$ $\qquad A = 0.999999$

Thus, intuitively,

$$\lim_{t \to \infty} \int_1^t \frac{1}{x^2}\, dx = 1$$

This example suggests the following definition.

IMPROPER INTEGRALS

If $f(x)$ is continuous over the interval of integration and if the limit exists, then

1. $\displaystyle\int_a^\infty f(x)\, dx = \lim_{t \to \infty} \int_a^t f(x)\, dx$; see Figure 7(a).

2. $\displaystyle\int_{-\infty}^b f(x)\, dx = \lim_{t \to -\infty} \int_t^b f(x)\, dx$; see Figure 7(b).

3. $\displaystyle\int_{-\infty}^\infty f(x)\, dx = \int_{-\infty}^c f(x)\, dx + \int_c^\infty f(x)\, dx$ [see Figure 7(c)], where c is any real number provided that both improper integrals exist.

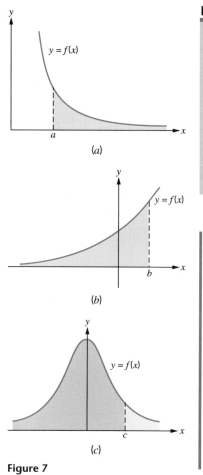

(a)

(b)

(c)

Figure 7

Calculator Note

With your calculator using fnInt, find values for

$$\int_0^t \frac{dx}{\sqrt{x^4 + 1}}$$

by using increasingly large values for t. What do your calculations tell you about

$$\lim_{t \to \infty} \int_0^t \frac{dx}{\sqrt{x^4 + 1}}?$$

Be careful here. Some calculators may give incorrect information if t gets too large. The answer is 1.85407, and if you use 10,000 as the upper limit you get very close to this answer. However, if you use 90,000,000, the answer given is incorrect. Choose your upper limits carefully and use enough limits to see a pattern.

If the indicated limit of an improper integral exists, then the improper integral is said to **converge**; if the limit does not exist, then the improper integral is said to **diverge**.

EXAMPLE 15   Find

$$\int_1^\infty e^{-x}\, dx$$

if the value of this integral exists. With your calculator draw the graph of $y = e^{-x}$ over [0, 10] (see Figure 8).

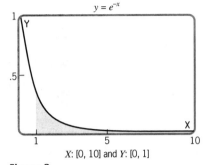

$$y = e^{-x}$$

X: [0, 10] and Y: [0, 1]

Figure 8

SOLUTION

$$\int_1^\infty e^{-x}\,dx \quad \text{equals the value of} \quad \int_1^t e^{-x}\,dx$$

as t becomes very large, so

$$\int_1^t e^{-x}\,dx = -e^{-x}\Big|_1^t = -e^{-t} + e^{-1}$$

As t becomes very large, e^{-t} approaches 0; hence, $-e^{-t} + e^{-1}$ approaches e^{-1}. Thus

$$\int_1^\infty e^{-x}\,dx = e^{-1} = \frac{1}{e}$$

EXAMPLE 16. Find

$$\int_{-\infty}^{-3} \frac{dx}{(1-x)^{3/2}}$$

if it exists. Draw the graph of $y = 1/(1-x)^{3/2}$ over $[-9, 1]$ (see Figure 9).

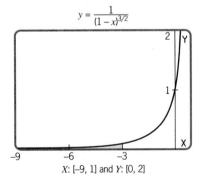

$$y = \frac{1}{(1-x)^{3/2}}$$

X: [-9, 1] and Y: [0, 2]

Figure 9

SOLUTION

$$\int_{-\infty}^{-3} \frac{dx}{(1-x)^{3/2}} = \lim_{t \to -\infty} \int_{t}^{-3} \frac{dx}{(1-x)^{3/2}}$$

$$= \lim_{t \to -\infty} \int_{t}^{-3} (1-x)^{-3/2} \, dx$$

$$= \lim_{t \to -\infty} \left[\frac{-(1-x)^{-1/2}}{-\frac{1}{2}} \right] \Bigg|_{t}^{-3} \qquad \int u^n \, du = \frac{u^{n+1}}{n+1}$$

$$= \lim_{t \to -\infty} [2(4)^{-1/2} - 2(1-t)^{-1/2}]$$

$$= \lim_{t \to -\infty} \left[\frac{2}{\sqrt{4}} - \frac{2}{\sqrt{1-t}} \right]$$

$$= \lim_{t \to -\infty} \left[1 - \frac{2}{\sqrt{1-t}} \right] = 1 \qquad \lim_{t \to -\infty} \frac{2}{\sqrt{1-t}} = 0 \quad \blacksquare$$

Calculator Note

Evaluating improper integrals is not as easy to do on your calculator as definite integrals. The numbers you may have to use as the limits of integration to get a fair estimate of the integral may be very large. With this in mind, it is really better to use the properties of integrals for calculating an improper integral and use the calculator for drawing the graph and perhaps getting an estimate of the answer. However, you should be aware that even numbers as large as 250,000 used as a limit of integration may not give a really accurate answer. Also, you will need to see the graph and use care in working with integrals that are divergent. It may not be readily apparent from the graph that the integral diverges. For example, fnInt($1/x$, x, 1, 1000) = 6.9077 but fnInt($1/x$, x, 1, 1,000,000) = 13.816.

EXAMPLE 17   Find

$$\int_{1}^{\infty} \frac{x \, dx}{(1+x^2)^2}$$

if it exists. Use your graphing calculator to draw the graph of $y = x/(1+x^2)^2$ over [1, 11] (see Figure 10).

SOLUTION

$$\lim_{t \to \infty} \int_{1}^{t} \frac{x \, dx}{(1+x^2)^2} = \lim_{t \to \infty} \int_{1}^{t} (1+x^2)^{-2} x \, dx$$

Now let's find the indefinite integral $\int (1+x^2)^{-2} x \, dx$, which can be obtained by a change of variable. Let $u = 1 + x^2$. Then

$$du = 2x \, dx \quad \text{and} \quad x \, dx = \tfrac{1}{2} \, du$$

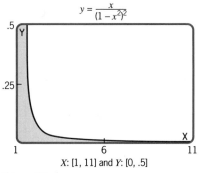

$$y = \frac{x}{(1-x^2)^2}$$

X: [1, 11] and Y: [0, .5]

Figure 10

Substituting gives

$$\int (1 + x^2)^{-2}x \, dx = \int u^{-2}\left(\frac{1}{2}\right) du = \frac{1}{2}\int u^{-2} \, du$$

$$= -\frac{1}{2}u^{-1} + C = -\frac{1}{2(1 + x^2)} + C$$

Hence

$$\lim_{t\to\infty} \int_1^t (1 + x^2)^{-2}x \, dx = \lim_{t\to\infty} \frac{-1}{2(1 + x^2)}\bigg|_1^t = \lim_{t\to\infty}\left[\frac{-1}{2(1 + t^2)} + \frac{1}{4}\right]$$

Since $-1/[2(1 + t^2)]$ approaches 0 as t becomes very large,

$$\int_1^\infty \frac{x \, dx}{(1 + x^2)^2} = 0 + \frac{1}{4} = \frac{1}{4}$$

Practice Problem 1 Evaluate each of the following.

(a) $\displaystyle\int_2^\infty \frac{dx}{(1 + x)^{3/2}}$ (b) $\displaystyle\int_{-\infty}^2 e^{2x} \, dx$

ANSWER

(a) $\dfrac{2}{\sqrt{3}}$ (b) $\dfrac{e^4}{2}$

EXAMPLE 18 Graph $f(x) = 1/x$ and find the area of the region under this curve and above the x-axis to the right of $x = 1$, if this area exists.

SOLUTION

$$A = \int_1^\infty \frac{dx}{x}$$

$$\int_1^t \frac{dx}{x} = \ln x \bigg|_1^t = \ln t - \ln 1 \qquad \text{See Figure 11.}$$

Now, $\ln 1 = 0$; hence,

$$\int_1^t \frac{dx}{x} = \ln t$$

$$\int_1^\infty \frac{dx}{x} = \lim_{t \to \infty} \int_1^t \frac{dx}{x}$$

$$\int_1^\infty \frac{dx}{x} = \lim_{t \to \infty} (\ln t)$$

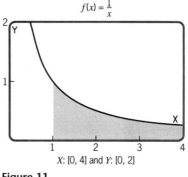

X: [0, 4] and Y: [0, 2]

Figure 11

Now with your calculator (see Figure 11), you can substitute values and intuitively determine that $\ln t$ becomes large as t becomes large and the limit does not exist. That is,

$$\int_1^\infty \frac{dx}{x}$$

does not exist or the integral diverges. Be careful, however, when you use the calculator for a problem like this. Some large upper bounds require a lot of time for your calculator to compute. Numerical integration without some thought given to the problem can be deceiving. ■

Calculator Note

You can approximate infinite limits with your graphing calculator using the numerical integration function. Look at

$$\int_1^\infty \frac{dx}{\sqrt{x^2 + x + 4}}$$

Take the upper bound to be increasingly larger numbers and use this to determine whether

$$\lim_{t \to \infty} \int_1^\infty \frac{dx}{\sqrt{x^2 + x + 4}}$$

converges or diverges. What do you see as you take the upper bound from $x = 10$ to $x = 1000$?

We have already discussed that the present value of a continuous income stream over a fixed number of years can be found by integrating a definite integral. When the number of years is extended indefinitely, this present value is called a **capital value of a money flow** or the **present value of a perpetuity**.

CAPITAL VALUE OF MONEY FLOW

$$\text{Capital value} = \int_0^\infty f(t)e^{-rt}\, dt$$

where $f(t)$ is the annual rate of flow at time t, and r is the annual interest rate, compounded continuously.

EXAMPLE 19 How much capital must you invest to provide a continuous money flow with an annual flow given by $f(t) = 10{,}000$ if the interest rate is 8% compounded continuously?

SOLUTION The capital value of the fund is given by

$$\int_0^\infty 10{,}000e^{-0.08t}\, dt = \lim_{b \to \infty} \int_0^b 10{,}000e^{-0.08t}\, dt$$

$$= \lim_{b \to \infty} \frac{10{,}000e^{-0.08t}}{-0.08} \bigg|_0^b$$

$$= \lim_{b \to \infty} \left[\frac{10{,}000e^{-0.08b}}{-0.08} - \frac{10{,}000}{-0.08} \right] \qquad e^0 = 1$$

$$= \lim_{b \to \infty} \left[\frac{10{,}000}{0.08} \left(1 - \frac{1}{e^{0.08b}} \right) \right]$$

$$= \frac{10{,}000}{0.08} = \$125{,}000 \qquad \lim_{b \to \infty} \left(\frac{1}{e^{0.08b}} \right) = 0$$

Do you see why

$$\lim_{b \to \infty} \left(\frac{1}{e^{0.08b}} \right) = 0?$$

If not, you can show that this is a reasonable answer with your calculator. ■

Probability Density Functions

A study of improper integrals in this chapter enables us to consider probability density functions of a continuous variable. These functions are very important in the study of statistics. A continuous probability density function is used to measure probabilities for a continuous random variable. For example, the normal distribution is the foundation for many statistical tests. First we look at five characteristics of probability density functions, commonly called a **pdf of a distribution**. The continuous variable is called a **random variable**. Then we consider problems relative to these characteristics.

Characteristics of Probability Density Functions

If X is a continuous random variable and $f(x)$ is a probability density function, then:

1. $f(x) \geq 0$ for $(-\infty, \infty)$ The function is always nonnegative.

2. $\displaystyle\int_{-\infty}^{\infty} f(x)\, dx = 1$ The area under the curve is 1.

3. The probability of x in $[a, b]$ is

$$P(a \leq X \leq b) = \int_{a}^{b} f(x)\, dx$$

This is the area under the curve from $x = a$ to $x = b$.

4. The expected value or mean of x is defined to be

$$\mu = E(x) = \int_{-\infty}^{\infty} xf(x)\, dx$$

5. The variance of x (square of standard deviation σ) is defined to be

$$\sigma^2 = \int_{-\infty}^{\infty} (x - \mu)^2 f(x)\, dx \quad \text{or}$$

$$= \int_{-\infty}^{\infty} x^2 f(x)\, dx - \mu^2 \qquad \text{Measure of dispersion or scattering.}$$

Four useful probability density functions of a random variable X are the **uniform** probability density function, the **exponential** probability density funtion, the **normal** probability density function, and the **standard** normal density function (see Figure 12). μ is the mean of the distribution and δ is the standard deviation.

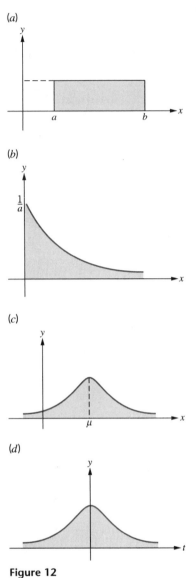

(a)

(b)

(c)

(d)

Figure 12

The **uniform probability density function,** (Figure 12(a)), is given by

$$f(x) = \begin{cases} \left(\dfrac{1}{b-a}\right), & a \le x \le b \\ 0, & \text{otherwise} \end{cases}$$

$$\mu = \frac{a+b}{2}$$

$$\sigma = \frac{b-a}{\sqrt{12}}$$

The **exponential probability density function,** (Figure 12(b)), is given by

$$f(x) = \begin{cases} \dfrac{1}{(a)} e^{-x/a}, & x \ge 0 \\ 0, & \text{otherwise} \end{cases}$$

$$\mu = a$$

$$\sigma = a$$

The **normal probability density function** (Figure 12(c)), is defined

$$f(x) = \frac{1}{\sigma\sqrt{2\pi}} e^{-(x-\mu)^2/2\sigma^2}$$

$$\mu = \mu \quad \text{(of the formula)}$$

$$\sigma = \sigma \quad \text{(of the formula)}$$

The standard normal density function, (Figure 12(d)), is defined by

$$f(x) = \frac{1}{\sqrt{2\pi}} e^{-t^2/2}$$

$$\mu = 0$$

$$\sigma = 1$$

In the normal probability density function and the standard normal density function, the value of μ gives the x-coordinate of the point at which there is a relative maximum and σ determines the spread of the curve.

EXAMPLE 20   Suppose that a random variable X has a pdf defined by

$$f(x) = \begin{cases} 1 - \frac{1}{2}x, & 0 \le x \le 2 \\ 0, & \text{elsewhere} \end{cases}$$

See Figure 13. Find   (a) $P(1 \le X \le 2)$ and   (b) $P(X \le 1)$.

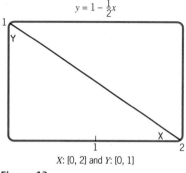

$$y = 1 - \frac{1}{2}x$$

X: [0, 2] and Y: [0, 1]

Figure 13

SOLUTION

(a) $P(1 \le X \le 2) = \int_1^2 \left(1 - \frac{1}{2}x\right) dx = \left(x - \frac{x^2}{4}\right)\Big|_1^2 = 1 - \frac{3}{4} = \frac{1}{4}$

(b) The random variable X assumes nonzero values only in the interval [0, 2] because the pdf is 0 outside of this interval. Therefore, we have

$$P(X \le 1) = P(0 \le X \le 1) = \int_0^1 \left(1 - \frac{1}{2}x\right) dx = \left(x - \frac{x^2}{4}\right)\Big|_0^1$$

$$= \left(\frac{3}{4} - 0\right) = \frac{3}{4}$$

Practice Problem 2 Using the pdf from Example 20, find $P(\frac{1}{2} \le x)$.

ANSWER

$$P\left(\frac{1}{2} \le x\right) = P\left(\frac{1}{2} \le X \le 2\right) = \int_{1/2}^2 \left(1 - \frac{1}{2}x\right) dx = \frac{9}{16}$$

EXAMPLE 21 Because of unpredictable variations in weather and air traffic, the 9:45 A.M. flight from New Orleans to Houston takes off at various times from 9:45 A.M. to 10:00 A.M. What is the probability that a passenger will experience a delay of at most 10 minutes?

SOLUTION We let X be the delay in minutes; therefore, X is in the interval [0, 15]. Thus X is uniformly distributed over the interval and its pdf is

$$f(x) = \begin{cases} \frac{1}{15}, & 0 \le x \le 15 \\ 0, & \text{elsewhere} \end{cases}$$

Thus, the probability of a delay of at most 10 minutes is

$$P(0 \le X \le 10) = \int_0^{10} \frac{1}{15} dx = \frac{1}{15} x \Big|_0^{10} = \frac{10}{15} - 0 = \frac{2}{3}.$$

EXAMPLE 22   A company has found that the average length of each business call is 7 minutes. If the length of a call has an exponential distribution

$$f(x) = \frac{1}{a} e^{-x/a}$$

find the following probabilities.

(a)   The probability that a call will be no more than 5 minutes.
(b)   The probability that a call will be longer than 5 minutes.

SOLUTION

(a)   Since the average call is 7 minutes, $a = 7$, so the pdf for X is $f(x) = \frac{1}{7} e^{-x/7}$, where $x \geq 0$. We now find the probability.

$$P(0 \leq X \leq 5) = \int_0^5 \frac{1}{7} e^{-x/7} \, dx$$

$$= -e^{-x/7} \Big|_0^5$$

$$= -e^{-5/7} - (-e^0)$$

$$= 1 - e^{-5/7}$$

$$\approx 0.510458$$

Therefore, approximately 51% of the calls are 5 minutes or less.

(b)   The probability that a call will be longer than 5 minutes is

$$P(x > 5) = \int_5^\infty \frac{1}{7} e^{-x/7} \, dx$$

However, from the information in part (a) and the fact that the total area under the curve is 1 (why?), we can find the probability as follows:

$$P(x > 5) = 1 - P(0 \leq X \leq 5) = 1 - 0.510458 = 0.489542 \qquad ∎$$

Practice Problem 3   Let X be an exponentially distributed random variable with an average value of 2. Find $P(1 \leq X \leq 4)$.

ANSWER

$$P(1 \leq X \leq 4) = \int_1^4 \frac{1}{2} e^{-x/2} \, dx = -e^{-x/2} \Big|_1^4 = 0.4712$$

SUMMARY

Improper integrals allow us to solve a number of important applications in all areas, from cash flow to probability distributions. Remember, it is not always possible to integrate an improper integral. If a limit can be found, the integral converges, otherwise it is divergent.

Exercise Set 6.3

Find the value of the following improper integrals if they exist.

1. $\int_2^\infty \dfrac{dx}{x^2}$

2. $\int_{-\infty}^{-1} \dfrac{dx}{x^2}$

3. $\int_1^\infty \dfrac{dx}{x^2}$

4. $\int_1^\infty \dfrac{dx}{x^{3/2}}$

5. $\int_1^\infty \dfrac{dx}{(1+x)^{3/2}}$

6. $\int_3^\infty \dfrac{dx}{(1+x)^{3/2}}$

7. $\int_0^\infty e^x\, dx$

8. $\int_{-\infty}^0 e^x\, dx$

9. $\int_0^\infty e^{-x}\, dx$

10. $\int_{-\infty}^{-1} e^{-x}\, dx$

11. $\int_0^\infty e^{-2x}\, dx$

12. $\int_{-\infty}^0 e^{3x}\, dx$

13. $\int_1^\infty \dfrac{x\, dx}{1+x^2}$

14. $\int_{-\infty}^{-1} \dfrac{x\, dx}{1+x^2}$

15. $\int_0^\infty xe^{-x^2}\, dx$

16. $\int_{-\infty}^0 xe^{-2x}\, dx$

17. $\int_2^\infty \dfrac{x\, dx}{(1+x^2)^2}$

18. $\int_{-\infty}^{-1} \dfrac{x\, dx}{(1+x^2)^2}$

19. $\int_0^\infty x^2 e^{-x^3}\, dx$

20. $\int_{-\infty}^{-1} 2x^2 e^{x^3}\, dx$

21. $\int_{-\infty}^\infty \dfrac{x\, dx}{(1+x^2)^2}$

Determine whether the following integrals are convergent or divergent. If convergent, find the value. Check your answers with your graphing calculator.

22. $\int_{-\infty}^\infty x\, dx$

23. $\int_{-\infty}^\infty \dfrac{2x\, dx}{1+x^2}$

24. $\int_{-\infty}^\infty xe^{-x^2}\, dx$

25. $\int_0^\infty xe^{-x}\, dx$

26. $\int_{-\infty}^0 e^{3x}\, dx$

27. $\int_{-\infty}^0 xe^{-x^2}\, dx$

Use your graphing calculator to find each limit, if it exists.

28. $\displaystyle\lim_{t\to\infty} \int_1^t \dfrac{dx}{x^4}$

29. $\displaystyle\lim_{t\to\infty} \int_1^t \dfrac{dx}{\sqrt[3]{x}}$

30. $\displaystyle\lim_{t\to\infty} \int_1^t \dfrac{dx}{\sqrt{x}}$

Applications (Business and Economics)

31. **Perpetuity Trust.** Find the present value of a perpetuity trust that pays a yearly amount given by

$$f(t) = \$2000e^{-0.06t}$$

if interest is compounded continuously.

32. **Perpetuity Trust.** What is the value of a trust that pays $10,000 a year in perpetuity (forever) if interest is compounded continuously at a rate of 6% per year? See Exercise 31.

33. **Capital Value of Perpetuity.** A donor wishes to provide a gift for a university that will provide a continuous money flow of $15,000 per year for scholarships. If money is worth 8% compounded continuously, find the capital value of this gift.

34. **Production.** Suppose that the rate of production of an oil company per month is given in thousands of barrels as

$$\dfrac{dP}{dt} = 26te^{-0.04t}$$

How much oil is produced during the first 12 months of operation? If the production is continued indefinitely, what would be the total amount produced?

35. **Investment.** An investment gives

$$\dfrac{dA}{dt} = 2000e^{-0.12t}$$

dollars per year. Assume that the investment is continued indefinitely, and find the total amount A returned from this investment.

36. **Capital Value.** How much capital must you invest to provide a continuous money flow of $f(t) = 10,000 + 500t$, if the interest rate is 8% compounded continuously?

37. ***Present Value of a Perpetuity.*** If the continuous money flow for a perpetuity is given by $f(t) = R$ dollars per year, show that the present value of the perpetuity is R/r, where r is the rate compounded continuously.

38. ***pdf.*** Show that

$$\int_{-\infty}^{\infty} f(x)\, dx = 1$$

for the uniform density function.

39. ***pdf.*** Show that

$$\int_{-\infty}^{\infty} f(x)\, dx = 1$$

for the exponential density function.

40. ***pdf.*** Using the definition, find the mean of the uniform density function.

41. ***pdf.*** Using the definition, find the standard deviation of the uniform density function.

42. ***pdf.*** Using the definition, find the mean of the exponential density function.

43. ***pdf.*** Using the definition, find the standard deviation of the exponential density function.

44. ***pdf.*** If $a = 5$ and $b = 35$, find $P(10 \le X \le 20)$ for the uniform density function.

45. ***pdf.*** If $a = 10$ for the exponential density function, find $P(X \ge 20)$.

46. ***pdf.*** In Exercise 45, find $P(1 \le X \le 20)$.

47. ***pdf.*** In Exercise 45, find $P(X \le 10)$.

48. ***pdf.*** For $f(x)$ to be a probability density function,

$$\int_{-\infty}^{\infty} f(x)\, dx$$

must equal 1. Find k so that this condition is satisfied for

$$f(x) = \begin{cases} kx^2, & \text{for } 0 < x < 2 \\ 0, & \text{elsewhere.} \end{cases}$$

49. ***pdf.*** Find k so that

$$\int_{-\infty}^{\infty} f(x)\, dx = 1 \quad \text{for } f(x) = \begin{cases} ke^{-x/2}, & \text{for } x > 0 \\ 0, & \text{elsewhere.} \end{cases}$$

50. ***pdf.*** Let X be uniformly distributed over the interval [1, 6]. Find each of the following.
(a) $P(2 \le X \le 3)$ (b) $P(X \ge 2)$
(c) $P(X \le 4)$ (d) $P(X \ge 1)$
(e) Find the expected value (mean).

51. ***pdf.*** The voltage in a 220-volt line varies randomly between 210 and 230 volts.
(a) What is the probability that the voltage will be between 215 and 225 volts?
(b) If a power surge can cause damage to an electrical system if the voltage is more than 225 volts, what percentage of the time will there be a danger of damage?
(c) What is the expected voltage?

52. ***pdf.*** Let X be an exponentially distributed random variable with an average value of 5.
(a) Find the pdf for X.
(b) Find $P(X \le 5)$.
(c) Find $P(0 \le X \le 3)$.
(d) Use the result in part (b) to find $P(X > 5)$.
(e) Find $P(X > 5)$ by evaluating an appropriate improper integral.

53. ***pdf.*** Suppose that the average distance between barges on a river is 300 feet and the distance is exponentially distributed.
(a) What is the probability that the distance between two successive barges will be no more than 150 feet?
(b) What is the probability that the distance will be between 250 and 350 feet?

54. Given the pdf

$$f(x) = \begin{cases} 1.5x - 0.75x^2, & \text{for } 0 \le x \le 2 \\ 0, & \text{elsewhere} \end{cases}$$

use your graphing calculator to graph the function and approximate to three decimal places the necessary value of x to make the equation true.
(a) $P(0 \le X \le x) = 0.3$
(b) $P(x \le X \le 2) = 0.5$

55. Given the pdf

$$f(x) = \begin{cases} 0.75 - 0.375\sqrt{x}, & \text{for } 0 \le x \le 4 \\ 0, & \text{elsewhere} \end{cases}$$

use your graphing calculator to graph the function and approximate to three decimal places the necessary value of x to make the equation true.

(a) $P(0 \le X \le x) = 0.5$

(b) $P(x \le X \le 4) = 0.3$

Applications (Social and Life Sciences)

56. **Drug Elimination.** When a person takes a drug, the body does not absorb all of the drug. One way to determine the amount of drug absorbed is to measure the rate at which the drug is eliminated from the body. A doctor finds that the rate of elimination of a drug in milliliters per minute, t, is given by

$$\frac{dE}{dt} = te^{-0.3t}$$

Assume that the elimination continues indefinitely, and find the total amount eliminated, E.

Review Exercises

Find the following integrals.

57. $\displaystyle \int \frac{5e^x \, dx}{\sqrt[3]{e^x + 2}}$

58. $\displaystyle \int 3xe^{-2x} \, dx$

59. $\displaystyle \int (2x - 3)e^{2x} \, dx$

60. $\displaystyle \int 4x^2 \ln x \, dx$

61. $\displaystyle \int x^4 \ln x \, dx$

62. $\displaystyle \int 3x\sqrt{x + 1} \, dx$

6.4 INTEGRATION USING TABLES

OVERVIEW   We have studied several techniques for finding the antiderivatives of a function. You have probably noticed that finding antiderivatives is not as straightforward and is somewhat more difficult than finding derivatives. Because of this, formulas for integration have been compiled into tables of integrals in most mathematics handbooks. Our goal in this section is to practice properly matching a given integral with a formula found in a table. To accomplish this goal, we have included a very abbreviated table of integrals. It is customary in integral tables to omit the constant of integration. Therefore, you must remember to include it in your answer. In this section, we

- Practice recognizing formulas
- Use substitution to match an integral formula
- Use reduction formulas

In our abbreviated table of integrals we have not included formulas that have already been given. The table is divided into four sections:

Integrals involving $au + b$
Integrals involving $\sqrt{au + b}$
Integrals involving $u^2 \pm a^2$ and $a \pm u^2$
Integrals involving e^{au} and $\ln u$.

This is not, by any means, an exhaustive table, but it will serve to introduce you to the technique. Remember to add the constant of integration. There are handbooks that contain extensive tables of integration.

A Brief Table of Integrals

Integrals Involving $au + b$

1. $\displaystyle\int \frac{u}{au + b}\, du = \frac{u}{a} - \frac{b}{a^2} \ln |au + b|$

2. $\displaystyle\int \frac{u^2}{au + b}\, du = \frac{1}{a^3}\left[\frac{1}{2}(au + b)^2 - 2b(au + b) + b^2 \ln |au + b|\right]$

3. $\displaystyle\int \frac{u}{(au + b)^2}\, du = \frac{1}{a^2}\left[\frac{b}{au + b} + \ln |au + b|\right]$

4. $\displaystyle\int \frac{1}{(au + b)(cu + d)}\, du = \frac{1}{bc - ad} \ln \left|\frac{cu + d}{au + b}\right| \quad (bc - ad > 0)$

5. $\displaystyle\int \frac{1}{u^2(au + b)}\, du = -\frac{1}{bu} + \frac{a}{b^2} \ln \left|\frac{au + b}{u}\right|$

6. $\displaystyle\int \frac{1}{u(au + b)^2}\, du = \frac{1}{b(au + b)} + \frac{1}{b^2} \ln \left|\frac{u}{au + b}\right|$

Integrals Involving $\sqrt{au + b}$

7. $\displaystyle\int u\sqrt{au + b}\, du = \frac{2(3au - 2b)(au + b)^{3/2}}{15a^2}$

8. $\displaystyle\int \frac{u}{\sqrt{au + b}}\, du = \frac{2}{3a^2}(au - 2b)\sqrt{au + b}$

9. $\displaystyle\int \frac{u^n}{\sqrt{au + b}}\, du = \frac{2u^n\sqrt{au + b}}{a(2n + 1)} - \frac{2bn}{a(2n + 1)}\int \frac{u^{n-1}}{\sqrt{au + b}}\, du \quad (n \geq 2)$

Integrals Involving $u^2 \pm a^2$ and $a^2 \pm u^2 \quad (a > 0)$

10. $\displaystyle\int \frac{du}{u^2 - a^2} = \frac{1}{2a} \ln \left|\frac{u - a}{u + a}\right| \quad (u^2 \neq a^2)$

11. $\displaystyle\int \frac{du}{a^2 - u^2} = \frac{1}{2a} \ln \left|\frac{a + u}{a - u}\right| \quad (u^2 \neq a^2)$

12. $\displaystyle\int \frac{du}{\sqrt{u^2 \pm a^2}} = \ln |u + \sqrt{u^2 \pm a^2}|$

13. $\displaystyle\int \frac{du}{u\sqrt{a^2 \pm u^2}} = -\frac{1}{a} \ln \left|\frac{a + \sqrt{a^2 \pm u^2}}{u}\right|$

14. $\displaystyle\int \frac{du}{u^2\sqrt{a^2 \pm u^2}} = -\frac{\sqrt{a^2 \pm u^2}}{a^2 u}$

15. $\displaystyle\int \frac{du}{u^2\sqrt{u^2 - a^2}} = \frac{\sqrt{u^2 - a^2}}{a^2 u}$

16. $\displaystyle\int \sqrt{u^2 \pm a^2}\, du = \frac{1}{2}(u\sqrt{u^2 \pm a^2} \pm a^2 \ln |u + \sqrt{u^2 \pm a^2}|)$

17. $\displaystyle\int u^2\sqrt{u^2 \pm a^2}\, du = \frac{1}{8}[u(2u^2 \pm a^2)\sqrt{u^2 \pm a^2} - a^4 \ln |u + \sqrt{u^2 \pm a^2}|]$

18. $\displaystyle\int \frac{\sqrt{a^2 \pm u^2}}{u}\, du = \sqrt{a^2 \pm u^2} - a \ln \left| \frac{a + \sqrt{a^2 \pm u^2}}{u} \right|$

19. $\displaystyle\int \frac{\sqrt{u^2 \pm a^2}}{u^2}\, du = -\frac{\sqrt{u^2 \pm a^2}}{u} + \ln \left| u + \sqrt{u^2 \pm a^2} \right|$

20. $\displaystyle\int \frac{u^2}{\sqrt{u^2 \pm a^2}}\, du = \frac{1}{2}\left(u\sqrt{u^2 \pm a^2} \mp a^2 \ln \left| u + \sqrt{u^2 \pm a^2} \right| \right)$

Integrals Involving e^{au} and $\ln u$

21. $\displaystyle\int ue^{au}\, du = \frac{e^{au}}{a^2}(au - 1)$

22. $\displaystyle\int u^n e^{au}\, du = \frac{u^n e^{au}}{a} - \frac{n}{a}\int u^{n-1} e^{au}\, du \qquad n > 0$

23. $\displaystyle\int u^n \ln u\, du = \frac{u^{n+1}}{(n+1)^2}\left[(n+1)\ln u - 1\right]$

24. $\displaystyle\int \frac{1}{u \ln u}\, du = \ln |\ln u|$

EXAMPLE 23   Find $\displaystyle\int \frac{5\, dx}{x^2 - 36}$.

SOLUTION

$$\int \frac{5\, dx}{x^2 - 36} = 5 \int \frac{dx}{x^2 - 36}$$

The denominator seems to involve $u^2 - a^2$. Letting $u = x$ and $du = dx$ and comparing with formula 10, we have

$$\int \frac{du}{u^2 - a^2} = \frac{1}{2a} \ln \left| \frac{u - a}{u + a} \right| + C \qquad u^2 \ne a^2,\ a^2 = 36$$

$$\int \frac{dx}{x^2 - 36} = \frac{1}{2 \cdot 6} \ln \left| \frac{x - 6}{x + 6} \right| + C \qquad \text{for } x^2 \ne 36$$

$$\text{So} \quad \int \frac{5\, dx}{x^2 - 36} = \frac{5}{12} \ln \left| \frac{x - 6}{x + 6} \right| + C$$

To check this result with your calculator, calculate

$$\int \frac{5\, dx}{x^2 - 36}$$

as a definite integral, say

$$\int_1^3 \frac{5\, dx}{x^2 - 36}$$

Use the answer you got for the integral and substitute $x = 1$ and $x = 3$. Thus, we have

$$\frac{5}{12} \ln \left| \frac{x - 6}{x + 6} \right| \Big|_1^3 = \frac{5}{12} \ln \left| \frac{3 - 6}{3 + 6} \right| - \frac{5}{12} \ln \left| \frac{1 - 6}{1 + 6} \right| \approx -0.317558$$

Likewise, using the numeric integration on the calculator, you get

$$\text{fnInt}(5/(x^2 - 36), x, 1, 3) \approx -.317558$$

Calculator Note

For each example in this section you should continue to check your work as we did in the last example.

EXAMPLE 24 Find $\displaystyle\int \frac{dx}{6x^2 + x - 15}$.

SOLUTION There does not appear to be a formula in the table for this expression. Note, however, that the denominator may be factored.

$$\int \frac{dx}{6x^2 + x - 15} = \int \frac{dx}{(3x + 5)(2x - 3)}$$

It seems that formula 4 can be used with $u = x$ and $du = dx$.

$$\int \frac{1}{(au + b)(cu + d)} \, du = \frac{1}{bc - ad} \ln \left| \frac{cu + d}{au + b} \right| + C \qquad \text{Formula 4}$$

We will use $a = 3$, $b = 5$, $c = 2$, and $d = -3$.

$$\int \frac{1}{(3x + 5)(2x - 3)} \, dx = \frac{1}{\underbrace{5 \cdot 2 - (3)(-3)}_{bc - ad}} \ln \left| \frac{2x - 3}{3x + 5} \right| + C \qquad bc - ad = 19 > 0$$

$$= \frac{1}{19} \left[\ln |2x - 3| - \ln |3x + 5| \right] + C$$

EXAMPLE 25 Find $\displaystyle\int \frac{2x^2}{\sqrt{25x^2 - 36}} \, dx$.

SOLUTION To put this integral into a form involving $\sqrt{u^2 - a^2}$, let $u = 5x$. Then $x = \frac{1}{5}u$ and $dx = \frac{1}{5} \, du$.

$$\int \frac{2x^2}{\sqrt{25x^2 - 36}} \, dx = \int \frac{2(\frac{1}{25})u^2}{\sqrt{u^2 - 36}} \frac{1}{5} \, du = \frac{2}{125} \int \frac{u^2}{\sqrt{u^2 - 36}} \, du$$

Using formula 20 gives

$$\frac{2}{125} \int \frac{u^2}{\sqrt{u^2 - 36}} \, du$$

$$= \left(\frac{2}{125}\right) \left(\frac{1}{2}\right) (u\sqrt{u^2 - 36} + 36 \ln |u + \sqrt{u^2 - 36}|) + C$$

Thus

$$\int \frac{2x^2}{\sqrt{25x^2 - 36}} \, dx = \frac{1}{125} (5x\sqrt{25x^2 - 36} + 36 \ln |5x + \sqrt{25x^2 - 36}|) + C \quad \blacksquare$$

EXAMPLE 26 Find $\int x^2 e^{4x} \, dx$.

SOLUTION Formula 22 can be used as a first step in this problem. Let $u = x$ and $du = dx$.

$$\int u^n e^{au} \, du = \frac{u^n e^{au}}{a} - \frac{n}{a} \int u^{n-1} e^{au} \, du + C \qquad \text{Formula 22}$$

$$\int x^2 e^{4x} \, dx = \frac{x^2 e^{4x}}{4} - \frac{2}{4} \int x^{2-1} e^{4x} \, dx + C \qquad n = 2, a = 4$$

Now using formula 21, let $u = x$ and $du = dx$.

$$\int u e^{au} \, du = \frac{e^{au}}{a^2} (au - 1) \qquad\qquad \text{Formula 21}$$

$$\int x e^{4x} \, dx = \frac{e^{4x}}{4^2} (4x - 1) \qquad\qquad a = 4$$

$$\text{So} \quad \int x^2 e^{4x} \, dx = \frac{x^2 e^{4x}}{4} - \frac{2}{4} \left[\frac{e^{4x}}{16} (4x - 1)\right] + C$$

$$= e^{4x} \left[\frac{x^2}{4} - \frac{1}{8} x + \frac{1}{32}\right] + C$$

$$= e^{4x} \left[\frac{8x^2 - 4x + 1}{32}\right] + C \qquad \blacksquare$$

Formulas such as formula 22 used in the preceding example are called **reduction formulas**. These may be applied repeatedly until the integral is in a form that you can integrate.

Practice Problem 1 Find $\displaystyle\int \frac{x^2}{3x + 2}\, dx$.

ANSWER

$$\frac{1}{27}\left[\frac{1}{2}(3x + 2)^2 - 4(3x + 2) + 4 \ln |3x + 2|\right] + C$$

EXAMPLE 27 Find

$$\int_9^{12} \frac{1}{x^2 \sqrt{225 - x^2}}\, dx$$

Use your calculator to check your work.

SOLUTION Use formula 14 with $u = x$ and $a = 15$. Then

$$\int_9^{12} \frac{1}{x^2 \sqrt{225 - x^2}}\, dx = \int_9^{12} \frac{1}{u^2 \sqrt{225 - u^2}}\, du$$

$$= -\frac{\sqrt{225 - u^2}}{225u}\bigg|_{u=9}^{u=12}$$

$$= -\frac{\sqrt{225 - 144}}{(225)(12)} + \frac{\sqrt{225 - 81}}{(225)(9)}$$

$$= -\frac{9}{(225)(12)} + \frac{12}{(225)(9)}$$

$$= -\frac{9}{2700} + \frac{16}{2700} = \frac{7}{2700}$$

Practice Problem 2 Using a calculator, graph

$$y = \frac{1}{x^2 \sqrt{225 - x^2}}$$

for $9 \le x \le 12$ and determine whether the answer in Example 27 can be considered as an area.

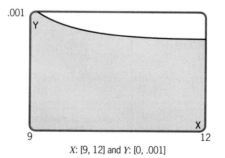

X: [9, 12] and Y: [0, .001]

Figure 14

ANSWER   Yes. (see Figure 14).

EXAMPLE 28   Find the consumers' surplus at a price level of $8 for the price–demand equation

$$p = D(x) = \frac{20x - 5000}{x - 400}$$

SOLUTION

1. The first step is to find the production for a price level of $8.

$$8 = \frac{20x_1 - 5000}{x_1 - 400} \qquad p_1 = 8$$

$$8x_1 - 3200 = 20x_1 - 5000 \qquad \text{Multiply by } x_1 - 400.$$

$$12x_1 = 1800$$

$$x_1 = 150$$

2. The second step is to show the consumers' surplus by a shaded region on a graph (see Figure 15).

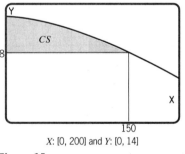

X: [0, 200] and Y: [0, 14]

Figure 15

3. The consumers' surplus, the shaded region of the graph, is the area between

$$p_1 = 8 \quad \text{and} \quad p = D(x) = \frac{20x - 5000}{x - 400}$$

Thus

$$CS = \int_0^{x_1} [D(x) - p_1]\, dx$$

$$= \int_0^{150} \left[\frac{20x - 5000}{x - 400} - 8 \right] dx$$

$$= \int_0^{150} \left[\frac{12x - 1800}{x - 400} \right] dx$$

$$= 12 \int_0^{150} \frac{x}{x - 400} \, dx - 1800 \int_0^{150} \frac{dx}{x - 400}$$

$$= 12[x + 400 \ln |x - 400|] \Big|_0^{150} - 1800 \ln |x - 400| \Big|_0^{150} \qquad \text{Formula 1}$$

$$= 12[150 + 400 \ln 250 - 400 \ln 400]$$
$$\qquad - 1800 \ln 250 + 1800 \ln 400$$
$$= 1800 + 3000 \ln 250 - 3000 \ln 400$$
$$\approx 389.99$$

The consumers' surplus is $389.99.

SUMMARY

Integration tables can be extensive. We have included only a few entries to introduce you to the use of an integration table. Using a table requires practice, since you must become familiar with the various forms that the table has and how a formula may fit the problem you are trying to solve.

Exercise Set 6.4

Using a table of integrals, find each of the following.

1. $\int \dfrac{dx}{x^2 - 25}$

2. $\int \dfrac{3\,dx}{x^2 - 36}$

3. $\int \dfrac{dx}{36 - x^2}$

4. $\int \dfrac{4\,dx}{16 - x^2}$

5. $\int \dfrac{dx}{\sqrt{x^2 + 36}}$

6. $\int \dfrac{dx}{\sqrt{x^2 - 36}}$

7. $\int \dfrac{dx}{x\sqrt{x^2 + 25}}$

8. $\int \dfrac{dx}{x\sqrt{49 - x^2}}$

9. $\int \dfrac{dx}{x^2\sqrt{49 + x^2}}$

10. $\int \dfrac{dx}{x^2\sqrt{49 - x^2}}$

11. $\int \dfrac{dx}{x^2\sqrt{x^2 - 49}}$

12. $\int \sqrt{x^2 - 49}\,dx$

13. $\int x^2\sqrt{x^2 + 25}\,dx$

14. $\int \dfrac{\sqrt{x^2 + 25}}{x^2}\,dx$

15. $\int \dfrac{\sqrt{x^2 + 25}}{x}\,dx$

16. $\int \dfrac{x^2}{\sqrt{x^2 - 16}}\,dx$

17. $\int \dfrac{x^2}{\sqrt{x^2 + 49}}\,dx$

18. $\int \dfrac{\sqrt{x^2 + 16}}{x^2}\,dx$

19. $\int \dfrac{dx}{(x + 3)(x - 2)}$

20. $\int \dfrac{dx}{(2x + 3)(3x - 2)}$

21. $\int x^2 e^{x/2}\,dx$

22. $\int x^3 \ln x\,dx$

23. $\int \dfrac{x}{\sqrt{2x + 3}}\,dx$

Find the value of each of the following and check your work with your calculator.

24. $\int_3^5 \dfrac{1}{x^2 - 1}\,dx$

25. $\int_4^6 \dfrac{1}{\sqrt{x^2 - 9}}\,dx$

26. $\int_2^4 \dfrac{1}{x^2(2x + 1)}\,dx$

Find the following integrals.

27. $\int \dfrac{dx}{\sqrt{25x^2 + 16}}$

28. $\int \dfrac{3\,dx}{\sqrt{16x^2 - 25}}$

29. $\displaystyle\int \frac{8\,dx}{x\sqrt{25-4x^2}}$

30. $\displaystyle\int \frac{5\,dx}{x^2\sqrt{16+9x^2}}$

31. $\displaystyle\int 3x^2\sqrt{9x^2+16}\,dx$

32. $\displaystyle\int \frac{\sqrt{16-25x^2}}{3x}\,dx$

33. $\displaystyle\int \frac{\sqrt{16+25x^2}}{3x^2}\,dx$

34. $\displaystyle\int \frac{4x^2}{\sqrt{36x^2-9}}\,dx$

35. $\displaystyle\int \frac{dx}{6x^2-5x-6}$

36. $\displaystyle\int \frac{2\,dx}{8x^2-10x-3}$

Applications (Business and Economics)

37. **Producers' Surplus.** Find the producers' surplus for a price level of \$10 given the price–demand equation

$$p = S(x) = \frac{5x}{30-x}$$

Show the producers' surplus on a graph.

38. **Consumers' Surplus.** Find the consumers' surplus for

$$D(x) = \frac{360}{x^2+4x+3} \quad \text{and} \quad S(x) = \frac{5x}{x+3}$$

39. **Cash Reserves.** The cash reserves of a company for a year are given in thousands of dollars by

$$c = 2 + \frac{x^2\sqrt{1+x^2}}{144}, \qquad 0 \le x \le 12$$

where x represents the number of months. Find the average cash reserve for (a) the first quarter, (b) the first half of the year, and (c) the last quarter.

Applications (Social and Life Sciences)

40. **Learning Rates.** A rate of learning is given by

$$\frac{dL}{dt} = \frac{64}{\sqrt{t^2+36}}, \qquad t \ge 0$$

where L is the number of items learned in t hours. Find the number of items learned in 8 hours.

41. **Pollution.** The radius R of an oil slick is increasing at the rate

$$\frac{dR}{dt} = \frac{144}{\sqrt{t^2+16}}, \qquad t \ge 0$$

where t is time in minutes. Find the radius after 5 minutes if $R = 0$ when $t = 0$.

Review Exercises

42. Suppose that the rate of production of an oil company per month is given in thousands of barrels as

$$\frac{dP}{dt} = 24te^{-0.06t}$$

How much oil is produced during the first 6 months? If production is continued indefinitely, what would be the total amount produced?

Find the value of the improper integrals if they exist.

43. $\displaystyle\int_2^\infty 3x^{-3/2}\,dx$

44. $\displaystyle\int_2^\infty \frac{4\,dx}{(1+x)^{3/2}}$

45. $\displaystyle\int_0^\infty 3e^{-2x}\,dx$

46. $\displaystyle\int_{-\infty}^\infty e^{2x}\,dx$

6.5 APPROXIMATE INTEGRATION

OVERVIEW   We have defined the definite integral $\int_a^b f(x)\,dx$ over the closed interval $[a, b]$ to be $F(b) - F(a)$, where $F(x)$ is an antiderivative of $f(x)$. Sometimes it is difficult or maybe impossible to find $F(x)$. For example,

$$\int_a^b e^{-x^2}\,dx$$

cannot be evaluated using antiderivatives because no antiderivatives exist in

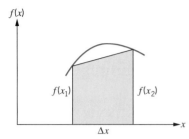

$f(x)$

$f(x_1)$ $f(x_2)$

Δx

Figure 16

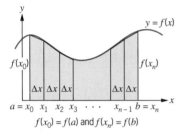

y

$y = f(x)$

$f(x_0)$ $f(x_n)$

Δx Δx Δx $\quad$ Δx Δx

$a = x_0 \ x_1 \ x_2 \ x_3 \ \cdots \ x_{n-1} \ b = x_n$

$f(x_0) = f(a)$ and $f(x_n) = f(b)$

Figure 17

terms of functions we have studied. Numerical procedures exist to help us evaluate such integrals. In this section we study two techniques:

- Trapezoidal rule
- Simpson's rule

The Riemann sum procedures that we have used to define a definite integral are sometimes the procedures we use to approximate the value of the definite integral. The two methods of numerical integration that we consider (the trapezoidal rule and Simpson's rule) are based on the interpretation of an integral as an area. For example, one can approximate the area under a curve as a trapezoid instead of as a rectangle, as seen in Figure 16. The area of the trapezoid is

$$\left[\frac{f(x_1) + f(x_2)}{2} \right] \Delta x$$

An approximation for the definite integral $\int_a^b f(x)\, dx$ can be found by calculating the area of n trapezoids. In Figure 17, the area under the curve from $x = a$ to $x = b$ is approximately

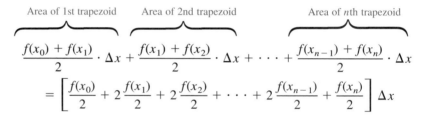

Area of 1st trapezoid $\quad$ Area of 2nd trapezoid $\quad$ Area of nth trapezoid

$$\frac{f(x_0) + f(x_1)}{2} \cdot \Delta x + \frac{f(x_1) + f(x_2)}{2} \cdot \Delta x + \cdots + \frac{f(x_{n-1}) + f(x_n)}{2} \cdot \Delta x$$

$$= \left[\frac{f(x_0)}{2} + 2\frac{f(x_1)}{2} + 2\frac{f(x_2)}{2} + \cdots + 2\frac{f(x_{n-1})}{2} + \frac{f(x_n)}{2} \right] \Delta x$$

This allows us to employ the trapezoidal rule for approximating the value of a definite integral.

TRAPEZOIDAL RULE

Let $f(x)$ be a continuous function on $[a, b]$. Then

$$\int_a^b f(x)\, dx \approx \left[\frac{f(x_0)}{2} + f(x_1) + f(x_2) + \cdots + f(x_{n-1}) + \frac{f(x_n)}{2} \right] \Delta x$$

where

$$\Delta x = \frac{b - a}{n}, \quad x_0 = a, \quad x_1 = x_0 + \Delta x, \quad x_2 = x_1 + \Delta x,$$

$$\ldots, \quad x_n = x_{n-1} + \Delta x = b$$

The number of approximations will affect the accuracy of the approximation. As n increases, the approximation becomes more accurate.

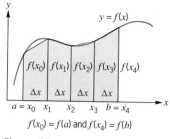

Figure 18

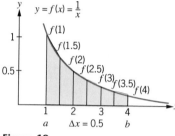

Figure 19

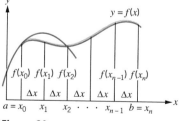

Figure 20

EXAMPLE 29   Write the trapezoidal rule to approximate the area under the curve defined by $y = f(x)$ in Figure 18 from $x = a$ to $x = b$ using $n = 4$ subdivisions.

SOLUTION   In the formula

$$\Delta x = \frac{b - a}{n} = \frac{b - a}{4}, \quad a = x_0, \quad b = x_4$$

$$\int_a^b f(x) \, dx \approx \left[\frac{f(x_0)}{2} + f(x_1) + f(x_2) + f(x_3) + \frac{f(x_4)}{2} \right] \Delta x$$

EXAMPLE 30   Use the trapezoidal rule with six intervals to approximate

$$\int_1^4 \frac{1}{x} \, dx$$

Compare this answer to the answer you get by numeric integration on your calculator (see Figure 19).

SOLUTION   We first find Δx and identify all other parts to be used in the formula.

$$\Delta x = \frac{b - a}{n} = \frac{4 - 1}{6} = 0.5 \qquad x_0 = a = 1$$

$$x_1 = 1 + 0.5 = 1.5 \qquad\qquad x_2 = 1.5 + 0.5 = 2$$

$$x_3 = 2 + 0.5 = 2.5 \qquad\qquad x_4 = 2.5 + 0.5 = 3$$

$$x_5 = 3 + 0.5 = 3.5 \qquad\qquad x_6 = b = 4 \qquad \text{See Figure 19}$$

$$\int_1^4 \frac{1}{x} \, dx \approx \left[\frac{f(1)}{2} + f(1.5) + f(2) + f(2.5) + f(3) + f(3.5) + \frac{f(4)}{2} \right] (0.5)$$

$$= \left[\frac{1}{2} + \frac{1}{1.5} + \frac{1}{2} + \frac{1}{2.5} + \frac{1}{3} + \frac{1}{3.5} + \frac{1}{8} \right] (0.5)$$

$$= 1.41$$

With a calculator, we get the answer 1.386. How could we get the answer using the trapezoidal rule to be closer to the answer we get with the calculator?

When we use the trapezoidal rule, the approximation of the area under $f(x)$ is being done, in effect, by a linear function. Intuition suggests that the accuracy of a numerical integration procedure can be improved by replacing a linear approximation with something that would better fit the curve, a second-degree polynomial (Figure 20).

It can be shown (but will be omitted here) that the area under the parabola between $x = x_0$ and $x = x_2$ is given by

$$\frac{\Delta x}{3} [f(x_0) + 4f(x_1) + f(x_2)]$$

This argument can be repeated for $[x_2, x_4]$, $[x_4, x_6]$, and so on. Therefore, if n is even, the area under the curve from $x = a$ to $x = b$ is approximately the sum of the areas under $n/2$ approximating parabolas and we obtain Simpson's rule.

SIMPSON'S RULE

Let $f(x)$ be a continuous function on $[a, b]$. Then,

$$\int_a^b f(x)\, dx \approx \frac{\Delta x}{3} [f(x_0) + 4f(x_1) + 2f(x_2) + 4f(x_3)$$
$$+ \cdots + 2f(x_{n-2}) + 4f(x_{n-1}) + f(x_n)]$$

where $\Delta x = (b - a)/n$ is the width of each subinterval, n is even, and

$$x_0 = a, \quad x_1 = x_0 + \Delta x, \quad x_2 = x_1 + \Delta x, \quad \ldots, \quad x_n = b = x_{n-1} + \Delta x$$

EXAMPLE 31   Approximate the definite integral

$$\int_0^1 e^{-x^2}\, dx$$

using Simpson's rule with $n = 4$ subintervals.

SOLUTION   For this problem $a = 0$, $b = 1$, $n = 4$, and $f(x) = e^{-x^2}$.

$$\Delta x = \frac{b - a}{n} = \frac{1 - 0}{4} = 0.25 \qquad \text{Width of the subintervals}$$

$$x_0 = a = 0 \qquad\qquad\qquad x_1 = 0 + 0.25 = 0.25$$

$$x_2 = 0.25 + 0.25 = 0.50 \qquad\qquad x_3 = 0.50 + 0.25 = 0.75$$

$$x_4 = b = 1$$

$$\int_0^1 e^{-x^2}\, dx \approx \frac{0.25}{3} [f(x_0) + 4f(x_1) + 2f(x_2) + 4f(x_3) + f(x_4)]$$

$$= \frac{0.25}{3} [f(0) + 4f(0.25) + 2f(0.50) + 4f(0.75) + f(1)]$$

$$= \frac{0.25}{3} [e^0 + 4e^{-(0.25)^2} + 2e^{-(0.50)^2} + 4e^{-(0.75)^2} + e^{-(1)^2}]$$

$$\approx \frac{0.25}{3} [1 + 3.7577 + 1.5576 + 2.2791 + 0.3679]$$

$$\approx 0.7469$$

You will find that the answer above is the same as the one you will get with numeric integration on your calculator to 4 decimal places. In fact most calculators use Simpson's rule for numeric integration, but use more terms. ■

Practice Problem 1   Using Simpson's rule with eight intervals to integrate

$$\int_1^5 \left(\frac{1}{x^2}\right) dx$$

find the following.

(a) Δx (b) $f(x_0)$ (c) $4f(x_1)$

ANSWER

(a) $\Delta x = \dfrac{5 - 1}{8} = \dfrac{1}{2}$ (b) $x_0 = 1, f(x_0) = \dfrac{1}{1^2} = 1$

(c) $x_1 = 1 + \dfrac{1}{2} = \dfrac{3}{2}, f(x_1) = \dfrac{4}{9}, 4f(x_1) = \dfrac{16}{9} = 1.7778$

NOTE: In more advanced courses, it is proven that the error in using the trapezoidal rule to approximate the integral $\int_a^b f(x)\, dx$ with n subdivisions has the following upper bound:

$$|\text{Error}| \leq \frac{(b - a)^3 M}{12n^2}$$

where M is the maximum value of $|f''(x)|$ on $[a, b]$.
For Simpson's rule it is

$$|\text{Error}| \leq \frac{(b - a)^5 M}{180n^4}$$

where M is the maximum value of $|f^{(4)}(x)|$ on $[a, b]$.
In Example 30, the maximum error is calculated as follows:

$$f(x) = \frac{1}{x}, \quad f'(x) = -\frac{1}{x^2}, \quad \text{and} \quad f''(x) = \frac{2}{x^3}$$

The maximum value of $\dfrac{2}{x^3}$ on $[1, 4]$ is 2; so $M = 2$.

$$|\text{Error}| \leq \frac{(4 - 1)^3 \cdot 2}{12 \cdot 6^2} = \frac{1}{8} \qquad n = 6$$

SUMMARY

For the trapezoidal rule, errors are approximately proportional to $1/n^2$ and for a given n, the size of the error depends on the size of f''. For Simpson's rule, errors are approximately proportional to $1/n^4$, and, for a given n, the size of the error depends on the size of the fourth derivative, $f^{(4)}$. Thus, we can see that the definite integral $\int_a^b f(x)\, dx$ can be computed fairly quickly and quite accurately in most cases using Simpson's rule. Difficulties will arise when f' or a higher derivative of f does not exist or gets very large over the interval $a \le x \le b$.

Exercise Set 6.5

Evaluate the given integrals using the trapezoidal rule. Compare your answer with the answer you get by numeric integration on your graphing calculator.

1. $\displaystyle\int_1^2 (x^2 + 1)\, dx; \quad n = 4$

2. $\displaystyle\int_1^2 (x^3 - 4)\, dx; \quad n = 4$

3. $\displaystyle\int_0^1 \sqrt{x + 1}\, dx; \quad n = 4$

4. $\displaystyle\int_0^1 x(x^2 + 1)^2\, dx; \quad n = 6$

Evaluate the given integrals using Simpson's rule. Compare your answer with the answer you get by numeric integration on your graphing calculator.

5. $\displaystyle\int_1^2 (x^2 + 1)\, dx; \quad n = 4$

6. $\displaystyle\int_1^2 (x^3 - 4)\, dx; \quad n = 4$

7. $\displaystyle\int_0^1 \sqrt{x + 1}\, dx; \quad n = 4$

8. $\displaystyle\int_0^1 x(x^2 + 1)^2\, dx; \quad n = 6$

9. Using the trapezoidal rule, find the maximum error in Exercise 3.

10. Using Simpson's rule, find the maximum error in Exercise 7.

Applications (Business and Economics)

11. **Total Cost.** A corporation determines its marginal costs at various production levels to be as follows:

Production (units of 100)	Marginal Costs ($)
20	400
25	420
30	440
35	450
40	460

Use the trapezoidal rule to approximate the cost of increasing production from 20 units to 40 units.

12. **Production.** The rate of production of an oil company per month is given in thousands of barrels as

$$\frac{dp}{dt} = 30e^{-0.06t}$$

How much oil is produced during the first 6 months of operation? Approximate the answer using the trapezoidal rule with $n = 6$.

13. **Consumers' Surplus.** For a demand function

$$p = D(x) = 85 - e^{0.01x^2}$$

find the consumers' surplus using Simpson's rule with six subintervals when the price level is 30.

Applications (Social and Life Sciences)

14. **Learning Function.** The number of words learned in a minute is given at the following times.

Time After Start (minutes)	Number of Words Learned
4	15
5	14.5
6	13.2
7	12
8	11
9	10

Using the trapezoidal rule, find the number of words learned from $t = 4$ to $t = 9$.

Review Exercises

15. Find each integral if it exists.

(a) $\displaystyle\int_{-\infty}^{2} xe^{-3x}\, dx$ (b) $\displaystyle\int_{-\infty}^{\infty} \frac{2x}{\sqrt{x^2 + 1}}\, dx$

16. Find each indefinite integral.

(a) $\displaystyle\int \frac{-3\sqrt{\ln x}}{x}\, dx$ (b) $\displaystyle\int \frac{1 + x}{2 - x}\, dx$

Chapter Review

Review to make certain that you understand the following concepts.

Integration by parts   Improper integrals
Simpson's rule   Trapezoidal rule
Integration by use of tables   Convergent integral
Divergent integral   Probability density function

Make sure that you can use the following formulas.

$$\int u\, dv = uv - \int v\, du$$

$$\int_{a}^{\infty} f(x)\, dx = \lim_{t \to \infty} \int_{a}^{t} f(x)\, dx$$

$$\int_{-\infty}^{a} f(x)\, dx = \lim_{t \to -\infty} \int_{t}^{a} f(x)\, dx$$

$$\int_{-\infty}^{\infty} f(x)\, dx = \lim_{t \to -\infty} \int_{t}^{c} f(x)\, dx + \lim_{t \to \infty} \int_{c}^{t} f(x)\, dx$$

Chapter Test

1. Determine whether $\displaystyle\int_{0}^{\infty} \frac{dx}{\sqrt{x}}$ is convergent or divergent.

2. Find $\displaystyle\int x\sqrt{x + 2}\, dx$.

3. Use $\displaystyle\int \frac{du}{u^2 - a^2} = \frac{1}{2a} \ln \left| \frac{u - a}{u + a} \right| + C$ to find $\displaystyle\int \frac{dx}{25x^2 - 36}$.

4. Evaluate $\int_0^2 xe^{4x}\, dx$ and check with your graphing calculator.

5. Approximate

$$\int_0^2 x^3\, dx$$

by using Simpson's rule with four subintervals and compare this answer to the one you get by using your graphing calculator.

6. Evaluate $\int_2^4 x^2 \ln x\, dx$. Check your work with your graphing calculator.

7. Approximate

$$\int_0^4 x^2\, dx$$

using the trapezoidal rule with four subintervals and compare to the answer given by your graphing calculator.

8. Evaluate $\int x^2 e^x\, dx$ and check with your graphing calculator.

9. Evaluate $\int_0^1 15x^2 \sqrt{5x^3 + 4}\, dx$ using substitution.

10. Let X be an exponentially distributed random variable with an average value of 7. Find the pdf for X and find $P(X \le 8)$.

Multivariable Calculus

T he goal of this chapter is to extend the theory of calculus to real-valued functions that involve more than one independent variable. Examples of such functions abound in both business and economics and the social and life sciences. In economics and business, one encounters revenue as a function of both price, p, and the number of items sold, x. That is, $R = f(p, x)$. You need to know the number of items sold and the price per item to calculate the revenue. In this chapter, we study three-dimensional coordinates, functions of several variables, partial derivatives, relative maxima and minima for functions of two or more variables, and Lagrange multipliers.

7.1   FUNCTIONS OF SEVERAL VARIABLES

OVERVIEW   We have already encountered many examples of functions of several variables. In the formula for simple interest, $A = P(1 + rt)$, A is a function of P, r, and t, and in the formula for compound interest, $A = P(1 + i)^n$, A is a function of P, i, and n. In an earlier exercise, the cost function $C(x)$ for the number x of items produced was given as

$$C(x) = 0.25x + 70$$

Suppose now that the manufacturer decides to expand and produce an additional, different item. Let y represent the number of these additional items produced. The cost function $C(x, y)$ for producing both items would depend upon both x and y, and hence is a function of two variables. For example, suppose that

$$C(x, y) = 130 + 0.25x + 0.30y$$

If the manufacturer decides to produce even more items, the cost function would involve more variables. We could define functions of three variables, four variables, or, in general, n variables. In this section, we consider

- Functions of two or more independent variables
- Applications of functions of several variables
- The three-dimensional coordinate system
- Traces of graphs in three dimensions

A company produces pens and pencils. The cost of producing a pen is $2.40 and a pencil is $1.80 with fixed costs of $130. The cost function for producing x pens and y pencils is

$$C(x, y) = 130 + 2.40x + 1.80y$$

We say that C is a function of x and y. The domain of (x, y) is the set of ordered pairs where the elements are positive integers. To understand this concept, we need the following definition of a function with two independent variables.

FUNCTION OF TWO INDEPENDENT VARIABLES

The function $f(x, y)$ describes a function of two independent variables if for each ordered pair (x, y) from the domain of f there is one and only one value of $f(x, y)$ in the range of f. Unless stated otherwise, the domain of $f(x, y)$ is the set of all ordered pairs of real numbers (x, y) such that $f(x, y)$ is also a real number.

An equation such as

$$f(x, y) = 2x^2 + xy + 3y^2$$

defines f as a function of two independent variables x and y. The domain of this function is the set of ordered pairs (a, b), where a and b are any real numbers. The range of f is the set of real numbers.

EXAMPLE 1   Let $f(x, y) = 2x^2 + xy + 3y^2$. Then

$$f(0, 0) = 2(0)^2 + (0)(0) + 3(0)^2 = 0$$

$$f(1, 2) = 2(1)^2 + (1)(2) + 3(2)^2 = 16$$

$$f(-1, -2) = 2(-1)^2 + (-1)(-2) + 3(-2)^2 = 16$$

$$f(h, k + 2) = 2(h)^2 + h(k + 2) + 3(k + 2)^2$$
$$= 2h^2 + hk + 2h + 3k^2 + 12k + 12$$

Practice   Problem   1 Let $f(x, y) = 2x^3 - 3x^2y + 2xy^2 + y^3$.   Find (a) $f(0, 0)$, (b) $f(-1, -2)$, and (c) $f(1, 2)$.

ANSWER

(a) 0

(b) -12

(c) 12

Similarly, a function defined by an equation such as

$$f(x, y, z) = 3x^2 - 2xy + y^2 - 3xyz + 2z^2$$

is a function of three variables. The domain of such a function is the set of ordered triples (a, b, c), where a, b, and c are real numbers.

EXAMPLE 2   If $f(x, y, z) = 3x^2 - 2xy + y^2 - 3xyz + 2z^2$, find (a) $f(1, 2, 3)$ and (b) $f(-1, 2, -3)$.

SOLUTION

(a) $f(1, 2, 3) = 3(1)^2 - 2(1)(2) + (2)^2 - 3(1)(2)(3) + 2(3)^2$
$$= 3 - 4 + 4 - 18 + 18$$
$$= 3$$

(b) $f(-1, 2, -3) = 3(-1)^2 - 2(-1)(2) + (2)^2 - 3(-1)(2)(-3) +$
$$2(-3)^2$$
$$= 3 + 4 + 4 - 18 + 18$$
$$= 11$$

Practice Problem 2 For $f(x, y, z) = 3x^2 - 2xy + y^2 - 3xyz + 2z^2$, find (a) $f(1, -1, 2)$ and (b) $f(h, h + 2, \Delta h)$.

ANSWER

(a)   20

(b) $2h^2 + 4 - 3h^2 \, \Delta h - 6h \, \Delta h + 2(\Delta h)^2$

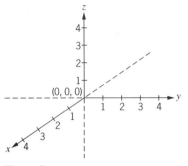

Figure 1

Functions with two independent variables can be represented on a coordinate system in much the same manner as functions with one independent variable. A function with one independent variable is drawn on a two-dimensional coordinate system. For a function with two independent variables, a three-dimensional coordinate system is needed. A three-dimensional Cartesian coordinate system is shown in Figure 1. In this figure the x-axis, the y-axis, and the z-axis intersect and are perpendicular to one another at the origin, (0, 0, 0). The dashed lines indicate the negative portion of each of these coordinate lines. Any ordered triple (a, b, c) can be plotted as a point using this coordinate system (a is the x-coordinate, b is the y-coordinate, and c is the z-coordinate).

The plane formed by the x-axis and the y-axis is called the **xy-coordinate plane**; the plane formed by the x-axis and the z-axis is called the **xz-coordinate plane**; and the plane formed by the y-axis and the z-axis is called the **yz-coordinate plane**. In a three-dimensional system, the three axes divide the space into eight parts; each part is called an **octant**. In octant 1, the x-, y-, and z-coordinates are all positive.

The point (3, 2, 4) is located by starting at the origin, (0, 0, 0), and moving 3 units in the direction of the positive x-axis, 2 units in the direction of the positive y-axis, and 4 units in the direction of the positive z-axis [see Figure 2(a)]. Any point can be plotted in this manner. First move along the x-axis, then move parallel to the y-axis, and then parallel to the z-axis.

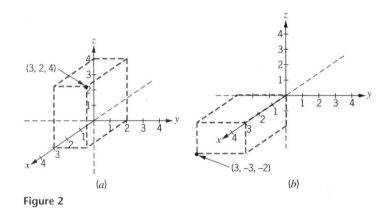

(a)   (b)

Figure 2

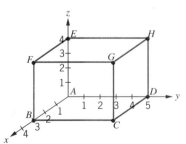

Figure 3

EXAMPLE 3   Plot the point $(3, -3, -2)$ on the coordinate system in Figure 2(b).

SOLUTION   First move 3 units in a positive direction on the x-axis. Then move 3 units in a negative direction parallel to the y-axis, and finally move 2 units in a negative direction parallel to the z-axis and mark the point $(3, -3, -2)$.

Practice Problem 3 Find the coordinates of the vertices (A to H) of the rectangular box shown in Figure 3.

ANSWER $A(0, 0, 0)$, $B(3, 0, 0)$, $C(3, 5, 0)$, $D(0, 5, 0)$, $E(0, 0, 4)$, $F(3, 0, 4)$, $G(3, 5, 4)$, $H(0, 5, 4)$

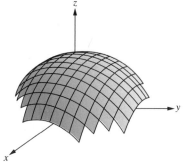

Figure 4

The graph of a function $z = f(x, y)$ consists of all the points $(x, y, f(x, y))$ in a three-dimensional coordinate system, in other words, all points (x, y, z) where (x, y) is in the domain of f and $z = f(x, y)$. The graph of all such points is a **surface** in space. It is difficult or even impossible to sketch the graph of some three-variable equations by hand. Computers and math programs have the capabilities of graphing these surfaces. Figure 4 shows a graph generated by a computer. We will not try to sketch difficult figures, but we will consider some techniques for graphing equations with three variables.

To draw graphs of functions in three dimensions, we make use of traces. The intersections of a graph of a function with the coordinate planes of the three-dimensional coordinate system are called **traces** of the graph of the function. Since in each coordinate plane one variable is always zero, the equations of the traces can be obtained by setting $x = 0$, $y = 0$, and $z = 0$ one at a time.

TRACES OF FUNCTIONS OF TWO VARIABLES

Let $z = f(x, y)$ be a function of two variables.

(a)   If we set $x = 0$, the curve $z = f(0, y)$ is the trace of the graph of f in the yz-plane.

(b)   If we set $y = 0$, the curve $z = f(x, 0)$ is the trace of the graph of f in the xz-plane.

(c)   If we set $z = 0$, the curve $f(x, y) = 0$ is the trace of the graph of f in the xy-plane.

One type of surface in space is a plane. The general equation of a plane is $Ax + By + Cz = D$. By forming the intersection of a plane with each coordinate plane, we can see that the resulting traces are lines.

For example, to graph the plane $3y + 4z + 2x = 16$, we draw three traces.

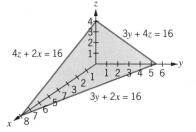

Figure 5

1.   Setting $z = 0$ results in the equation $3y + 2x = 16$. This is the trace on the xy-plane. The x-intercept is the point $(8, 0, 0)$ and the y-intercept is the point $(0, \frac{16}{3}, 0)$.

2.   Setting $y = 0$ results in the equation $4z + 2x = 16$, which is the trace on the xz-plane. The x-intercept is the point $(8, 0, 0)$ and the z-intercept is the point $(0, 0, 4)$.

3.   Setting $x = 0$ results in the equation $3y + 4z = 16$, which is the trace on the yz-plane. The y- and z-intercepts are the same as those found above, $(0, \frac{16}{3}, 0)$ and $(0, 0, 4)$.

We draw the three traces of this plane as shown in Figure 5 and shade the region bounded by the traces. The shaded region shows a portion of the plane in the first octant—the octant where x, y, and z are all positive.

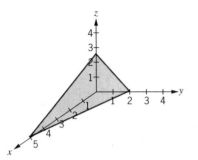

Figure 6

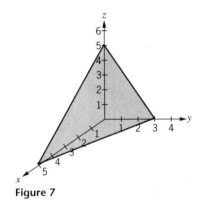

Figure 7

EXAMPLE 4 Sketch the traces of $2x + 5y + 4z = 10$ by finding the intercepts, and then shade the plane.

SOLUTION By letting $y = 0$ and $z = 0$, $2x = 10$ or $x = 5$ (the x-intercept). From $z = 0$ and $x = 0$, $5y = 10$ or $y = 2$ (the y-intercept). The z-intercept is obtained by setting $y = 0$ and $x = 0$; that is, $4z = 10$, so $z = \frac{5}{2}$. The three intercepts are plotted in Figure 6; the lines connecting these points are the traces of the plane, the first octant of which is shaded. ∎

Practice Problem 4 Sketch the traces of $3x + 5y + 3z = 15$ by finding the intercepts, and then shade the plane.

ANSWER See Figure 7.

Special planes are obtained when some of the variables are missing in the three-dimensional linear equation. Consider the surface that is the graph of $z = 4$. In this case x and y can have any value but z is always 4. Thus $z = 4$ is the equation of a plane parallel to the xy-plane and four units above the xy-plane (see Figure 8).

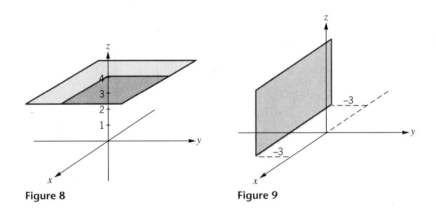

Figure 8   **Figure 9**

EXAMPLE 5 Sketch the plane $y = -3$ on a three-dimensional coordinate system.

SOLUTION This plane is parallel to the xz-plane and 3 units in a negative direction from the xz-plane (see Figure 9). Remember: $y = -3$ *regardless of the values of x and z.* ∎

The planes shown in Figures 8 and 9 have only one intercept. It is possible for a plane to have two intercepts. Let's look at such a plane in the next example.

EXAMPLE 6 Sketch the graph of $2x + y = 2$ on a three-dimensional coordinate system.

SOLUTION   Since the variable z is missing in the equation, z can assume any value. The trace in the xy-plane is $2x + y = 2$. The x-intercept is 1 and the y-intercept is 2. In Figure 10, vertical lines are drawn to the intercepts to show that z can assume any value. ∎

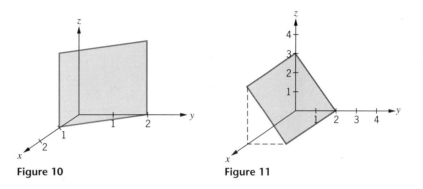

Figure 10   **Figure 11**

Practice Problem 5 Sketch the plane $3y + 2z = 6$ in three-space.

ANSWER

See Figure 11. Note that x can be anything.

Contour Maps

Your calculator is valuable in visualizing a three-dimensional surface using a **contour map**. To illustrate, let's consider the graph of the surface defined by $z = x^2 + y^2$. Suppose that we let $z = c$, where $c \neq 0$. The equation $z = c$ is a plane parallel to the xy-plane. If we draw the graph of $x^2 + y^2 = c$, we will have the trace of the surface on the $z = c$ plane. By considering a number of such traces, as z takes on various values, we can often visualize the surface that such traces would yield. These traces projected onto the xy-plane constitute a contour map. With your calculator, make a contour map for $z = x^2 + y^2$, considering traces when $z = 4, 9, 16,$ and 25.

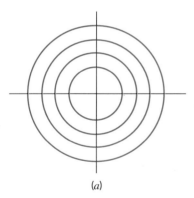

(a)

For $z = 4$, let $Y_1 = \sqrt{4 - x^2}$ and $Y_2 = -Y_1$.
For $z = 9$, let $Y_3 = \sqrt{9 - x^2}$ and $Y_4 = -Y_3$.
For $z = 16$, let $Y_5 = \sqrt{16 - x^2}$ and $Y_6 = -Y_5$.
For $z = 25$, let $Y_7 = \sqrt{25 - x^2}$ and $Y_8 = -Y_7$.

You will get a contour plot that looks something like that in Figure 12(a). In Figure 12(b) the graph shown is a contour plot done by a computer program.

We can visualize the traces as z takes on five values as we draw the sketch of the surface in Figure 13(a). In Figure 13(b), a computer generated plot of $z = x^2 + y^2$ is shown.

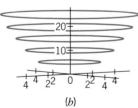

(b)

Figure 12

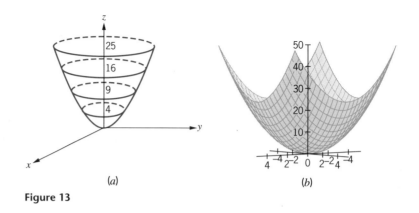

(a)

(b)

Figure 13

SUMMARY

When evaluating functions of more than one independent variable, the same techniques that we use with functions of one independent variable apply. Graphing, however, requires that we use a three-dimensional axis. Your calculator can help you with contour maps and then you can try to visualize the three-dimensional figure. In addition, traces on the coordinate planes help to visualize three-dimensional figures.

Exercise Set 7.1

Find the values of the following functions at the indicated points of their domain.

$$f(x, y) = 3x^2 - 2xy + y^3$$

$$g(x, y) = 7 - 3x + 2y$$

$$h(x, y) = 4x^3 - 3xy + 2y^2$$

1. $f(0, 1)$　　　　　　　　2. $f(1, 0)$
3. $f(-1, 0)$　　　　　　　4. $f(-1, 1)$
5. $f(2, 1)$　　　　　　　　6. $f(1, 2)$
7. $g(0, 0)$　　　　　　　　8. $g(3, 2)$
9. $g(2, 3)$　　　　　　　10. $g(-3, -1)$
11. $h(1, 2)$　　　　　　　12. $h(2, -1)$
13. Plot the following points in a three-space coordinate system.
　　(a) $(2, 4, 3)$　　　　　(b) $(3, -1, -2)$
　　(c) $(-2, 0, 1)$　　　　(d) $(-1, 0, -2)$
　　(e) $(-1, 2, -3)$　　　(f) $(0, 0, -1)$

Obtain the equations of the traces, find the coordinate intercepts, and then shade parts of the following planes.

14. $x + y + z = 3$　　　　15. $3x + 7y + 2z = 14$
16. $x - y + z = 4$　　　　17. $5x + 2y + 3z = 15$
18. $-x + y - z = 3$　　　19. $2x - 3y - 4z = 12$
20. Classify the following statements as true or false.
　　(a) $3x + y = 7$ is a line in three-space.
　　(b) In three-space, $y = 0$ is the equation of the x-axis.
　　(c) One trace of $x - 3y + 4z = 6$ is $3y - 4z = -6$.
　　(d) In three-space, $z = 0$ represents the xz-plane.
　　(e) In three-space, every linear equation represents a plane.
　　(f) The trace of $6z + 4x = 12$ in the yz-plane is $z = 2$.
　　(g) The x-intercept of $y + 6z + 4x = 12$ is 3.
　　(h) The point $(1, 3, -2)$ is a point on the plane $x - y + 2z = -6$.

(i) The graph of $y = 4$ in three-space is a plane perpenicular to the y-axis.

(j) The plane $2x + 3y = 6$ never intersects the z-axis.

Shade the following planes in three-space.

21. $z = 5$ 22. $x = -2$

23. $y = -4$ 24. $x + 3y = 6$

25. $2x - y = 4$ 26. $3y - 4x = 8$

For Exercises 27–31, make contour maps (possibly using a graphing calculator). (Optional: sketch the surface.)

27. $z = 4x^2 + y^2$ 28. $y = 4z^2 + x^2$

29. $z = x^2 - y^2$ 30. $y = x^2 - z^2$

Applications (Business and Economics)

31. **Profit Function.** A profit function $P(x, y)$ is given by

$$P(x, y) = 2x^2 - xy + 3y^2 + 4x + 2y + 4$$

Find (a) $P(3, 2)$, (b) $P(4, 1)$, and (c) $P(3, 4)$.

32. **Revenue and Cost Functions.** Suppose that the revenue and cost functions for a firm producing two items are given in units of thousands by

$$R(x, y) = 3x + 4y$$

$$C(x, y) = x^2 - 3xy + 3y^2 + 2x + 4y + 7$$

Find each of the following.
(a) $P(x, y) = R(x, y) - C(x, y)$ (b) $R(3, 2)$
(c) $C(2, 3)$ (d) $P(3, 2)$

33. **Prince–Earnings Ratio.** The price–earnings ratio of a stock is given by

$$R(P, E) = \frac{P}{E}$$

where P is the price of a stock, and E is the earnings per share for a year. Find and interpret $R(110, 10)$ for IBM.

34. **Stock Yield.** The yield of a stock is given by

$$Y(d, P) = \frac{d}{P}$$

where d is the yearly dividends and P is the price. Find and interpret $Y(33, 110)$ for IBM.

35. **Production Model.** The production of a given company is given by

$$P = 0.8l^{1/2}C^{1/2}$$

where P is the number of units produced, l is the number of hours of labor available, and C is the number of dollars of capital available. Find the production when $l = 400$ and $C = \$10,000$.

36. **Production Model.** Find the production for $P = 10l^{3/4}C^{1/4}$ if $l = 1296$ work-hours and $C = \$625$ in capital.

Applications (Social and Life Sciences)

37. **Scuba Diving.** The time of a scuba dive is estimated by

$$T(v, d) = \frac{35v}{d + 35}$$

where v is the volume of air in the diver's tanks and d is the depth in feet of the dive. Find (a) $T(965,30)$ and (b) $T(75, 45)$.

38. **Intelligence Quotient.** The function for determining the intelligence quotient is given as

$$I(M, C) = \frac{100M}{C}$$

where I represents the intelligence quotient, M represents mental age, and C represents chronological age.
(a) If $M = 14$ and $C = 10$, find $I(14, 10)$.
(b) If $M = 11$ and $C = 10$, find $I(11, 10)$.
(c) If $M = 10$ and $C = 10$, find $I(10, 10)$.

7.2 PARTIAL DERIVATIVES

OVERVIEW   A function of more than one variable may have derivatives with respect to each independent variable. These derivatives in turn may have derivatives with respect to each independent variable. Such derivatives are called **partial derivatives.** We consider first a function of two variables. Since we already know how to differentiate a function of a single variable, it would seem reasonable to reduce our function of two variables to a function of one variable by holding one variable as constant. This leads to the definition of a partial derivative. In this section we

- Introduce partial derivatives
- Evaluate partial derivatives at points
- Interpret partial derivatives as slopes of tangents to surfaces.

The definition of a partial derivative is introduced for a function with two independent variables.

PARTIAL DERIVATIVE OF $z = f(x, y)$

The **partial derivative of** $z = f(x, y)$ **with repect to** x—denoted by $\partial z/\partial x$, f_x, or $f_x(x, y)$—is defined to be

$$\frac{\partial z}{\partial x} = \lim_{h \to 0} \frac{f(x + h, y) - f(x, y)}{h} \qquad \text{y is held constant.}$$

The **partial derivative of** $z = f(x, y)$ **with respect to** y—denoted by $\partial z/\partial y$, f_y, or $f_y(x, y)$—is defined to be

$$\frac{\partial z}{\partial y} = \lim_{k \to 0} \frac{f(x, y + k) - f(x, y)}{k} \qquad \text{x is held constant.}$$

if these limits exist.

NOTE: This definition states that partial derivatives of a function of two variables are determined by treating one variable as a constant. For example, for $z = f(x, y)$, to determine $\partial z/\partial x$, y is treated like a constant in the differentiation process and x is treated as a variable.

Since partial derivatives are simply ordinary derivatives with respect to one variable while keeping the other variable constant, partial derivatives can be found by using our previously obtained differentiation formulas instead of using the limit process.

EXAMPLE 7   If $z = f(x, y) = 2x^2 + y$, find $\partial z/\partial x$ and $\partial z/\partial y$.

SOLUTION   To find $\partial z/\partial x$, we differentiate with respect to x while holding y constant.

$$\frac{\partial z}{\partial x} = 4x \qquad \text{Since y is a constant, $(\partial/\partial x)(y) = 0$} \\ \text{and $(\partial/\partial x)(2x^2) = 4x$.}$$

Likewise, to find $\partial z/\partial y$, we differentiate with respect to y while holding x constant.

$$\frac{\partial z}{\partial y} = 1 \qquad \text{Since } x \text{ is a constant, } (\partial/\partial y)(2x^2) = 0 \\ \text{and } (\partial/\partial y)(y) = 1.$$

■

Practice Problem 1 For $z = f(x, y) = 4x^3 + 3x - y^2 + 2y$, find $\partial z/\partial x$ and $\partial z/\partial y$.

ANSWER $\partial z/\partial x = 12x^2 + 3$ and $\partial z/\partial y = -2y + 2$

EXAMPLE 8 If $z = e^{x^2+y^2}$, find $\partial z/\partial y$.

SOLUTION

$$\frac{\partial z}{\partial y} = 2ye^{x^2+y^2} \qquad \text{Since } x \text{ is a constant, } \frac{\partial}{\partial y}(x^2 + y^2) = 2y.$$

■

In the next example, the function has both independent variables in the same term. In this situation, you still treat one variable as a constant and differentiate as before.

EXAMPLE 9 If $z = f(x, y) = 3x^2 - 2xy + 5xy^2 - 3y$, find each of the following.

(a) $\dfrac{\partial z}{\partial y}$ (b) $f_y(3, 2)$

SOLUTION

(a) $\dfrac{\partial z}{\partial y} = -2x + 10xy - 3$ Hold x constant.

(b) $\begin{aligned} f_y(3, 2) &= -2(3) + 10(3)(2) - 3 \\ &= -6 + 60 - 3 \\ &= 51 \end{aligned}$

■

Practice Problem 2 Given $z = f(x, y) = e^{xy}$, find $\partial f/\partial x$ and $\partial f/\partial y$.

ANSWER $\dfrac{\partial f}{\partial x} = ye^{xy}$ and $\dfrac{\partial f}{\partial y} = xe^{xy}$

EXAMPLE 10 If $z = \ln(3x^2 + 2y)$, find each of the following.

(a) $\dfrac{\partial z}{\partial x}$ (b) $f_x(1, 0)$

(c) $\dfrac{\partial z}{\partial y}$ (d) $f_y(1, 0)$

SOLUTION

(a) $\dfrac{\partial z}{\partial x} = \dfrac{6x}{3x^2 + 2y}$ Hold y constant.

(b) $f_x(1, 0) = \dfrac{6(1)}{3(1)^2 + 2(0)} = \dfrac{6}{3 + 0} = 2$

(c) $\dfrac{\partial z}{\partial y} = \dfrac{2}{3x^2 + 2y}$ Hold x constant.

(d) $f_y(1, 0) = \dfrac{2}{3(1)^2 + 2(0)} = \dfrac{2}{3 + 0} = \dfrac{2}{3}$

Practice Problem 3 For $z = \ln(5x^2y - 2x)$ find $\partial z/\partial x$ and $\partial z/\partial y$.

ANSWER

$$\frac{\partial z}{\partial x} = \frac{10xy - 2}{5x^2y - 2x} \quad \text{and} \quad \frac{\partial z}{\partial y} = \frac{5x^2}{5x^2y - 2x}$$

The definition of partial derivatives can be extended for functions of more than two independent variables. To be specific, if z is a function of more than two independent variables, such as $z = f(w, x, y) = x^2yw + ye^w$, consider all independent variables except one as fixed and take the derivative with respect to this one variable. For example, to find $\partial z/\partial w = f_w(w, x, y)$, treat both x and y as constants and w as a variable.

EXAMPLE 11 Given $z = x^2yw + ye^w$, find $\partial z/\partial x$, $\partial z/\partial y$, and $\partial z/\partial w$.

SOLUTION

$$\frac{\partial z}{\partial x} = 2xyw \qquad w, y \text{ treated as constants.}$$

$$\frac{\partial z}{\partial y} = x^2w + e^w \qquad w, x \text{ treated as constants.}$$

$$\frac{\partial z}{\partial w} = x^2y + ye^w \qquad x, y \text{ treated as constants.}$$

Practice Problem 4 Given $f(x, y, z) = (x - y + 2z)^2$, find f_x, f_y, and f_z.

ANSWER $f_x = 2(x - y + 2z)$

$f_y = -2(x - y + 2z)$

$f_z = 4(x - y + 2z)$

In Section 7.1 we considered $z = f(x, y)$ as a surface in three dimensions. The partial derivative of z with respect to x at a point (x_1, y_1, z_1) may be thought of as the slope of the tangent to the surface in the direction of the x-axis at the point

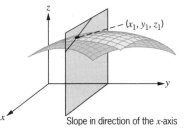

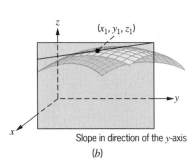

Slope in direction of the x-axis

(a)

Slope in direction of the y-axis

(b)

Figure 14

(x_1, y_1, z_1) on the surface. That is, if $\partial z/\partial x$ is evaluated at (x_1, y_1, z_1), then it represents the slope of the tangent to the surface at (x_1, y_1, z_1) parallel to the x-axis [Figure 14(a)]. Similarly, if $\partial z/\partial y$ is evaluated at (x_1, y_1, z_1), then it represents the slope of the tangent to the surface at (x_1, y_1, z_1) parallel to the y-axis [Figure 14(b)].

EXAMPLE 12   Find the slope of the tangent line to $z = x^2 + 2y^2$ at $(1, -1, 3)$ (a) in the direction of the x-axis and (b) in the direction of the y-axis.

SOLUTION

(a)   We have

$$\frac{\partial z}{\partial x} = 2x \quad \text{and} \quad \frac{\partial z}{\partial x}(1, -1, 3) = 2 \cdot 1 = 2$$

Thus the slope of the tangent line to the surface in the direction of the x-axis at the point $(1, -1, 3)$ is 2 [see Figure 15(a)].

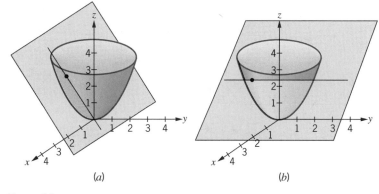

(a)   (b)

Figure 15

(b)   We have

$$\frac{\partial z}{\partial y} = 4y \quad \text{and} \quad \frac{\partial z}{\partial y}(1, -1, 3) = -4$$

Thus the slope of the tangent line to the surface in the direction of the y-axis at $(1, -1, 3)$ is -4 [see Figure 15(b)].

SUMMARY

Evaluating partial derivatives for functions of more than one independent variable involves the same rules that were used in earlier chapters involving functions with only one independent variable. The difference is that when evaluating a partial derivative, only one variable is differentiated at a time and the others are held constant. If z is defined to be a function of two variables, x and y, $z = f(x, y)$, then $\partial z/\partial x$ is found by taking y to be constant while we differentiate with respect to x. Likewise, $\partial z/\partial y$ is found by differentiating z with respect to y while x is held constant. This work can be extended to a function, z, of any number of variables. To differentiate z with respect to any variable, you just hold all the others constant while differentiating with respect to that variable.

Exercise Set 7.2

Find $\partial z/\partial x$, $\partial z/\partial y$, $f_x(3, 2)$, and $f_y(3, 2)$ for the following functions.

1. $z = f(x, y) = 7 + 2x$
2. $z = f(x, y) = 7 - 3y$
3. $z = f(x, y) = 7 + 3x - 2y$
4. $z = f(x, y) = 5 - 2x + 3y$
5. $z = f(x, y) = 10 - 3x^2 - 2xy + y^3$
6. $z = f(x, y) = 4xy^2 - 3y^2$
7. $z = f(x, y) = 3e^x - 2e^y$
8. $z = f(x, y) = 3xy^2 - 2e^y$
9. $z = f(x, y) = 3 \ln (xy + x^2)$
10. $z = f(x, y) = 2 \ln (3x^2 + 2y^2)$

Find z_x, z_y, and z_w if they exist.

11. $z = e^{xyw}$
12. $z = e^{3x^2yw^2}$
13. $z = \ln (x^2 + xy + w^2)$
14. $z = \ln (x^2 + w^2 + yw)$

Find the slope of the tangent line in an x direction and in a y direction at the point given.

15. $z = x^3 + y^2$, $(1, 1, 2)$
16. $z = 4x^2 + 6y^2$, $(1, 1, 10)$
17. $z = xy$, $(2, \frac{1}{2}, 1)$
18. $z = x^2 + xy$, $(2, 1, 6)$

Applications (Business and Economics)

19. **Compound Interest.** If a principal P is invested at an interest rate i compounded annually for n years, then the amount after n years is given by $A =$ $P(1 + i)^n$. Find the rate of change of A with respect to i with P and n held constant.

20. **Revenue and Cost Functions.** The revenue and cost functions for a company are given by

$$R(x, y) = 3x + 5y$$

$$C(x, y) = x^2 - 2xy + 3y^2 + 3x + 2y + 5$$

where x and y represent the number of items produced. Find each of the following.
   (a) $P(x, y) = R(x, y) - C(x, y)$
   (b) $R_y(3, 2)$ (c) $C_x(2, 3)$
   (d) $C_y(2, 3)$ (e) $P_x(2, 3)$

21. **Profit Function.** A profit function is given in thousands of dollars as

$$P(x, y) = 3x^2 - xy + 2y^3 + 2x + 3y + 3$$

where x and y represent the number of items produced. Find each of the following.
   (a) $P(1, 2)$ (b) $P(2, 1)$
   (c) $P_x(1, 2)$ (d) $P_y(1, 2)$

22. **Production Model.** The production model of a given company is given by $P = 0.8l^{1/2}C^{1/2}$, where P is the number of items produced, l is the number of hours of available labor, and C is the number of dollars of capital available. Find $\partial P/\partial l$ and $\partial P/\partial C$.

23. **Production Model.** For the production model in Exercise 22, find $\partial P/\partial l$ and $\partial P/\partial C$ when $l = 400$ hours and $C = \$10,000$.

Applications (Social and Life Sciences)

24. **Safety.** The length L of skid marks is given by

$$L(w, s) = 0.000014ws^2$$

where w is the weight of the car and s is the speed of the car. Find and interpret the results for each of the following.

(a) $L(2000, 50)$ (b) $L(3000, 55)$
(c) $L_w(3000, 55)$ (d) $L_s(3000, 55)$

25. **Intelligence Quotient.** IQ, represented by I, is given by the following equation:

$$I(M, C) = \frac{100M}{C}$$

where M represents mental age and C represents chronological age.

(a) Find $I(13, 10)$.
(b) Find $I(10, 13)$.
(c) Find $I_M(13, 10)$.
(d) Interpret your results.

26. **Scuba Diving.** The time of a scuba dive is estimated by

$$T(v, d) = \frac{36v}{d + 36}$$

where v is the volume of air in the diver's tanks and d is the depth of the dive.

(a) Find $T(65, 30)$.
(b) Find $T_v(65, 30)$.
(c) Find $T_d(65, 30)$.
(d) Interpret your results.

Review Exercises

Sketch the graphs of the following functions.

27. $z = 4z + 2y + 6$
28. $z = 6x - y + 12$

7.3 APPLICATION OF PARTIAL DERIVATIVES

OVERVIEW   Recall that partial derivatives represent the instantaneous rate of change of the dependent variable with respect to one independent variable (the others are held fixed). As an independent variable increases, a positive partial derivative indicates an increase in the dependent variable and a negative partial derivative indicates a decrease in the dependent variable. Keep these concepts in mind as we study marginal analysis (rates of change) with two or more independent variables. In this section we consider

- Marginal productivity
- Marginal cost, revenue, and profit for joint cost functions
- Marginal demand functions using related products

Companies that produce more than one product can be classified as multicommodity firms. Although most of these companies produce several products, for mathematical simplicity we discuss the cost, revenue, and profit in terms of two products, item 1 and item 2. We will discuss $C(x_1, x_2)$, $R(x_1, x_2)$, and $P(x_1, x_2)$, for x_1 units of item 1 and x_2 units of item 2. The following partial derivatives give marginal costs (C_{x_1} and C_{x_2}), marginal revenues (R_{x_1} and R_{x_2}), and marginal profits (P_{x_1} and P_{x_2}), which are useful approximations for the management of a multicommodity company.

When more than one product is produced by a company, the cost function involves all the products and is called a joint cost function.

C_{x_1}: Cost of making one *additional* unit of item 1 for given values of x_1 and x_2.

C_{x_2}: Cost of making one *additional* unit of item 2 for given values of x_1 and x_2.

R_{x_1}: Increase in revenue from selling one *additional* unit of item 1 for given values of x_1 and x_2.

R_{x_2}: Increase in revenue from selling one *additional* unit of item 2 for given values of x_1 and x_2.

P_{x_1}: Change in profit from selling one *additional* unit of item 1 for given values of x_1 and x_2.

P_{x_2}: Change in profit from selling one *additional* unit of item 2 for given values of x_1 and x_2.

EXAMPLE 13   If the joint cost function of a multicommodity firm is

$$C(x_1, x_2) = 100 + x_1^2 + 4x_1x_2 + 2x_2^2$$

find the following.

(a) C_{x_1} at $x_1 = 20$, $x_2 = 30$
(b) C_{x_2} at $x_1 = 20$, $x_2 = 30$

SOLUTION

(a) $C_{x_1} = 2x_1 + 4x_2$; $C_{x_1}(20, 30) = 2 \cdot 20 + 4 \cdot 30 = 160$
(b) $C_{x_2} = 4x_1 + 4x_2$; $C_{x_2}(20, 30) = 4 \cdot 20 + 4 \cdot 30 = 200$

Practice Problem 1   A revenue function is given by

$$R(x_1, x_2) = 200x_1 + 200x_2 - 4x_1^2 - 8x_1x_2 - 4x_2^2$$

If $x_1 = 4$ and $x_2 = 12$, find (a) R_{x_1} and (b) R_{x_2}.

ANSWER

(a) $R_{x_1} = 200 - 8x_1 - 8x_2$; $R_{x_1}(4, 12) = 72$
(b) $R_{x_2} = 200 - 8x_1 - 8x_2$; $R_{x_2}(4, 12) = 72$

Practice Problem 2   If a cost function is defined as

$$C(x_1, x_2) = 32\sqrt{x_1 x_2} + 175x_1 + 205x_2 + 1050$$

find the marginal costs, C_{x_1} and C_{x_2}, when $x_1 = 80$ and $x_2 = 20$.

ANSWER $C_{x_1}(80, 20) = 183$; $C_{x_2}(80, 20) = 237$

Production Model

Production depends on many factors, such as labor, capital, type of machinery, age of machinery, and so on. However, for the purpose of simplicity, we consider production (or output) as a function of only two variables, labor and capital. A function

$$P = f(l, C)$$

that gives the output P for l units of labor (such as the number of hours of labor) and C units of capital is called a **production function.**

$\dfrac{\partial P}{\partial l}$ gives the marginal productivity with respect to labor units.

$\dfrac{\partial P}{\partial C}$ gives the marginal productivity with respect to capital.

EXAMPLE 14   A furniture manufacturing company can produce

$$P = 0.8l^{1/2}C^{1/2}$$

custom chairs per week. Here l is the number of hours of labor and C is the number of dollars of capital. For example, if 400 work-hours and \$10,000 in capital are available, the production would be

$$P(400, 10,000) = 0.8(400)^{1/2}(10,000)^{1/2}$$
$$= 1600 \text{ chairs}$$

Then the marginal productivities are

$$P_l = \frac{0.4C^{1/2}}{l^{1/2}}$$

$$P_l(400, 10,000) = \frac{0.4(10,000)^{1/2}}{(400)^{1/2}} = 2$$

$$P_C = \frac{0.4l^{1/2}}{C^{1/2}}$$

$$P_C(400, 10,000) = \frac{0.4(400)^{1/2}}{(10,000)^{1/2}} = 0.08$$

Since marginals give an approximation of the change in a function, $P_l = 2$ tells us that if capital remains fixed at \$10,000 per week, the availability of an *additional* work-hour each week would result in an *approximate* increase in production of 2 chairs. On the other hand, $P_C = 0.08$ tells us that if labor remains fixed at 400 work-hours per week, the availability of one additional dollar of capital would increase production by *approximately* 0.08 of a chair. ■

This example introduces what is known as the Cobb–Douglas production function that is used extensively in business and economics to calculate the number of units produced by different amounts of capital and labor.

COBB–DOUGLAS PRODUCTION FUNCTION

The Cobb–Douglas production function is given by

$$P(l, C) = Al^aC^{1-a}, \qquad A > 0, 0 < a < 1$$

where P is the number of units produced with l units of labor and C units of capital.

EXAMPLE 15   A company has the following production model for a product:

$$P(l, C) = 10l^{3/4}C^{1/4}$$

(a)   Find the production that will result from 1296 work-hours of labor and $625 in capital.
(b)   Find the marginal productivity of labor and the marginal productivity of capital.
(c)   Evaluate the marginal productivity of labor and capital when $l = 64$ work-hours and $C = \$256$.
(d)   Interpret each result from part (c).

SOLUTION

(a)   When $l = 1296$ and $C = 625$,

$$P(l, C) = 10(1296)^{3/4}(625)^{1/4} = 10,800$$

Production is 10,800 units.

(b)   Given $P(l, C) = 10l^{3/4}C^{1/4}$, the marginal productivity of labor and the marginal productivity of capital are, respectively,

$$\frac{\partial P}{\partial l} = \frac{7.5C^{1/4}}{l^{1/4}} \quad \text{and} \quad \frac{\partial P}{\partial C} = \frac{2.5l^{3/4}}{C^{3/4}}$$

(c)   We have

$$\frac{\partial P}{\partial l}(64,256) = \frac{7.5(256)^{1/4}}{(64)^{1/4}} = 10.6066$$

and

$$\frac{\partial P}{\partial C}(64,256) = \frac{2.5(64)^{3/4}}{(256)^{3/4}} = 0.8839$$

(d)   From part (c), $\partial P/\partial l = 10.6066$ tells us that if capital is kept fixed at
$256, the availability of an additional hour of labor would *increase* pro-
duction by approximately 10.6066 units when $l = 64$ hours. Furthermore,
$\partial P/\partial C = 0.8839$ tells us that if work-hours are kept constant, the availa-
bility of an extra dollar of capital would *increase* production by approximately 0.8839 units when $C = \$256$ and $l = 64$ hours.   ■

Joint Demand Functions

Sometimes two products are so related that a change in the price of one affects the
demand for the other. Typical examples are pork and beef, chicken and fish, IBM
computers and MacIntosh computers. If such a relationship exists, then the
demand for each product is a function of the price of *both* products. Let p_1 and q_1
and p_2 and q_2 be the prices and numbers demanded, respectively, for two products. Then

$$q_1 = f(p_1, p_2)$$

$$q_2 = g(p_1, p_2)$$

We can find four partial derivatives.

$\dfrac{\partial q_1}{\partial p_1}$:   Marginal demand for the first product with respect to the price of the first
product (rate of change in demand for product 1 as the price of product 1
changes).

$\dfrac{\partial q_1}{\partial p_2}$:   Marginal demand for the first product with respect to the price of the second
product (rate of change in demand for product 1 as the price of product 2
changes).

$\dfrac{\partial q_2}{\partial p_1}$:   Marginal demand for the second product with respect to the price of the first
product (rate of change in demand for product 2 as the price of product 1
changes).

$\dfrac{\partial q_2}{\partial p_2}$:   Marginal demand for the second product with respect to the price of the
second product (rate of change in demand for product 2 as the price of
product 2 changes).

Two products are said to be **competitive** if a decrease in the quantity demanded
of one product can lead to an increase in the quantity demanded of the other
product. For example, IBM computers and MacIntosh computers are competitive.
On the other hand, if a decrease in the quantity demanded of one product leads to a
decrease in the quantity demanded of a second product, then the products are said
to be **complementary.** For example, motorboats and motorboat engines are complementary. The marginal demands can be used to classify products as competitive or complementary.

COMPETITIVE OR COMPLEMENTARY OR NEITHER

If p_1 and p_2 and q_1 and q_2 represent, respectively, the price and demand for two products, then the two products are said to be **competitive** if

$$\frac{\partial q_1}{\partial p_2} > 0 \quad \text{and} \quad \frac{\partial q_2}{\partial p_1} > 0$$

and **complementary** if

$$\frac{\partial q_1}{\partial p_2} < 0 \quad \text{and} \quad \frac{\partial q_2}{\partial p_1} < 0$$

All other cases are neither competitive nor complementary.

EXAMPLE 16 Given the following two demand functions for product 1 and product 2, classify the two products as competitive or complementary.

$$q_1 = 1000 - 6p_1 + 8p_2$$
$$q_2 = 600 + 12p_1 - 10p_2$$

SOLUTION

$$\frac{\partial q_1}{\partial p_2} = 8 \quad \text{and} \quad \frac{\partial q_2}{\partial p_1} = 12$$

Since $\partial q_1/\partial p_2$ and $\partial q_2/\partial p_1$ are both positive, the two products are competitive.

SUMMARY

We can see that the concepts and uses of partial derivatives relative to marginal analysis are much the same as we encountered in Chapter 3 when we dealt with only one independent variable. Remember that these ideas can be used with any number of independent variables, and that often, in real cases, there are numerous independent variables. Partial derivatives are very useful in the discussion of production models such as the Cobb–Douglas model.

Exercise Set 7.3

Applications (Business and Economics)

1. **Marginal Profit.** A corporation makes and sells x units of product A and y units of product B. The cost and revenue functions are given by

$$C(x, y) = 400 + 30x + 20y$$

and

$$R(x, y) = 50x + 30y$$

Find and interpret each of the following.
(a) Marginal cost with respect to x
(b) Marginal cost with respect to y
(c) Marginal revenue with respect to x
(d) Marginal revenue with respect to y
(e) Marginal profit with respect to x
(f) Marginal profit with respect to y

2. **Marginal Cost.** Suppose that a company produces x units of product A and y units of product B with a cost function of

$$C(x, y) = 1000 + 40x + 30y + 2x^2 + y^2$$

Find and interpret $C_x(8, 10)$ and $C_y(8, 10)$.

3. **Marginal Revenue.** Relative to the company in Exercise 2, the revenue function is

$$R(x, y) = 100x + 60y - 0.02xy$$

Find and interpret $R_x(8, 10)$ and $R_y(8, 10)$.

4. **Marginal Profit.** Relative to the company in Exercises 2 and 3, find and interpret the following for the profit function: $P_x(8, 10)$ and $P_y(8, 10)$.

5. **Marginal Profit.** A company uses the following cost and revenue functions:

$$C(x, y) = 10x^2 + 4y^2 + 2xy$$

and

$$R(x, y) = 12x^2 + 6y^2$$

Find and interpret the marginal profit functions P_x and P_y.

Analyze the demand functions in Exercises 6–11 to determine whether product 1 and product 2 are competitive, complementary, or neither. p_1 and p_2 are the prices and q_1 and q_2 are the demands for products 1 and 2, respectively.

6. $q_1 = 160 - 0.10p_1 - 0.05p_2$
   $q_2 = 200 - 0.08p_1 - 0.01p_2$

7. $q_1 = 250 - 0.04p_1 + 0.01p_2$
   $q_2 = 200 - 0.03p_1 + 0.005p_2$

8. $q_1 = 6000 - \dfrac{400}{p_1 + 2} - 40p_2$

   $q_2 = 8000 + 50p_1 - \dfrac{400}{p_2 + 2}$

9. $q_1 = \dfrac{200}{p_1\sqrt{p_2}}, \quad q_2 = \dfrac{400}{p_2\sqrt{p_1} + 3}$

10. $q_1 = \dfrac{60\sqrt{p_1}}{p_2}, \quad q_2 = \dfrac{40p_2}{p_1}$

11. $q_1 = 600 - 0.04p_1{}^2 - 0.02p_2{}^2$
    $q_2 = 400 - 0.01p_1{}^2 - 0.04p_2{}^2$

12. **Production Function.** The production function for a given product of an electronics company is given by

$$f(l, C) = 400l^{1/3}C^{2/3}$$

where l is labor in terms of work-hours per week, C is capital expenditures in dollars spent per week, and the output f is in units manufactured per week. Find each of the following and interpret your results.
(a) $f(512, 27{,}000)$
(b) $f_l(512, 27{,}000)$
(c) $f_C(512, 27{,}000)$

13. **Production Function.** Suppose that the production function for a product is

$$f = 60l^{3/4}C^{1/4}$$

where l is in work-hours and C is capital expenditures in dollars.
(a) Find the marginal productivity of l.
(b) Find the marginal productivity of C.

Review Exercises

Find $\partial z/\partial x$, $\partial z/\partial y$, $f_x(1, 2)$, and $f_y(1, 2)$ for the following functions.

14. $z = f(x, y) = 3x^2 + 2xy - 5y^2$
15. $z = f(x, y) = 4x^2y - 3xy^3$
16. $z = f(x, y) = 4x^2y + 2e^x$
17. $z = f(x, y) = 4xy + 5e^{-y}$

7.4 HIGHER-ORDER PARTIAL DERIVATIVES

OVERVIEW   When working with functions of one variable, we found that it was easy to extend the concept of the first derivative to obtain a second derivative, a third derivative, and so on. Although such an extension with partial derivatives is somewhat more involved, in this section we

- Extend the partial derivative concept to higher-order partial derivatives
- Find mixed partial derivatives

Just as with ordinary derivatives, it is possible to take the second partial derivative of a function. We can also take the third, fourth, and so on partial derivatives. The definition is given for second-order partial derivatives but can be extended to higher derivatives as needed.

SECOND-ORDER PARTIAL DERIVATIVES

Suppose that $z = f(x, y)$. Then

$$\frac{\partial^2 z}{\partial x^2} = \frac{\partial}{\partial x}\left(\frac{\partial z}{\partial x}\right) \qquad \text{Take the partial derivative with respect to } x \text{ of } \frac{\partial z}{\partial x}.$$

$$\frac{\partial^2 z}{\partial y^2} = \frac{\partial}{\partial y}\left(\frac{\partial z}{\partial y}\right) \qquad \text{Take the partial derivative with respect to } y \text{ of } \frac{\partial z}{\partial y}.$$

$$\frac{\partial^2 z}{\partial x \partial y} = \frac{\partial}{\partial x}\left(\frac{\partial z}{\partial y}\right) \qquad \text{Take the partial derivative with respect to } x \text{ of } \frac{\partial z}{\partial y}.$$

The following notations are for second-order derivatives.

NOTATION OF SECOND-ORDER PARTIAL DERIVATIVES

Suppose that $z = f(x, y)$; then

$$\frac{\partial^2 z}{\partial x^2} = \frac{\partial}{\partial x}\left(\frac{\partial z}{\partial x}\right) = f_{xx}(x, y) = f_{xx}$$

$$\frac{\partial^2 z}{\partial y \partial x} = \frac{\partial}{\partial y}\left(\frac{\partial z}{\partial x}\right) = f_{xy}(x, y) = f_{xy}$$

$$\frac{\partial^2 z}{\partial x \partial y} = \frac{\partial}{\partial x}\left(\frac{\partial z}{\partial y}\right) = f_{yx}(x, y) = f_{yx}$$

$$\frac{\partial^2 z}{\partial y^2} = \frac{\partial}{\partial y}\left(\frac{\partial z}{\partial y}\right) = f_{yy}(x, y) = f_{yy}$$

EXAMPLE 17   If $z = f(x, y) = 2x^3 - 3xy^2 + y^3 - 2$, find each of the following.

(a) $\dfrac{\partial^2 z}{\partial x^2}$ (b) $\dfrac{\partial^2 z}{\partial y \partial x}$ (c) $\dfrac{\partial^2 z}{\partial x \partial y}$ (d) $\dfrac{\partial^2 z}{\partial y^2}$

(e) $f_{xx}(1, 2)$ (f) $f_{xy}(1, 2)$ (g) $f_{yx}(2, 1)$ (h) $f_{yy}(5, 2)$

SOLUTION

(a) $\dfrac{\partial z}{\partial x} = 6x^2 - 3y^2$ (b) $\dfrac{\partial z}{\partial x} = 6x^2 - 3y^2$

$\dfrac{\partial^2 z}{\partial x^2} = 12x$ $\dfrac{\partial^2 z}{\partial y \partial x} = -6y$

(c) $\dfrac{\partial z}{\partial y} = -6xy + 3y^2$ (d) $\dfrac{\partial z}{\partial y} = -6xy + 3y^2$

$\dfrac{\partial^2 z}{\partial x \partial y} = -6y$ $\dfrac{\partial^2 z}{\partial y^2} = -6x + 6y$

(e) $f_{xx}(x, y) = \dfrac{\partial^2 z}{\partial x^2} = 12x$ (f) $f_{xy}(x, y) = \dfrac{\partial^2 z}{\partial y \partial x} = -6y$

$f_{xx}(1, 2) = 12(1) = 12$ $f_{xy}(1, 2) = -6(2) = -12$

(g) $f_{yx}(x, y) = \dfrac{\partial^2 z}{\partial x \partial y} = -6y$ (h) $f_{yy}(x, y) = \dfrac{\partial^2 z}{\partial y^2} = -6x + 6y$

$f_{yx}(2, 1) = -6(1) = -6$ $f_{yy}(5, 2) = -6(5) + 6(2)$

$= -30 + 12 = -18$ ■

Practice Problem 1 For $f(x, y) = 7x^2y^3 + 3x^2y - 2y + 4x^4$, find f_{xx}, f_{xy}, f_{yy}, and f_{yx}.

ANSWER $f_{xx} = 48x^2 + 6y + 14y^3$, $f_{xy} = 6x + 42xy^2$, $f_{yy} = 42x^2y$, and $f_{yx} = 6x + 42xy^2$

EXAMPLE 18 Find $f_{xx}(1, 2)$, $f_{yy}(1, 2)$, and $f_{xy}(1, 2)$ for $f(x, y) = e^{x^2+y^2}$.

SOLUTION

$$f(x, y) = e^{x^2+y^2}$$

$$f_x = e^{x^2+y^2}(2x) \qquad \dfrac{\partial(x^2 + y^2)}{\partial x} = 2x$$

$$f_{xx} = 2e^{x^2+y^2} + e^{x^2+y^2}(4x^2) \qquad \text{Product formula}$$
$$= (4x^2 + 2)e^{x^2+y^2}$$

$$f_y = e^{x^2+y^2}(2y) \qquad \dfrac{\partial(x^2 + y^2)}{\partial y} = 2y$$

$$f_{yy} = 2e^{x^2+y^2} + e^{x^2+y^2}(4y^2) \qquad \text{Product formula}$$
$$= (4y^2 + 2)e^{x^2+y^2}$$

$$f_{xy} = e^{x^2+y^2}(4xy) \qquad \dfrac{\partial(f_x)}{\partial y} = \dfrac{\partial[e^{x^2+y^2}(2x)]}{\partial y}$$

$$f_{xx}(1, 2) = [4(1)^2 + 2]e^{1^2+2^2} = 6e^5$$

$$f_{yy}(1, 2) = [4(2)^2 + 2]e^{1^2+2^2} = 18e^5$$

$$f_{xy}(1, 2) = 4(1)(2)e^{1^2+2^2} = 8e^5$$

■

Practice Problem 2 Given $f = e^{xy}$, find f_{xx}, f_{yy}, and f_{xy}.

ANSWER $f_{xx} = y^2 e^{xy}, f_{yy} = x^2 e^{xy}, f_{xy} = e^{xy} + xye^{xy}$

You may have already discovered in many examples that

$$\frac{\partial^2 z}{\partial y \partial x} = \frac{\partial^2 z}{\partial x \partial y}$$

Actually, all of the functions with which we shall be dealing satisfy this relationship.

There are four ways to find a second-order partial derivative of $z = f(x, y)$.

1. f_{xx} Differentiate twice with respect to x.
2. f_{yy} Differentiate twice with respect to y.
3. f_{xy} Differentiate with respect to x and then with respect to y.
4. f_{yx} Differentiate with respect to y and then with respect to x.

The derivatives f_{xy} and f_{yx} are called **mixed partial derivatives.**

We now extend the concept of a second-order partial derivative to a higher-order partial derivative. We use the following notation:

$$\frac{\partial^3 f}{\partial x^3} = f_{xxx} = \frac{\partial (f_{xx})}{\partial x}$$

$$\frac{\partial^3 f}{\partial y^3} = f_{yyy} = \frac{\partial (f_{yy})}{\partial y}$$

$$\frac{\partial f}{\partial x^2 \partial y} = f_{xxy} \quad \text{or} \quad f_{xyx} \quad \text{or} \quad f_{yxx}$$

$$f_{xxy} = \frac{\partial (f_{xx})}{\partial y}$$

$$\text{and} \quad f_{xyx} = \frac{\partial (f_{xy})}{\partial x}$$

$$\text{and} \quad f_{yxx} = \frac{\partial (f_{yx})}{\partial x}$$

EXAMPLE 19   If $f(x, y) = x^4 y^3 + x^6 + y^4$, find f_{xxy}, f_{xyx}, and f_{xxx}

SOLUTION

$$f(x, y) = x^4 y^3 + x^6 + y^4$$

$$f_x = 4x^3 y^3 + 6x^5$$

$$f_{xx} = 12x^2 y^3 + 30x^4$$

$$f_{xxy} = 36x^2 y^2 \qquad\qquad f_{xxy} = \frac{\partial (f_{xx})}{\partial y}$$

$$f_{xy} = 12x^3y^2 \qquad f_{xy} = \frac{\partial(f_x)}{\partial y}$$

$$f_{xyx} = 36x^2y^2 \qquad f_{xyx} = \frac{\partial(f_{xy})}{\partial x}$$

$$f_{xxx} = 24xy^3 + 120x^3 \qquad f_{xxx} = \frac{\partial(f_{xx})}{\partial x}$$

■

Practice Problem 3 For $f(x, y) = x^4 - 3x^2y^2 + y^4$, find f_{xxy}, f_{xyx}, and f_{xxx}.

ANSWER $f_{xxy} = -12y, f_{xyx} = -12y, f_{xxx} = 24x$

SUMMARY

In this section, we have expanded what we previously learned about partial derivatives. Partial derivatives can be expanded to any level to give second, third, or higher derivatives. These derivatives may all be with respect to the same variable, or they can be mixed.

Exercise Set 7.4

Find $\partial^2z/\partial x^2$, $\partial^2z/\partial y^2$, and $f_{xy}(1, 3)$ for the following functions.

1. $z = f(x, y) = 7 + 5x + 3y$
2. $z = f(x, y) = 5x^2 - 3y^3$
3. $z = f(x, y) = 10x^2y^3$
4. $z = f(x, y) = 5x^3y^2$
5. $z = f(x, y) = 5x^2y - 3xy^3$
6. $z = f(x, y) = 4xy^2 - 3x^2y^3$
7. $z = f(x, y) = 3e^x - 2e^y$
8. $z = f(x, y) = 4xe^x - 2xe^y$
9. $z = f(x, y) = 3e^{xy}$
10. $z = f(x, y) = \ln(3x + y)$
11. $z = f(x, y) = x \ln(3x^2 + 2y^2)$
12. $z = f(x, y) = 3e^{x^2} + y^2$
13. $z = f(x, y) = e^x \ln(3x + 2y)$
14. $z = f(x, y) = (e^x + e^y) \ln x$
15. $z = f(x, y) = \sqrt{2x + 3y}$

16. $z = f(x, y) = \sqrt[3]{x^2 + 3y^2}$
17. $z = f(x, y) = \dfrac{x}{y}$
18. $z = f(x, y) = \dfrac{1}{2x + 3y^2}$
19. $z = f(x, y) = e^{x-y}(y^2)$
20. $z = f(x, y) = xe^{x^2-y^2}$

Find f_{xxx}, f_{yyy}, f_{xyx}, and f_{yxy} for each of the following:

21. $f(x, y) = x^3y + y^3x$
22. $f(x, y) = y^2x + x^2y$
23. $f(x, y) = e^{x+y}$
24. $f(x, y) = e^{x^2 + 3y^2}$
25. $f(x, y) = \ln(x + y)$
26. $f(x, y) = \ln(x^2 + y)$

Applications (Business and Economics)

27. ***Production Function.*** The production function for a given product of an electronics company is given by

$$f(l, C) = 400l^{1/3}C^{2/3}$$

where l is labor in terms of work-hours per week, C is the dollars spent per week, and the output f is in units manufactured per week. Find f_{ll}, f_{CC}, and f_{lC}.

28. **Production Function.**   Suppose that the production function for a product is

$$f(l, C) = 60l^{3/4}C^{1/4}$$

where l is in work-hours and C is capital expenditures in dollars. Find f_{ll}, f_{CC}, and f_{Cl}.

Review Exercises

Find $\partial z/\partial x$, $\partial z/\partial y$, $f_x(1, 2)$, and $f_y(1, 2)$.

29. $z = f(x, y) = 3xy + \ln(x^2 + 2y^2)$

30. $z = f(x, y) = e^{xy} + \ln(2x + y)$

7.5   RELATIVE MAXIMA AND MINIMA FOR FUNCTIONS OF TWO VARIABLES

OVERVIEW   In this section, we show how partial derivatives may be used to find relative maxima or relative minima for functions of two variables. The development of the theory is somewhat similar to the development of the theory for functions of one variable. Since we have demonstrated that there are two first partial derivatives and three or more second partial derivatives for functions of two variables, you may expect the procedure of finding and testing critical points to be more complicated. To simplify the discussion, we are going to assume that all higher-order partial derivatives of $z = f(x, y)$ exist in some circular region of the xy-plane. In this section we

- Define a relative maximum and relative minimum for two independent variables
- Define a saddle point
- Find necessary conditions for a relative maximum and relative minimum
- Find sufficient conditions for a relative maximum, relative minimum, and a saddle point

The following definition extends the definition of relative maximum and relative minimum to a function of two variables.

RELATIVE MAXIMUM AND RELATIVE MINIMUM FOR FUNCTIONS OF TWO VARIABLES

(a)   If there exists a circular region with (a, b) as the center in the domain of $z = f(x, y)$ such that

$$f(a, b) \geq f(x, y)$$

for all (x, y) in the region, then $f(a, b)$ is a **relative maximum**.

(b)   If there exists a circular region with (a, b) as the center in the domain of $z = f(x, y)$ such that

$$f(a, b) \leq f(x, y)$$

for all (x, y) in the region, then $f(a, b)$ is a **relative minimum**.

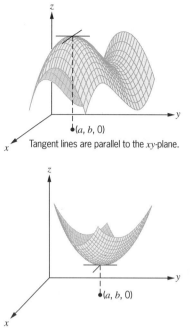

Tangent lines are parallel to the xy-plane.

Figure 16

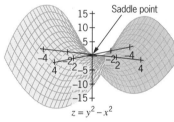

$z = y^2 - x^2$

Figure 17

Recall that at a relative maximum or relative minimum for a function of one variable, the first derivative is 0. For a function of two variables, Figure 16 suggests that both partial derivatives must be 0 at a relative maximum or relative minimum. This is true and is stated as a *necessary* condition for a relative maximum or relative minimum.

A NECESSARY CONDITION FOR A RELATIVE MAXIMUM OR RELATIVE MINIMUM

If f_x and f_y exist at (a, b), then a *necessary* condition for $f(a, b)$ to be a relative maximum or a relative minimum is that

$$f_x(a, b) = f_y(a, b) = 0$$

The point (a, b) is called a **critical point** of the domain.

The converse of the theorem is not true. If $f_x(a, b) = f_y(a, b) = 0$, then $f(a, b)$ may or may not be a relative maximum or relative minimum. Figure 17 shows a point (a, b) at which both partial derivatives equal 0, but the point is not a relative maximum or a relative minimum; this is called a **saddle point**.

The following test is similar to the second derivative test in one variable and gives *sufficient* conditions for a relative minimum, a relative maximum, or a saddle point.

Sufficient Conditions for a Relative Minimum, a Relative Maximum, or a Saddle Point

Assume that $z = f(x, y)$ and all higher-order partial derivatives exist in some circular region with (a, b) as the center. Suppose that $f_x(a, b) = f_y(a, b) = 0$ and let

$$A = f_{xx}(a, b), \quad B = f_{xy}(a, b), \quad \text{and} \quad C = f_{yy}(a, b)$$

1. If $AC - B^2 > 0$, and $A > 0$, then $f(a, b)$ is a relative minimum.
2. If $AC - B^2 > 0$, and $A < 0$, then $f(a, b)$ is a relative maximum.
3. If $AC - B^2 < 0$, then $f(x, y)$ has a saddle point at (a, b).
4. If $AC - B^2 = 0$, then no conclusion can be reached about $f(a, b)$.

We now outline the procedure that should be followed in locating points where there are relative maxima, relative minima, or saddle points.

> ## Four-Step Procedure for Finding Relative Maxima, Relative Minima, or Saddle Points
>
> Step 1. Find f_x and f_y and set each equal to zero.
> Step 2. Solve the equations in step 1 to get the critical points, which we denote as (a, b).
> Step 3. Compute $A = f_{xx}(a, b)$, $B = f_{xy}(a, b)$, and $C = f_{yy}(a, b)$.
> Step 4. Evaluate $AC - B^2$ and determine which sufficient condition applies.

Step 2 may require solving the pair of simultaneous equations

$$\frac{\partial z}{\partial x} = 0$$

$$\frac{\partial z}{\partial y} = 0$$

EXAMPLE 20   Find all critical points and classify them for $z = f(x, y) = x^2 + y^2$.

SOLUTION   We will follow the four-step process outlined above.

Step 1.   Find f_x and f_y and set each equal to zero.

$$f_x(x, y) = 2x$$
$$f_y(x, y) = 2y$$

Setting both of these equal to 0 gives

$$f_x(x, y) = 2x = 0$$
$$f_y(x, y) = 2y = 0$$

Step 2.   Solve the equations in step 1 to get the critical points, which we denote as (a, b). In this step, we may obtain two equations in two unknowns in which we desire all common solutions. For this example, the only common solution is $x = 0$ and $y = 0$ and $(0, 0)$ is the only critical point of the domain.

Step 3.   Compute $A = f_{xx}(a, b)$, $B = f_{xy}(a, b)$, and $C = f_{yy}(a, b)$.

$$f_{xx}(x, y) = 2 \qquad \text{Generally, second partials are not constants.}$$
$$f_{xy}(x, y) = 0$$
$$f_{yy}(x, y) = 2$$

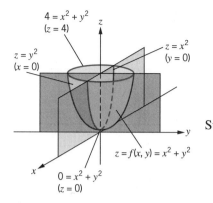

$4 = x^2 + y^2$
$(z = 4)$

$z = y^2$
$(x = 0)$

$z = x^2$
$(y = 0)$

$z = f(x, y) = x^2 + y^2$

$0 = x^2 + y^2$
$(z = 0)$

Figure 18

Therefore,

$$A = f_{xx}(0, 0) = 2$$
$$B = f_{xy}(0, 0) = 0$$
$$C = f_{yy}(0, 0) = 2$$

Step 4. Evaluate $AC - B^2$ and determine which sufficient condition applies. $AC - B^2 = 2(2) - (0)^2 = 4 > 0$ and $A = 2 > 0$; so, condition 1 applies and $f(0, 0)$ is a relative minimum. The minimum is $f(0, 0) = 0$. This fact agrees with the graph shown in Figure 18. ∎

EXAMPLE 21 Find and classify all critical points of

$$z = f(x, y) = x^2 + xy + y^2 - 7x - 8y + 10$$

SOLUTION
Step 1.

$$f_x(x, y) = 2x + y - 7 = 0$$
$$f_y(x, y) = x + 2y - 8 = 0$$

Step 2. Solving this system gives $(2, 3)$ as the only critical point of the domain.
Step 3.

$$f_{xx}(x, y) = 2 = f_{xx}(2, 3) = A$$
$$f_{xy}(x, y) = 1 = f_{xy}(2, 3) = B$$
$$f_{yy}(x, y) = 2 = f_{yy}(2, 3) = C$$

Step 4. $AC - B^2 = (2)(2) - (1)^2 = 3 > 0$. Since $A = 2 > 0$, condition 1 tells us that $f(2, 3)$ is a relative minimum. The minimum is $f(2, 3) = -9$. ∎

Practice Problem 1 Find and classify all critical points for

$$z = f(x, y) = 2x^2 + y^2 + 8x - 6y + 20$$

ANSWER   Relative minimum at $(-2, 3)$; $f(-2, 3) = 3$

Practice Problem 2 Find and classify all critical points for $z = x - \frac{1}{3}x^3 - \frac{1}{2}y^2$.

ANSWER   Relative maximum at $(-1, 0)$; $f(-1, 0) = -\frac{2}{3}$; saddle point at $(1, 0)$

EXAMPLE 22 Find all critical points and classify them for

$$z = f(x, y) = xy - \frac{x^4}{4} - \frac{y^2}{2} + 10$$

SOLUTION

Step 1.

$$f_x(x, y) = y - x^3 = 0$$
$$f_y(x, y) = x - y = 0$$

Step 2. To solve these two equations, we obtain $y = x$ from the second equation and substitute into the first equation to give

$$x - x^3 = 0$$
$$x(1 - x^2) = 0$$
$$x(1 + x)(1 - x) = 0$$
$$x = 0 \quad \text{or} \quad x = -1 \quad \text{or} \quad x = 1$$

There are three critical points of the domain: $(0, 0)$, $(-1, -1)$, and $(1, 1)$. We will test each of these by repeating steps 3 and 4.

Step 3.

$$f_{xx}(x, y) = -3x^2$$
$$f_{xy}(x, y) = 1$$
$$f_{yy}(x, y) = -1$$

For $(0, 0)$ we have

$$A = f_{xx}(0, 0) = -3(0)^2 = 0$$
$$B = f_{xy}(0, 0) = 1$$
$$C = f_{yy}(0, 0) = -1$$

Step 4. $AC - B^2 = 0(-1) - (1)^2 = -1 < 0$. Condition 3 tells us that $f(x, y)$ has a saddle point at $(0, 0)$.

Step 3. For $(-1, -1)$ we have

$$A = f_{xx}(-1, -1) = -3(-1)^2 = -3$$
$$B = f_{xy}(-1, -1) = 1$$
$$C = f_{yy}(-1, -1) = -1$$

Step 4. $AC - B^2 = (-3)(-1) - (1)^2 = 2 > 0$. Since $A = -3 < 0$, condition 2 tells us that $f(-1, -1)$ is a relative maximum.

Step 3.   For (1, 1) we have

$$A = f_{xx}(1, 1) = -3(1)^2 = -3$$
$$B = f_{xy}(1, 1) = 1$$
$$C = f_{yy}(1, 1) = -1$$

Step 4. $AC - B^2 = (-3)(-1) - (1)^2 = 2 > 0$. Since $A = -3 < 0$, condition 2 tells us that $f(1, 1)$ is a relative maximum.   ∎

Practice Problem 3 Find the points at which there are relative maxima, relative minima, or saddle points for $z = 3xy - x^3 - y^3$.

ANSWER   Relative maximum of 1 at (1, 1); saddle point at (0, 0)

There are times when a restriction is placed on a function. These restrictions may be given as equations and are called **constraint equations**.

EXAMPLE 23   Find three numbers x, y, and z such that $x + y + z = 10$ and $x^2 yz$ is a relative maximum. (The equation $x + y + z = 10$ is a constraint equation.)

SOLUTION   Although $x^2 yz$ is not a function of two variables, we can solve $x + y + z = 10$ for z to obtain $z = 10 - x - y$. Substituting this value of z gives a function of two variables.

$$f(x, y) = x^2 y(10 - x - y) = 10x^2 y - x^3 y - x^2 y^2$$

Step 1.

$$f_x(x, y) = 20xy - 3x^2 y - 2xy^2 = xy(20 - 3x - 2y) = 0$$
$$f_y(x, y) = 10x^2 - x^3 - 2x^2 y = x^2(10 - x - 2y) = 0$$

Step 2.   Although $(0, y)$ is a critical point for any y, it is obvious that none of these yield a maximum because for each of these $f(0, y) = 10(0)^2 y - (0)^3 y - (0)^2 y^2 = 0$. An additional critical point $(5, \frac{5}{2})$ is obtained from solving the system

$$20 - 3x - 2y = 0$$
$$10 - x - 2y = 0$$

Step 3.

$$f_{xx}(x, y) = 20y - 6xy - 2y^2$$
$$f_{xy}(x, y) = 20x - 3x^2 - 4xy$$
$$f_{yy} = (x, y) = -2x^2$$

For $(5, \frac{5}{2})$ we have

$$A = f_{xx}(5, \frac{5}{2}) = 20(\frac{5}{2}) - 6(5)(\frac{5}{2}) - 2(\frac{5}{2})^2 = 50 - 75 - \frac{25}{2} = -\frac{75}{2}$$

$$B = f_{xy}(5, \frac{5}{2}) = 20(5) - 3(5)^2 - 4(5)(\frac{5}{2}) = 100 - 75 - 50 = -25$$

$$C = f_{yy}(5, \frac{5}{2}) = -2(5)^2 = -50$$

Step 4. $AC - B^2 = -\frac{75}{2}(-50) - (-25)^2 = 1250 > 0$. Since $A = -\frac{75}{2} < 0$, condition 2 tells us that

$$f(5, \frac{5}{2}) = 10(5)^2(\frac{5}{2}) - (5)^3(\frac{5}{2}) - (5)^2(\frac{5}{2})^2 = \frac{625}{4}$$

is a relative maximum. So x^2yz is a maximum at $(5, \frac{5}{2}, \frac{5}{2})$.   ■

SUMMARY

To locate relative extrema on the surface $z = f(x, y)$, we begin by finding the critical points. The critical points are found by solving the pair of simultaneous equations

$$\frac{\partial z}{\partial x} = 0 \quad \text{and} \quad \frac{\partial z}{\partial y} = 0$$

In this section we stated the necessary and sufficient conditions for a relative maximum, relative minimum, and a saddle point. This can be used in marginal analysis to determine maximum profit, minimum cost, and maximum revenue.

Exercise Set 7.5

In Exercises 1–5, find the critical points of each function.

1. $f(x, y) = x^2 + y^2 - 2y + 4$
2. $f(x, y) = x^2 + 2y^2 - 2xy - 3x + 5y + 8$
3. $f(x, y) = x^2 - 2x + y^4 - 2y^2 + 6$
4. $f(x, y) = 4x^2 - 4xy + 3y^2 + 8x - 4y + 6$
5. $f(x, y) = x^2 - 2xy + 2y^2 + 3x + 8$

In Exercises 6–10, classify each critical point as a relative maximum, relative minimum, or saddle point.

6. Exercise 1   7. Exercise 2

8. Exercise 3   9. Exercise 4
10. Exercise 5

Find all critical points and classify each as a relative maximum, relative minimum, or saddle point.

11. $f(x, y) = x^2 + 2y^2$
12. $f(x, y) = 3x^2 + y^2$
13. $f(x, y) = 4 - x^2 - y^2$
14. $f(x, y) = 9 - x^2 - y^2$
15. $f(x, y) = 10 - x^2 - 2y^2$
16. $f(x, y) = y + x^2 + 3y^2$
17. $f(x, y) = x^2 - y^2$
18. $f(x, y) = xy$

19. $f(x, y) = 5 - x^2 + 4x - y^2$

20. $f(x, y) = x^2 + y^2 - 6x + 2y + 10$

21. $f(x, y) = xy - 3x + 2y - 3$

22. $f(x, y) = xy + 5x - 3y + 7$

23. $f(x, y) = x^2 + y^2 - 6xy$

24. $f(x, y) = 8xy - x^2 - y^2$

25. $f(x, y) = x^2 + xy + 2y^2 - 3x + 2y + 2$

26. $f(x, y) = x^2 + 4xy + y^2 + 6y + 1$

27. $f(x, y) = -2x^2 + xy - y^2 + 10x + y - 3$

28. $f(x, y) = 3x^2 - 2xy + 2y^2 - 8x - 4y + 10$

29. $f(x, y) = 3xy - x^2y - xy^2$

30. $f(x, y) = e^{xy}$

31. $f(x, y) = x^3 + y^3 - 6x^2 - 3y^2 - 9y$

32. $f(x, y) = 2x^3 - x^2 - 4x + y^2 - 4y + 2$

33. Divide a straight line of length L into three parts such that the sum of the squares of their lengths is a minimum.

Applications (Business and Economics)

34. **Profit Function.** A company has the following profit function (in thousands of dollars):

$$P(x, y) = 2xy - x^2 - 2y^2 - 4x + 12y - 5$$

where x is the number of thousands of item I and y is the number of thousands of item II produced. How many of each type should be produced to maximize (relative maximum) profit? What is the maximum profit?

35. **Cost Function.** A cost function (in thousands of dollars) is given by

$$C(x, y) = 3x^2 + 2xy + 2y^2 - 18x - 16y + 180$$

Find the critical point and the minimum (relative minimum) cost if x and y represent the number of items produced.

36. **Maximum Revenue.** The revenue function (in thousands of dollars) associated with Exercise 35 is

$$R(x, y) = -5x^2 + 42x - 8y^2 - 2xy + 102y$$

Find the point that will give maximum (relative maximum) revenue and find the maximum revenue.

37. **Maximum Profit.** Using the cost and revenue functions of Exercises 35 and 36, find the point at which there is a maximum (relative maximum) profit and find the maximum profit.

Review Exercises

38. Find f_{xx}, f_{ww}, f_{xy}, and f_{yw} for $f(x, y, w) = e^{xyw}$.

39. Find f_{xx} and f_{xy} for $f(x, y) = \ln(x^2 + 3y^2)$.

7.6 LAGRANGE MULTIPLIERS

OVERVIEW   In Example 23 of Section 7.5, we found three numbers x, y, and z, which made $f(x, y, z) = x^2yz$ a maximum subject to the constraint $x + y + z = 10$. In the solution to that example, we solved the constraint equation for z in terms of x and y, that is,

$$z = 10 - x - y$$

and substituted this value of z into the function x^2yz to obtain a function of two variables,

$$x^2yz = x^2y(10 - x - y)$$

We then found the critical points for this function and used the critical points to find the relative maximum.

Sometimes it is difficult or even impossible to solve a constraint equation for one variable in terms of the others. In that case, we may try a method called the **method of Lagrange multipliers**. In this method, the introduction of another variable allows us to solve the constrained optimization problem without first solving the constraint equation for one of the variables. The method can be used for functions of two variables, three variables, four variables, or more. For simplicity, we state the method for two independent variables. The method was invented by the French mathematician Joseph Louis Lagrange (1736–1813). In this section we

- Learn to use Lagrange multipliers
- Use Lagrange multipliers to solve optimization problems

The method of Lagrange multipliers is used to solve constrained optimization problems. For example, a soft drink distributor may wish to design the least expensive can that will hold a certain number of ounces, or a company may need to maximize production subject to a strict budget. The method of Lagrange multipliers helps to solve these problems.

We will consider problems such as, find the points that maximize or minimize the function $f(x, y)$ subject to the constraint $g(x, y) = 0$. The example we mentioned in the overview could be stated as follows: Find the point on the line $x + y = 10$ for which $f(x, y) = x^2 y$ is a maximum. That is, maximize $f(x, y) = x^2 y$ subject to $x + y = 10$.

The following procedure is very helpful in solving such problems.

THE METHOD OF LAGRANGE MULTIPLIERS

To find the relative maxima or relative minima of a function $f(x, y)$ subject to a constraint $g(x, y) = 0$, introduce a new variable, λ (called a Lagrange multiplier). Then:

Step 1.   Form

$$F(x, y, \lambda) = f(x, y) + \lambda \cdot g(x, y)$$

Step 2.   Form the system

$$F_x(x, y, \lambda) = 0$$

$$F_y(x, y, \lambda) = 0$$

$$F_\lambda(x, y, \lambda) = 0$$

We assume all indicated partial derivatives exist.

Step 3.   Solve the system found in step 2 for values of x, y, and λ that satisfy the system. The desired extrema will be found among the points (x, y) that satisfy the system.

The steps in solving a problem by the method of Lagrange multipliers are illustrated in the next example.

EXAMPLE 24   Find the minimum value of $f(x, y) = x^2 + y^2$ subject to $2x + y = 5$.

SOLUTION   Note that $f(x, y) = x^2 + y^2$ and $g(x, y) = 2x + y - 5 = 0$.

Step 1.   Form

$$F(x, y, \lambda) = f(x, y) + \lambda \cdot g(x, y)$$

$$F(x, y, \lambda) = \underbrace{x^2 + y^2}_{\text{Objective function}} + \underbrace{\lambda \cdot (2x + y - 5)}_{\text{Constraint function}}$$

Step 2.   Form the system.

$$F_x(x, y, \lambda) = 2x + 2\lambda = 0$$

$$F_y(x, y, \lambda) = 2y + \lambda = 0$$

$$F_\lambda(x, y, \lambda) = 2x + y - 5 = 0$$

Step 3.   The system may be solved by any of several methods, but we choose to use the substitution method for this particular example. The first equation gives $x = -\lambda$. The second equation gives $y = -\frac{1}{2}\lambda$. Substituting these values into the third equation gives

$$2(-\lambda) + \left(-\frac{1}{2}\lambda\right) - 5 = 0$$

$$-\frac{5}{2}\lambda = 5$$

$$\lambda = -2$$

Hence $x = -\lambda = -(-2) = 2$

and $y = -\frac{1}{2}\lambda = -\frac{1}{2}(-2) = 1$

The point $(x, y) = (2, 1)$ gives a relative minimum of $f(2, 1) = 2^2 + 1^2 = 5$ for the function subject to the constraint $2x + y = 5$.   ∎

Note that in the preceding example we did not prove that $f(2, 1) = 5$ is a relative minimum subject to the constraint. Sufficient conditions may be found in textbooks on mathematical analysis, but you can usually judge by substituting some point close to the critical point and comparing the values obtained.

Practice Problem 1   Minimize $f(x, y) = x^2 + y^2$ subject to $x + 2y = 10$.

ANSWER   The minimum is 20, which occurs at $x = 2$ and $y = 4$.

Practice Problem 2 Use the method of Lagrange multipliers to find two numbers whose sum is 76 and whose product is a maximum.

ANSWER $x = 38$ and $y = 38$

The method of Lagrange multipliers can be applied to functions of three or more variables, as the following example illustrates.

EXAMPLE 25   Use the method of Lagrange multipliers to find a maximum of $f(x, y, z) = x^2yz$ subject to the constraint $x + y + z = 10$. (This is a Lagrange multiplier solution to the example mentioned in the Overview.)

SOLUTION

Step 1.   Form

$$F(x, y, z, \lambda) = x^2yz + \lambda(x + y + z - 10)$$

Step 2.   Form the system.

$$F_x(x, y, z, \lambda) = 2xyz + \lambda = 0$$
$$F_y(x, y, z, \lambda) = x^2z + \lambda = 0$$
$$F_z(x, y, z, \lambda) = x^2y + \lambda = 0$$
$$F_\lambda(x, y, z, \lambda) = x + y + z - 10 = 0$$

Step 3.   From the first equation,

$$\lambda = -2xyz$$

Substituting in the second equation gives

$$x^2z - 2xyz = 0$$

which gives

$$xz(x - 2y) = 0$$
$$x = 0, \quad z = 0, \quad \text{or} \quad x = 2y$$

Substituting in the third equation gives

$$x^2y - 2xyz = 0$$

Therefore,

$$xy(x - 2z) = 0$$
$$x = 0, \quad y = 0, \quad \text{or} \quad x = 2z$$

From the fourth equation,

$$z = 10 - x - y$$

If $x = 0$, $y = 0$, or $z = 0$, the function $f(x, y, z) = 0$ and is not a maximum. So,

$$x = 2y$$
$$x = 2z$$
$$z = 10 - x - y$$

or

$$\frac{x}{2} = 10 - x - \frac{x}{2} \qquad \left(z = \frac{x}{2}, y = \frac{x}{2} \right)$$
$$2x = 10$$
$$x = 5, \quad y = \tfrac{5}{2}, \quad \text{or} \quad z = \tfrac{5}{2}$$

Thus, $f(5, \tfrac{5}{2}, \tfrac{5}{2}) = (5)^2(\tfrac{5}{2})(\tfrac{5}{2}) = \tfrac{625}{4}$ is a relative maximum subject to the constraint. This is the same value that we obtained in Example 23 of the last section. ∎

Practice Problem 3 Using the method of Lagrange multipliers, maximize the function $f(x, y) = x^2 y$ subject to $x + y = 10$.

ANSWER $x = \tfrac{20}{3}, y = \tfrac{10}{3}, f(\tfrac{20}{3}, \tfrac{10}{3}) = \tfrac{4000}{27}$

EXAMPLE 26 A manufacturing company has a production function

$$P(l, C) = 200l^{1/2}C^{1/2}$$

where l denotes the number of units of labor and C denotes the number of units of capital. In addition, there is a budget constraint of \$30,000 where each unit of labor costs \$10 and each unit of capital \$100. Find the l and C that will maximize production.

SOLUTION We want to maximize

$$P(l, C) = 200l^{1/2}C^{1/2}$$

subject to the constraint

$$10l + 100C = 30,000$$

$$F(l, \lambda) = 200l^{1/2}C^{1/2} + \lambda(10l + 100C - 30,000)$$

$$F_l = \frac{100C^{1/2}}{l^{1/2}} + 10\lambda$$

$$F_C = \frac{100l^{1/2}}{C^{1/2}} + 100\lambda$$

$$F_\lambda = 10l + 100C - 30,000$$

Set each partial equal to zero.

$$\frac{100C^{1/2}}{l^{1/2}} + 10\lambda = 0$$

$$\frac{100l^{1/2}}{C^{1/2}} + 100\lambda = 0$$

$$10l + 100C - 30,000 = 0$$

Solving for λ in the first two equations gives

$$\lambda = \frac{-10C^{1/2}}{l^{1/2}} = \frac{-l^{1/2}}{C^{1/2}}$$

Multiplying by $l^{1/2}C^{1/2}$ gives

$$-10C = -l \quad \text{or} \quad l = 10C$$

Substituting this value in the third equation gives

$$10(10C) + 100C = 30,000$$

$$C = 150$$

So $l = 10(150) = 1500$ and we see that 1500 units of labor and 150 units of capital maximize production. The maximum production is

$$P(1500, 150) = 200\sqrt{150 \cdot 1500} = 94,868 \quad \text{units of production}$$

At $C = 120$, $l = 1800$ and production is 92,952.
At $C = 180$, $l = 1200$ and production is 92,952.

It seems that 94,868 is indeed a maximum.

EXAMPLE 27   A utility function is used to measure the satisfaction, called the utility, that one gets from using products in a given time. Suppose that $U(x, y)$ is such a function for products A and B. Let p_1 and p_2 be the prices of products A and B, respectively. Now suppose that there is a limit or budget that restricts the amount I that can be spent on products A and B. For example, suppose that

$$I = p_1 x + p_2 y$$

We are interested in maximizing the function $U(x, y)$ subject to the budget constraint $I = p_1 x + p_2 y$. For example, if $U = x^2 y^2$, find the maximum satisfaction if $40 = 2x + 3y$.

SOLUTION

$$F(x, y, \lambda) = x^2 y^2 + \lambda(2x + 3y - 40)$$

$$F_x = 2xy^2 + 2\lambda$$

$$F_y = 2x^2 y + 3\lambda$$

$$F_\lambda = 2x + 3y - 40$$

Setting each partial equal to zero yields

$$2xy^2 + 2\lambda = 0$$

$$2x^2 y + 3\lambda = 0$$

$$2x + 3y - 40 = 0$$

From the first two equations we get $\lambda = -xy^2$ and $\lambda = -\frac{2}{3}x^2 y$ or $-xy^2 = -\frac{2}{3}x^2 y$. We now have

$$2x^2 y - 3xy^2 = xy(2x - 3y) = 0$$

so

$$x = 0, \quad y = 0, \quad \text{or} \quad x = \frac{3y}{2}$$

Either $x = 0$ or $y = 0$ makes $U(x, y) = x^2 y^2$ equal to 0. A maximum value of $U(x, y)$ does not occur at either $x = 0$ or $y = 0$. [Note that $U(x, y)$ is always greater than or equal to 0.]

Substituting $x = 3y/2$ in $2x + 3y - 40 = 0$ yields

$$2\left(\frac{3y}{2}\right) + 3y - 40 = 0$$

$$6y = 40 \quad \text{or} \quad y = \frac{20}{3}$$

and $2x + 3(\frac{20}{3}) = 40$ or $x = 10$. Then

$$U\left(10, \frac{20}{3}\right) = 10^2 \left(\frac{20}{3}\right)^2 = \frac{40{,}000}{9}$$

Use your calculator to evaluate $U(x, y)$ at two or three points as evidence that $\frac{40{,}000}{9}$ is a maximum.

At $(5, 10)$, which satisfies $2x + 3y = 40$,

$$U(5, 10) = 5^2(10)^2 = 2500 < \frac{40{,}000}{9}$$

At $(11, 6)$, which again satisfies $2x + 3y = 40$,

$$U(11, 6) = 11^2(6)^2 = 4356 < \frac{40{,}000}{9}$$ ■

SUMMARY

In this section we learned that it is often necessary to maximize or minimize a function subject to a certain constraint. This is especially true in business. To find a relative maximum or minimum value of $f(x, y, z)$ subject to the constraint $g(x, y, z) = 0$, we form

$$F(x, y, z, \lambda) = f(x, y, z) + \lambda g(x, y, z)$$

We then set all four partial derivatives of F equal to zero and solve for x, y, and z. The partial derivatives give

$$F_x(x, y, z, \lambda) = \frac{\partial f}{\partial x} + \lambda \frac{\partial g}{\partial x} = 0$$

$$F_y(x, y, z, \lambda) = \frac{\partial f}{\partial y} + \lambda \frac{\partial g}{\partial y} = 0$$

$$F_z(x, y, z, \lambda) = \frac{\partial f}{\partial z} + \lambda \frac{\partial g}{\partial z} = 0$$

$$g(x, y, z) = 0$$

Exercise Set 7.6

Use the method of Lagrange multipliers to solve the following exercises.

1. Find a relative maximum of $f(x, y) = 3xy$ subject to $x + y = 10$.

2. Find a relative maximum of $f(x, y) = 4xy$ subject to $x + y = 8$.

3. Find a relative maximum of $f(x, y) = 4xy$ subject to $x + 2y = 6$.

4. Find a relative maximum of $f(x, y) = 2xy + 3$ subject to $x + y = 12$.

5. Find a relative maximum of $f(x, y) = 3xy + 2$ subject to $3x + y = 4$.

6. Find x and y such that $x + y = 22$ and xy^2 is a maximum.

7. Find x and y such that $x + y = 34$ and x^2y is a maximum.

8. Find a relative maximum of $f(x, y) = 3x^2y$ subject to $x + 2y = 5$.

9. Find a relative maximum of $f(x, y) = 2xy^2$ subject to $2x + 3y = 5$.

10. Find two numbers whose sum is 26 and whose product is a maximum.

11. Find two numbers whose sum is 94 and whose product is a maximum.

12. Find a relative minimum of $f(x, y) = 3x^2 + 2y^2 - xy$ subject to $x + y = 4$.

13. Find a relative minimum of $f(x, y) = x^2 + 3y^2 - 2xy$ subject to $2x + y = 5$.

14. Find three numbers whose sum is 80 and whose product is a maximum.

15. Find three numbers whose sum is 146 and whose product is a maximum.

Applications (Business and Economics)

16. **Production.** If x thousand dollars are spent on labor and y thousand dollars are spent on equipment, the production of a factory would be

$$P(x, y) = 40x^{1/3}y^{2/3}$$

How should $100,000 be allocated to obtain the maximum production?

17. **Change in Production.** Use the method of Lagrange multipliers to estimate the change in the maximum production that would occur if the money available was increased by $1000 in Exercise 16.

18. **Cost Function.** The cost function for x units of item I and y units of item II is

$$C(x, y) = 8x^2 + 14y^2$$

If it is necessary that $x + y = 99$, how many of each item should be produced for minimum cost? What is the minimum cost?

19. **Maximum Utility.** If $U(x, y) = 14x - x^2 + 20y - y^2$ is a utility function with constraint $5x + 4y = 100$, where $p_1 = \$5$, $p_2 = \$4$, and $I = \$100$, find the maximum utility.

20. **Maximum Utility.** If the utility function for two products is $U(x, y) = xy^2$, with $p_1 = \$4$, $p_2 = \$6$, and $I = \$60$, find the maximum utility or satisfaction.

21. **Warehouse.** An electric utility needs a warehouse that contains 1 million cubic feet. It is estimated that the floor and ceiling will cost $3 per square foot and the walls will cost $7 per square foot. Find the dimensions of the most economical building.

Applications (Social and Life Sciences)

22. **Feed Mixture.** A feed company mixed three feeds (called A, B, and C) to obtain a desired feed for calves. Let x be the number of tons of A in the mixture, y the number of tons of B, and z the number of tons of C in the mixture. To obtain the desired amount of mixture (due to cost)

$$x + 2y + 3z = 140$$

In addition, the company desires to maximize the units of iron in the mixture. If 2% of A are units containing iron, 4% of B, and 6% of C, determine the desired mixture to maximize the number of units of iron.

23. **Farming**. A farmer has 300 feet of fence. Find the dimensions of the rectangular field of maximum area he can enclose.

24. Redo Exercise 23 if the farmer does not need to fence one side because of a building.

25. **Construction**. A rectangular box with no top is to be built with 600 square feet of material. Find the dimensions of the box that would have maximum volume.

26. Redo Exercise 25 if the box has a top.

Review Exercises

Find the relative maxima and relative minima for the following functions.

27. $f(x, y) = 4x^2 + y^2 - 4y$

28. $f(x, y) = 4x^2 + y^2 + 8x - 2y - 1$

29. $f(x, y) = x^2 + xy + y^2 - 6x - 2y$

30. $f(x, y) = x^3 - 3x - y^2$

Chapter Review

Review to ensure that you understand the following concepts.

Function of two variables	Saddle point
Function of several variables	Partial derivative
Relative maxima	Second-order partial derivative
Relative minima	Lagrange multiplier
Critical point	Lagrange method

Can you use the following formulas?

$$\frac{\partial^2 z}{\partial x^2} = \frac{\partial}{\partial x}\left(\frac{\partial z}{\partial x}\right)$$

$$\frac{\partial^2 z}{\partial y^2} = \frac{\partial}{\partial y}\left(\frac{\partial z}{\partial y}\right)$$

$$\frac{\partial^2 z}{\partial x\, \partial y} = \frac{\partial}{\partial y}\left(\frac{\partial z}{\partial x}\right) \quad \text{or} \quad \frac{\partial}{\partial x}\left(\frac{\partial z}{\partial y}\right)$$

Chapter Test

1. If $f(x, y, z) = x^3 + y^3 + z^3 - xyz$, find $f(1, 2, -1)$.
2. Plot the following points on a three-dimensional coordinate system.
   (a) $(0, 2, 1)$ (b) $(-3, 2, -1)$
3. Sketch the graph of $z = 4x + 3y + 12$.
4. If $z = x^2 + 4xy^2$, find $\partial^2 z/\partial y^2$.
5. Given $z = 3 \ln(x^2 + 3y^2)$, find $\partial^2 z/\partial x^2$.
6. If $f = e^{xy}$, find f_y.
7. If $z = x^2 y + 4x + xy^2$, find $\partial^2 z/\partial x\, \partial y$.

8. Sketch the three dimensional graph of $x + y = 4$.

9. Find all critical points for

$$f(x, y) = x^2 + xy - x + \frac{3y^2}{2} + 7y$$

10. In Problem 9, classify the critical points as relative maxima, relative minima, or neither.

11. Make a contour map for $z = 4x^2 + y^2$ and then draw a rough sketch of the surface defined by $z = 4x^2 + y^2$.

12. Use the method of Lagrange multipliers to find a maximum of $f(x, y, z) = 9xyz$ subject to $6x + 4y + 3z = 24$.

CHAPTER 8

Trigonometric Functions

S o far in this book we have restricted our attention to polynomial functions, rational functions, exponential functions, and functions involving radicals. We turn our attention now to another collection of functions called **trigonometric functions**. Each of the trigonometric functions that we define is periodic, which means that its values repeat on regular intervals. Trigonometric functions are important because they model a wide variety of cyclical phenomena such as business cycles, blood pressure in the aorta, and atmospheric pollution.

It is hoped that the reader has some knowledge of trigonometry; however, we begin with the most elementary concepts, so the information in this chapter can be learned without prior knowledge of the subject.

8.1 ANGLES AND TRIGONOMETRIC FUNCTIONS

OVERVIEW   Before we define trigonometric functions, we must first understand the meaning of an angle. We express the measure of an angle in two different ways, both of which are used with trigonometric functions. In application problems the measure of an angle is usually expressed in terms of radians. For example, a particular sales function is given by $S(x) = 200 + 400 \sin(\pi x/6)$. For $x = 3$, $S(3) = 200 + 400 \sin(\pi/2)$. The measure of the angle in the sine function is $\pi/2$ radians. In this section we define

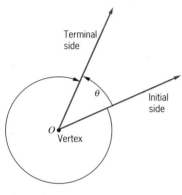

- Degree measure
- Radian measure
- Trigonometric functions
- Relationships between trigonometric functions

Figure 1

The ancient Babylonian civilization introduced angle measurement in terms of degrees, minutes, and seconds, and these units are still used in certain engineering applications. However, it is more convenient in calculus to measure angles in terms of radians. We will discuss both measurements.

We describe the measurement of an angle in terms of a rotating ray (see Figure 1). The beginning ray, called the **initial side** of the angle, is rotated about its endpoint, called the **vertex** of the angle until it assumes the position of the **terminal side**. An angle can be obtained by a counterclockwise rotation or a clockwise rotation, as shown in Figure 1.

An angle is in **standard position** when it is positioned on a rectangular coordinate system with its vertex at the origin and its initial side on the positive x-axis. The angles in Figure 2 are in standard position. The measurement of an angle determined by one complete rotation in a counterclockwise direction is defined to be 360°. One-half of this rotation is a 180° angle, and one-fourth is a 90° angle. When an angle is rotated clockwise, its measure in degrees is negative.

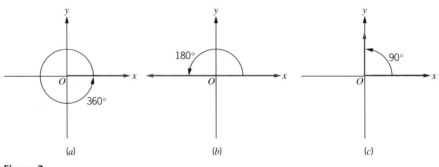

Figure 2

To define the measurement of an angle in radians, we consider a circle with unit radius and measure angles in terms of distances around the circumference of the circle. The angle determined by an arc of length 1 along the circumference has a measure of 1 **radian** [see Figure 3(a)]. Since the circumference of a circle of radius 1 has a length of 2π, there are 2π radians in one complete rotation [Figure 3(b)]; there are π radians in one-half revolution [Figure 3(c)]; and $\pi/2$ radians in one-fourth revolution [Figure 3(d)].

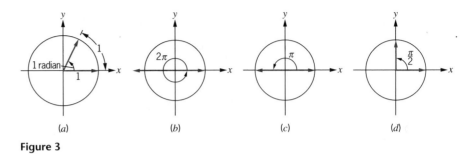

(a)   (b)   (c)   (d)

Figure 3

The following relations between angle measure in degrees and in radians are very important.

$$360° = 2\pi \text{ radians}$$

$$270° = \frac{3\pi}{2} \text{ radians}$$

$$180° = \pi \text{ radians}$$

$$90° = \frac{\pi}{2} \text{ radians}$$

For $180° = \pi$ radians, divide both sides of the equation by 180 to obtain

$$1° = \frac{\pi}{180} \text{ radians}$$

Thus,

$$x° = x \cdot \frac{\pi}{180} \text{ radians}$$

Conversely, for π radians $= 180°$, divide both sides of the equation by π to obtain

$$1 \text{ radian} = \frac{180}{\pi} \text{ degrees}$$

Thus,

$$x \text{ radians} = x \cdot \frac{180}{\pi} \text{ degrees}$$

We will usually omit the word *radian* when measuring angles in this book because all of our angles will be in radians unless otherwise specified.

EXAMPLE 1   Convert the following in degree measure to radians and in radian measure to degrees.

(a)   45°   (b)   60°   (c)   120°

(d) $\dfrac{\pi}{6}$ (e) $\dfrac{\pi}{3}$ (f) $\dfrac{4\pi}{3}$

SOLUTION

(a)   Since $1° = \pi/180$ radians,

$$45° = 45 \cdot \frac{\pi}{180} = \frac{\pi}{4} \text{ radians}$$

(b) $60° = 60 \cdot \dfrac{\pi}{180} = \dfrac{\pi}{3}$ radians

(c) $120° = 120 \cdot \dfrac{\pi}{180} = \dfrac{2\pi}{3}$ radians

(d)   Since 1 radian $= 180/\pi$ degrees,

$$\frac{\pi}{6} \text{ radians} = \frac{\pi}{6} \cdot \frac{180}{\pi} = 30°$$

(e) $\dfrac{\pi}{3}$ radians $= \dfrac{\pi}{3} \cdot \dfrac{180}{\pi} = 60°$

(f) $\dfrac{4\pi}{3}$ radians $= \dfrac{4\pi}{3} \cdot \dfrac{180}{\pi} = 240°$ ∎

An angle greater than 360° or 2π radians in standard form can be reduced by multiples of 360° or multiples of 2π radians, and the terminal side of the two angles are the same. For example, a 540° angle has the same terminal side as a $540° - 360° = 180°$ angle. Likewise, an angle that measures $7\pi/2$ has the same terminal side as

$$\frac{7\pi}{2} - 2\pi = \frac{3\pi}{2}$$

The Trigonometric Functions

We define the six basic trigonometric functions in terms of an angle, denoted by a Greek letter such as θ, α, or β, or by capital letters such as A, B, C, or D in standard position. As shown in Figure 4, we choose an arbitrary point $P(x, y)$ on

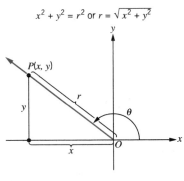

Figure 4

the terminal side of the angle. The distance from P to the origin is denoted by r, which is always a positive number. Using the Pythagorean theorem,

$$x^2 + y^2 = r^2 \quad \text{or} \quad r = \sqrt{x^2 + y^2}$$

The six trigonometric functions are defined as follows.

TRIGONOMETRIC FUNCTIONS

Let (x, y) be a point other than the origin on the terminal side of an angle θ in standard position. Let r be the distance from the point to the origin. Then

$$\text{Sine } \theta = \sin \theta = \frac{y}{r} \qquad\qquad \text{Cosecant } \theta = \csc \theta = \frac{r}{y}, y \neq 0$$

$$\text{Cosine } \theta = \cos \theta = \frac{x}{r} \qquad\qquad \text{Secant } \theta = \sec \theta = \frac{r}{x}, x \neq 0$$

$$\text{Tangent } \theta = \tan \theta = \frac{y}{x}, x \neq 0 \qquad \text{Cotangent } \theta = \cot \theta = \frac{x}{y}, y \neq 0$$

EXAMPLE 2   Find $\sin \alpha$, $\cos \alpha$, and $\tan \alpha$ where the following points lie on the terminal side of angle α.

(a) $(3, 4)$ (b) $(-8, 15)$ (c) $(12, -5)$

SOLUTION

(a)   We have $r = \sqrt{(3)^2 + (4)^2} = 5$. Then

$$\sin \alpha = \frac{4}{5}, \quad \cos \alpha = \frac{3}{5}, \quad \text{and} \quad \tan \alpha = \frac{4}{3}$$

(b)   We have $r = \sqrt{(-8)^2 + (15)^2} = 17$. Then

$$\sin \alpha = \frac{15}{17}, \quad \cos \alpha = -\frac{8}{17}, \quad \text{and} \quad \tan \alpha = -\frac{15}{8}$$

(c)   We have $r = \sqrt{(12)^2 + (-5)^2} = 13$. Then

$$\sin \alpha = -\frac{5}{13}, \quad \cos \alpha = \frac{12}{13}, \text{and} \quad \tan \alpha = -\frac{5}{12}$$

EXAMPLE 3   Find the values of the six trigonometric functions if $\theta = \pi/2$.

SOLUTION

First we select one point on the terminal side of the angle of measure $\pi/2$. One such point is $(0, 1)$. Then we have $r = \sqrt{0^2 + 1^2} = 1$. Consequently, the values of the six trigonometric functions are

$$\sin \frac{\pi}{2} = \frac{1}{1} = 1 \qquad\qquad \csc \frac{\pi}{2} = \frac{1}{1} = 1$$

$$\cos \frac{\pi}{2} = \frac{0}{1} = 0 \qquad\qquad \sec \frac{\pi}{2} = \frac{1}{0} \quad \text{(undefined)}$$

$$\tan \frac{\pi}{2} = \frac{1}{0} \quad \text{(undefined)} \qquad \cot \frac{\pi}{2} = \frac{0}{1} = 0$$

Use your calculator to verify the six trigonometric functions for the angles with measures of 0, $\pi/2$, and $3\pi/2$. First make certain the calculator is in radian mode. To display the ANGLE menu, press $\boxed{\text{2nd}}$ $\boxed{\text{ANGLE}}$ and select 1: for degree notation and 3: for radian notation. To get the values for csc θ, sec θ, and cot θ in Table 1, note that these are the reciprocals of sin θ, cos θ, and tan θ, respectively. Also note that

$$\tan \theta = \frac{y}{x} = \frac{y/r}{x/r} = \frac{\sin \theta}{\cos \theta}$$

TABLE 1

θ (in radians)	θ (in degrees)	sin θ	cos θ	tan θ	cot θ	sec θ	csc θ
0	0°	0	1	0	Undefined	1	Undefined
$\pi/2$	90°	1	0	Undefined	0	Undefined	1
π	180°	0	-1	0	Undefined	-1	Undefined
$3\pi/2$	270°	-1	0	Undefined	0	Undefined	-1
2π	360°	0	1	0	Undefined	1	Undefined

We can obtain the trigonometric functions of certain frequently used angles such as $\pi/6$ radians (30°), $\pi/3$ radians (60°), and $\pi/4$ radians (45°) from a calculator. For example, obtain the following trigonometric functions on your calculator:

$$\sin \frac{\pi}{4} = 0.7071067812 \quad \text{and} \quad \cos \frac{\pi}{4} = 0.7071067812$$

The point (0.7071067812, 0.7071067812) could serve as a point on the terminal side of an angle of measure $\pi/4$ because

$$\sqrt{(0.7071067812)^2 + (0.7071067812)^2} \approx 1$$

from which

$$\sin \frac{\pi}{4} = \frac{0.7071067812}{1}, \quad \cos \frac{\pi}{4} = \frac{0.7071067812}{1}, \quad \text{and} \quad \tan \frac{\pi}{4} = 1$$

The exact point is $(\sqrt{2}/2, \sqrt{2}/2)$, where $\sqrt{2}/2 \approx 0.7071067812$. In practice, the trigonometric functions are given as

$$\sin \frac{\pi}{4} = \frac{\sqrt{2}}{2}, \quad \cos \frac{\pi}{4} = \frac{\sqrt{2}}{2}, \quad \text{and} \quad \tan \frac{\pi}{4} = 1$$

In a similar manner we can use a calculator to develop the following trigonometric functions of $\pi/3$ and $\pi/6$:

$$\sin \frac{\pi}{3} = \frac{\sqrt{3}}{2} \qquad \sin \frac{\pi}{6} = \frac{1}{2}$$

$$\cos \frac{\pi}{3} = \frac{1}{2} \qquad \cos \frac{\pi}{6} = \frac{\sqrt{3}}{2}$$

$$\tan \frac{\pi}{3} = \frac{\sqrt{3}}{1} \qquad \tan \frac{\pi}{6} = \frac{1}{\sqrt{3}}$$

In later sections of this chapter we will need to find an angle when we know a trigonometric function of the angle. We accomplish this goal by defining the inverse functions for sin x, cos x, and tan x. To define the inverse functions we must restrict the domains.

INVERSE TRIGONOMETRIC FUNCTIONS

(a) The **inverse sine function**, denoted $y = \sin^{-1} x$, is the function with a domain of $[-1, 1]$ and a range of $-\pi/2 \le x \le \pi/2$ that satisfies the relation $\sin y = x$.

(b) The **inverse cosine function**, denoted by $y = \cos^{-1} x$ is the function with a domain of $[-1, 1]$ and a range of $[0, \pi]$, which satisfies the relation $\cos y = x$.

(c) The **inverse tangent function**, denoted by $y = \tan^{-1} x$ is the function with a domain of $(-\infty, \infty)$ and a range of $(-\pi/2, \pi/2)$, which satisfies the relation $\tan y = x$.

Using the inverse keys $\boxed{\sin^{-1}}$, $\boxed{\cos^{-1}}$, or $\boxed{\tan^{-1}}$ on most calculators gives an angle in the range of definition. For example, when $\tan x = -1$, x could be an angle in either quadrants II or IV. Because of the definition of the inverse function, only the angle in quadrant IV is read from the calculator.

EXAMPLE 4 If $\cos\theta = 0.45$, find θ.

SOLUTION We have $\theta = \cos^{-1} 0.45 \approx 1.10403$ radians. Since π radians is approximately 3.1416 radians,

$$\theta = \frac{1.10403}{3.1416}\,\pi \approx \frac{\pi}{2.85}$$

Since $\pi/3$ is a $60°$ angle, this angle is slightly larger than $60°$.

EXAMPLE 5 For the point $(-2, 3)$ on the terminal side of an angle, find the angle in radians.

SOLUTION The point $(-2, 3)$ is in quadrant II, so we use the cosine function. Since $r = \sqrt{(-2)^2 + (3)^2} \approx 3.60555$, we have

$$\cos\theta = \frac{x}{r} = \frac{-2}{3.60555} = -0.554700$$

$$\theta = 2.1588$$

Practice Problem 1

Find x in radians from a calculator for the following functions.

a) $\sin x = -0.3$
b) $\cos x = -0.3$
c) $\tan x = -0.6$

ANSWER

a) $x = -0.3047$ radians
b) $x = 1.8755$ radians
c) $x = -0.5404$ radians

SUMMARY

In this section we introduced the concept and measure of an angle as background for the differentiation of trigonometric functions in later sections. In this book the most frequently used trigonometric functions are

$$\sin\theta = \frac{y}{r}, \quad \cos\theta = \frac{x}{r}, \quad \text{and} \quad \tan\theta = \frac{y}{x}$$

where $r = \sqrt{x^2 + y^2}$ and (x, y) is a point on the terminal side of the angle θ. We use these concepts as we discuss the graphs of periodic functions in the next section and as we use calculus to describe the characteristics of such functions in the remainder of the chapter.

Exercise Set 8.1

In Exercises 1–6, determine the quadrant of the terminal side of each angle in standard position.

1. $260°$
2. $342°$
3. $-280°$
4. $-340°$
5. $-120°$
6. $287°$

In Exercises 7–9, determine the sign of (a) sin θ, (b) cos θ and (c) tan θ, where θ is in standard position.

7. $286°$
8. $-300°$
9. $-126°$

Find the equivalent radian measure for each angle.

10. $45°$
11. $120°$
12. $305°$
13. $81°$
14. $164°$
15. $342°$

Find the equivalent degree measure for each angle.

16. $\dfrac{3\pi}{4}$
17. $\dfrac{5\pi}{6}$
18. $\dfrac{-2\pi}{3}$
19. 0.60
20. 2.33
21. -1.62

In Exercises 22–27, write out the values of sin θ, cos θ, *and* tan θ *without using a calculator.*

22. $45°$
23. $60°$
24. $270°$
25. π radians
26. $\dfrac{\pi}{6}$ radians
27. $-\dfrac{\pi}{2}$ radians

In Exercises 28–33, use a calculator to find the values of sin θ, cos θ, *and* tan θ.

28. $26°$
29. $154°$
30. 1.6
31. 4.2
32. $\dfrac{\pi}{18}$
33. $\dfrac{5\pi}{7}$

In Exercises 34–39, find sin θ, cos θ, tan θ, csc θ, sec θ, *and* cot θ, *where $P(x, y)$ is a point on the terminal side of θ in standard position.*

34. $P(x, y) = (3, 4)$
35. $P(x, y) = (-5, 12)$
36. $P(x, y) = (2, -3)$
37. $P(x, y) = (-1, 0)$
38. $P(x, y) = (0, -4)$
39. $P(x, y) = (-2, -3)$

Find a value of θ in the range of inverse functions for the following trigonometric equations.

40. $\sin \theta = \dfrac{\sqrt{3}}{2}$
41. $\cos \theta = \dfrac{1}{2}$
42. $\tan \theta = \sqrt{3}$
43. $\tan \theta = -1$
44. $\sin \theta = \dfrac{-\sqrt{3}}{2}$
45. $\cos \theta = \dfrac{-1}{2}$
46. $\sin \theta = 0.6$
47. $\cos \theta = 0.32$
48. $\tan \theta = 1.32$
49. $\cos \theta = -0.41$
50. $\tan \theta = -3$
51. $\sin \theta = -0.7$

Applications (Business and Economics)

52. **Business cycle.** An ice cream company has revenues from sales for 1 year given approximately by

$$R(x) = 0.6 - 0.5 \sin \frac{\pi x}{12} \qquad 0 \le x \le 12$$

where $R(x)$ is revenue in millions of dollars for a month of sales x months after July 1.
(a) Find the value of $R(0$ months$)$, $R(4$ months$)$, and $R(6$ months$)$ without using a calculator.
(b) Find the value of $R(1$ month$)$, $R(5$ months$)$, and $R(9$ months$)$ using a calculator.

53. **Sales cycle.** A garden shop finds that sales per month for the past year were cyclical with sales of lawn mowers given by

$$S(x) = 100 + 200 \sin \frac{\pi x}{12}$$

where S is the number of lawn mowers sold and x is the time in months measured from January 1. How many lawn mowers were sold (a) in June, (b) in July, (c) in September, and (d) in November?

8.2 GRAPHS OF TRIGONOMETRIC FUNCTIONS AND TRIGONOMETRIC EQUATIONS

OVERVIEW   In this section we use a calculator to explore the characteristics of graphs of trigonometric functions. These graphs provide excellent examples in the study of periodic functions. Knowing the length of a business cycle or the period of a breathing cycle is useful. Also, graphical exploration leads to horizontal shift identities. In addition, we will validate other trigonometric identities in the next section by showing that the graph of one side of the identity is identical to the other side. In this section we

- Find and use periods of trigonometric functions
- Discuss the characteristics of the graphs of these functions, including amplitude
- Solve trigonometric equations

Many calculators have the capability to graph the sine, cosine, and tangent functions. Furthermore, we can graph the cosecent, secant, and cotangent functions as reciprocals of the sine, cosine, and tangent functions, respectively. We use this capability to discuss the characteristics of trigonometric functions.

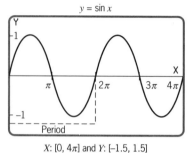

$y = \sin x$

X: [0, 4π] and Y: [−1.5, 1.5]

Figure 5

EXAMPLE 6 Using X: $[0, 4\pi]$ and Y: $[-1.5, 1.5]$, graph $y = \sin x$ with your calculator.

SOLUTION The graph is shown in Figure 5.

In Figure 5 notice that there is a repetition of values of the function starting at $x = 2\pi$. From $x = 2\pi$ to $x = 4\pi$, the graph is exactly the same as from $x = 0$ to $x = 2\pi$. Therefore, $y = \sin x$ is a periodic function with period 2π.

PERIODIC FUNCTION

A function f is a **periodic function** if there is a positive real number k such that $f(x + k) = f(x)$ for every value of x in the domain of f. The smallest value of k for which this is true is called the **period** of f. The period corresponds to the length of one cycle on the graph of f.

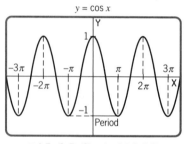

$y = \cos x$

X: [−7π/2, 7π/2] and Y: [−1.5, 1.5]

Figure 6

Now let's graph $y = \cos x$ on X: $[-7\pi/2, 7\pi/2]$ and Y: $[-1.5, 1.5]$ (see Figure 6). Note that the period of $y = \cos x$ is also 2π. In addition, note for both the sine and cosine functions that the maximum value is 1 and the minimum value is -1.

To discuss the characteristics of $y = \tan x$, we graph this function on X: $[-2\pi, 2\pi]$ and Y: $[-10, 10]$ (see Figure 7).

The graph in Figure 7 suggests that the period of $y = \tan x$ is π. Note that as $x \rightarrow (\pi/2)^-$, $\tan x \rightarrow \infty$; and as $x \rightarrow (\pi/2)^+$, $\tan x \rightarrow -\infty$.

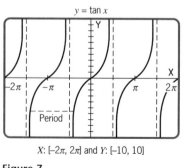

$y = \tan x$

X: [−2π, 2π] and Y: [−10, 10]

Figure 7

EXAMPLE 7 On the same coordinate system graph $y = \sin x$, $y = 2 \sin x$, $y = \frac{1}{2} \sin x$, and $y = -\sin x$. How do these graphs differ? What are the maximum and minimum values of each function?

(a) $y_1 = 2 \sin x$
(b) $y_2 = \sin x$
(c) $y_3 = \frac{1}{2}\sin x$
(d) $y_4 = -\sin x$

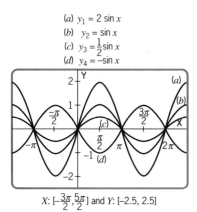

$X: [-\frac{3\pi}{2}, \frac{5\pi}{2}]$ and $Y: [-2.5, 2.5]$

Figure 8

SOLUTION   From Figure 8 we note that each graph has the same period. However, the maximum and minimum values of the functions are different. The function $y = \sin x$ has a maximum value of 1, $y = 2 \sin x$ has a maximum value of 2, and $y = \frac{1}{2}\sin x$ has a maximum value of $\frac{1}{2}$. One value of x for which these maximum values occur is $x = \pi/2$. At $x = \pi/2$, $y = -\sin x$ has a minimum value of -1. In fact, the graph of $y = -\sin x$ is a reflection of the graph of $y = \sin x$ across the x-axis. Similarly, $y = \sin x$ has a minimum of -1, $y = 2 \sin x$ has a minimum of -2, and $y = \frac{1}{2}\sin x$ has a minimum of $-\frac{1}{2}$ at $x = 3\pi/2$. ■

The discussion in the preceding example suggests the following definition.

AMPLITUDE OF SINE AND COSINE

> The amplitude of the sine function (or cosine function) is defined to be half of the difference between its maximum and minimum values. For $y = a \sin x$ and $y = a \cos x$, the amplitude is $|a|$.

Using the preceding definition, the amplitude of $y = 2 \sin x$ is 2, the amplitude of $y = \frac{1}{2} \cos x$ is $\frac{1}{2}$, and the amplitude of $y = -\sin x$ is 1.

What happens to the graphs of the sine and cosine functions when the angle changes from x to $3x$ or from x to $x/2$?

We investigate these changes in the following example.

EXAMPLE 8   Draw the graphs of $y = \sin x$, $y = \sin 3x$, and $y = \sin x/2$ on the same coordinate system. Are the amplitudes the same? Are the periods the same?

(a) $y_1 = \sin 3x$
(b) $y_2 = \sin x$
(c) $y_3 = \sin \frac{x}{2}$

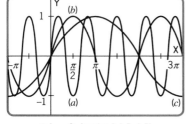

$X: [-\pi, 3\pi]$ and $Y: [-1.5, 1.5]$

Figure 9

SOLUTION   In Figure 9 note that the amplitudes are the same. The period of $y = \sin x$ is 2π, the period of $y = \sin x/2$ is 4π, and the period of $y = \sin 3x$ is $2\pi/3$. ■

The graphs in Figure 9 suggest the following definition.

PERIOD OF SINE AND COSINE FUNCTIONS

> The period for $y = \sin bx$ and $y = \cos bx$ is
>
> $$\frac{2\pi}{|b|}$$

Similarly, with calculator exploration we can show that the period for $y = \tan bx$ is $\pi/|b|$. The function $y = \cos x + k$ (or $y = \sin x + k$) shifts the graph of $y = \cos x$ (or $y = \sin x$) k units in a positive y direction. If k is negative, the shift is in a negative y direction. Likewise, $y = \cos(x - a)$ [or $y = \sin(x - a)$] shifts the graph of $y = \cos x$ (or $y = \sin x$) a units in a positive direction along the horizontal axis.

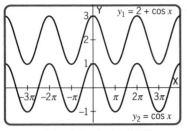

X: [-4π, 4π] and Y: [-1.5, 3.5]

Figure 10

(a) $y_1 = \cos x$
(b) $y_2 = \cos \left(x - \frac{\pi}{2}\right)$

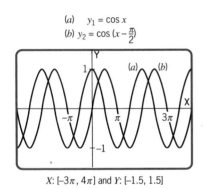

X: [-3π, 4π] and Y: [-1.5, 1.5]

Figure 11

(a) $y_1 = \sin x$
(b) $y_2 = \sin \left(x - \frac{\pi}{2}\right)$

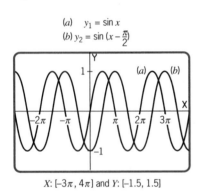

X: [-3π, 4π] and Y: [-1.5, 1.5]

Figure 12

EXAMPLE 9   Draw the graph of $y = \cos x$ and $y = 2 + \cos x$ on the same coordinate system and discuss the relationship between the two graphs. Use X: $[-4\pi, 4\pi]$ and Y: $[-1.5, 3.5]$.

SOLUTION   As shown in Figure 10, the graph of $y = \cos x$ is shifted 2 units upward to get the graph of $y = 2 + \cos x$. ■

EXAMPLE 10   Draw the graphs of $y = \cos x$ and $y = \cos(x - \pi/2)$ on the same coordinate system and discuss the relationship between the two graphs.

SOLUTION   As shown in Figure 11, the graph of $y = \cos x$ is shifted $\pi/2$ units to the right to obtain the graph of $y = \cos(x - \pi/2)$. ■

EXAMPLE 11   Draw the graph of $y = \sin x$ and the graph of $y = \sin(x - \pi/2)$ and discuss the relationship between the two graphs. Use X: $[-3\pi, 4\pi]$ and Y: $[-1.5, 1.5]$.

SOLUTION   As shown in Figure 12, the graph of $y = \sin x$ is shifted $\pi/2$ units to the right to obtain the graph of $y = \sin(x - \pi/2)$. ■

Using Figure 11, draw the graph of $y = \sin x$ along with the other two graphs. What do you see? Are you surprised that the graph of $y = \sin x$ coincides with the graph of $y = \cos(x - \pi/2)$? This result shows that $\sin x = \cos(x - \pi/2)$. In a similar manner, using Figure 12 you can show that $\cos x = -\sin(x - \pi/2)$.

The preceding trigonometric functions often occur in trigonometric equations such as the following.

EXAMPLE 12   Use the inverse trigonometric keys on your calculator to solve $(\cos 2x + 0.8)(\cos [x - \pi/2]) = 0$

SOLUTION   The solutions are obtained when factors are set equal to 0.

$$\text{From } \cos(2x) + 0.8 = 0$$
$$\cos(2x) = -0.8$$
$$2x = \cos^{-1}(-0.8)$$
$$2x = 2.4981$$
$$x = 1.2490$$
$$\text{Also } \cos(x - \pi/2) = 0$$
$$x - \pi/2 = \cos^{-1} 0$$
$$x - \pi/2 = 1.5708$$
$$x = 3.1416$$

■

Practice Problem 1 Solve $2 \sin^2 x + \sin x = 0$

ANSWER $2 \sin^2 x + \sin x = \sin x (2 \sin x + 1) = 0$

$$x = 0 \quad \text{and} \quad x = -\pi/6$$

SUMMARY

The graphing techniques that we discussed in this section for trigonometric functions will be of much value in the last two sections where we study derivatives and integrals of trigonometric functions. Knowing the period and amplitude of a graph should be helpful in discussing slopes and curvature and in finding areas. The trigonometric identities should assist in simplifying answers.

Exercise Set 8.2

Find the period and amplitude of the graph of the given function without drawing the graph.

1. $y = 0.5 \cos 2x$

2. $y = 4 \cos 0.5x$

3. $y = \sin \dfrac{\pi x}{2}$

4. $y = 2 \cos \dfrac{2\pi x}{3}$

Find the period of the graph of a given function without drawing the graph.

5. $y = \sec 2x$

6. $y = \cot 3x$

7. $y = 4 \tan \dfrac{x}{2}$

8. $y = 2 \csc \dfrac{x}{3}$

9. $y = \cot \dfrac{\pi x}{2}$

10. $y = \tan \dfrac{2\pi x}{3}$

Draw the graphs of the given functions for $-2\pi \le x \le 2\pi$ and then verify the answers obtained previously.

11. Exercise 1

12. Exercise 2

13. Exercise 3, $-2 \le x \le 2$

14. Exercise 4, $-2 \le x \le 2$

15. Exercise 5

16. Exercise 6

17. Exercise 8

18. Exercise 7

Sine and cosine, tangent and cotangent, and secant and cosecant are called cofunctions. If x is any acute angle, a trigonometric function of x is equal to the cofunction of $(\pi/2) - x$. Verify this by drawing a graph of the function and a graph of the cofunction for $0 \le x \le \pi/2$. If the graphs do not coincide, the identities are not valid.

19. $\sin x = \cos \left(\dfrac{\pi}{2} - x \right)$

20. $\cos x = \sin \left(\dfrac{\pi}{2} - x \right)$

21. $\tan x = \cot \left(\dfrac{\pi}{2} - x \right)$

22. $\sec x = \csc \left(\dfrac{\pi}{2} - x \right)$

Find the period, amplitude, maximum, and minimum values of the graph of each function and then graph each function on $-2\pi \le x \le 2\pi$ to verify your answer.

23. $y = 4 \sin 2x + 3$

24. $y = -4 \cos 5x - 2$

25. $y = 0.6 \cos 2x - 0.5$

26. $y = 0.5 \sin 0.5x + 3$

27. $y = 0.6 \cos 0.2x - 5$

28. $y = 3 \sin(x - 2) - 1$

Solve for x. (Find one solution.)

29. $\sin 2x = 0.4$

30. $\cos 3x = -0.6$

31. $4 \cos 2x = 1$

32. $2 \sin \dfrac{x}{2} = -0.5$

33. $4 \sin 0.5x = 3$

34. $2 \cos 0.3x = 1$

Solve the given equation. (Find two solutions.)

35. $2 \cos^2 x = 1$

36. $\tan^2 x - \tan x = 0$

37. $\sin x = \cos x$

38. $\tan^2 x = 3$

39. $1 - \sin^2 x = 1 - \sin x.$

Applications (Business and Economics)

40. **Business cycle.**   An ice cream company has reve-
    nues from sales given approximately by

$$R(x) = 0.6 - 0.5 \sin \frac{\pi x}{12}, \quad 0 \le x \le 12$$

where $R(x)$ is revenue in millions of dollars for a
month of sales x months after July 1. Draw the graph
of this revenue frunction.

41. **Sales cycle.**   A garden shop finds that sales per
    month for the past year are cyclical with sales of
    lawn mowers given by

$$S(x) = 100 + 200 \sin \frac{\pi x}{12}$$

where S is the number of lawn mowers sold and x is
the time in months measured from January 1. Draw a
graph of sales for $0 \le x \le 12$.

Applications (Social and Life Sciences)

42. **Breathing cycle.**   The breathing cycle for a person
    is given by

$$v = 0.9 \sin \frac{2 \pi t}{5}$$

where v is in liters per second and t is in seconds.
Graph one breathing cycle of this person.

43. **Blood pressure.**   A person's blood pressure is ap-
    proximated by

$$P = 76 + 10 \sin 72 \pi t$$

Graph one cycle of this person's blood pressure.

Review Exercises

Find the value of each expression.

44. $\cos^{-1} 0.5$

45. $\sin^{-1} (-0.5)$

46. $\tan^{-1} (-1)$

47. $\sin^{-1} (0.16)$

Find the values of $\sin \theta$, $\cos \theta$, *and* $\tan \theta$ *for* θ *as
given.*

48. $90°$ 49. $\dfrac{3\pi}{4}$ 50. 1.5 51. -0.8

8.3   TRIGONOMETRIC IDENTITIES

An equation that includes trigonometric functions and is true for all possible
values of the variable is called a **trigonometric identity**. Trigonometric identities
are useful in simplifying expressions involving trigonometric functions, in developing differentiation formulas for trigonometric functions, and as substitutions to
perform integrations. We consider the first two of these uses in this chapter. In
this section we

- Verify trigonometric identities using the Pythogorean Theorem
- Verify trigonometric identities using a calculator
- Simplify expressions using trigonometric identities

For any value of θ in standard position, a point $P(x, y)$ is defined where $x^2 +
y^2 = r^2$, as discussed in the preceding section. This expression is valid for all x
and y and the corresponding r. Dividing both sides of this expression by r^2 gives

$$\frac{x^2}{r^2} + \frac{y^2}{r^2} = 1$$

But $\sin \theta = \dfrac{y}{r}$ and $\cos \theta = \dfrac{x}{r}$. Therefore $\cos^2 \theta + \sin^2 \theta = 1$.

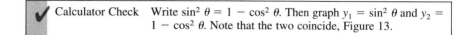

Calculator Check   Write $\sin^2 \theta = 1 - \cos^2 \theta$. Then graph $y_1 = \sin^2 \theta$ and $y_2 = 1 - \cos^2 \theta$. Note that the two coincide, Figure 13.

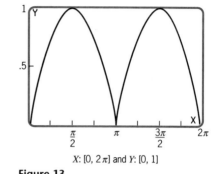

X: $[0, 2\pi]$ and Y: $[0, 1]$

Figure 13

Next divide both sides of $\cos^2 \theta + \sin^2 \theta = 1$ by $\cos^2 \theta$, where $\cos \theta \neq 0$.

$$1 + \frac{\sin^2 \theta}{\cos^2 \theta} = \frac{1}{\cos^2 \theta}$$

$$1 + \tan^2 \theta = \sec^2 \theta \qquad \tan \theta = \frac{\sin \theta}{\cos \theta} \text{ and } \sec \theta = \frac{1}{\cos \theta}$$

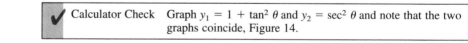

Calculator Check   Graph $y_1 = 1 + \tan^2 \theta$ and $y_2 = \sec^2 \theta$ and note that the two graphs coincide, Figure 14.

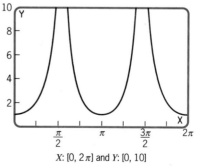

X: $[0, 2\pi]$ and Y: $[0, 10]$

Figure 14

In a similar manner divide both sides of $\cos^2 \theta + \sin^2 \theta = 1$ by $\sin^2 \theta$ where $\sin \theta \neq 0$.

$$\frac{\cos^2 \theta}{\sin^2 \theta} + 1 = \frac{1}{\sin^2 \theta}$$

$$\cot^2 \theta + 1 = \csc^2 \theta$$

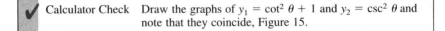 **Calculator Check** Draw the graphs of $y_1 = \cot^2 \theta + 1$ and $y_2 = \csc^2 \theta$ and note that they coincide, Figure 15.

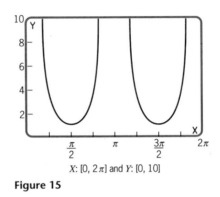

$X: [0, 2\pi]$ and $Y: [0, 10]$

Figure 15

The fundamental identities are stated as follows.

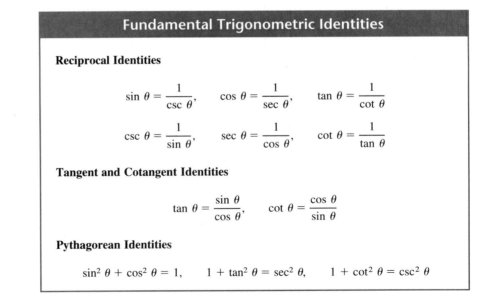

Fundamental Trigonometric Identities

Reciprocal Identities

$$\sin \theta = \frac{1}{\csc \theta}, \qquad \cos \theta = \frac{1}{\sec \theta}, \qquad \tan \theta = \frac{1}{\cot \theta}$$

$$\csc \theta = \frac{1}{\sin \theta}, \qquad \sec \theta = \frac{1}{\cos \theta}, \qquad \cot \theta = \frac{1}{\tan \theta}$$

Tangent and Cotangent Identities

$$\tan \theta = \frac{\sin \theta}{\cos \theta}, \qquad \cot \theta = \frac{\cos \theta}{\sin \theta}$$

Pythagorean Identities

$$\sin^2 \theta + \cos^2 \theta = 1, \qquad 1 + \tan^2 \theta = \sec^2 \theta, \qquad 1 + \cot^2 \theta = \csc^2 \theta$$

To show that an equation is an identity, we must show that the equation is true for all values of the variable for which all expressions in the equation are defined. To verify that an equation is an identity we follow one of two approaches: simplify one side of the equation until it is identical to the other side or simplify each side separately until both sides are identical.

EXAMPLE 13   Verify that $\dfrac{\tan \theta}{\sec \theta} = \sin \theta$ in two ways: graphically and by using trigonometric identities.

SOLUTION   Graph $y_1 = \tan \theta / \sec \theta$ and $y_2 = \sin \theta$ and note that the two graphs coincide, Figure 16.

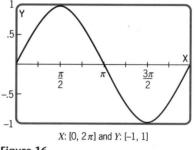

X: [0, 2π] and Y: [−1, 1]

Figure 16

Algebraically,

$$\frac{\tan \theta}{\sec \theta} = \tan \theta \cdot \frac{1}{\sec \theta}$$

$$\frac{\tan \theta}{\sec \theta} = \frac{\sin \theta}{\cos \theta} \cos \theta \qquad \cos \theta = 1/\sec \theta$$

$$\frac{\tan \theta}{\sec \theta} = \sin \theta$$

Practice Problem 1 Show that $\dfrac{1 + \cos \theta}{\cos \theta} = \sin \theta$ is not an identity.

ANSWER   Graphing $y_1 = \dfrac{1 + \cos \theta}{\cos \theta}$ and $y_2 = \sin \theta$, we note that the two graphs

do not coincide. Therefore $\dfrac{1 + \cos \theta}{\cos \theta} = \sin \theta$ is not an identity,
Figure 17.

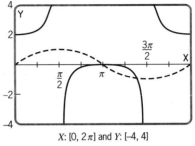

X: [0, 2π] and Y: [–4, 4]

Figure 17

EXAMPLE 14   Show that $\dfrac{\sin x}{1 - \cos x} = \dfrac{1 + \cos x}{\sin x}$ is an identity and then
verify this fact graphically.

SOLUTION   The important step in this problem is to multiply the numerator
and denominator of the first fraction by something so that the denominator will
simplify. Thus we multiply the numerator and denominator by $1 + \cos x$.

$$\frac{\sin x}{1 - \cos x} = \frac{\sin x}{1 - \cos x} \cdot \frac{1 + \cos x}{1 + \cos x}$$
$$= \frac{\sin x \,(1 + \cos x)}{1 - \cos^2 x}$$
$$= \frac{\sin x \,(1 + \cos x)}{\sin^2 x}$$
$$= \frac{1 + \cos x}{\sin x}$$

The given expression is an identity for $\sin x \neq 0$ and $1 - \cos x \neq 0$.

 Calculator Check   Graphing $y_1 = \dfrac{\sin x}{1 - \cos x}$ and $y_2 = \dfrac{1 + \cos x}{\sin x}$, we note that
the two graphs coincide, Figure 18.

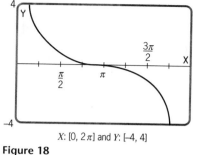

$X: [0, 2\pi]$ and $Y: [-4, 4]$

Figure 18

Practice Problem 2 Classify the following expressions as identities or not identities.

(a) $\sec x \sin x = \tan x$

(b) $\dfrac{\tan^2 x}{\sin^2 x} = \sec x$

(c) $\sec^2 x - 1 = \csc^2 x$

(d) $\dfrac{\sin^2 x}{1 - \cos x} = 1 + \cos x$

ANSWER (a) Identity (b) Not an identity (c) Not an identity
(d) Identity

In Exercise Set 14.3 we will draw graphs to show that $\sin(-\theta) = -\sin\theta$, $\cos(-\theta) = \cos\theta$, and $\tan(-\theta) = -\tan\theta$.

For θ in quadrant I let $P = (a, b)$ be the point on the terminal side of the angle θ that is on the unit circle. (See Figure 19.) The point Q on the terminal side of the angle $-\theta$ that is on the unit circle will have coordinates $(a, -b)$. Then

$$\sin\theta = b \quad \cos\theta = a \quad \sin(-\theta) = -b \quad \cos(-\theta) = a$$

so that

$$\sin(-\theta) = -\sin\theta \quad \cos(-\theta) = \cos\theta$$

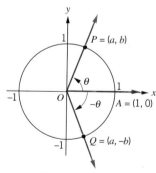

Figure 19

The same reasoning can be used to verify these identities for θ in quadrants II, III, and IV.

Now, using these results and some of the fundamental identities, we have

$$\tan(-\theta) = \frac{\sin(-\theta)}{\cos(-\theta)} = \frac{-\sin\theta}{\cos\theta} = -\tan\theta \quad \cot(-\theta) = \frac{1}{\tan(-\theta)} = \frac{1}{-\tan\theta} = -\cot\theta$$

$$\sec(-\theta) = \frac{1}{\cos(-\theta)} = \frac{1}{\cos\theta} = \sec\theta \quad \csc(-\theta) = \frac{1}{\sin(-\theta)} = \frac{1}{-\sin\theta} = -\csc\theta$$

We discussed the horizontal shift identities listed below in the preceding section using graphs.

Horizontal Shift Identities

For all values of x,

1. $\cos x = -\sin\left(x - \dfrac{\pi}{2}\right) = \sin\left(\dfrac{\pi}{2} - x\right)$

2. $\sin x = \cos\left(x - \dfrac{\pi}{2}\right) = \cos\left(\dfrac{\pi}{2} - x\right)$

These identities can be obtained easily using the trigonometric function of the sum and difference of two angles. There are six sum and different formulas. These are stated without proof.

SUM AND DIFFERENCE FORMULAS

1. $\sin(\alpha + \beta) = \sin\alpha\cos\beta + \cos\alpha\sin\beta$
2. $\sin(\alpha - \beta) = \sin\alpha\cos\beta - \cos\alpha\sin\beta$
3. $\cos(\alpha + \beta) = \cos\alpha\cos\beta - \sin\alpha\sin\beta$
4. $\cos(\alpha - \beta) = \cos\alpha\cos\beta + \sin\alpha\sin\beta$

5. $\tan(\alpha + \beta) = \dfrac{\tan\alpha + \tan\beta}{1 - \tan\alpha\tan\beta}$

6. $\tan(\alpha - \beta) = \dfrac{\tan\alpha - \tan\beta}{1 + \tan\alpha\tan\beta}$

COMMON ERROR   Students will often write $\sin(\alpha + \beta) = \sin\alpha + \sin\beta$. This is not a valid identity for all angles.

There are a variety of uses for sum and difference formulas. One use is to find exact values of trigonometric functions.

EXAMPLE 15

Find the exact value of $\sin\dfrac{\pi}{12}$.

SOLUTION

We can rewrite the angle $\dfrac{\pi}{12} = \dfrac{\pi}{3} - \dfrac{\pi}{4}$ and use the difference formula for the sine function.

$$\sin\left(\frac{\pi}{3} - \frac{\pi}{4}\right) = \sin\frac{\pi}{3}\cos\frac{\pi}{4} - \cos\frac{\pi}{3}\sin\frac{\pi}{4}$$

$$= \frac{\sqrt{3}}{2} \cdot \frac{\sqrt{2}}{2} - \frac{1}{2} \cdot \frac{\sqrt{2}}{2}$$

$$= \frac{\sqrt{6} - \sqrt{2}}{4}$$

✔ Calculator Check: $\sin\dfrac{\pi}{12} \approx 0.25883$ and $\dfrac{\sqrt{6} - \sqrt{2}}{4} \approx 0.25882$.

EXAMPLE 16   Verify the identity: $\cos(\pi + x) = -\cos x$

SOLUTION   We will use the sum formula for cosine.

$$\cos(\pi + x) = \cos\pi\cos x - \sin\pi\sin x$$

$$= (-1)\cos x - (0)\sin x$$

$$= -\cos x$$

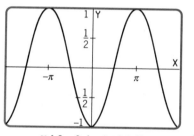

X: [−2π, 2π] and Y: [−1, 1]

Figure 20

✔ Calculator Check   Graphing both $y_1 = \cos(\pi + x)$ and $y_2 = -\cos x$, we see that the graphs coincide in Figure 20.

From the sum formulas, we can develop double angle formulas.

Double-Angle Formulas
7. $\sin 2\alpha = 2\sin\alpha\cos\alpha$
8. $\cos 2a = \cos^2\alpha - \sin^2\alpha = 2\cos^2\alpha - 1 = 1 - 2\sin^2\alpha$
9. $\tan 2\alpha = \dfrac{2\tan\alpha}{1 - \tan^2\alpha}$

EXAMPLE 17

Verify the identity: $\sin 2\alpha = 2\sin\alpha\cos\alpha$

SOLUTION

$$\sin 2\alpha = \sin(\alpha + \alpha) = \sin \alpha \cos \alpha + \cos \alpha \sin \alpha = 2 \sin \alpha \cos \alpha$$

The other double-angle formulas can be proved in a similar manner.

EXAMPLE 18

Verify the identity: $\cos^2 2\alpha - \sin^2 2\alpha = \cos 4\alpha$

SOLUTION

$$\cos^2 2\alpha - \sin^2 2\alpha = \cos^2 2\alpha - (1 - \cos^2 2\alpha)$$
$$= 2 \cos^2 2\alpha - 1$$
$$= \cos 2(2\alpha)$$
$$= \cos 4\alpha$$

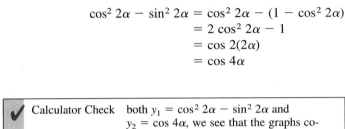

✔ Calculator Check   both $y_1 = \cos^2 2\alpha - \sin^2 2\alpha$ and $y_2 = \cos 4\alpha$, we see that the graphs co-incide in Figure 21.

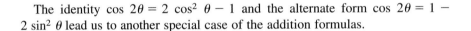

X: $[-\pi, \pi]$ and Y: $[-1, 1]$

Figure 21

The identity $\cos 2\theta = 2 \cos^2 \theta - 1$ and the alternate form $\cos 2\theta = 1 - 2 \sin^2 \theta$ lead us to another special case of the addition formulas.

Half-Angle Formulas
10. $\sin \dfrac{\theta}{2} = \pm \sqrt{\dfrac{1 - \cos \theta}{2}}$
11. $\cos \dfrac{\theta}{2} = \pm \sqrt{\dfrac{1 + \cos \theta}{2}}$
12. $\tan \dfrac{\theta}{2} = \dfrac{\sin \theta}{1 + \cos \theta} = \dfrac{1 - \cos \theta}{\sin \theta}$

NOTE: The symbol $\pm$ is used for formulas 10 and 11 because it is necessary to select the proper sign after you determine the quadrant where $\dfrac{\theta}{2}$ is located. Do not use both symbols in a particular problem.

EXAMPLE 19

Verify the identity: $\tan \dfrac{\theta}{2} = \dfrac{\sin \theta}{1 + \cos \theta}$

SOLUTION

$$\frac{\sin \theta}{1 + \cos \theta} = \frac{\sin 2 \left(\frac{\theta}{2}\right)}{1 + \cos 2 \left(\frac{\theta}{2}\right)}$$

$$= \frac{2 \sin \left(\frac{\theta}{2}\right) \cos \left(\frac{\theta}{2}\right)}{1 + [2 \cos^2 \left(\frac{\theta}{2}\right) - 1]}$$

$$= \frac{2 \sin \left(\frac{\theta}{2}\right) \cos \left(\frac{\theta}{2}\right)}{2 \cos^2 \left(\frac{\theta}{2}\right)}$$

$$= \frac{\sin \left(\frac{\theta}{2}\right)}{\cos \left(\frac{\theta}{2}\right)}$$

$$= \tan \left(\frac{\theta}{2}\right)$$

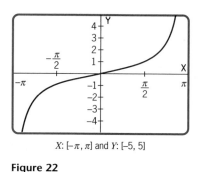

X: [−π, π] and Y: [−5, 5]

Figure 22

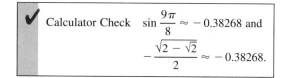 **Calculator Check**   Graphing both $y_1 = \tan \dfrac{\theta}{2}$ and $y_2 = \dfrac{\sin \theta}{1 + \cos \theta}$, we see that the graphs coincide, Figure 22.

EXAMPLE 20

Find exact value of $\sin \dfrac{9\pi}{8}$.

SOLUTION

We can rewrite the angle $\dfrac{1}{2} \cdot \dfrac{9\pi}{4} = \dfrac{\frac{9\pi}{4}}{2}$. Since $\dfrac{\frac{9\pi}{4}}{2}$ is in quadrant 3, the sine will be negative.

$$\sin \frac{9\pi}{8} = \sin \frac{\frac{9\pi}{4}}{2}$$

$$= -\sqrt{\frac{1 - \cos \frac{9\pi}{4}}{2}}$$

$$= -\sqrt{\frac{1 - \frac{\sqrt{2}}{2}}{2}}$$

$$= -\frac{\sqrt{2 - \sqrt{2}}}{2}$$

 **Calculator Check** $\sin \dfrac{9\pi}{8} \approx -0.38268$ and

$$-\frac{\sqrt{2 - \sqrt{2}}}{2} \approx -0.38268.$$

SUMMARY

In this section we studied the following trigonometric identities both algebraically and graphically:

(a)   Reciprocal identities
(b)   Tangent and cotangent identities
(c)   Pythagorean identities
(d)   Horizontal shift identities
(e)   Sum and difference formulas
(f)   Double angle formulas
(g)   Half-angle formulas

Exercise Set 8.3

In Exercises 1 to 6, simplify each expression using the fundamental identities.

1. $\tan \theta \cos \theta$

2. $\sec \theta \cos \theta$

3. $\dfrac{1 - \sin^2 \theta}{\cos \theta}$

4. $\dfrac{1 + \tan^2 \theta}{\csc^2 \theta}$

5. $\dfrac{1 + \tan \theta}{1 + \cot \theta}$

6. $\dfrac{1 + \sin \theta}{1 - \cos \theta}$

In Exercises 7–22, verify that each is an identity. Do a calculator check using graphs.

7. $\csc^2 x(1 - \cos^2 x) = 1$

8. $\sin^2 x - \cos^2 x = 1 - 2 \cos^2 x$

9. $1 + \cot^2 x = \csc^2 x$

10. $\dfrac{\sin x - \cos x}{\cos x} + 1 = \tan x$

11. $1 - 2 \sin^2 x = 2 \cos^2 x - 1$

12. $\sin x + \cos x \cot x = \csc x$

13. $\dfrac{1 - \cos x}{\sin x} = \dfrac{\sin x}{1 + \cos x}$

14. $\dfrac{\tan x}{\sec x - 1} = \dfrac{\sec x + 1}{\tan x}$

15. $\dfrac{1 + \cos x}{1 - \cos x} = \dfrac{\sec x + 1}{\sec x - 1}$

16. $\dfrac{\sin x}{1 + \cos x} + \dfrac{1 + \cos x}{\sin x} = 2 \csc x$

17. $\tan (2\pi - \theta) = - \tan \theta$

18. $\dfrac{\sin(\alpha + \beta)}{\cos \alpha \cos \beta} = \tan \alpha + \tan \beta$

19. $\sin (\alpha - \beta) \sin (\alpha + \beta) = \sin^2 \alpha - \sin^2 \beta$

20. $\dfrac{\cos (\alpha + \beta)}{\cos \alpha \cos \beta} = 1 - \tan \alpha \tan \beta$

21. $\tan \frac{\theta}{2} = \csc \theta - \cot \theta$

22. $\csc^2 (\theta/2) = \csc \theta - \cot \theta$

Find the exact value for each term.

23. $\tan (7\pi/8)$ 24. $\sin (- \pi/8)$

25. $\cos (7\pi/12)$ 26. $\tan (19\pi/12)$

27. In Section 4 we use $\sin (\frac{\pi}{2} - x)$. Show that $\sin (\frac{\pi}{2} - x) = \cos x$.

28. Show that $\sin x = \cos (\frac{\pi}{2} - x)$.

Review Exercises

29. Find a value of θ for $\sin \theta = - \frac{1}{2}$.

30. Find $\sin \theta$, $\cos \theta$, and $\tan \theta$ for angle θ that has the point $(- 3, - 4)$ on the terminal side.

31. Find the period, amplitude, maximum, and minimum values of the graph of $y = - 2 \sin (2\pi x - \pi/2)$. Sketch the graph over the interval $[- 1, 1]$.

8.4   DIFFERENTIATION OF TRIGONOMETRIC FUNCTIONS

OVERVIEW   Phenomena such as earthquakes, tides, radio waves, weather, and heart rhythms are periodic. Since trigonometric functions are also periodic, they are important in obtaining information about such phenomena. Continuous periodic functions can be expressed in terms of sines and cosines; therefore, the derivatives of sine and cosine functions play a vital role in predicting and describing changes in tides, heart rhythms, and so on.

In this section we

- Learn the rules for differentiating the six trigonometric functions
- Study applications of these derivatives

To develop the rule for the derivative of the sine function, we must first look at two limits:

$$\lim_{x \to 0} \frac{\sin x}{x} \quad \text{and} \quad \lim_{x \to 0} \frac{\cos x - 1}{x}$$

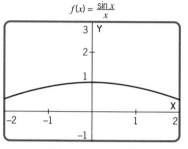

$f(x) = \dfrac{\sin x}{x}$

X: [–2, 2] and Y: [–1, 3]

Figure 23

Figure 23 shows the graph of the function $f(x) = (\sin x)/x$. Using the window shown in the figure, have your calculator draw the function and use the $\boxed{\text{TRACE}}$ key to determine that

$$\lim_{x \to 0} \frac{\sin x}{x} = 1$$

Make certain that your calculator is in radian mode.

To determine that

$$\lim_{x \to 0} \frac{\cos x - 1}{x} = 0$$

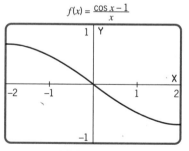

$f(x) = \dfrac{\cos x - 1}{x}$

X: [–2, 2] and Y: [–1, 1]

Figure 24

have your calculator draw the graph of

$$f(x) = \frac{\cos x - 1}{x}$$

and use the $\boxed{\text{TRACE}}$ key. Use the window shown in Figure 24.

Now let's use the two limits just found to develop the rule for the derivative of the sine function.

For $y = \sin x$,

$$\frac{dy}{dx} = \lim_{h \to 0} \frac{\sin (x + h) - \sin x}{h} \qquad \text{Definition of derivative}$$

$$= \lim_{h \to 0} \frac{\sin x \cos h + \cos x \sin h - \sin x}{h} \qquad \text{Sum formula for sine}$$

$$= \lim_{h \to 0} \frac{\sin x (\cos h - 1) + \cos x \sin h}{h}$$

$$= \lim_{h \to 0} \sin x \cdot \lim_{h \to 0} \frac{\cos h - 1}{h} + \lim_{h \to 0} \cos x \cdot \lim_{h \to 0} \frac{\sin h}{h} \qquad \text{Properties of limits}$$

$$= \sin x \cdot 0 + \cos x \cdot 1 \qquad\qquad \lim_{h \to 0} \frac{\cos h - 1}{h} = 0$$

$$\lim_{h \to 0} \frac{\sin h}{h} = 1$$

$$= \cos x$$

The chain rule can be used to generalize the derivative of the sine function. If u is a differentiable function of x, and if $y = \sin u$, applying the chain rule yields

$$\frac{dy}{dx} = \frac{dy}{du} \cdot \frac{du}{dx} = (\cos u) \left(\frac{du}{dx} \right)$$

DERIVATIVES OF THE SINE FUNCTION

(a)   If $y = \sin x$, then

$$\frac{dy}{dx} = \cos x$$

(b)   If u is a differentiable function of x, and if $y = \sin u$, then

$$\frac{dy}{dx} = \cos u \cdot \frac{du}{dx}$$

Calculator Note

The rule for the derivative of the sine can be illustrated on your calculator by graphing $y = \cos x$ and $y = $ nDeriv $(\sin x, x, x)$. As your calculator draws the graphs, note that they are the same.

We now use these rules to differentiate functions containing the sine function. Since the sine function is differentiable and continuous, all rules of differentiation can be applied.

EXAMPLE 21   Differentiate each function.

(a) $y = -4x^3 \sin x$ (b) $y = \sin (2x^2 - 4x)$
(c) $y = [\sin (x^3 - 4x)]^4$

SOLUTION

(a)   We use the product rule.

$$y = -4x^3 \sin x$$

$$\frac{dy}{dx} = \frac{d}{dx}(-4x^3) \cdot \sin x + (-4x^3) \cdot \frac{d}{dx}(\sin x) \qquad \text{Product rule}$$

$$= -12x^2 \sin x + (-4x^3) \cos x$$

(b)   This function requires the second rule:

$$\frac{dy}{dx} = \cos u \cdot \frac{du}{dx}$$

Thus,

$$y = \sin (2x^2 - 4x)$$

$$\frac{dy}{dx} = \cos (2x^2 - 4x) \cdot \frac{d}{dx}(2x^2 - 4x) \qquad u = 2x^2 - 4x$$

$$= \cos (2x^2 - 4x) \cdot (4x - 4)$$

(c)   This problem requires the general power rule for differentiation and the rule

$$\frac{dy}{dx} = \cos u \cdot \frac{du}{dx}$$

Thus,

$$y = [\sin (x^3 - 4x)]^4$$

$$\frac{dy}{dx} = 4[\sin (x^3 - 4x)]^3 \cdot \frac{d}{dx}[\sin (x^3 - 4x)] \qquad \text{General power rule:} \\ y' = nu^{n-1}u'$$

$$= 4[\sin (x^3 - 4x)]^3 \cdot [\cos (x^3 - 4x)](3x^2 - 4)$$

Calculator Note

You can check your answers in the preceding example by having your calculator graph your answer and then use the nDeriv function to graph the derivative. The graphs should be the same. If your calculator has table capabilities, you can use the table to check many values. Although this is not a proof that the two are equal, it provides a good check.

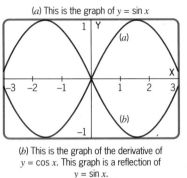

(a) This is the graph of $y = \sin x$

(b) This is the graph of the derivative of $y = \cos x$. This graph is a reflection of $y = \sin x$.

X: [-3, 3] and Y: [-1, 1]

Figure 25

Before we develop the rule for the derivative of the cosine function, let's look at a graphic display that will give us some insight into the derivative of the cosine function. On your graphing calculator, enter the functions $Y_1 = \text{nDeriv}(\cos x, x, x)$ and $Y_2 = \sin x$. Have your calculator draw Y_1 and Y_2. These graphs, shown in Figure 25, suggest that

$$\frac{d}{dx} \cos x = -\sin x$$

As discussed in Section 8.3, the sine and cosine functions have the relationship:

$$\cos x = \sin \left(\frac{\pi}{2} - x \right) \quad \text{and} \quad \sin x = \cos \left(\frac{\pi}{2} - x \right)$$

We use these identities to find the rule for the derivative of the cosine function. For $y = \cos x$,

$$\frac{dy}{dx} = \frac{d}{dx} \cos x$$

$$= \frac{d}{dx} \sin \left(\frac{\pi}{2} - x \right)$$

$$= \cos \left(\frac{\pi}{2} - x \right) \cdot \frac{d}{dx} \left(\frac{\pi}{2} - x \right) \qquad \text{Derivative of sine function}$$

$$= (\sin x)(-1) \qquad\qquad \cos \left(\frac{\pi}{2} - x \right) = \sin x$$

$$= -\sin x$$

We can generalize this rule using the chain rule as we did for the sine function. If u is a differentiable function of x and $y = \cos u$, applying the chain rule yields

$$\frac{dy}{dx} = \frac{dy}{du} \cdot \frac{du}{dx} = (-\sin u) \left(\frac{du}{dx} \right)$$

DERIVATIVE OF THE COSINE FUNCTION

(a) If $y = \cos x$, then

$$\frac{dy}{dx} = -\sin x$$

(b) If u is a differentiable function of x, and if $y = \cos u$, then

$$\frac{dy}{dx} = -\sin u \cdot \frac{du}{dx}$$

EXAMPLE 22   Differentiate each function.

(a) $y = x \cos \sqrt{x}$ (b) $y = \sin^2 x + \cos^2 x$
(c) $y = \cos (2x^3 - 5x)$

SOLUTION

(a)   Use the product rule.

$$y = x \cos \sqrt{x}$$

$$\frac{dy}{dx} = \frac{d}{dx}(x) \cdot (\cos \sqrt{x}) + (x) \cdot \frac{d}{dx}(\cos \sqrt{x})$$

$$= (1) \cos \sqrt{x} + (x)(-\sin \sqrt{x}) \cdot \frac{d}{dx}(\sqrt{x})$$

$$= (1) \cos \sqrt{x} + (x)(-\sin \sqrt{x})\left(\frac{1}{2} x^{-1/2}\right)$$

$$= \cos \sqrt{x} - \frac{\sqrt{x} \sin \sqrt{x}}{2}$$

(b)   Since $\sin^2 x = (\sin x)^2$ and $\cos^2 x = (\cos x)^2$, we use the general power rule to differentiate this function.

$$y = \sin^2 x + \cos^2 x$$

$$\frac{dy}{dx} = \frac{d}{dx}(\sin x)^2 + \frac{d}{dx}(\cos x)^2$$

$$= 2(\sin x)\frac{d}{dx}(\sin x) + 2(\cos x)\frac{d}{dx}(\cos x) \qquad \text{General power rule}$$

$$= 2(\sin x)(\cos x) + (2 \cos x)(-\sin x)$$

$$= 2(\sin x)(\cos x) - (2 \cos x)(\sin x)$$

$$= 0$$

(c)   We use the chain rule.

$$y = \cos(2x^3 - 5x)$$

$$\frac{dy}{dx} = -\sin(2x^3 - 5x) \cdot \frac{d}{dx}(2x^3 - 5x) \qquad \text{Chain rule: } \frac{d}{dx}(\cos u) = -\sin u \cdot \frac{du}{dx}$$

$$= (-\sin(2x^3 - 5x))(6x^2 - 5)$$

Practice Problem 1   Differentiate each function.

(a) $y = 4 - x^2 \sin x$ (b) $y = x \sin x + \cos x$

(c) $y = x^2 \cos \dfrac{1}{x}$ (d) $y = \dfrac{x - \sin x}{1 + \cos x}$

ANSWER:

(a) $\dfrac{dy}{dx} = -x^2 \cos x - 2x \sin x$

(b) $\dfrac{dy}{dx} = x \cos x$

(c) $\dfrac{dy}{dx} = 2x \cos \dfrac{1}{x} + \sin \dfrac{1}{x}$

(d) $\dfrac{dy}{dx} = \dfrac{(1 + \cos x)(1 - \cos x) - (x - \sin x)(-\sin x)}{(1 + \cos x)^2}$

$\qquad = \dfrac{x \sin x}{(1 + \cos x)^2}$

EXAMPLE 23   Find an equation of the tangent line to the curve $f(x) = x \sin x$ at $x = 4$. Support your answer graphically.

SOLUTION   For $f(x) = x \sin x$, we use the product rule to determine $f'(x) = x \cos x + \sin x$. So, at $x = 4$, the point on the graph where the tangent is to be drawn is $(4, 4 \sin 4) = (4, -3.03)$ and the slope of the tangent line is $f'(4) = 4 \cos 4 + \sin 4 = -3.37$. This answer for slope can be verified on your calculator using nDeriv($x \sin x, x, 4$) = -3.371376165. The equation of the tangent line is

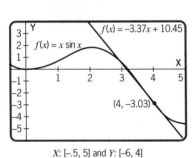

$f(x) = -3.37x + 10.45$

$f(x) = x \sin x$

$(4, -3.03)$

X: [-.5, 5] and Y: [-6, 4]

Figure 26

$$y - (-3.03) = -3.37(x - 4) \quad \text{or} \quad y = -3.37x + 10.45$$

This can be checked on your calculator by graphing both the function $f(x) = x \sin x$ and the line $y = -3.37x + 10.45$. The line should appear tangent to the curve at $x = 4$. Check the value of each function at $x = 4$; they should be the same. Figure 26 shows these functions. ∎

EXAMPLE 24   Find the derivative of $f(x) = \sqrt{3 - \sin x}$ and use the derivative and the graph of the function to find the maximum and minimum values of $f(x)$ over the interval $[0, 2\pi)$.

SOLUTION   The first step in solving the problem is to find the derivative. Then we will find the critical values. Since $f(x) = \sqrt{3 - \sin x} = (3 - \sin x)^{1/2}$, we use the general power rule for differentiation to determine that

$$f'(x) = \frac{1}{2}(3 - \sin x)^{-1/2} \frac{d}{dx}(3 - \sin x)$$

$$= \frac{1}{2}(3 - \sin x)^{-1/2}(-\cos x) = \frac{-\cos x}{2\sqrt{3 - \sin x}}$$

Setting this to zero and solving the equation, we find the critical values in the interval $[-\frac{\pi}{4}, 2\pi)$:

$$f'(x) = 0$$

$$\frac{-\cos x}{2\sqrt{3 - \sin x}} = 0$$

$$-\cos x = 0 \qquad \text{Only the numerator can be 0.}$$

$$\cos x = 0$$

$$x = \frac{\pi}{2} \quad \text{or} \quad x = \frac{3\pi}{2}$$

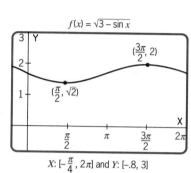

$$f(x) = \sqrt{3 - \sin x}$$

$$X: [-\tfrac{\pi}{4}, 2\pi] \text{ and } Y: [-.8, 3]$$

Figure 27

Looking at the graph of the function in Figure 27, we see that there is a relative minimum at $x = \pi/2$, which yields the point $(\pi/2, \sqrt{2})$, and a relative maximum at $x = 3\pi/2$, which yields the point $(3\pi/2, 2)$. ◾

EXAMPLE 25   With a graphing calculator, find any relative extrema for $f(x) = \sin x + 2 \cos x$ over the interval $(-\pi, \pi)$ using the graphs of both the function and the derivative.

SOLUTION   The graphs of the function and its derivative over the interval $[-\pi, \pi]$ are shown in Figure 28. We determine that $f'(x) = 0$ at $x = -2.674$ and $x = 0.468$. We also determine that the relative minimum is -2.236 at $(-2.674, -2.236)$ and the relative maximum is 2.236 at $(0.468, 2.236)$. ◾

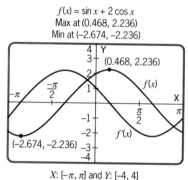

$$f(x) = \sin x + 2 \cos x$$
Max at (0.468, 2.236)
Min at (−2.674, −2.236)

$$X: [-\pi, \pi] \text{ and } Y: [-4, 4]$$

Figure 28

Practice Problem 2   Find any relative extrema for $y = \cos x - \sin x$ over the interval $[-\pi, \pi]$.

ANSWER   Relative maximum is $\sqrt{2}$ at $(-\pi/4, \sqrt{2})$; relative minimum is $-\sqrt{2}$ at $(3\pi/4, -\sqrt{2})$.

Since the sine and cosine functions are differentiable functions, the other trigonometric functions—tangent, cotangent, secant, and cosecant—are differentiable for all values where they are defined because they can all be written in terms of the sine and cosine functions. The derivatives of these functions are given as follows.

DERIVATIVES OF OTHER TRIGONOMETRIC FUNCTIONS

1.   For $y = \tan x$, $\dfrac{dy}{dx} = \sec^2 x$

2.   For $y = \cot x$, $\dfrac{dy}{dx} = -\csc^2 x$

3.   For $y = \sec x$, $\dfrac{dy}{dx} = \sec x \tan x$

4.   For $y = \csc x$, $\dfrac{dy}{dx} = -\csc x \cot x$

If u is a differentiable function of x, then we have the following.

5.   For $y = \tan u$, $\dfrac{dy}{dx} = \sec^2 u \, \dfrac{du}{dx}$

6.   For $y = \cot u$, $\dfrac{dy}{dx} = -\csc^2 u \, \dfrac{du}{dx}$

7.   For $y = \sec u$, $\dfrac{dy}{dx} = \sec u \tan u \, \dfrac{du}{dx}$

8.   For $y = \csc u$, $\dfrac{dy}{dx} = -\csc u \cot u \, \dfrac{du}{dx}$

EXAMPLE 26   Differentiate each function.

(a) $y = \cot \sqrt{x}$ (b) $y = \csc(x^2 + 1)$
(c) $y = \sec x \csc x$ (d) $y = 3x + x \tan x$

SOLUTION

(a)   Use rule 6.

$$y = \cot x^{1/2}$$

$$\frac{dy}{dx} = -\csc^2(x^{1/2}) \, \frac{d}{dx}(x^{1/2}) \qquad \frac{d}{dx}\cot u = -\csc^2 u \cdot \frac{dw}{dx}$$

$$= [-\csc^2(x^{1/2})] \cdot \frac{1}{2} x^{-1/2}$$

$$= \frac{-\csc^2 \sqrt{x}}{2\sqrt{x}}$$

(b)   Use rule 8.

$$y = \csc(x^2 + 1)$$

$$\frac{dy}{dx} = -\csc(x^2 + 1) \cot(x^2 + 1) \cdot \frac{d}{dx}(x^2 + 1)$$

$$\frac{d}{dx}\csc u = -\csc u \cot u \, \frac{du}{dx}$$

$$= (-2x) \csc(x^2 + 1) \cot(x^2 + 1)$$

(c) Use the product rule.

$$y = \sec x \csc x$$

$$\frac{dy}{dx} = (\csc x) \cdot \frac{d}{dx} (\sec x) + (\sec x) \cdot \frac{d}{dx} (\csc x)$$

$$= (\csc x)(\sec x)(\tan x) + (\sec x)(-\csc x)(\cot x)$$

$$= \sec^2 x - \csc^2 x \qquad \text{Rewrite in terms of } \sin x \text{ and } \cos x \text{ and reduce.}$$

(d) Use the product rule.

$$y = 3x + x \tan x$$

$$\frac{dy}{dx} = \frac{d}{dx} (3x) + (x) \cdot \frac{d}{dx} (\tan x) + (\tan x) \cdot \frac{d}{dx} (x)$$

$$= 3 + (x) \sec^2 x + \tan x$$

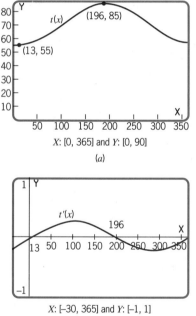

X: [0, 365] and Y: [0, 90]

(a)

X: [−30, 365] and Y: [−1, 1]

The value x = 195.5 is rounded to 196.

(b)

Figure 29

EXAMPLE 27 The average daily Fahrenheit temperature at a golf resort can be approximated by this function

$$t(x) = 70 - 15 \cos \left[\frac{2\pi(x - 13)}{365} \right]$$

where $x = 1$ corresponds to January 1. Draw both the function and its derivative to determine on what day of the year that the temperature average is the highest and on what day it is the lowest.

SOLUTION To graph the function, enter $Y_1 = 70 - 15 \cos ((2\pi(x - 13))/ 365)$ and then enter $Y_2 = \text{nDeriv}(Y_1, x, x)$. The values that the derivative takes on are very small compared to the function. Therefore, have the calculator draw only the derivative in one window and only the function in another window, as shown in Figure 29. Notice that the windows are different. Using ROOT on your calculator, determine from the graph that the derivative of the temperature function is zero at the points $x = 13$ and $x = 195.5$, which we round to $x = 196$. Using these values of x as we view the graph of the temperature function, we see that on the 13th day of the year the average daily temperature reaches its yearly minimum of 55°F, and on the 196th day the average daily temperature reaches its maximum of 85°F.

Practice Problem 3 Using the function given in Example 27, find where the function is increasing and where it is decreasing over the interval [0, 365].

ANSWER Since the derivative is negative over the intervals (0, 13) and (196, 365) (rounding 195.5 to 196), the average temperatures are decreasing during the corresponding days. The function is increasing over the interval (13, 196), which is the interval of days where the average daily temperature is increasing.

SUMMARY

In this section, we introduced the derivatives of the six trigonometric functions. Be sure to learn these rules of differentiation, keeping in mind that only the sine and cosine functions are defined for all real numbers.

Exercise Set 8.4

Differentiate each function. Check your answers with your graphing calculator.

1. $y = \sin 7x$

2. $y = \tan 3x^2$

3. $y = 2x \cos \pi x$

4. $y = \cot (x^2 - 3)$

5. $y = -3 \sin 2x - 5 \cos 3x$

6. $y = x^2 \sec x$

7. $y = \dfrac{1 + \cos 2x}{\sin x}$

8. $y = \tan e^x$

9. $y = e^{\tan x^2}$

10. $y = \ln(\sin x^2 - 2x)$

11. $y = \ln(3x + \cos 2x)$

12. $y = e^{\csc 2x}$

13. $y = e^{2x} \sec(3x + 1)$

14. $y = \csc \sqrt{x}$

15. $y = \sqrt{\cot x}$

16. $y = \dfrac{2 + \cos x}{3x - \sin x}$

17. $y = \dfrac{\sin 3x}{2x}$

18. $y = \sin x \sec x$

19. $y = \dfrac{-5}{\cos x}$

20. $y = x^2 \cot x$

21. $y = (\cos x^2 - 2x)^3$

22. $y = \cos x(1 + \sec x)$

Find an equation for the tangent line for $y = f(x)$ at the given value of x.

23. $y = \sin x$, at $x = 1$

24. $y = \cos 2x$, at $x = 2$

25. $y = \dfrac{2 + \tan x}{x}$, at $x = 2.5$

26. $y = 3 \cos^2 x$, at $x = -1$

Graph each function and determine if there are any horizontal tangents in the interval $0 \le x \le 2\pi$. If there are none, state why.

27. $y = x - 2 \cos x$

28. $y = x - \sin x$

29. $y = \dfrac{1}{2}x + \sin x$

Use a graphing calculator to graph the second derivative of each function and determine where there are inflection points over the interval $[0, 2\pi]$.

30. $y = \sin x + \cos x$

31. $y = 2 \sin x - \sin 2x$

32. $y = x - \cos x$

33. $y = \dfrac{1}{2} \sin \pi x - 2 \cos x$

Applications (Business and Economics)

34. **Revenue.** The monthly revenue (in thousands of dollars) of a golf shop is given by

$$R(x) = 10 - 8 \cos \frac{\pi x}{6}, \qquad 0 \le x \le 12$$

During what month is the revenue the greatest?

35. **Cost.** A manufacturer knows that the cost (in hundreds of dollars) of making x items (in thousands) is given by

$$C(x) = x \left[200 + 20 \sin \left(\frac{\pi x}{12} \right) \right]$$

What is the marginal cost at $x = 15{,}000$?

36. **Profit.** A company finds that its yearly profits can be found using the model

$$P(x) = 50 + 40 \sin \pi x, \qquad 0 \le x \le 5$$

where x is in years and $P(x)$ is in thousands of dollars. Find the marginal profit at $x = 2$.

37. Profit. A company's weekly profits (in hundreds of dollars) from the sale of running shoes x weeks after January 1 is approximated by the model

$$P(x) = 6 - 6 \cos \frac{\pi x}{26}, \qquad 0 \le x \le 52$$

(a) What is the rate of change of profit as a function of x?

(b) Calculate $P'(10)$ and $P'(32)$ and analyze the results.

(c) Find all relative extrema over the interval $(0, 52)$.

38. Revenue. The monthly revenue (in thousands of dollars) x months after January 1 can be approximated by

$$R(x) = 5 - 4 \cos \frac{\pi x}{6}, \qquad 0 \le x \le 12$$

(a) What is the rate of change of revenue as a function of x?

(b) Calculate $R'(2)$ and $R'(10)$ and interpret the results.

(c) During what month does maximum revenue occur?

(d) What is maximum revenue?

Applications (Social and Life Sciences)

39. Chemical Wastes. A chemical company began putting waste into a river January 1, 1995. The amount of daily waste (in hundreds of gallons) is given by the model

$$W(t) = 12 + 10 \sin \frac{\pi}{6} t$$

where t is the number of months after the company began. What is the rate of change of waste being dumped at $t = 10$?

40. Respiration. The number of liters of air inhaled at t seconds by a person during respiration is given by the function

$$v(t) = \frac{6}{5\pi} - \frac{6}{5\pi} \cos \frac{\pi t}{2}$$

At what value of t is the volume a maximum? What is the maximum volume?

41. Body Temperature. The temperature of a person during a 10-day illness is given by

$$T(t) = 100.8° + 3 \sin \frac{\pi t}{8}, \qquad 0 \le t \le 10$$

Find $T'(6)$ and interpret the result.

42. Roller Coaster. A roller coaster x meters from the starting point is $h(x)$ meters above the ground where

$$h(x) = 15 + 15 \sin \frac{\pi x}{50}$$

Find $h'(x)$.

Review Exercises

Find the period, amplitude, and maximum and minimum values of the graph of each function. Then graph each function over the interval $-2\pi \le x \le 2\pi$.

43. $y = -3 \cos(2x - \pi)$ 44. $y = 0.5 \sin \dfrac{x}{2}$

Graphically determine whether the given equation is an identity.

45. $\sin x \csc x - \cos^2 x = \sin^2 x$

46. $\tan x \cot x - \cos^2 x = \sin^2 x$

8.5 INTEGRATION OF TRIGONOMETRIC FUNCTIONS

OVERVIEW   We have seen that integrals are used in many types of problems, such as calculating the total amount of a revenue stream, finding the area between two curves, or finding the average value of a function over an interval. We can also use integrals to predict the future size of a population. We continue discussing these applications using trigonometric functions. In this section we

- Integrate trigonometric functions
- Study applications involving integrals

In the last section, rules of differentiation were given. These can be integrated to produce integration formulas.

TRIGONOMETRIC INTEGRATION FORMULAS

For u a function of x:

1. $\displaystyle\int \sin u \, du = -\cos u + C$

2. $\displaystyle\int \cos u \, du = \sin u + C$

3. $\displaystyle\int \sec^2 u \, du = \tan u + C$

4. $\displaystyle\int \csc^2 u \, du = -\cot u + C$

5. $\displaystyle\int \sec u \tan u \, du = \sec u + C$

6. $\displaystyle\int \csc u \cot u \, du = -\csc u + C$

Below is a list of some additional trigonometric integration formulas.

MORE TRIGONOMETRIC INTEGRATION FORMULAS

7. $\displaystyle\int \tan x \, dx = -\ln |\cos x| + C$

8. $\displaystyle\int \sec x \, dx = \ln |\sec x + \tan x| + C$

9. $\displaystyle\int \csc x \, dx = \ln |\csc x - \cot x| + C$

10. $\displaystyle\int \cot x \, dx = \ln |\sin x| + C$

Remember that integration problems can be checked with differentiation. It will often be necessary to use substitution when integrating indefinite integrals that involve trigonometric functions.

EXAMPLE 28   Find each integral.

(a) $\displaystyle\int \sin 3x \, dx$ (b) $\displaystyle\int \sin^3 x \cos x \, dx$

(c) $\displaystyle\int (\sin x)e^{\cos x} \, dx$ (d) $\displaystyle\int 3 \sec^3 x \tan x \, dx.$

SOLUTION

(a)   We use the substitution $u = 3x$.

$$\int \sin 3x \, dx = \int (\sin u)\left(\frac{1}{3} \, du\right) \qquad u = 3x: \, du = 3 \, dx \Longrightarrow \frac{1}{3} \, du = dx$$

$$= \frac{1}{3} \int \sin u \, du$$

$$= -\frac{1}{3} \cos u + C$$

$$= -\frac{1}{3} \cos 3x + C$$

Check by differentition:

$$\frac{d}{dx}\left[\frac{-1}{3} \cos 3x\right] = \left(-\frac{1}{3}\right)(-\sin 3x)(3) = \sin 3x$$

(b)   The substitution for this problem is $u = \sin x$.

$$\int \sin^3 x \cos x \, dx = \int u^3 \, du \qquad u = \sin x: \, du = \cos x \, dx$$

$$= \frac{u^4}{4} + C$$

$$= \frac{\sin^4 x}{4} + C$$

Check by differentiation:

$$\frac{d}{dx}\left[\frac{\sin^4 x}{4}\right] = \left(\frac{1}{4}\right)(4 \sin^3 x)(\cos x) = \sin^3 x \cos x$$

(c)   This integral seems to be of the form $\displaystyle\int e^u \, du$, so we use the substitution $u = \cos x$.

$$\int (\sin x)e^{\cos x}\, dx = \int (e^{\cos x})(\sin x \, dx) \qquad u = \cos x\colon du = -\sin x \, dx \Rightarrow$$
$$-du = \sin x \, dx$$

$$= \int e^u(-\, du)$$

$$= -\int e^u \, du$$

$$= -e^u + C$$

$$= -e^{\cos x} + C$$

Check by differentiation:

$$\frac{d}{dx}[-e^{\cos x}] = -(-\sin x)e^{\cos x} = (\sin x)e^{\cos x}$$

(d)   To integrate this function, we need to rewrite the integral and then use substitution.

$$\int 3 \sec^3 x \tan x \, dx = 3 \int (\sec^2 x)(\sec x \tan x \, dx) \qquad \text{Rewrite the integrand.}$$

$$= 3 \int u^2 \, du \qquad\qquad u = \sec x$$
$$du = \sec x \tan x \, dx$$

$$= 3 \cdot \frac{u^3}{3} + C$$

$$= \sec^3 x + C$$

Check by differentiation:

$$\frac{d}{dx}[\sec^3 x] = (3 \sec^2 x)(\sec x \tan x) = 3 \sec^3 x \tan x \qquad\blacksquare$$

Practice Problem 1 Find each integral.

(a) $\displaystyle\int \cos 3x \, dx$ (b) $\displaystyle\int \csc \pi x \cot \pi x \, dx$

(c) $\displaystyle\int \cos 3x \sqrt{1 - 2 \sin 3x}\, dx$

ANSWER

(a) $\dfrac{1}{3} \sin 3x + C$ (b) $\dfrac{-1}{\pi} \csc \pi x + C$

(c) $\dfrac{-1}{9} (1 - 2 \sin 3x)^{3/2} + C$

Integrals can be used to find the area between two curves.

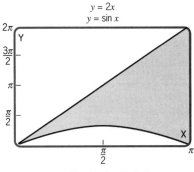

$y = 2x$
$y = \sin x$

Figure 30 X: $[0, \pi]$ and Y: $[0, 2\pi]$

EXAMPLE 29 Find the area of the region bounded by the graphs of $y = 2x$ and $y = \sin x$ from $x = 0$ to $x = \pi$.

SOLUTION The graphs of the functions and the area to be calculated are shown in Figure 30. To calculate the area, we use a definite integral.

$$\int_0^{\pi} (2x - \sin x)\, dx = \frac{2x^2}{2} - (-\cos x)\Big|_0^{\pi}$$

$$= x^2 + \cos x \Big|_0^{\pi}$$

$$= (\pi^2 + \cos \pi) - (0^2 + \cos 0)$$

$$= \pi^2 - 1 - 1 \approx 7.8696$$

We can check this answer with a graphing calculator by entering

$$\text{fnInt}(2x - \sin x, x, 0, \pi) = 7.8696 \qquad \blacksquare$$

EXAMPLE 30 A business determines that its monthly revenue function is

$$R(x) = 8750 + 500 \sin \frac{x}{2}$$

Use a graphing calculator to find the average monthly revenue over 1 year, that is, over the interval [0, 12].

SOLUTION

$$(1/12)\ \text{fnInt}(8750 + 500 \sin x/2, x, 0, 12) = 8753.32$$

The average monthly revenue for the year was \$8753.32. $\qquad \blacksquare$

Practice Problem 2 Find the total revenue for 1 year for the function given in Example 30:

$$R(x) = 8750 + 500 \sin \frac{x}{2}$$

ANSWER

$$\int_0^{12} \left(8750 + 500 \sin \frac{x}{2}\right) dx = 105{,}039.83$$

SUMMARY

Using the rules for integrals of trigonometric functions requires practice. It will usually be necessary to use substitution. If you are calculating a definite integral, you can use the fnInt function on your graphing calculator.

Exercise Set 8.5

Find each integral.

1. $\displaystyle\int \cos 5x\, dx$

2. $\displaystyle\int \cos(-8x)\, dx$

3. $\displaystyle\int \sin \frac{x}{4}\, dx$

4. $\displaystyle\int 2 \sin \frac{x}{3}\, dx$

5. $\displaystyle\int \sin(1 + 3x)\, dx$

6. $\displaystyle\int \sin(2 - 4x)\, dx$

7. $\displaystyle\int \sec^2 3x\, dx$

8. $\displaystyle\int \sec 4x \tan 4x\, dx$

9. $\displaystyle\int (\sin x)^5 \cos x\, dx$

10. $\displaystyle\int \sqrt[3]{\cos x} \sin x\, dx$

11. $\displaystyle\int \frac{\cos x}{\sqrt{\sin x}}\, dx$

12. $\displaystyle\int x^3 \cos x^4\, dx$

13. $\displaystyle\int \csc^2\left(\frac{x}{2}\right) dx$

14. $\displaystyle\int (1 + \sec^2 x)\, dx$

15. $\displaystyle\int (1 - \csc x)^2\, dx$

16. $\displaystyle\int \frac{(1 + \cos x)^2\, dx}{\cos x}\, dx$

17. $\displaystyle\int (\cos^3 x)(-\sin x)\, dx$

18. $\displaystyle\int x \sin x^2\, dx$

19. $\displaystyle\int e^x \sin(e^x)\, dx$

20. $\displaystyle\int e^{\sin x} \cos x\, dx$

21. $\displaystyle\int \frac{1 + \cos x}{x + \sin x}\, dx$

22. $\displaystyle\int e^x \cos(e^x)\, dx$

Evaluate each integral and check with a graphing calculator.

23. $\displaystyle\int_{-\pi}^{\pi} \cos x\, dx$

24. $\displaystyle\int_{-\pi/2}^{\pi/2} \sin x\, dx$

25. $\displaystyle\int_{0}^{2} \sin x\, dx$

26. $\displaystyle\int_{1}^{2} \cos x\, dx$

27. $\displaystyle\int_{\pi/2}^{\pi} 3 \sin x\, dx$

28. $\displaystyle\int_{-3\pi}^{\pi} \cos 5x\, dx$

Use your calculator and calculus techniques to find the area between the x-axis and the graph of f over the interval stated.

29. $f(x) = \cos x, \quad \left[\dfrac{\pi}{6}, \dfrac{5\pi}{6}\right]$

30. $f(x) = \tan x, \quad \left[0, \dfrac{\pi}{4}\right]$

31. $f(x) = \sin x, \quad \left[\dfrac{\pi}{2}, \dfrac{3\pi}{2}\right]$

Use your calculator to find the area between the curves given over the stated interval.

32. $y = 2 + \cos x, \ y = \sin x; \quad [0, \pi]$

33. $y = x + 1, \ y = \sin x; \quad [0, \pi]$

34. $y = 2x, \ y = -\cos x; \quad [0, 2\pi]$

35. $y = \sin x, \ y = -\cos\left(x - \frac{\pi}{2}\right); \quad [0, 2\pi]$

Find the average value for each function over the given interval.

36. $f(x) = \cos^2 x, \quad [-\pi, \pi]$

37. $f(x) = x \sin x, \quad [0, 5]$

38. $f(x) = e^x \sin e^x, \quad [0, 2]$

39. $f(x) = -\cos^2 x \sin x, \quad [0, 2\pi]$

Applications (Business and Economics)

40. **Revenue.** A sporting goods store's weekly revenue (in thousands of dollars) from tennis racket sales x weeks after January 1 is approximated by

$$R(x) = 1.5 + 1.4 \cos \frac{\pi x}{20}, \qquad 0 \le x \le 52$$

(a) Find the total revenue for the year.
(b) Find the average weekly revenue for the year.

41. **Revenue.** A clothing distributor has found that his monthly revenue (in thousands of dollars) from the sales of a certain shirt x months after January 1 can be approximated by

$$R(x) = 3 + 2 \cos \frac{\pi x}{6}, \qquad 0 \le x \le 12$$

(a) Find the total revenue received for the year from sales of the shirt.
(b) Find the average monthly revenue for the year from sales of the shirt.

42. **Cash Flow.** A manufacturer has determined that his daily cash flow is

$$C(x) = 100 \sin \frac{2\pi x}{30} + 100$$

on day x of a 30-day month.
 (a) Find the total amount of money earned during the month.
 (b) Find the average daily cash flow for the month.

43. **Dow Averages.** The weekly closing price of a stock in week x is approximated by the function

$$P(x) = 75 + 3x \cos \frac{\pi x}{6}, \qquad 0 \le x \le 26$$

where $P(x)$ is the price (in dollars) per share. Find the average weekly closing price over the 26-week period.

Applications (Social and Life Sciences)

44. **Roller Coaster.** A roller coaster x meters from the starting point is $h(x)$ meters above the ground where

$$h(x) = 15 + 15 \sin \frac{\pi}{50} x$$

Find the area under the roller coaster 110 meters from the starting point.

45. **Population.** The population of coyotes in a certain region of Texas over a 2-year period is approximated by the model

$$P(m) = 400 + 50 \sin \left(\frac{\pi m}{12} \right)$$

in month m. Find the average population over the first year and over the second year.

Review Exercises

Differentiate each function.

46. $f(x) = \sin 3x$
47. $f(x) = e^{-x} \tan 2x$
48. $f(x) = \ln(\cos^2 x)$
49. $f(x) = (1 - \sec 3x)^3$

Chapter Review

Important Terms

Degree	Quadrant
Radian	Periodic
Angle	For $y = a \sin bx$ and $y = a \cos bx$,
Standard position	Amplitude $= |a|$ and Period $= \dfrac{2\pi}{|b|}$

Trigonometric Functions

(x, y) is a point on the terminal side of angle θ and $r = \sqrt{x^2 + y^2}$.

$$\sin \theta = \frac{y}{r} \qquad\qquad \csc \theta = \frac{r}{y}, \quad y \neq 0$$

$$\cos \theta = \frac{x}{r} \qquad\qquad \sec \theta = \frac{r}{x}, \quad x \neq 0$$

$$\tan \theta = \frac{y}{x}, \quad x \neq 0 \qquad \cot \theta = \frac{x}{y}, \quad y \neq 0$$

Conversions

$$1° = \frac{\pi}{180} \text{ radian} \qquad 1 \text{ radian} = \frac{180°}{\pi}$$

Trigonometric Identities

1. $\sin(-x) = -\sin x$

2. $\cos(-x) = \cos x$

3. $\sin^2 x + \cos^2 x = 1$

4. $1 + \tan^2 x = \sec^2 x$

5. $1 + \cot^2 x = \csc^2 x$

6. $\sin 2x = 2 \sin x \cos x$

7. $\cos 2x = \cos^2 x - \sin^2 x$

8. $\tan 2x = \dfrac{2 \tan x}{1 - \tan^2 x}$

Derivatives of Trigonometric Functions

For u a differentiable function of x:

1. $\dfrac{d}{dx} \sin u = \cos u \cdot \dfrac{du}{dx}$

2. $\dfrac{d}{dx} \cos u = -\sin u \cdot \dfrac{du}{dx}$

3. $\dfrac{d}{dx} \tan u = \sec^2 u \dfrac{du}{dx}$

4. $\dfrac{d}{dx} \cot u = -\csc^2 u \dfrac{du}{dx}$

5. $\dfrac{d}{dx} \sec u = \sec u \tan u \dfrac{du}{dx}$

6. $\dfrac{d}{dx} \csc u = -\csc u \cot u \dfrac{du}{dx}$

Trigonometric Integration Formulas

For u a function of x:

1. $\displaystyle\int \sin u \, du = -\cos u + C$

2. $\displaystyle\int \cos u \, du = \sin u + C$

3. $\displaystyle\int \sec^2 u \, du = \tan u + C$

4. $\displaystyle\int \csc^2 u \, du = -\cot u + C$

5. $\displaystyle\int \sec u \tan u \, du = \sec u + C$

6. $\displaystyle\int \csc u \cot u \, du = -\csc u + C$

7. $\displaystyle\int \tan u \, du = -\ln |\cos u| + C$

8. $\displaystyle\int \sec u \, du = \ln |\sec u + \tan u| + C$

9. $\displaystyle\int \csc u \, du = \ln |\csc u - \cot u| + C$

10. $\displaystyle\int \cot u \, du = \ln |\sin u| + C$

Chapter Test

1. Find the equivalent radian measure for $-275°$.

2. Find the equivalent degree measure for $7\pi/11$ radians.

3. Given that the point $(-3, 4)$ is on the terminal side of an angle θ in standard position, find the value of the six trigonometric functions for θ.

4. Find the period, amplitude, and maximum and minimum values of the graph of each function and sketch the graph over $-2\pi \le x \le 2\pi$ to verify your answer.
    (a) $y = 0.8 \cos 0.3x + 2$ (b) $y = 3 \sin 4x - 2$

5. Solve each equation over the interval $[0, 2\pi)$.
    (a) $\cos^2 x - 1 = 0$ (b) $\sin 2x = \dfrac{1}{2}$

6. Verify each identity algebraically.
    (a) $\dfrac{1 - \sin x}{\cos x} = \dfrac{\cos x}{1 + \sin x}$
    (b) $(\sin x + \cos x)^2 + (\sin x - \cos x)^2 = 2$

7. Differentiate each function.
    (a) $y = \sin 2x^2 - \cos 3x$ (b) $y = \ln (2x^3 - 3 \cos x^2)$
    (c) $y = x^3 \tan x$ (d) $y = e^{2x-3} \csc x^2$

8. Determine any inflection points for $y = 2 \sin x - 3 \cos x$ over the interval $[0, 2\pi]$.

9. A store's weekly revenue (in thousands of dollars) from the sale of golf clubs x weeks after January 1 is approximated by

    $$R(x) = 1.5 + 1.4 \cos \frac{\pi x}{20}, \qquad 0 \le x \le 52$$

    Find the maximum weekly revenue.

10. Find each integral.
    (a) $\displaystyle\int \sin (4x - 3)\, dx$ (b) $\displaystyle\int (\cos x)^4 \sin x\, dx$

    (c) $\displaystyle\int e^{\cos x} \sin x\, dx$ (d) $\displaystyle\int (\sin 2x \cos 2x\, dx)^2\, dx$

11. Evaluate each integral.
    (a) $\displaystyle\int_{-\pi}^{\pi} - 2 \sin 3x\, dx$ (b) $\displaystyle\int_{0}^{2\pi} \cos^2 x \sin x\, dx$

12. Find the area between $y = 2x + 1$ and $y = 2 \sin x$ over the interval $[0, 2\pi]$.

13. The yearly profits of a company can be found using the model

    $$P(x) = 50 + 40 \sin \pi x, \qquad 0 \le x \le 4$$

    where x is in years and $P(x)$ is in thousands of dollars. Find the total profit over the 4-year period.

14. Algebraically and graphically show that $\dfrac{\sin \theta}{1 - \cos \theta} = \dfrac{1 + \cos \theta}{\sin \theta}$.

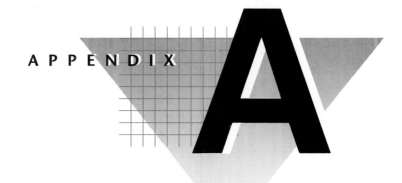

A P P E N D I X

Using Graphing Calculators

INSTRUCTIONS FOR TI-82

Graphing Functions

The first step in graphing a function is to press the $\boxed{\text{Y} =}$ key. The screen shown in Figure 1(*a*) will appear in the calculator window. Press the down arrow to see the remaining $\text{Y} =$ locations [Figure 1(*b*)].

Suppose that you want to graph $y = -x + 4$. With the arrow keys ($\boxed{\blacktriangledown}$, $\boxed{\blacktriangleleft}$, $\boxed{\blacktriangleright}$, $\boxed{\blacktriangle}$) you can move the cursor to any $\text{Y} =$ location you choose. Move the cursor to the Y_1 position. Press the negative key $\boxed{(-)}$ (grey key, row 10, column 4), the $\boxed{\text{X, T, } \theta}$ key (row 3, column 2), the addition operation sign $\boxed{+}$, and the number 4. On the screen you have $\text{Y}_1 = -\text{X} + 4$ [Figure 2(*a*)]. You could have entered the X with the $\boxed{\text{ALPHA}}$ key and the $\boxed{\text{X}}$ key (row 9, column 1), but the $\boxed{\text{X, T, } \theta}$ key is easier. Note also that we used the negative key $\boxed{(-)}$ (grey key, row 10, column 4) instead of the subtraction key $\boxed{-}$ (blue key, row 8, column 5).

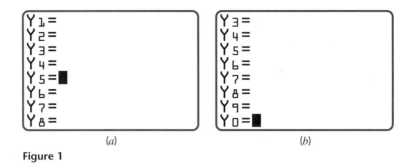

Figure 1

When you graph a function with paper and pencil, you must draw coordinate axes and decide what values you want to see on the axes. Similarly, with a graphing calculator you must decide what viewing window to use. Press WINDOW (row 1, column 2). The Ymin, Ymax, Xmin, and Xmax define what is called a viewing window. You can change the size of your viewing window by changing the window settings. Move the arrow down to Xmin and input -6. Each time you press ENTER or the down arrow the cursor moves down one line. Let Xmin $= -6$, Xmax $= 6$, Xscl $= 1$, Ymin $= -6$, Ymax $= 6$, and Yscl $= 1$ [Figure 2(b)]. When you have input all of the entries, press Y = (to make certain of the function to be graphed) and then press GRAPH. When you press GRAPH, you obtain the graph shown in Figure 2(c).

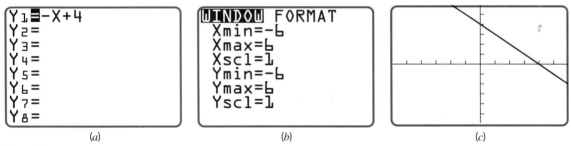

Figure 2

The hash lines (called tick marks) on both the *x*-axis and *y*-axis represent 1 unit. If we count the tick marks in Figure 2(c), we see that the maximum *y* value on the screen is 6 and the minimum *y* value is -6. Press the WINDOW key and change the minimum and maximum values of *x* and *y* to -10 and 10 and examine the graph in Figure 3(a). Note this time that there are 10 hash lines, each representing 1 unit. This particular range setting is the built-in window setting on the T1-82 that is attained by pressing ZOOM and then choosing 6:ZStandard [Figure 3(b)]. To clear the screen of a graph, simply press CLEAR when the graph is on the screen. This will return you to the home screen.

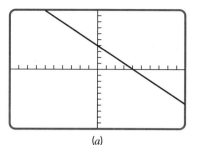

(a)

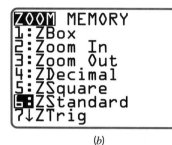

(b)

Figure 3

Press $\boxed{\text{Y =}}$ to input the function

$$y = \frac{320 - 4x}{3}$$

[see Figure 4(*b*)]. Note that we need to use parentheses:

Now when you press $\boxed{\text{GRAPH}}$ the screen is blank. What is the difficulty? The viewing window is too small. When $x = 0$, y is 320/3 or 106.6. The range of y values is too small. Suppose that we are interested in the graph for only positive values of x and large values of y. We change our window to Xmin = 0, Xmax = 8, Xscl = 2, Ymin = 80, Ymax = 120, and Ysci = 10 [Figure 4(*a*)]. The graph is shown in Figure 4(*c*).

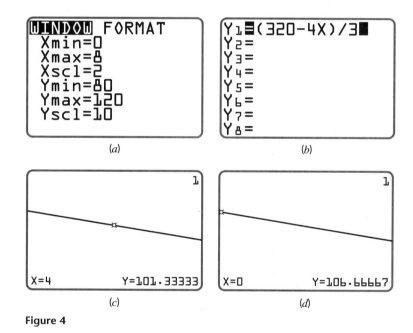

(a)

(b)

(c)

(d)

Figure 4

Trace Features

There are two cursors that can be moved to different positions on the screen. The free-moving cursor (+) is activated by pressing one of the arrow keys ($\boxed{\blacktriangledown}$, $\boxed{\blacktriangleleft}$, $\boxed{\blacktriangleright}$, $\boxed{\blacktriangle}$). The x- and y-coordinates of the position of the cursor are given at the bottom of the screen. The trace cursor ($\boxtimes$) is activated by pressing the $\boxed{\text{TRACE}}$ key.

The trace cursor moves along graphs of functions that have been activated and placed on the screen. The trace cursor gives the current x- and the $f(x)$-coordinates of the selected function along the bottom of the screen. A small number at the upper right side of the screen indicates which function is being traced [see Figures 4(c) and 4(d)]. The up and down arrows ($\boxed{\blacktriangle}$, $\boxed{\blacktriangledown}$) move between Y = functions, and the left and right arrows ($\boxed{\blacktriangleleft}$, $\boxed{\blacktriangleright}$) move the cursor along the selected function. To deactivate the trace cursor, press $\boxed{\text{CLEAR}}$ or press $\boxed{\text{GRAPH}}$.

To graph $Y_1 = 2X + 1$ and $Y_2 = -X + 7$, enter the functions on the Y = screen [Figure 5(a)]. Press the $\boxed{\text{ZOOM}}$ key, select 6:ZStandard, and then press $\boxed{\text{TRACE}}$ [Figure 5(b)]. Now press $\boxed{\blacktriangleright}$ and move the cursor to what seems to be the intersection of the two lines [Figure 5(c)]. The actual intersection is $x = 2$ and $y = 5$. However, the most accurate answer we can obtain by moving the cursor is $x = 2.1276596$ and $y = 4.8723404$ or $x = 1.9148936$ and $y = 4.8297872$.

The answer is not exact because the cursor has width and length and moves a pixel distance; thus, it cannot be placed exactly on the intersection of the two lines. The viewing window 8:ZInteger [Figure 5(d)] under the $\boxed{\text{ZOOM}}$ menu makes the pixel distance such that the cursor will be on or close to the x- and y-coordinates. This procedure enables us to obtain a more accurate answer for the coordinates of intersection, especially if they are integral coordinates. Press the $\boxed{\text{ZOOM}}$ key, choose 8:ZInteger, and then press $\boxed{\text{ENTER}}$. The calculator gives you a chance to change the center of the coordinates to something other than (0, 0). For now, let's use (0, 0), so press $\boxed{\text{ENTER}}$ again. Press $\boxed{\text{TRACE}}$; notice now that the

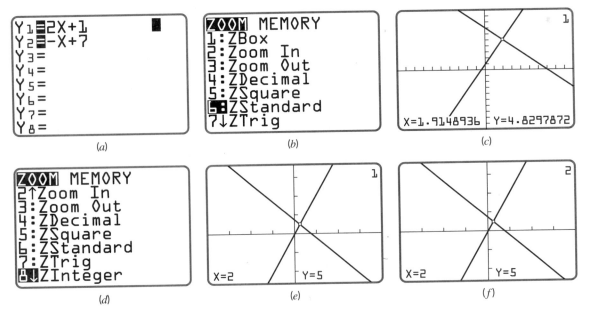

Figure 5

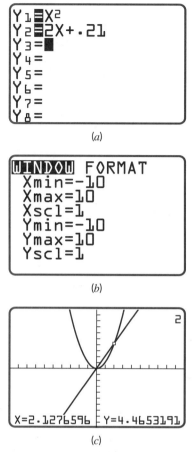

(a)

(b)

(c)

Figure 6

trace cursor can be moved to exactly $x = 2$ and $y = 5$ (left and right arrows) whether you specify function 1 [Figure 5(e)] or change to view function 2 (up and down arrows) [Figure 5(f)].

The point of intersection can also be found by using the CALC menu. This will be discussed later in this appendix. Using the TABLE, we can also show where the two functions have the same x and y values. This will also be discussed later in this appendix.

Friendly Windows

A "friendly window" is one where the TRACE key moves in units of .1 for the x-values. You can get a friendly window by using the ZDecimal option on the ZOOM menu (ZOOM 4:). When you use this key the WINDOW is set at XMin $= -4.7$, XMax $= 4.7$, XScl $= 1$, YMin $= -3.1$, YMax $= -3.1$. Graph $y = x^2 - 2$ and use this window. Now use the TRACE key and you will see that the values of x change in increments of .1.

The incremental movement of the cursor is set by a variable, ΔX. ΔX is evaluated as $\Delta X = (XMax - XMin)/94$. By setting the values of XMin and XMax appropriately, you can have $\Delta X = .1$. For example if you set XMax $= 5$, then XMin $= -4.4$. With some practice you can set these values so that you have a friendly window covering the area of the graph you wish to examine.

There is another variable, $\Delta Y = (YMax - YMin)/62$, but this variable does not affect the TRACE. Both ΔX and ΔY can be found using VARS : WINDOW. ΔX is VARS 1 : 7 and ΔY is VARS 1 : 8. You can store a value in these variables, but the resulting range on the graph screen may not be suitable.

Box Command

Suppose that we wish to graph $y = x^2$ and $y = 2x + 0.21$ on the standard screen. To graph $y = x^2$, press , select Y_1, press $\boxed{X,T,\theta}$, and $\boxed{x^2}$. Similarly graph $Y_2 = 2X + .21$ [Figure 6(a)]. Using the standard viewing rectangle [Figure 6(b)], the point of intersection is close to $x = 2.13$, $y = 4.47$ [Figure 6(c)].

To improve the answer, place a box around the intersection and then change the screen to that which is in the box; that is, enlarge the picture of the intersection. Press $\boxed{ZOOM}$, select 1 : ZBox [Figure 7(a)], and move the cursor to a point where the upper left corner of the box could be, for example, $x = 1.4893617$ and $y = 5.8064516$ [Figure 7(b)]. Press $\boxed{ENTER}$ to fix the corner. Then move the cursor to the lower right corner, for example, $x = 2.9787234$ and $y = 3.2258065$. A box is formed that visually contains the intersection [Figure 7(c)]. Press $\boxed{ENTER}$ to obtain the interior of the box as the image on the screen [Figure 7(d)]. Repeat the process to obtain the answer $x = 2.10$, $y = 4.41$ to the accuracy desired [Figures 8(a), 8(b), and 8(c)]. (Note that we could also use the intersect option under the CALC menu or the ΔTbl increment using the TblSet and TABLE to find the same solution.)

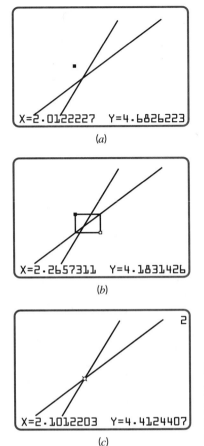

(a)

(b)

(c)

Figure 8

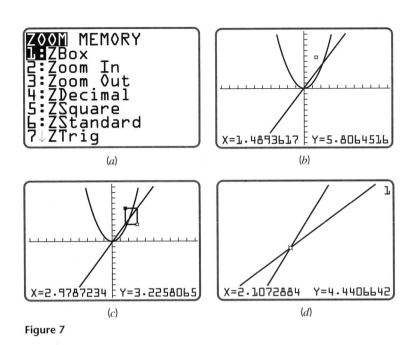

(a)

(b)

(c)

(d)

Figure 7

Solving Equations

In this section we illustrate two techniques for solving equations. The first technique gives an approximation of the answer to the desired accuracy. The second technique is helpful when the solution is an integer. Find the solution to $2x - 2 = 4$. Perform the necessary operations to get everything on one side of the equation, that is, $0 = 2x - 6$. The value of x that will make this expression equal to 0 is the value that will make y equal to 0 in $y = 2x - 6$. Of course, $y = 0$ is where the graph crosses the x-axis. This point is called the x-intercept. The first method consists of putting a box around the x-intercept on the graph and then viewing the x-intercept when the box becomes the screen. That is, enlarge the picture so that the coordinates of the x-intercept can be read easily. Graph $y = 2x - 6$ in the Y_1 position [Figure 9(a)]. Press ZOOM, select 6:ZStandard and then press ZOOM; next 1:ZBox [Figure 9(b)], move the cursor to the upper left corner, and press ENTER to fix one corner of the box. Next move the cursor to the lower right corner to fix the box. Make sure that the box contains the x-intercept. Then press ENTER and the new graphing screen becomes the region that we enclosed in the box [Figure 9(c)]. Now move the cursor to the position that seems to be the location of the x-intercept [Figure 9(d)]. Continue this process until you are satisfied with the accuracy of the answer. The x-intercept is at $x = 3$; that is, the solution to the equation $2x - 2 = 4$ is $x = 3$.

Now let's solve the same equation, $2x - 2 = 4$, by considering two equations made up of the two sides of our given equation. Let $Y_1 = 2X - 2$ (left side) and

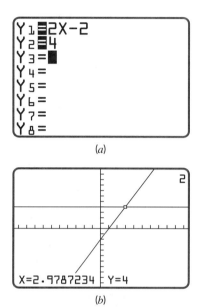

(a)

(b)

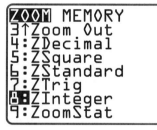

(c)

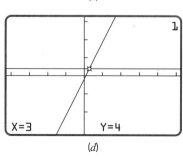

(d)

Figure 10

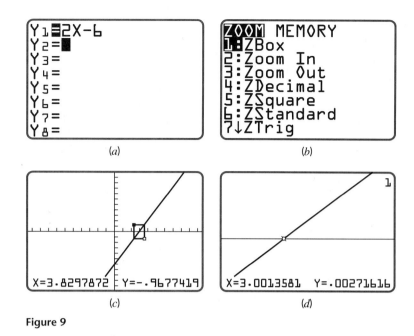

(a)

(b)

(c)

(d)

Figure 9

$Y_2 = 4$ (right side) [Figure 10(a)]. Now press $\boxed{\text{ZOOM}}$, choose 6:ZStandard, and graph the two equations. This time let's use a different method to locate the point of intersection. Press $\boxed{\text{TRACE}}$ and move the cursor to what seems to be the point of intersection. Notice at the bottom of the screen that $x = 2.9787234$, $Y = 4$ [Figure 10(b)]. To determine if these are integral values, press $\boxed{\text{ZOOM}}$, select 8:ZInteger, and then press $\boxed{\text{ENTER}}$ and $\boxed{\text{TRACE}}$ [Figure 10(c)]. Move the cursor until you see that the coordinates of the intersection are $x = 3$, $y = 4$ [Figure 10(d)]. The solution to the equation is $x = 3$.

This problem could also be solved using the TABLE key or the CALC intersect key.

$\boxed{\text{TABLE}}$ Key

The $\boxed{\text{TABLE}}$ key displays the numerical values similar to the Graph command but in table format. Let's return to the earlier example of finding the intersection of $Y_1 = 2X + 1$ and $Y_2 = -X + 7$ [Figure 11(a)], but this time we will use the $\boxed{\text{TABLE}}$ key. Press $\boxed{\text{TblSet}}$ using $\boxed{\text{2nd}}$ $\boxed{\text{WINDOW}}$ [Figure 11(b)]. Set TblMin $= -3$ and ΔTbl $= 1$, and then press $\boxed{\text{TABLE}}$ using $\boxed{\text{2nd}}$ $\boxed{\text{GRAPH}}$. Notice that in the table at $x = 2$, both Y_1 and Y_2 have values of 5 [Figure 11(c)].

Table Setup

The selections we make on the TABLE SETUP screen determine which cells contain values when we press the $\boxed{\text{TABLE}}$ key.

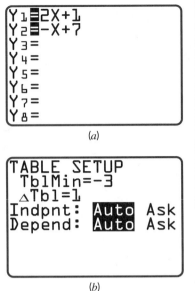

(a)

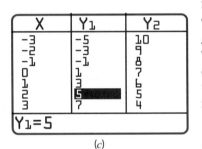

(b)

(c)

Figure 11

Indpnt: Auto Depend: Auto	Values appear in all cells in the table automatically.
Indpnt: Ask Depend: Auto	Table is empty. When a value is entered for the independent variable, the dependent values are calculated automatically.
Indpnt: Auto Depend: Ask	Values appear for the independent variable. To generate a value for a dependent variable, move to the specific cell and press ENTER.
Indpnt: Ask Depend: Ask	Table is empty. Enter values for independent variable. To generate a value for the dependent variable, move to the specific cell and press ENTER.

We use the functions in Y_1 and Y_2 and change the TblSet. Notice the effect of changing from Auto to Ask in each of the screens in Figure 12.

Now let's return to another example: $Y_1 = X^2$ and $Y_2 = 2X + .21$. Input the functions using the Y= key [Figure 13(a)]. Then press the TblSet key and set the TblMin to -1 and ΔTbl to 1 [Figure 13(b)]. Press 2nd TABLE; notice that at $x = 2$, both Y_1 and Y_2 have values that seem close [Figure 13(c)]. Now set TblMin to 1.9 and ΔTbl to .1 [Figure 13(d)]. At $x = 2.1$, the y values agree exactly [Figure 13(e)]. If need be, we could continue to change the minimum value of the table and the increment of change in the table until the two y values reached an approximation close to the desired accuracy.

CALC Key

We can use the value command to calculate a function value for a given x [see Figure 14(c)]. Let $Y_1 = X^2$ and $Y_2 = 2X + .21$. Press the GRAPH key to see the graphs of the functions [Figure 14(d)]. Press 2nd TRACE. Use 1:Value. Input an x value (in this case, 3) and press ENTER. Notice that the x and y values are displayed at the bottom of the screen [Figure 14(e)]. Press the up arrow or the down arrow and the x and y values for the other function will be displayed [Figure 14(f)].

To solve for the intersection of $Y_1 = X^2$ and $Y_2 = 2X + .21$ using the CALC key, graph the functions using Y= [Figure 15(a)] and GRAPH. To obtain the CALC menu press 2nd TRACE and select 5:intersect [Figure 15(b)]. In response to the question ''First curve?'' press ENTER if the cursor is on curve 1 close to the point of intersection [Figure 15(c)]. In response to the question ''Second curve?'' press ENTER if the cursor is on curve 2 close to the point of intersection [Figure 15(d)]. In response to ''Guess?'' move the cursor close to the point of intersection and press ENTER [Figure 15(e)]. The calculator will respond on the bottom of the screen with Intersection X = 2.1, Y = 4.41 [Figure 15(f)].

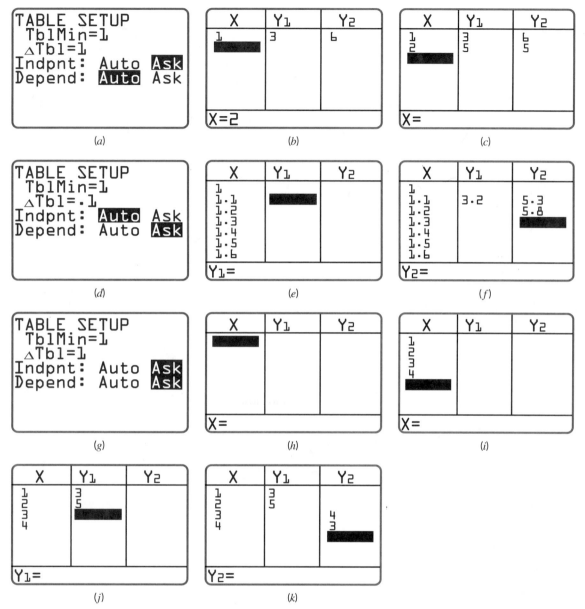

Figure 12

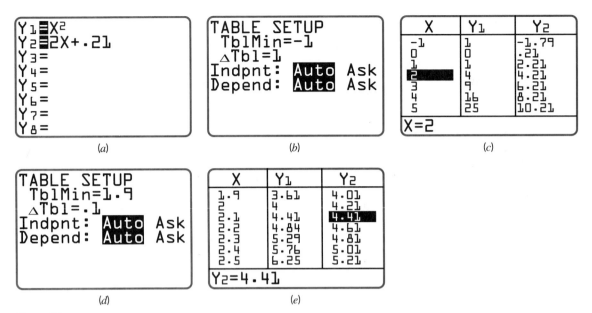

Figure 13

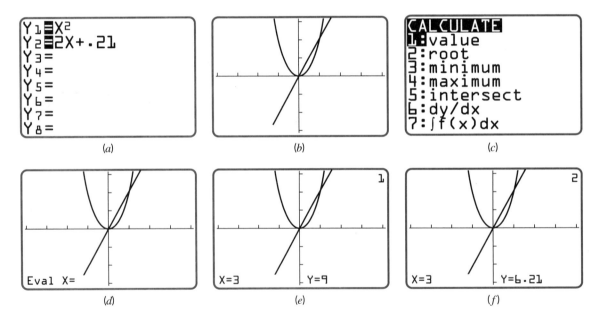

Figure 14

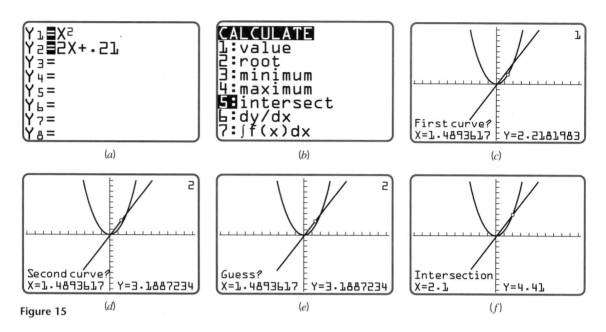

Figure 15

(a)   (b)   (c)

(d)   (e)   (f)

Calculus

The use of the TI-82 in the study of calculus is explained in the first five chapters on calculus.

INSTRUCTIONS FOR THE CASIO 7000G and 7700 GB

Two types of graphs can be generated using this calculator: built-in function graphs and user-generated graphs. The calculator contains the following built-in graphs:

$$\sin \quad \cos \quad \tan \quad x^2 \quad \log \quad \ln$$
$$10 \quad e \quad x^{-1} \quad \sqrt{} \quad \sqrt[3]{}$$

We can draw such graphs as follows:

$\boxed{\text{Graph}}\ \boxed{x^{-1}}\ \boxed{\text{EXE}}$ sketches $y = \dfrac{1}{x}$

$\boxed{\text{Graph}}\ \boxed{\text{SHIFT}}\ \boxed{e^x}\ \boxed{\text{EXE}}$ sketches $y = e^x$

Any time a built-in graph is executed, the ranges are set to their optimum values. User-generated graphs are not set automatically; instead, the user must decide on the range. The parts of a graph outside of the selected range do not appear on the display. The range contents are given in Figure 16(a); the maximum and minimum on each axis and the scale on each axis are given in Figure 16(b).

To change the range use the keys $\boxed{\Leftarrow}\ \boxed{\Rightarrow}\ \boxed{\Downarrow}\ \boxed{\Uparrow}$ to move the cursor to the number you wish to change. Then insert the appropriate numbers. After you

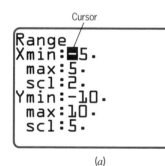

(a)

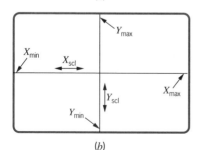

(b)

Figure 16

enter [RANGE], the cursor is located at Xmin. After making a change, the cursor moves automatically down the scale by keying [EXE] as follows:

$-6.$ [EXE]   Xmin
$6.$ [EXE]   Xmax
$1.$ [EXE]   Scl
$-6.$ [EXE]   Ymin
$6.$ [EXE]   Ymax
$1.$ [EXE]   Scl

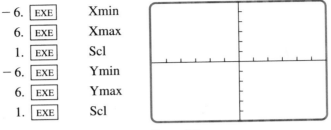

Figure 17

Now punch [G↔T] and you will see the coordinate system given in Figure 17. Again, punch [G↔T] to transfer the window from graphics back to text. Unless stated otherwise or unless built-in functions are used, the graphs in this appendix will use the above range of values. Enter [AC] and you are ready to draw a graph.

User-generated graphs can be drawn by entering the formula after entering [Graph]. An unknown is entered by using [ALPHA] [X]. (The X appears in red and is the [+] key.) For example,

$$y = 4x^3 - 2x^2$$

is entered as

[Graph] 4 [ALPHA] [X] [x^y] 3 [−] 2 [ALPHA] [X] [x²] [EXE]

You will use the following keys in working problems.

[AC]	Clears the screen
[SHIFT] [Cls] [EXE]	Clears the memory of previously used graphs
[RANGE] followed by [RANGE]	Will show your coordinate system, then return to text
[SHIFT] [▲] (key [:])	Enables one to key in one graph followed by another; press [EXE] twice

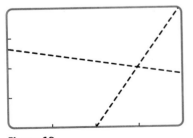

Figure 18

To assist you in using the graphing calculator as a tool throughout the book, the following program is given. To graph $y = -x + 4$, use [Graph] [(−)] [ALPHA] [X] [+] 4 [EXE] (see Figure 18).

Two graphs can be graphed on the screen at the same time and the calculator can be used to obtain an approximate solution of the point of intersection (see Figure 19). To find the intersection of $y = \dfrac{320 - 4x}{3}$ and $y = 20x$ first change the range on x to $2 \le x \le 6$ with a scale of 1, and on y to $80 \le y \le 120$ with a scale of 10:

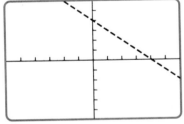

Figure 19

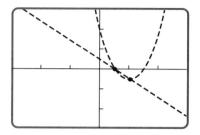

Figure 20

To read the point of intersection, use ⌷SHIFT⌷ ⌷Trace⌷. Locate the blinking pixel at the left of the screen. Use the arrow keys ⌷⇒⌷ and ⌷⇐⌷ to move it to the point of intersection. Then enter ⌷SHIFT⌷ ⌷X↔Y⌷ and read the y-coordinate of the point of intersection. For greater accuracy we will use an automatic zoom-in feature. Press ⌷SHIFT⌷ ⌷x̂⌷ (the ⌷×⌷ key). You can also zoom in by changing the range settings, but the preceding procedure is much faster. The first graph appears automatically. The second graph appears when you press ⌷EXE⌷. Again use ⌷SHIFT⌷ ⌷Trace⌷. Now, move the pixel to what seems to be the intersection and read x and y. This whole process can be repeated until you obtain the accuracy you desire. In two repetitions, it was found that $x = 5$ and $y = 100$.

For practice, let's obtain the intersections of $y = x^2 - 4x + 3$ and $y = -x + 1$. For $-6 \le x \le 6$ and $-6 \le y \le 6$,

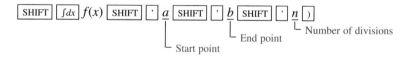

See Figure 20. To find a point of intersection for minimum y, use ⌷SHIFT⌷ ⌷Trace⌷. Use the arrow keys ⌷⇒⌷ and ⌷⇐⌷ to move the blinking pixel to the desired intersection. Read the x value of this intersection from the screen. ⌷SHIFT⌷ ⌷X↔Y⌷ gives the y value associated with the x value of the intersection. For greater accuracy, ⌷SHIFT⌷ ⌷x̂⌷ will zoom in on the intersection by telescoping the range. This procedure regraphs the first function. Use ⌷EXE⌷ to regraph the second function. Now start with ⌷SHIFT⌷ ⌷Trace⌷ and repeat the process. Continue to repeat the process until you have attained the desired accuracy ($x = 1.9997$, $y = -0.9997$).

Integration   The following is the input format for integrations using Simpson's rule:

INSTRUCTIONS FOR SHARP EL-9200/9300

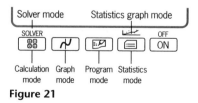

Figure 21

To turn on the calculator, press ⌷ON⌷. To turn the calculator off, press ⌷2ndF⌷ ⌷OFF⌷. The top row (Figure 21) contains keys that determine the operation mode of the calculator.

The simplest mode to begin with is calculation mode (real). To operate in calculation mode press $\boxed{88}$. Make sure the calculator is set to real mode and press $\boxed{\text{MENU}}$ $\boxed{1}$.

NOTE: Unless otherwise noted, all examples in this manual assume that you are in floating point display format.

Enter numbers using the number keys $\boxed{0}$ through $\boxed{9}$, the decimal point key and the change-sign key $\boxed{(-)}$.

Example: Enter the number 123.4.

Press: 1 2 3 $\boxed{.}$ 4 $\boxed{\text{ENTER}}$.

Result: 123.4

Graphing

The general procedure for graphing equations is as follows:

1. Enter graph mode.
2. Key-in one or more functions.
3. Specify the range of the x- and y-axis.
4. Press $\boxed{\sim}$ or $\boxed{\text{2ndF}}$ $\boxed{\text{AUTO}}$.
5. Press $\boxed{\text{SET UP}}$ $\boxed{\text{E}}$.
6. When the menu appears, select 1 for rectangular coordinates. Then press $\boxed{\text{ENTER}}$.

After entering graph mode, use the keyboard and $\boxed{\text{MATH}}$ menus to enter a function (an equation or a number) and then press $\boxed{\text{ENTER}}$. The cursor will move to start of the next function. If there are any previous functions, use $\boxed{\text{CL}}$ to clear them. If you want to move between functions, press $\boxed{\blacktriangledown}$ or $\boxed{\blacktriangle}$ (if you are in equation edit mode, you must press $\boxed{\text{2ndF}}$ $\boxed{\blacktriangledown}$ or $\boxed{\text{2ndF}}$ $\boxed{\blacktriangle}$). Pressing $\boxed{\text{MENU}}$ displays a menu that lets you jump directly to any equation.

Before you graph a function, you must set a range. The easiest way to set the range is to use the auto scaling feature. Press $\boxed{\text{RANGE}}$, enter a minimum, maximum, and scale value for x, and press $\boxed{\text{2ndF}}$ $\boxed{\text{AUTO}}$.

The following example shows some of the powerful graphing features: to graph $y = x^4$ and $y = x + 2$, press

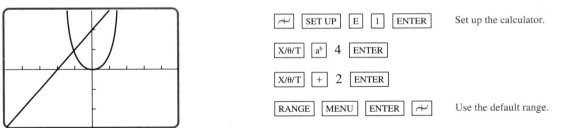

$\boxed{\sim}$ $\boxed{\text{SET UP}}$ $\boxed{\text{E}}$ $\boxed{1}$ $\boxed{\text{ENTER}}$ Set up the calculator.

$\boxed{\text{X/}\theta\text{/T}}$ $\boxed{a^b}$ 4 $\boxed{\text{ENTER}}$

$\boxed{\text{X/}\theta\text{/T}}$ $\boxed{+}$ 2 $\boxed{\text{ENTER}}$

$\boxed{\text{RANGE}}$ $\boxed{\text{MENU}}$ $\boxed{\text{ENTER}}$ $\boxed{\sim}$ Use the default range.

to get the graphs shown in Figure 22.

Figure 22

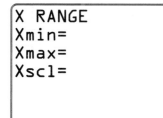

Figure 23

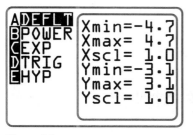

Figure 24

Press $\boxed{\text{RANGE}}$ to get the screen shown in Figure 23. If rectangular coordinates are selected, the X RANGE screen lets you change the minimum (Xmin) and maximum (Xmax) points of the displayed *x*-axis. Xscl (*x*-scale) sets the scale of the tick marks on the *x*-axis. You can change any of these settings. Move to the Y RANGE screen by pressing $\boxed{\blacktriangleright}$. The Y RANGE screen displays the minimum (Ymin) and maximum (Ymax) points of the *y*-axis. Yscl sets the scale of the tickmarks on the *y*-axis.

While in the range screen, press $\boxed{\text{MENU}}$ and the screen shown in Figure 24 appears. This menu lets you select predefined ranges that are appropriate for common functions graphed in rectangular coordinates. To select the default press $\boxed{\text{ENTER}}$.

Zooming

The zoom feature lets you change the range of the graph without entering specific numbers for each coordinate. Press $\boxed{\text{ZOOM}}$ and the menu shown in Figure 25 is displayed. $\boxed{1}$ BOX lets you draw a box around an area of interest. The boxed area then fills the entire screen, distorting the graph as necessary. After BOX is selected move the cursor to a starting point for the box and press $\boxed{\text{ENTER}}$. Move the cursor diagonally to draw a box and press $\boxed{\text{ENTER}}$.

$\boxed{2}$ IN zooms in on the graph by an amount determined by FACTOR. If you use the trace feature to select a point on the curve (before you zoom in), that point becomes the screen center.

$\boxed{3}$ OUT zooms out on the graph by an amount determined by FACTOR. If you use the trace feature to select a point on the curve (before you zoom out), that point becomes the screen center.

$\boxed{4}$ FACTOR lets you set the zoom factor (the number that the range is divided by when zooming in, or multiplied by when zooming out). The *x*-factor can differ from the *y*-factor.

$\boxed{5}$ AUTO performs the same function as pressing $\boxed{\text{2ndF}}$ $\boxed{\text{AUTO}}$.

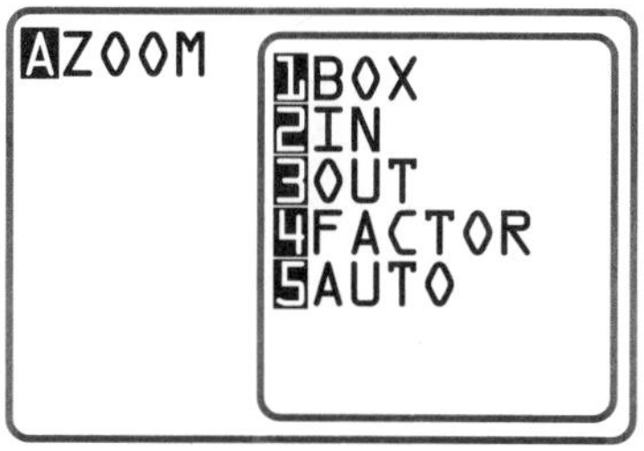

```
AZOOM   1BOX
        2IN
        3OUT
        4FACTOR
        5AUTO
```

Figure 25

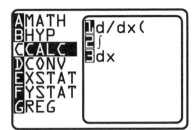

Figure 26

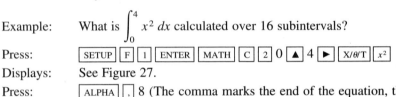

Figure 27

Calculus Functions

Your calculator can integrate and differentiate many types of functions using numerical estimations. Pressing $\boxed{\text{MATH}}$ $\boxed{\text{C}}$ displays the screen shown in Figure 26. $\boxed{1}$ d/dx(selects the derivative function. This function estimates the first derivative[3] of a function at a given value. The syntax of the derivative function is d/dx(*function*, x [*value*], Δx [*change in x*]). (Δx is optional.)

Example: If $F(x) = x^2 + x$, find $F'(4)$.

Press: $\boxed{\text{MATH}}$ $\boxed{\text{C}}$ $\boxed{1}$ $\boxed{\text{X/}\theta\text{/T}}$ $\boxed{x^2}$ $\boxed{+}$ $\boxed{\text{X/}\theta\text{/T}}$ $\boxed{\text{ALPHA}}$ $\boxed{,}$ 4 $\boxed{\text{ALPHA}}$ $\boxed{,}$ 0.00001 $\boxed{)}$ $\boxed{\text{ENTER}}$

Result: 9.

$\boxed{2}$ $\int$ selects the integrate function. This calculates the area under a curve between two points. The calculator uses Simpson's rule to partition the area into a number of even subintervals and estimate the answer. The calculator doubles the number of subintervals that you specify. The answer is an estimate, and therefore will not be exactly correct.

Example: What is $\displaystyle\int_{0}^{4} x^2 \, dx$ calculated over 16 subintervals?

Press: $\boxed{\text{SETUP}}$ $\boxed{\text{F}}$ $\boxed{1}$ $\boxed{\text{ENTER}}$ $\boxed{\text{MATH}}$ $\boxed{\text{C}}$ $\boxed{2}$ 0 $\boxed{\blacktriangle}$ 4 $\boxed{\blacktriangleright}$ $\boxed{\text{X/}\theta\text{/T}}$ $\boxed{x^2}$

Displays: See Figure 27.

Press: $\boxed{\text{ALPHA}}$ $\boxed{,}$ 8 (The comma marks the end of the equation, the 8 is one-half the number of desired subintervals.)

Answers

Exercise Set 1.1, page 9

1. -5

3. 1

5. -1

7. Not defined

9. 0

11.

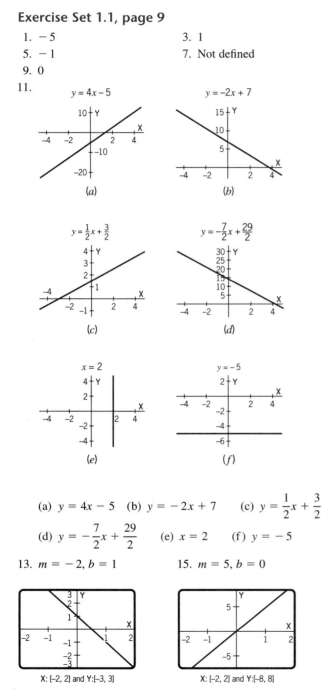

$y = 4x - 5$

(a)

$y = -2x + 7$

(b)

$y = \frac{1}{2}x + \frac{3}{2}$

(c)

$y = -\frac{7}{2}x + \frac{29}{2}$

(d)

$x = 2$

(e)

$y = -5$

(f)

(a) $y = 4x - 5$ (b) $y = -2x + 7$ (c) $y = \frac{1}{2}x + \frac{3}{2}$

(d) $y = -\frac{7}{2}x + \frac{29}{2}$ (e) $x = 2$ (f) $y = -5$

13. $m = -2, b = 1$

X: [-2, 2] and Y:[-3, 3]

15. $m = 5, b = 0$

X: [-2, 2] and Y:[-8, 8]

17. $m = 3, b = 1$

X: [-2, 2] and Y:[-3, 3]

19. $m = \frac{3}{2}, b = -\frac{5}{2}$

X: [-2, 3] and Y:[-5, 2]

21. $y = \frac{1}{2}x + \frac{5}{2}$

23. $y = -\frac{1}{3}x - \frac{7}{3}$

25. $y = 4x - 3$

27. $x = 1$

29. $y = -6$

31. $y = 0$

33. $a; b$

35. $b = -8$

37. (a) $P = 0.8x - 50$
    (b) 6.3   (c) 110

Exercise Set 1.2, page 18

1. (a), (b), (c), (d), (e), (g), (h), (j)

3. (a) Function   (b) Function
   (c) Not a function   (d) Function

5. $10, 4, -1$

7. $-1, 224, w^3 - 2w, 8z^3 - 4z, t^3 + 6t^3 + 10t + 4,$
   $27x^3 - 27x^2 + 3x + 1$

9. $[0, \infty)$

11. $(-\infty, -1), (-1,5), (5, \infty)$

13. $[-\sqrt{3}, \sqrt{3}]$ 15. $(-\infty, 0), (0, 3), (3, \infty)$

17. $(-\infty, -4], [0, \infty)$ 19. $(-\infty, -1), [1, \infty)$

21. $[-2, 1], [3, \infty)$ 23. $(-\infty, \infty)$

25. (a) 3   (b) 2   (c) $4(4 + h)$ (d) $4 + h$

(e) $\dfrac{\sqrt{2 + h} - \sqrt{2}}{h}$ (f) $\dfrac{-1}{2(2 + h)}$

27. (d)   29. (b)

31. $2, 3.50, 15.50, 36.50; 2 for first hour and $1.50 for
    each additional hour.

33. (a) $\dfrac{25}{2}$ (b) $\dfrac{15}{2}$ (c) 0

(d)

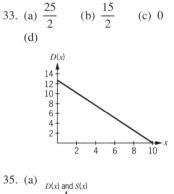

35. (a)

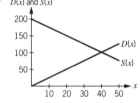

(b) (40,100)   (c) $p = 100$ (d) $x > 40$

37. $y = 19x$ for 19 mpg. More miles per gallon will cause the slope of the line to increase, and the line will be steeper.

Exercise Set 1.3, page 28

1. $y = 4x^2$ magnifies x^2 by 4. $y = 4(x + 1)^2$ shifts $y = x^2$ to the left 1 and magnifies it by 4. $y = 4(x - 1)^2$ shifts $y = x^2$ to the right 1 and magnifies it by 4.

(a) $y = 4x^2$
(b) $y = 4(x + 1)^2$
(c) $y = 4(x - 1)^2$

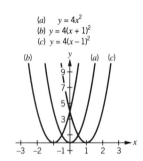

3. $y = -x^2$ reflects $y = x^2$ across the x-axis. $y = -(x - 3)^2$ shifts $y = x^2$ to the right 3 and reflects it across the x-axis. $y = -(x + 3)^2$ shifts $y = x^2$ to the left 3 and reflects it across the x-axis.

5. $y = x^2 - 2$ lowers x^2 down 2. $y = (x - 3)^2$ shifts $y = x^2$ to the right 3. $y = (x + 3)^2 + 2$ shifts $y = x^2$ to the right 3 and up 2.

(a) $y = -x^2$
(b) $y = -(x - 3)^2$
(c) $y = -(x + 3)^2$

(a) $y = x^2 - 2$
(b) $y = (x - 3)^2$
(c) $y = (x + 3)^2 + 2$

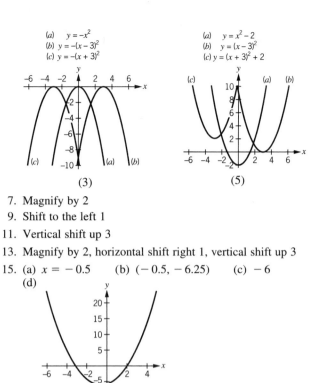

(3)   (5)

7. Magnify by 2

9. Shift to the left 1

11. Vertical shift up 3

13. Magnify by 2, horizontal shift right 1, vertical shift up 3

15. (a) $x = -0.5$ (b) $(-0.5, -6.25)$ (c) -6
(d)

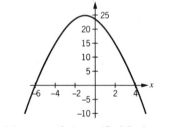

(e) $x = -3, 2$ (f) Minimum

17. (a) $x = -1$ (b) $(-1, 25)$ (c) 24
(d)

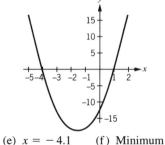

(e) $x = -6, 4$ (f) Maximum

19. (a) $x = -1.5$ (b) $(-1.5, -18.75)$ (c) -12
(d)

(e) $x = -4.1$ (f) Minimum

21. (a) $x = -3$ (b) $(-3, -9)$ (c) 0

(d)

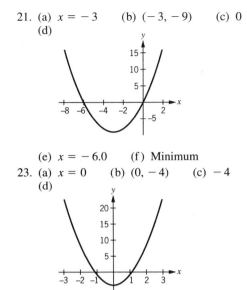

(e) $x = -6.0$ (f) Minimum

23. (a) $x = 0$ (b) $(0, -4)$ (c) -4

(d)

(e) $x = -1.1547, 1.1547$ (f) Minimum

25. $D = (-\infty, -4/3], [4/3, \infty), R = [0, \infty)$

27. $D = (-\infty, 0), (0, \infty), R = (-\infty, 1), (1, \infty)$

Exercise Set 1.4, page 35

1. $x = \pm 4$ 3. $x = \pm \sqrt{7}$

5. $x = 0, 5$ 7. $x = \pm 5$

9. $x = -1, 3$

11. $x = -4, 7$

13. $x = -2, 1$

15. $x = -3, 2$

17. $x = 1, -9$

19. $x = -5, -6$

21. $(2, 3), (3, 4)$

23. $(-1, 2), (2/3, 2)$

25. $(-\infty, -2], [2, \infty)$

27. $(-\infty, -3), (-2, \infty)$

29. $[-1/2, 2/3]$

31. $x = 0, -2, 2$; y is negative on $(-\infty, -2)$, positive on $(-2, 0)$, negative on $(0, 2)$, positive on $(2, \infty)$.

33. $x = 0, -1$; y is negative on $(-\infty, -1)$, negative on $(-1, 0)$, positive on $(0, \infty)$.

35. $x = -3, 0, 3$; y is positive on $(-\infty, -3)$, negative on $(-3, 0)$, negative on $(0, 3)$, positive on $(3, \infty)$.

37. $x = -2, 0, 1$; y is positive on $(-\infty, -2)$, negative on $(-2, 0)$, negative on $(0, 1)$, positive on $(1, \infty)$.

39. At $x = 193$, the maximum revenue is $37,249.

41. Equilibrium point is (8, 16) and the price is $16.

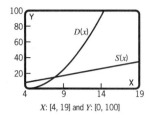

X: [4, 19] and Y: [0, 100]

43. Maximum profit occurs at $x = 14$ and is $242.

45.

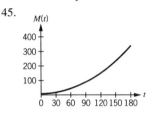

47. $r = 0$ and $r = 100$; $r = 76.83$

49. $y = -\dfrac{5}{9}x - \dfrac{1}{9}$

Exercise Set 1.5, page 44

1. 0.00137174   3. 123.144

5. 0.716531   7. -3.00417

9. 41.4666   11. 1, 0.707107, 4

13. 0; 2.16228; -0.99 15. 3; 2.12132; 12

17. 2; 2.25525; 7.52439   19.

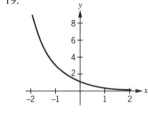

21.

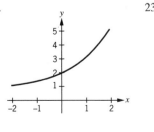

23.

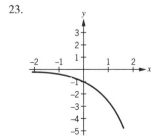

25. Range $(-\infty, 5)$

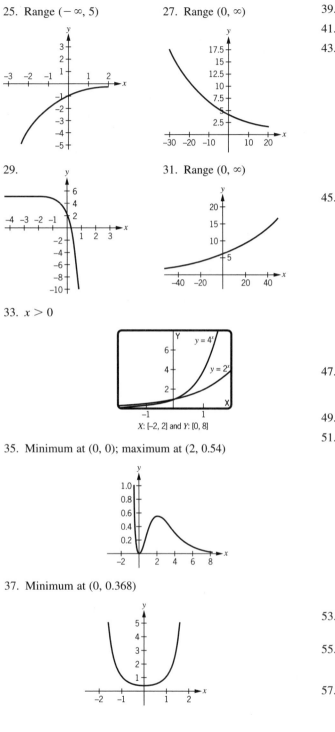

27. Range $(0, \infty)$

29.

31. Range $(0, \infty)$

33. $x > 0$

X: [-2, 2] and Y: [0, 8]

35. Minimum at $(0, 0)$; maximum at $(2, 0.54)$

37. Minimum at $(0, 0.368)$

39. $x = -0.2$

41. $x = -9$

43. Sales increase, but level off and are never above 600.

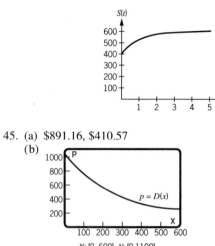

45. (a) $891.16, $410.57

(b)

X: [0, 600], Y: [0,1100]

Within the context of the problem the domain is $(0, \infty)$ and the range is $(0, 1125)$.

47. (a) $V(2) = 15{,}750$

(b) $V(5) = 6644.53$

(c) $V(t) \approx 14{,}000$ when $x = 2.4$ years

49. $x = 0.75$ hours

51. (a) 0

(b) 6.71399

(c) Around $x = 60$

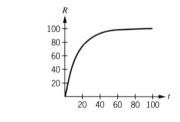

53. (a) 2412

(b) 3572

55. (a) $x = 4, -1$

(b) $x = 3, -0.5$

57. Vertex: $\left(\dfrac{2}{3}, \dfrac{4}{3}\right)$; x-intercepts: $0, \dfrac{4}{3}$; y-intercept: 0

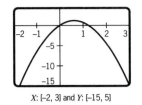

X: [−2, 3] and Y: [−15, 5]

59. Vertex: (1, 1); y-intercept: 3

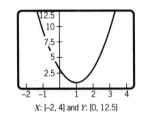

X: [−2, 4] and Y: [0, 12.5]

Chapter 1 Test, page 46

1. $y = x - 6$

2. $y = -\dfrac{1}{2}x + \dfrac{5}{2}$

3. $x = \pm \sqrt{7}$

4. $x = -3 \pm \sqrt{5}$

5. $x = -\dfrac{1}{2}, \dfrac{1}{3}$

6. $x = 0.333424$

7. $x = 4.420696, \ x = -0.7540291$

8. $x = -1.61803399, \ x = 0.61803399, \ x = 1$

9. (a) -7 (b) $2x + h + 2$

10. $D =$ all real values of x except -1 and 5

11. $D = [-5, 5]$

12. $D = (-\infty, \infty)$

13. $D = (-\infty, \infty)$

14. (a) $x = 1$ (b) Up   (c) minimum at $(1, -4)$
    (d) y-intercept $= -3$, x-intercepts $= -1, 3$
    (e) $y = (x - 1)^2 - 4$
    (f)

15.

16. Function

17. Not a function

18. $x = 44$

19. $D(100) = 599.19$

Exercise Set 2.1, page 61

1.

x	1.900	1.990	1.999	$\to 2 \leftarrow$	2.001	2.010	2.100
$f(x)$	0.800	0.980	0.998	$\to 1 \leftarrow$	1.002	1.020	1.200

3.

x	3.900	3.990	3.999	$\to 4 \leftarrow$	4.001	4.010	4.100
$f(x)$	7.900	7.990	7.999	$\to 8 \leftarrow$	8.001	8.010	8.100

5. $\displaystyle\lim_{x \to 1} \frac{3x^2 + 4x}{x + 2} = \frac{7}{3}$

7. $\displaystyle\lim_{x \to 4} -8 = -8$

9. $\displaystyle\lim_{x \to 3} 2x = 6$

11. $\displaystyle\lim_{x \to 2} \frac{3x}{2} = 3$

13. $\displaystyle\lim_{x \to 1} (x) = 3$

15. $\displaystyle\lim_{x \to 1^+} f(x) = 1$

17. $f(8) = 1$

19. $f(3)$ is not defined

21. $\displaystyle\lim_{x \to 3} f(x) = 2$

23. $\displaystyle\lim_{x \to 3} (2x^2 - 4) = 14$

25. $\lim\limits_{x\to 2} (3x^2 - 5)(x + 4) = 42$

27. $\lim\limits_{x\to -3} \dfrac{7x}{2x + 3} = 7$

29. $\lim\limits_{x\to 1} \dfrac{5x}{2 + x^2} = \dfrac{5}{3}$

31. $\lim\limits_{x\to 2} x^2(x^2 + 1)^3 = 500$

33. (a) $\lim\limits_{x\to 3} f(x) = 1$ (b) $\lim\limits_{x\to 3} f(x) = 4$
    (c) $\lim\limits_{x\to 3} f(x)$ does not exist
    (d) $\lim\limits_{x\to 3} f(x)$ does not exist
    (e) $\lim\limits_{x\to 3} f(x) = 3$ (f) $\lim\limits_{x\to 3} f(x) = 2$

35. $\lim\limits_{x\to 0} f(x) = 2$ 37. $\lim\limits_{x\to 4} f(x) = -2$

39. $\lim\limits_{x\to 2} g(x) = 1$ 41. $\lim\limits_{x\to 2} y$ does not exist

43. $\lim\limits_{x\to 2} y = -2$ 45. $\lim\limits_{x\to 3} \dfrac{x^3 - 27}{x - 3} = 27$

47. $\lim\limits_{x\to -3} \dfrac{x^2 - 9}{x + 3} = -6$

49.

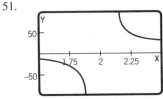

X: [1.5, 2.5] and Y: [9,16]

$\lim\limits_{x\to 2^-} f(x) = 12$; $\lim\limits_{x\to 2^+} f(x) = 12$; but $f(2)$ is not defined.

51.

X: [1.5, 2.5] and Y: [-100, 100]

$\lim\limits_{x\to 2} f(x)$ is undefined.

53. (a) $\lim\limits_{x\to 1} C(x) = 5$
    (b) $\lim\limits_{x\to 4} C(x)$ does not exist.
    (c) $\lim\limits_{x\to 6} C(x) = 6$
    (d) $\lim\limits_{x\to 8} C(x) = 8$

55. (a) $\lim\limits_{t\to 100} n(t) = 40{,}000$
    (b) $\lim\limits_{t\to 0} n(t) = 0$
    (c) $\lim\limits_{t\to -100} n(t)$ does not exist.

Exercise Set 2.2, page 72

1. Continuous at $-1, 0, 2$
3. Continuous at $-2, 0, 1$
5. Continuous at $-1, 0, 2$
7. Continuous at $-1, 2$; not continuous at 0
9. Continuous at $0, 1$; not continuous at -2
11. Continuous at $-1, 0$; not continuous at 2
13. Continuous at $-3, 0$; not continuous at 3
15. Continuous at -1; not continuous at $1, 0$
17. (a) Discontinuous because $\lim\limits_{x\to 2} f(x) = 1$ but $f(2) \neq 1$.
    (b) Continuous because $\lim\limits_{x\to 2} f(x) = 2$ and $f(2) = 2$.

19. $\lim\limits_{x\to 3} \dfrac{x^2 - 9}{x - 3} = 6$ 21. $\lim\limits_{x\to 0} \dfrac{4x^2 - 3x}{x} = -3$

23. $\lim\limits_{x\to 4} \dfrac{x - 4}{\sqrt{x} - 2} = 4$ 25. $\lim\limits_{x\to 0} \dfrac{(2/x) + 1}{3/x} = \dfrac{2}{3}$

27. Discontinuous at $x = 3$
29. Continuous everywhere
31. Discontinuous at $x = 0$
33. Discontinuous at $x = -2$
35. Discontinuous at $x = 2$ and $x = -2$
37. Discontinous at $x = 2$ and $x = -2$
39. (a) Discontinuous at $x = 2$
    (b) Continuous everywhere
41. Discontinuous at $x = 2$
43. (a) Discontinuous at $x = 4$ and $x = 6$
    (b) $\lim\limits_{x\to 2} C(x) = 5$ (c) $C(2) = 5$
    (d) $\lim\limits_{x\to 4} C(x)$ does not exist   (e) $C(4) = 5$
    (f) $\lim\limits_{x\to 7} C(x) = 15$ (g) $C(7) = 15$
45. Continuous on $(16, 96)$

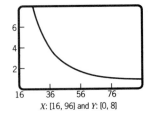

X: [16, 96] and Y: [0, 8]

47. (a) Discontinuous at $x = 2, 3, 5, 6$ (b) $N(1) = 4$
(c) $\lim\limits_{t \to 2^-} N(t) = 4$ (d) $\lim\limits_{t \to 2^+} N(t) = 8$

49. (a)

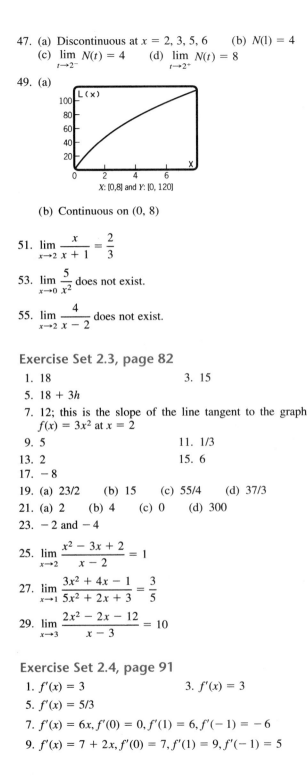

X: [0,8] and Y: [0, 120]

(b) Continuous on $(0, 8)$

51. $\lim\limits_{x \to 2} \dfrac{x}{x + 1} = \dfrac{2}{3}$

53. $\lim\limits_{x \to 0} \dfrac{5}{x^2}$ does not exist.

55. $\lim\limits_{x \to 2} \dfrac{4}{x - 2}$ does not exist.

Exercise Set 2.3, page 82

1. 18   3. 15

5. $18 + 3h$

7. 12; this is the slope of the line tangent to the graph of $f(x) = 3x^2$ at $x = 2$

9. 5   11. 1/3

13. 2   15. 6

17. -8

19. (a) 23/2   (b) 15   (c) 55/4   (d) 37/3

21. (a) 2   (b) 4   (c) 0   (d) 300

23. -2 and -4

25. $\lim\limits_{x \to 2} \dfrac{x^2 - 3x + 2}{x - 2} = 1$

27. $\lim\limits_{x \to 1} \dfrac{3x^2 + 4x - 1}{5x^2 + 2x + 3} = \dfrac{3}{5}$

29. $\lim\limits_{x \to 3} \dfrac{2x^2 - 2x - 12}{x - 3} = 10$

Exercise Set 2.4, page 91

1. $f'(x) = 3$ 3. $f'(x) = 3$

5. $f'(x) = 5/3$

7. $f'(x) = 6x, f'(0) = 0, f'(1) = 6, f'(-1) = -6$

9. $f'(x) = 7 + 2x, f'(0) = 7, f'(1) = 9, f'(-1) = 5$

11. $f'(x) = \dfrac{-4}{x^2}, f'(0)$ is not defined, $f'(1) = -4,$
$f'(-1) = -4$

13. $f'(x) = \dfrac{3}{2\sqrt{3x}}, f'(0)$ is not defined, $f'(1) = \dfrac{\sqrt{3}}{2}, f'(-1)$ is
not defined.

15. $f'(x) = \dfrac{-1}{2(x + 1)^{3/2}}, f'(0) = -1/2, f'(1) = -\dfrac{\sqrt{2}}{8}, f'(-1)$
is not defined.

17. $f'(x) = \dfrac{-4}{x^3}, f'(0)$ is not defined, $f'(1) = -4, f'(-1) = 4$

19. $f'(x) = 3x^2, f'(0) = 0, f'(1) = 3, f'(-1) = 3$

21. $y' = 2x, y'(1) = 2$

X: [-3, 3] and Y: [-3, 3]

23. $m = 6, y = 6x - 2$

25. (a) $f'(x) = 2x$ (b) $f'(2) = 4$
(c) $y = 4x - 3$
(d)

X: [-3, 3] and Y: [0, 6]

27. (a) $C'(x) = 300 + 2x$
(b) $C'(2) = 304$
(c) $C'(3) = 306$

29. (a) $p'(x) = -2x$
(b) $p'(3) = -6$
(c) $p'(5) = -10$

31. (a) $P'(x) = 10 - 10x$
(b) $P'(2) = -10$
(c) $P'(3) = -20$

33. (a) $N(2) = 2000$
(b) $N'(2) = -10$
(c) $N(4) = 1960$
(d) $N'(4) = -30$

35. $\lim\limits_{x \to -2/3} \dfrac{3x^2 + 4x - 1}{3x + 2}$ is not defined.

Exercise Set 2.5, page 103

1. $f'(x) = 0$

3. $\dfrac{dy}{dx} = -1$

5. $D_x\,[3x^2 + 5] = 6x$

7. $y' = 15x^2 - 6x$

9. $\dfrac{d}{dx}(2x^{-4}) = -8x^{-5}$

11. $f'(x) = -x^{-3/2} = \dfrac{-1}{x^{3/2}}$

13. $f'(x) = 24x^3 - 3, f'(0) = -3, f'(2) = 189,$
$f'(-3) = -651$

15. $f'(x) = \dfrac{-3}{x^2} + 4, f'(0)$ is not defined, $f'(2) = \dfrac{13}{4},$
$f'(-3) = \dfrac{11}{3}$

17. $f'(x) = 8x + \dfrac{4}{x^2}, f'(0)$ is not defined, $f'(2) = 17,$
$f'(-3) = \dfrac{-212}{9}$

19. $f'(x) = 2x^{-1/2}, f'(0)$ is not defined, $f'(2) = \sqrt{2}, f'(-3)$ is not defined.

21. $f'(x) = 4x^{-1/3} + 2x^{-1/2},$ $f'(0)$ is not defined, $f'(2) = 2^{5/3} + 2^{1/2} = 4.58902; f'(-3)$ is not defined.

23. $f'(x) = 6x + 2x^{-3/2}, f'(0)$ is not defined;
$f'(2) = 12 + 1/\sqrt{2} = 12.7071; f'(-3)$ is not defined.

25. $f'(x) = (7/2)x^{-1/2} + 3x^{-2/3}, f'(0)$ is not defined,
$f'(2) = \dfrac{7}{2^{3/2}} + \dfrac{3}{2^{2/3}} = 4.36476; f'(-3)$ is not defined.

27. $x = (0, 0)$

29. $(1/8, -1/16)$

31. $y = 6x - 1$

33. $y = 0.16x + 0.02$

35. (c) $8 + 0.02x$

37. (a) $C'(x) = 400 - 2x$ (b) $C'(1) = 398$
(c) $C'(2) = 396$ (d) $\bar{C}(2) = 798$
(e) $C'(1) = 398, C'(2) = 396, \bar{C}(2) = 798$

39. (a) $R'(x) = 20 - \dfrac{x}{250}$ (b) $x = 5000$

(c) $R(5000) = 50,000$ (d) $\bar{R}(x) = 20 - \dfrac{x}{500}$

41. (a) $C'(4) = 6; C'(10) = 6$ (b) Always 6
(c) $R'(5) = 39.9$ (d) $P'(4) = 33.92$

(e) $P(3) = 1.91$ (f) $\bar{P}(x) = 34 - 0.01x + \dfrac{100}{x}$

43. (a) $P'(x) = -0.4\,x^{-3}$ (b) $P'(1) = -0.4$
(c) $P'(2) = -0.05$

45. (a) $f'(x) = 24x^{-2/3}$ (b) $f'(1) = 24$
(c) $f'(8) = 6$ (d) $f(8) = 144$
(e) $f'(1) = 24, f'(8) = 6, f(8) = 144$

47. $\lim\limits_{x\to 2} \dfrac{x^2 + x + 4}{x - 2}$ is not defined.

49. 19.4

51. $y' = 8x + 3$

Exercise Set 2.6, page 114

1. $\dfrac{dy}{dx} = 6x - 7$

3. $\dfrac{dy}{dx} = 4x^3$

5. $\dfrac{dy}{dx} = 3x^2$

7. $\dfrac{dy}{dx} = \dfrac{-15}{(2x - 3)^2}$

9. $\dfrac{dy}{dx} = \dfrac{-13}{(2x - 3)^2}$

11. $\dfrac{dy}{dx} = 36x^2 + 14x - 6, \left.\dfrac{dy}{dx}\right|_{x=1} = 44$

13. $\dfrac{dy}{dx} = \dfrac{-10x}{(x^2 + 1)^2}, \left.\dfrac{dy}{dx}\right|_{x=1} = -\dfrac{5}{2}$

15. $\dfrac{dy}{dx} = \dfrac{-3x^2 + 4x - 9}{(x^2 - 3)^2}, \left.\dfrac{dy}{dx}\right|_{x=1} = -2$

17. $\dfrac{dy}{dx} = \dfrac{-9x^2 - 12}{2\sqrt{x}\,(x^2 - 4)^2}, \left.\dfrac{dy}{dx}\right|_{x=1} = -\dfrac{21}{18}$

19. $y' = \dfrac{20x^2 + 40x - 40}{(x + 1)^2}$

21. $y' = \dfrac{6x^3 - 6x^2 - 6}{(3x - 2)^2}$

23. $y' = -6(2x + 5)^{-2} + (9/2)x^{1/2} - (5/2)x^{-1/2}$
$= \dfrac{-6}{(2x + 5)^2} + \dfrac{9\sqrt{x}}{2} + \dfrac{5}{2\sqrt{x}}$

25. $y' = \dfrac{-2x^4 + 3x^2 + 4x}{(x^3 + 1)^2} - \dfrac{1}{2\sqrt{x^3}}$

27. $y = \dfrac{15x}{32} + \dfrac{25}{16}$

29. (a) $p'(x) = \dfrac{4000(x - 50)}{x^3}$ (b) $p'(2) = -24,000$

(c) $\bar{p}'(x) = \dfrac{4000\,[2x - 75]}{x^4}$

31. (a) $R'(x) = 1000(2x) = 2000x$
(b) $\bar{R}'(x) = \dfrac{1000x^2 + 3000}{x^2}$

33. (a) $L'(t) = \dfrac{6800}{(90t + 85)^2}$

(b) $L'(1) = \dfrac{272}{1225}$ (c) $L'(8) = \dfrac{272}{25,921}$

35. $\dfrac{dy}{dx} = 14x - 3$

37. $\dfrac{dy}{dx} = 6x - 3x^2 + 8x^3$

39. $\dfrac{dy}{dx} = 15x^2 + 2x^{-2}$

41. $\dfrac{dy}{dx} = 24x^3 + 10x^{-3}$

43. 12.3

45. $y' = \lim\limits_{h \to 0} \dfrac{(6xh + 3h^2)}{h} = 6x$

47. $f'(x) = x^{-1/2} + 6x = \dfrac{1}{\sqrt{x}} + 6x$

49. $f'(x) = x^{-1/2} + 2x^{-3/2} = \dfrac{2 + x}{x^{3/2}}$

51. $f'(x) = -2x^{-3/2} + 4x^{-5/3}$

Exercise Set 2.7, page 125

1. $f[g(x)] = x^2 - 2x + 4,\ g[f(x)] = x^2 + 2$

3. $f[g(x)] = \sqrt{x^3 - 7},\ g[f(x)] = (x - 7)^{3/2}$

5. $y = x^4 + 2x^2 + 1$

7. $y = \dfrac{1}{\sqrt{x^4 + 3}}$

9. $y = u^5,\ u = 3x + 4$

11. $y = 1/u,\ u = x + 8$

13. $\dfrac{dy}{dx} = 9$

15. $\dfrac{dy}{dx} = -24x$

17. $\dfrac{dy}{dx} = 72x + 120$

19. $\dfrac{dy}{dx} = 18x + 24$

21. $\dfrac{dy}{dx} = -12x(6 - 2x^2)^2$

23. $\dfrac{dy}{dx} = \dfrac{-2}{(2x + 4)^2}$

25. $\dfrac{dy}{dx} = \dfrac{2x^2}{(2x^3 - 1)^{2/3}}$

27. $\dfrac{dy}{dx} = (3x^2 - 2x + 1)(12x - 4)$

29. $\dfrac{dy}{dx} = -30(2x + 1)^2$

31. $\dfrac{dy}{dx} = \dfrac{(x - 1)^2(7x - 1)}{2\sqrt{x}}$

33. $\dfrac{dy}{dx} = \dfrac{-2x}{(2x^2 - 1)^{3/2}}$

35. $f'(x) = 36x^2 - 32x - 33$

37. $f'(x) = \dfrac{8x + 2}{\sqrt{2x^2 + x - 1}}$

39. $f'(x) = \dfrac{(3x - 1)(30x - 2)}{\sqrt{x}}$

41. $f'(x) = \dfrac{-24x}{(2x - 3)^3}$

43. $f'(x) = 6x(x + 3)^3(x + 1)$

45. $f'(x) = \dfrac{10\,(x - 3)}{(x + 2)^3}$

47. (a) $C'(x) = 18x - 36$
    (b) $C'(3) = -6$
    (c) $C'(4) = 12$
    (d) $\overline{C}'(x) = \dfrac{9x^2 - 124}{x^2}$

49. (a) $L'(1) = 72$
    (b) $L'(14) = 72$

51. $y' = \dfrac{-19}{(3x - 2)^2}$

53. $y' = 36x^2 + 16x - 6$

55. $y' = 18(2x + 5)^2$

Chapter 2 Test, page 127

1. $f'(x) = \lim\limits_{h \to 0} \dfrac{f(x + h) - f(x)}{h}$

2. $f'(x) = 6x$

3. 14

4. $f'(0) = 4;\ f'(-1) = 10$

5. $y' = 0$

6. $\dfrac{dy}{dx} = -16x^3$

7. $\overline{C}'(2) = -249$

8. -2

9. $\lim\limits_{x \to 1} \dfrac{x^2 - 3x + 3}{x - 1}$ does not exist.

10. $f'(x) = \dfrac{-3}{x^2} + \dfrac{2}{\sqrt{x}}$

11. $p'(x) > 0$ for $x > -4$
    $p'(x) < 0$ for $x < -4$

12. $y' = 6x\sqrt{x^2 - 3};\ y'(4) = 24\sqrt{13};\ y'(4) = 86.5332$

13. $y'(x) = \dfrac{18x - 1}{2\sqrt{3x + 2}};\ y'(1) = \dfrac{17}{2\sqrt{5}}$

14. $y' = \dfrac{-5x - 2}{(x^2 - 1)^{3/2}}; \; y'(2) = \dfrac{-12}{3^{3/2}}; \; y'(2) = -2.3094$

15. (a) $p'(x) = \dfrac{6000(x - 40)}{x^3}$

(b) $\bar{p}'(x) = \dfrac{12{,}000\,x - 360{,}000}{x^4}$

Exercise Set 3.1, page 138

1. $y' = 0$; no critical point; increasing, none

3. $f'(x) = 3$; no critical point; increasing, $(-\infty, \infty)$

5. $f'(x) = 2x$; critical point at $(0, 0)$; increasing, $(0, \infty)$

7. $y' = 2x + 2$; critical point at $(-1, -1)$; increasing, $(-1, \infty)$

9. $y' = 8x + 3$; critical point at $\left(-\dfrac{3}{8}, -\dfrac{41}{16}\right)$; increasing, $\left(-\dfrac{3}{8}, \infty\right)$

11. $y' = 6x + 12$; critical point at $(-2, -17)$; increasing, $(-2, \infty)$

13. $y' = 8x^{-1/3} + 1$; critical points at $(0, 0) \, (-512, 256)$; increasing, $(-\infty, -512), (0, \infty)$

15. Increasing, $(2, \infty)$; decreasing, $(-\infty\; 2)$

17. Increasing, $(-\infty, 4), (10, \infty)$; decreasing, $(4, 10)$

19. Answers will vary, but should contain the points given and have a relative minimum at the point $(1, 1)$, the only extremum.

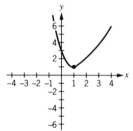

21. Answers will vary, but should contain the points given and have a relative maximum at the point $(3, 11)$, the only extremum.

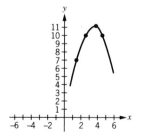

23. (a) $(a, c), (e, f), (i, k)$ (b) $(c, e), (f, i)$
   (c) c, e, f, i

25. Critical point, $(0, 0)$; increasing, $(0, \infty)$; decreasing $(-\infty, 0)$

27. Critical point, $(0, 3)$; increasing, $(-\infty, 0)$; decreasing, $(0, \infty)$

29. Critical point, $\left(-\dfrac{1}{2}, \dfrac{5}{4}\right)$; increasing, $\left(-\dfrac{1}{2}, \infty\right)$; decreasing, $\left(-\infty, -\dfrac{1}{2}\right)$

31. Critical point, $(0, 0)$; increasing, $(-\infty, \infty)$

33. Critical point, none; increasing, $(-\infty, \infty)$

35. Critical point, $(0, 0), (-4, 32)$; increasing, $(-\infty, -4), (0, \infty)$; decreasing, $(-4, 0)$

37. Critical point, $(0, 0), (2, -8)$; increasing, $(-\infty, 0), (2, \infty)$; decreasing, $(0, 2)$

39. Critical point, $(3, 0), (1, 4)$; increasing, $(-\infty, 1), (3, \infty)$; decreasing, $(1, 3)$

41. Critical point, $(0, 0)$; increasing, $(-\infty, \infty)$

43. Critical point, $(0, 0)$; increasing, $(0, \infty)$

45. Critical point, $(1, -4)$; increasing, $(1, 3), (3, \infty)$; decreasing, $(-\infty, 1)$

47. Up to \$20 the profit increases; after \$20 it decreases.

49. Decreasing for x in the interval $(1, 3)$ and increasing for $x > 3$.

51. Since the critical value $x = c$ means that $P'(c) = 0$ and also, $P'(x) = R'(x) - C'(x)$; therefore, $R'(c) - C'(c) = 0$, so $R'(c) = C'(c)$.

53. Increasing on $(3, 8)$ and decreasing on $(0, 3)$

Exercise Set 3.2, page 148

1. Relative minimum of -2 at $x = -3$ and -2 at $x = 1$. Relative maximum of 3 at $x = -1$.

3. (a) Relative maximum   (b) Neither   (c) Relative minimum   (d) Neither   (e) Relative minimum   (f) Neither   (g) Relative maximum   (h) Neither

5. $x = e, x = i$

In 7–27 (a) contains critical points (b) relative extrema (c) where the function is increasing, and (d) where the function is decreasing. Then insert (a), (b), (c), (d) in each problem

7. (a) $(0, 0)$ (b) Min. at $(0, 0)$ (c) $(0, \infty)$
   (d) $(-\infty, 0)$

9. (a) $(0, 3)$ (b) Max. at $(0, 3)$ (c) $(-\infty, 0)$
   (d) $(0, \infty)$

11. (a) $\left(-\frac{1}{2}, \frac{5}{4}\right)$ (b) Min. at $\left(-\frac{1}{2}, \frac{5}{4}\right)$ (c) $\left(-\frac{1}{2}, \infty\right)$
   (d) $\left(-\infty, -\frac{1}{2}\right)$

13. (a) $(0, 0)$ (b) None   (c) $(-\infty, \infty)$ (d) None

15. (a) None   (b) None   (c) $(-\infty, \infty)$ (d) None

17. (a) $(0, 0)$, $(4, -32)$ (b) Max. at $(0, 0)$; Min. at $(4, -32)$ (c) $(-\infty, 0)$, $(4, \infty)$ (d) $(0, 4)$

19. (a) $\left(\dfrac{2}{3}, \dfrac{2\sqrt{3}}{9}\right)$ (b) Max. at $\left(\dfrac{2}{3}, \dfrac{2\sqrt{3}}{9}\right)$
    (c) $(-\infty, \frac{2}{3})$ (d) $(\frac{2}{3}, 1)$

21. (a) $(-2, 0)$ (b) Min. at $(-2, 0)$ (c) $(-2, \infty)$
    (d) $(-\infty, -2)$

23. (a) $\left(\dfrac{2}{3}, \dfrac{-16\sqrt{6}}{9}\right)$ (b) Min. at $\left(\dfrac{2}{3}, \dfrac{-16\sqrt{6}}{9}\right)$
    (c) $\left(\dfrac{2}{3}, \infty\right)$ (d) $\left(0, \dfrac{2}{3}\right)$

25. (a) None   (b) None   (c) None
    (d) $(-\infty, 1)$, $(1, \infty)$

27. (a) $(2, 4)$, $(-2, -4)$ (b) Min. at $(2, 4)$; max. at $(-2, -4)$ (c) $(-\infty, -2)$, $(2, \infty)$ (d) $(-2, 0)$, $(0, 2)$

29. (d)

31. Relative minimum at $(1, 0)$

33. Minimum cost of 182 at $x = 3$.

35. A price of \$20 yields \$100,000 maximum profit.

37. Minimum population of 20,000 at $t = 3$

39. Critical point, $(1, 2)$; decreasing on $(-\infty, 1)$; increasing on $(1, \infty)$

41. Critical point, $(1, 1)$; decreasing on $(-\infty, 1)$; increasing on $(1, \infty)$

Exercise Set 3.3, page 158

$\dfrac{dx}{dy}$	$\dfrac{d^2y}{dx^2}$	$\dfrac{d^3y}{dx^3}$
1. 0	0	0
3. 3	0	0
5. $70x - 27$	70	0
7. $6x^2 + 6x - 1$	$12x + 6$	12
9. $12x^5 + 9x^2 - 2$	$60x^4 + 18x$	$240x^3 + 18$
11. $2x^3 + x^2$	$6x^2 + 2x$	$12x + 2$
13. $x^{-2/3} = \dfrac{1}{x^{2/3}}$	$\dfrac{-2}{3}x^{-5/3} = \dfrac{-2}{3x^{5/3}}$	$\dfrac{10}{9}x^{-8/3} = \dfrac{10}{9x^{8/3}}$
15. $\dfrac{4}{3}x^{1/3}$	$\dfrac{4}{9}x^{2/3} = \dfrac{4}{9x^{2/3}}$	$\dfrac{-8}{27}x^{-5/3} = \dfrac{-8}{27x^{5/3}}$
17. $-3(x-1)^{-2}$	$6(x-1)^{-3}$	$-18(x-1)^{-4}$

19. (a) x_2, x_3, x_7, x_9 (b) x_3, x_6, x_8 (c) x_4, x_6
    (d) x_2, x_7 (e) x_5, x_9

21. (a) Positive, zero, negative   (b) Negative, zero, positive
    (c) Zero, positive, zero   (d) Positive, positive, positive
    (e) Positive, negative, zero   (f) Positive
    (g) Positive   (h) Positive   (i) Positive
    (j) (x_2, x_5), (x_7, x_9) (k) (x_1, x_3), (x_6, x_8)

23. f is increasing and concave down.

25. f is decreasing and concave down.

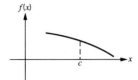

27. f is increasing for $x < c$ and decreasing for $x > c$ and concave down.

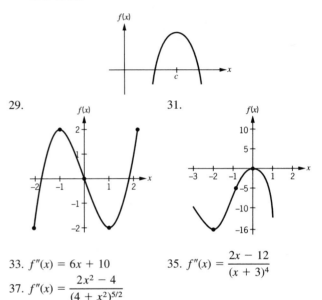

29.   31.

33. $f''(x) = 6x + 10$ 35. $f''(x) = \dfrac{2x - 12}{(x + 3)^4}$

37. $f''(x) = \dfrac{2x^2 - 4}{(4 + x^2)^{5/2}}$

In Exercises 39–51 (a) contains infection points, (b) intervals for concavity up (c) interval for concavity down and (d) relative extrema. Then insert (a) (b) (c) (d) in each

39. (a) None   (b) None   (c) $(-\infty, \infty)$
    (d) Max. at $(0, 0)$

41. (a) None   (b) None   (c) $(-\infty, \infty)$
    (d) Max. at $(0, 0)$

43. (a) None   (b) None   (c) $(-\infty, \infty)$
    (d) Max. at $(\frac{1}{3}, \frac{1}{3})$

45. (a) $(0, 0)$ (b) $(-\infty\ 0)$ (c) $(0, \infty)$
    (d) None

47. (a) $(-4/3, 4.128/27)$ (b) $(-4/3, \infty)$ (c) $(-\infty, -4/3)$
    (d) Min. at $(0, 0)$; max. at $(-8/3, 256/27)$

49. (a) None   (b) $(-\infty, \infty)$ (c) None
    (d) Min. at $(0, 0)$

51. (a) $(1, 6)$ (b) $(1, \infty)$ (c) $(-\infty, 1)$
    (d) Max. at $(-2, 4.76)$; min. at $(0, 0)$

53.

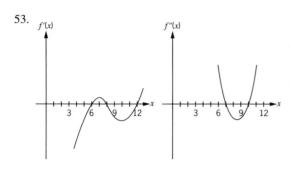

55.

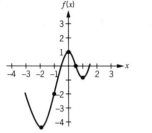

57. (a) $v(t) = t + 3t^2$, $a(t) = 1 + 6t$
(b) $v(8) = 200$, $a(8) = 49$
59. (a) $P'(7) = 8000 > 0$; profit is increasing.
(b) $(0, 15)$ (c) $S = 15$ (d) $112,500
61. (a) $x = 2664$ (b) $783,936
63. 324.15 million
65. Critical point, $(-1.5, -2.25)$; increasing, $(-1.50, \infty)$; decreasing, $(-\infty, -1.50)$

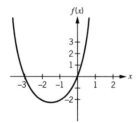

67. Critical points, $(1, 0)$, $(-1, 4)$; increasing, $(-\infty, -1)$, $(1, \infty)$; decreasing, $(-1, 1)$

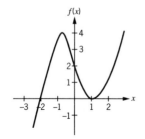

69. Critical point, $(1, -2)$; increasing $(1, \infty)$; decreasing, $(-\infty, 1)$

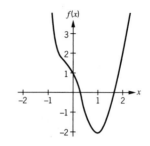

Exercise Set 3.4, page 168

1. Absolute max. at x_2; absolute min. at x_3
3. Absolute max. at x_4; absolute min. at x_1
5. (a) Absolute max. 0; absolute min. -0.25 (b) absolute max 6; absolute min. -0.25
(c) Absolute max. 6; absolute min. -0.25
7. (a) Absolute max. 2; absolute min. -2 (b) absolute max. 16; absolute min. -16
(c) Absolute max. 54; absolute min. -54
9. (a) Absolute max. 3; absolute min. 0   (b) absolute max. 24; absolute min. 0
(c) Absolute max. 99; absolute min. 0
11. $f(-1) = 2$, relative max.; $f(0) = 0$ is relative min.
13. Test fails, $f(1) = -1$, relative min.
15. $f(-5) = 12.5$, relative max.; $f\left(-\dfrac{4}{3}\right) = -12.15$, relative min.

In Exercises 17–25 (a) contains critical points, (b) relative extrema, (c) absolute extrema and (d) inflection points. Then insert (a) (b) (c) (d) in each

17. (a) $(\frac{1}{8}, -\frac{1}{16})$ (b) Min. $-\frac{1}{16}$ (c) Min. $-\frac{1}{16}$
(d) None
19. (a) $(0, 0)$ (b) None   (c) None   (d) $(0, 0)$
21. (a) $(0, 0), (2, -4)$ (b) Max. 0, min. -4 (c) None
(d) $(1, -2)$
23. (a) $(0, 0)$ (b) Min. 0   (c) Min. 0   (d) None
25. (a) $(1/\sqrt{2}, 2\sqrt{2}), (-1/\sqrt{2}, -2\sqrt{2})$
(b) Min. $2\sqrt{2}$, max. $-2\sqrt{2}$ (c) None   (d) None
27. Increasing, $(0, \infty)$; concave down, $(0, \infty)$; no points of inflection

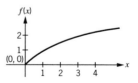

29. Increasing, $\left(-\infty, \dfrac{1}{3}\right)$, $(1, \infty)$; decreasing, $\left(\dfrac{1}{3}, 1\right)$; concave up, $\left(\dfrac{2}{3}, \infty\right)$; concave down, $\left(-\infty, \dfrac{2}{3}\right)$; max. at $\left(\dfrac{1}{3}, \dfrac{4}{27}\right)$; min at $(1, 0)$; inflection at $\left(\dfrac{2}{3}, \dfrac{2}{27}\right)$

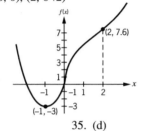

31. Increasing, $(-1, \infty)$; decreasing, $(-\infty, -1)$; concave up, $(-\infty, 0)$, $(2, \infty)$; concave down, $(0, 2)$; min. at $(-1, -3)$; inflection at $(0, 0)$, $(2, 6\sqrt[3]{2})$

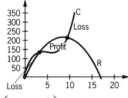

33. (c)   35. (d)
37. 20   39. $x = 5$; $\overline{P}(5) = \$0$
41. Profit is between $x = 3.28$ and $x = 9.5$. Loss is from $x = 0$ to $x = 3.28$ and for $x > 9.5$. Maximum profit is 73 when $x \approx 7$.

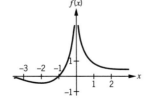

43. Critical Point, $\left(-2, -\dfrac{1}{4}\right)$; increasing, $(-2, 0)$; decreasing, $(-\infty, -2)$, $(0, \infty)$; min. at $\left(-2, -\dfrac{1}{4}\right)$, inflection point, $\left(-3, \dfrac{-2}{9}\right)$; concave up, $(-3, 0)$, $(0, \infty)$; concave down, $(-\infty -3)$

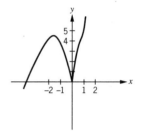

45. Critical points $(-3, 3)$, $(0, 0)$; increasing, $(-3, -2)$, $(-2, \infty)$; decreasing, $(-\infty, -3)$; relative min. at $(-3, 3)$; inflection point, $(0, 0)$; concave up, $(-\infty, -2)$, $(0, \infty)$; concave down, $(-2, 0)$

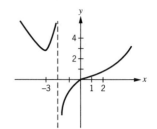

47. Critical points $(-2, 4.76)$, $(0, 0)$; increasing, $(-\infty, -2)$, $(0, \infty)$; decreasing, $(-2, 0)$; min. at $(0, 0)$; max. at $(-2, 4.76)$; inflection point, $(1, 6)$; concave up, $(1, \infty)$; concave down, $(-\infty, 1)$

Exercise Set 3.5, page 183

1. 0.333, 0.039, 0.00399, 0.0003999; $\lim\limits_{x \to \infty} f(x) = 0$
3. 8.33, 98.039, 998.0, 9998; $\lim\limits_{x \to \infty} f(x) \to \infty$
5. -0.5, -0.0408, -0.004008, -0.0004000; $\lim\limits_{x \to -\infty} f(x) = 0$
7. $-12.5, -102.04, -1002.0, -10002.0$; $\lim\limits_{x \to -\infty} f(x) \to -\infty$
9. $-20, -200, -2000, -20{,}000$; $\lim\limits_{x \to 1^-} f(x) \to -\infty$
11. 10, 100, 1000, 10,000; $\lim\limits_{x \to 1^-} f(x) \to \infty$
13. 20, 200, 2000, 20,000; $\lim\limits_{x \to 1^+} f(x) \to \infty$
15. 10, 100, 1000, 10,000; $\lim\limits_{x \to 1^+} f(x) \to \infty$
17. (a) $x = -2$ (b) $y = 0$ (c) 0   (d) 0
19. 0   21. 7
23. 1   25. $\dfrac{3}{4}$
27. $\to \infty$ 29. $\to \infty$
31. 0

33. Horizontal asymptote, $y = 1$; vertical asymptote, $x = -2$; no relative extrema; no inflection points

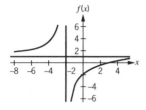

35. Horizontal asymptote, $y = 2$; vertical asymptote, $x = -1$; no relative extrema; no inflection points

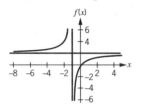

37. Horizontal asymptote, $y = 2$; vertical asymptote, $x = 1$; no relative extrema; no inflection points

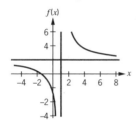

39. Horizontal asymptote, $y = 2$; vertical asymptote, $x = 1$; relative max. at $(3, 3)$; inflection point, $(4, 26/9)$

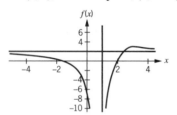

41. Horizontal asymptote, $y = 0$; vertical asymptote, $x = 0$; relative min. at $(12, -1/24)$; inflection point, $(18, -1/27)$

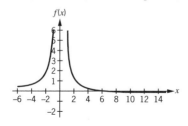

43. Horizontal asymptote, $y = 0$; vertical asymptote, $x = 0$; relative min. at $(1, -1)$; inflection point, $(3/2, -8/9)$

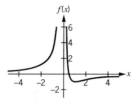

45. Vertical asymptotes, $x = 1$, $x = -1$

47. Horizontal asymptote, $y = 2$

49. $\to \infty$ 51. 3.50

53. $\to \infty$ 55. $\dfrac{1}{3}$

57. (a) At $x = a$, the rate of increase begins to go down.
    (b) This is the point of inflection where rate of increase changes from positive to negative
    (c) The horizontal asymptote.

59. Absolute max. $= 3$; absolute min. $= -75$

Exercise Set 3.6, page 200

1. 26 and 26   3. 7 and 7

5. 1000 cubic inches   7. 24 by 24

9. 30 by 45

11. 10 centimeters; 16,000 cubic centimeters

13. 300 by 250

15. (a) $C'(x) = 10 - 0.02x$

    (b) $\overline{C}(x) = 10 + \dfrac{30}{x} + 0.01x$

    (c) $\overline{C}'(x) = \dfrac{-30}{x^2} + 0.01$

    (d) Both at $x = \sqrt{3000} = 10\sqrt{30}$

17. 7 units

19. $\dfrac{50}{3}$ or 17 tables; profit $= \$233$

21. Maximum profit $= \$516{,}750$ at $x = 4050$

23. 110 people   25. ≈ 100 items

27. 18 orders per year of 22 cars

29. 6 weeks   31. 30 trees

33. 40 people   35. 5 more trees

37. (a) absolute minimum of 1.376 at $x = 6.4$
    (b) $x = 5$, max. at $N(5) = 10$

39. (a) $N'(t) = 3 - 3t^2$ (b) $t = 1$

41. Vertical symptote, $x = 2$; horizontal symptote, $y = 0$; no critical point; no extrema; decreasing, $(-\infty, 2)$, $(2, \infty)$; concave down, $(-\infty, 2)$; concave up, $(2, \infty)$

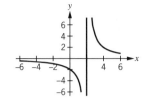

43. Relative max. at $(2\sqrt{2}, 4)$; relative min. at $(-2\sqrt{2}, -4)$; inflection point, $(0, 0)$; concave up, $(-4, 0)$; concave down, $(0, 4)$; increasing, $(-2\sqrt{2}, 2\sqrt{2})$; decreasing, $(-4, -2\sqrt{2})$, $(2\sqrt{2}, 4)$; no asymptotes

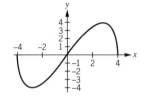

Exercise Set 3.7, page 211

1. (a) 1   (b) 3   (c) 3

3. $|E(1)| = \dfrac{2}{3}$; inelastic

5. $|E(3)| = 1$; unit elasticity

7. $|E(1)| = 2$; elastic

9. (a) $E(p) = \dfrac{p}{40 - p}$ (b) $|E(10)| = \dfrac{1}{3}$

   (c) Inelastic

   (d) Price increase of 5% at $p = 10$ would result in $\dfrac{1}{3} \cdot 0.05 = 1.67\%$ change in demand.

   (e) $|E(30)| = 3$ (f) Elastic

   (g) Price increase of 5% at $p = 30$ would result in $3 \cdot 0.05 = 15\%$ change in demand.

11. Elastic   13. Inelastic

15. Inelastic for $0 \le p < 20$, elastic for $20 < p \le 40$

17. Inelastic for $0 \le p < 33\frac{1}{3}$, elastic for $33\frac{1}{3} < p \le 10$

19. At $x = 4$, $|E(4)| = 0.17 < 1$, so the elasticity of cost is inelastic and a 10% increase in production produces a $(10\%)(0.17) = 1.7\%$ change in cost. At $x = 10$, $|E(10)| = 0.52 < 1$, so the elasticity of cost is inelastic and a 10% increase in production produces a $(10\%)(0.52) = 5.2\%$ change in cost. At $x = 20$, $|E(20)| = 1$, so there is unit elasticity and a 10% increase in production produces a 10% change in cost.

21. $R = p[100(100 - p)] = 10{,}000p - 100p^2$; increases and is inelastic over $(0, 50)$; decreases and is elastic over $(50, 100)$

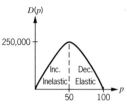

23. $R = 1{,}000{,}000p - 20{,}000p^2 + 100p^3$; increases and is inelastic over $\left(0, \dfrac{100}{3}\right)$; decreases and is elastic over $\left(\dfrac{100}{3}, 100\right)$

25. Critical point, $(0, 0)$, $(3, -9)$; relative minimum, $(3, -9)$; inflection points, $(0, 0)$, $\left(2, -\dfrac{16}{3}\right)$; decreasing, $(-\infty, 0)$, $(0, 3)$; increasing, $(3, \infty)$; concave up, $(-\infty, 0)$, $(2, \infty)$; concave down, $(0, 2)$; no horizontal or vertical asymptote

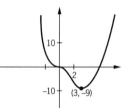

27. Critical points, $\left(\dfrac{2}{3}, \dfrac{64}{27}\right)$, $(2, 0)$; relative minimum, $(2, 0)$; relative maximum, $\left(\dfrac{2}{3}, \dfrac{64}{27}\right)$; inflection point, $\left(\dfrac{4}{3}, \dfrac{32}{27}\right)$; decreasing, $\left(\dfrac{2}{3}, 2\right)$; increasing, $\left(-\infty, \dfrac{2}{3}\right)$, $(2, \infty)$; concave up, $\left(\dfrac{4}{3}, \infty\right)$; concave down, $\left(-\infty, \dfrac{4}{3}\right)$; no horizontal or vertical asymptotes

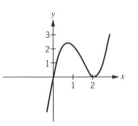

Chapter 3 Test, page 213

1. (a) (b, c), (d,e) (b) (a, b), (c, d) (e, f)
   (c) $x = c$, $x = e$
2. (a) $x = b$, $x = d$ (b) $x = b$
   (c) $x = a$
3. (a) $x = g$, $x = h$ (b) (a, g), (h, e) (e, f)
   (c) (g, h)
4. (a) (a, g), (h, f) (b) (g, h) (c) $x = e$
5. (a) Increasing, $(-\infty, 1)$; decreasing, $(1, \infty)$
   (b) Relative max. at $(1, 2)$
   (c)

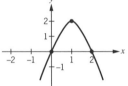

6. (a) Concave up, $(-\infty, 0)$, $(\frac{1}{2}, \infty)$; concave down, $(0, \frac{1}{2})$
   (b) Inflection points, $(0, 0)$, $(\frac{1}{2}, -\frac{1}{16})$
   (c) Relative min. at $(\frac{3}{4}, \frac{-27}{256})$
   (d)

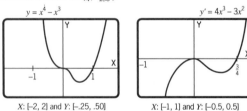

X: [-2, 2] and Y: [-.25, .50]   X: [-1, 1] and Y: [-0.5, 0.5]

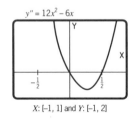

X: [-1, 1] and Y: [-1, 2]

7. (a) Increasing, $(-\infty, -1)$, $(1, \infty)$; decreasing, $(-1, 0)$, $(0, 1)$
   (b) Concave up $(0, \infty)$; concave down, $(-\infty, 0)$
   (c) Relative min. at $(1, 2)$; rel. max. at $(-1, -2)$

(d) Vertical asymptote, $x = 0$
(e)

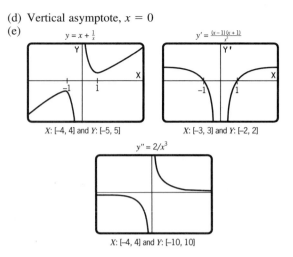

X: [-4, 4] and Y: [-5, 5]   X: [-3, 3] and Y: [-2, 2]

X: [-4, 4] and Y: [-10, 10]

8. Order 100 radios, 50 times per year.
9. 300 cars
10. $|E(2)| = \frac{1}{2} < 1$; therefore, the demand is inelastic and a 5% price increase will cause a 2.5% change in demand.

Exercise Set 4.1, page 222

1. $\dfrac{dy}{dx} = 4e^{4x}$

3. $\dfrac{dy}{dx} = 2e^{2x}$

5. $\dfrac{dy}{dx} = 4 + 3e^{-3x}$

7. $\dfrac{dy}{dx} = 4 - 3e^{3x}$

9. $\dfrac{dy}{dx} = 2xe^{x^2}$

11. $\dfrac{dy}{dx} = \dfrac{e^x}{2\sqrt{e^x + 4}}$

13. $\dfrac{dy}{dx} = 3e^{-x} + 2e^x$

15. $\dfrac{dy}{dx} = 12x^2e^{4x} + 6xe^{4x}$

17. $\dfrac{dy}{dx} = 3e^{3x} - e^x$

19. $\dfrac{dy}{dx} = \dfrac{2e^{2x} + e^x}{2\sqrt{e^{2x} + e^x}}$

21. $f''(x) = xe^{x-1} + 2e^{x-1}$
23. $f''(x) = e^{x^2+x}(4x^2 + 4x + 3)$

25. $y = 3x - 1$

27. $\dfrac{dy}{dx} = 3e^{3x} + 5e^x$

29. $\dfrac{dy}{dx} = 2xe^{2x} + e^{2x} + 4xe^{2x^2}$

31. $\dfrac{dy}{dx} = 3e^{3x} + 2xe^{x^2}$

33. (a) $\dfrac{dV}{dt} = -36,193.50$ (b) $\dfrac{dV}{dt} = -29,632.73$
    (c) $\dfrac{dV}{dt} = -14,715.18$

35. (a) $\dfrac{dS}{dt} = 7.2$ (b) $\dfrac{dS}{dt} = 2.2$ (c) $\dfrac{dS}{dt} = 0.3$

37.

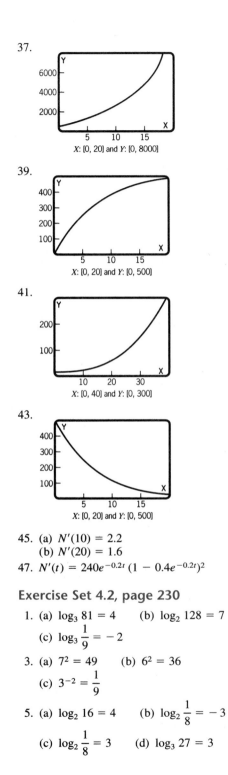

X: [0, 20] and Y: [0, 8000]

39.

X: [0, 20] and Y: [0, 500]

41.

X: [0, 40] and Y: [0, 300]

43.

X: [0, 20] and Y: [0, 500]

45. (a) $N'(10) = 2.2$
(b) $N'(20) = 1.6$

47. $N'(t) = 240e^{-0.2t}(1 - 0.4e^{-0.2t})^2$

Exercise Set 4.2, page 230

1. (a) $\log_3 81 = 4$ (b) $\log_2 128 = 7$
   (c) $\log_3 \dfrac{1}{9} = -2$

3. (a) $7^2 = 49$ (b) $6^2 = 36$
   (c) $3^{-2} = \dfrac{1}{9}$

5. (a) $\log_2 16 = 4$ (b) $\log_2 \dfrac{1}{8} = -3$
   (c) $\log_2 \dfrac{1}{8} = 3$ (d) $\log_3 27 = 3$

7. (a) $b = 4$ (b) $b = \dfrac{1}{2}$

9. (a) $x = 2$ (b) $x = 1000$

11. $2 \ln x + 4 \ln z - \ln y$

13. $\dfrac{2}{3} \log_b x - \dfrac{1}{3} \log_b y$

15. $\ln x + \dfrac{1}{2} \ln y - \dfrac{1}{2} \ln z$

17. 0.73916   19. 0.74055
21. 5.98348   23. $x = 0$
25. $x = 1.125$ 27. $x = e$ and $-e$
29. $x = -400$ 31. $x = 1$
33. $x = 1.9769$ 35. $x = 0.6442$
37. $x = -0.1567$ 39. 2002
41. 8.7 years
43. (a) $RS = 6$ (b) $I = 50,118,723 I_0$

45. $\dfrac{dy}{dx} = 15x^2 e^{5x^3}$

Exercise Set 4.3, page 235

1. $\dfrac{dy}{dx} = \dfrac{3}{x}$ 3. $\dfrac{dy}{dx} = \dfrac{2}{2x + 3}$

5. $\dfrac{dy}{dx} = \dfrac{12}{3x + 2}$ 7. $\dfrac{dy}{dx} = \dfrac{12x + 4}{3x^2 + 2x + 5}$

9. $\dfrac{dy}{dx} = 6x + 2 + \dfrac{2}{2x + 5}$

11. $\dfrac{dy}{dx} = \dfrac{9x^2}{3x + 2} + 6x \ln (3x + 2)$

13. $\dfrac{dy}{dx} = \dfrac{3}{x} (\ln x)^2$

15. $\dfrac{dy}{dx} = \dfrac{60x^3}{x^2 + 1} + 60x \ln (x^2 + 1)$

17. $\dfrac{dy}{dx} = \dfrac{14x + 1}{4x^2 + 2x}$

19. $\dfrac{dy}{dx} = \dfrac{(12x - 4) \ln (3x^2 - 2x + 5)}{3x^2 - 2x + 5}$

21. 15 items
23. 10,000 pounds

25. $\dfrac{dR}{dt} = \dfrac{0.67}{E \ln 10}$

27. $pH' = \dfrac{-1}{H^+ \ln 10}$. As H^+ increases, $1/H^+$ decreases as does $\ln (1/H^+)$, thus, pH decreases.

29. $x = 5^3 = 125$

31. $\dfrac{dy}{dx} = \dfrac{x + 3}{xe^{3/x}}$

33. $\dfrac{dy}{dx} = \dfrac{12e^{-3x} - 12e^{2x} - 10e^{-x}}{(3 + e^{-3x})^2}$

Exercise Set 4.4, page 240

1.
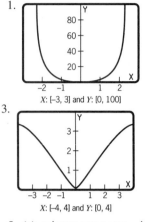

X: [-3, 3] and Y: [0, 100]

3.
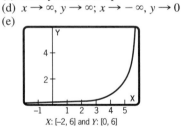

X: [-4, 4] and Y: [0, 4]

5. (a) x-intercept: none; y-intercept: $y = 1/e^4$
   (b) Domain: all real x (c) Always increasing
   (d) No relative extrema   (e) Always concave up
   (f) As $x \to \infty$, $y \to \infty$; as $x \to \infty$, $y \to 0$

7. (a) x-intercept: $x = 5$; y-intercept: none
   (b) Domain: $x > 4$ (c) Always increasing
   (d) No relative extrema   (e) Concave down on $(4, \infty)$
   (f) As $x \to \infty$, $y \to \infty$; as $x \to 4^+$, $y \to -\infty$

9. (a) x-intercept: none; y-intercept: $y = 1/e^4$
   (b) Always increasing; no relative extrema
   (c) Always concave up
   (d) $x \to \infty$, $y \to \infty$; $x \to -\infty$, $y \to 0$
   (e)

X: [-2, 6] and Y: [0, 6]

11. (a) x-intercept: $x = 5$; y-intercept: none
    (b) Always increasing $(4, \infty)$; no relative extrema
    (c) Concave down $(4, \infty)$
    (d) $x \to \infty$, $y \to \infty$; $x \to -\infty$, the function is not defined
    (e)
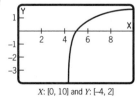

X: [0, 10] and Y: [-4, 2]

13. (a) x-intercept: $x = 0$; y-intercept: $y = 0$
    (b) Increasing for $x < 1$; decreasing for $x > 1$; at $x = 1$ is a relative maximum.
    (c) Concave down for $x < 2$; concave up for $x > 2$
    (d) As $x \to \infty$, $y \to 0$; as $x \to -\infty$, $y \to -\infty$
    (e)
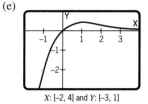

X: [-2, 4] and Y: [-3, 1]

15. (a) x-intercept: $x = 1$; y-intercept: none
    (b) Increasing for $x > 1/e^{1/2}$; decreasing for $x < 1/e^{1/2}$; there is a relative minimum at $x = 1/e^{1/2}$.
    (c) Concave down for $x < 1/e^{3/2}$; concave up for $x > 1/e^{3/2}$
    (d) As $x \to \infty$, $y \to \infty$; as $x \to 0$, $y \to 0$
    (e)
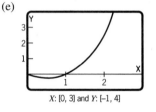

X: [0, 3] and Y: [-1, 4]

17. Relative minimum at $x = 0$; relative maximum at $x = \pm 1$

19.
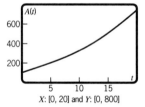

X: [0, 20] and Y: [0, 800]

21. (a) Relative maximum at $x = 0$; there are inflection points at $x = \pm 1$.

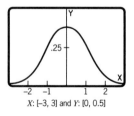

X: [-3, 3] and Y: [0, 0.5]

23. (a) Always increasing, $t > 0$
    (b) Always concave down, $t > 0$

(c) $N \to 1000$

(d)

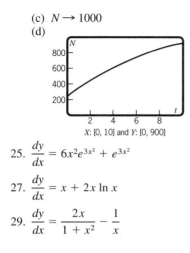

X: [0, 10] and Y: [0, 900]

25. $\dfrac{dy}{dx} = 6x^2 e^{3x^2} + e^{3x^2}$

27. $\dfrac{dy}{dx} = x + 2x \ln x$

29. $\dfrac{dy}{dx} = \dfrac{2x}{1 + x^2} - \dfrac{1}{x}$

Exercise Set 4.5, page 249

1. $\dfrac{dy}{dx} = 8x - 3$

3. $\dfrac{dy}{dx} = \dfrac{-1 - 6x}{2}$

5. $\dfrac{dy}{dx} = \dfrac{3 - 8x}{7}$

7. $\dfrac{dy}{dx} = \dfrac{3 - 2x}{5}$

9. $\dfrac{dy}{dx} = \dfrac{1 + 2x - 9x^2}{2}$

11. $\dfrac{dy}{dx} = \dfrac{8x - 3y}{3x}$

13. $\dfrac{dy}{dx} = \dfrac{3 - 4xy}{2x^2}$

15. $\dfrac{dy}{dx} = \dfrac{-3y}{1 + 3x}$

17. $\dfrac{dy}{dx} = \dfrac{-3 - 4xy}{1 + 2x^2}$

19. $\dfrac{dy}{dx} = \dfrac{2 + 6xy}{1 - 3x^2}$

21. $\dfrac{dy}{dx} = \dfrac{3}{2y + 1}$

23. $\dfrac{dy}{dx} = \dfrac{-8x}{6y - 1}$

25. $\dfrac{dy}{dx} = \dfrac{4 - y}{2y + x}$

27. $\dfrac{dy}{dx} = \dfrac{3y^2}{2 - 6xy}$

29. $\dfrac{dy}{dx} = \dfrac{3y - 2y^2 - 3x^2}{4xy - 3x}$

31. $y = -\dfrac{2}{3}x + \dfrac{5}{3}$

33. $y = -\dfrac{1}{7}x + \dfrac{9}{7}$

35. $y = \dfrac{3}{2}x + 5$

37. $\dfrac{dy}{dx} = \dfrac{-1}{2y - 5}\bigg|_{(2, 3)} = -1$

39. $\dfrac{dy}{dx} = \dfrac{-1 - 4x - 2y}{2x + y}\bigg|_{(-1, 3)} = -3$

41. $\dfrac{dy}{dx} = -\dfrac{y}{x}\bigg|_{(3, 2)} = -\dfrac{2}{3}$

43. $\dfrac{dy}{dx} = (10^{3x+4})(3 \ln 10)$

45. $\dfrac{dy}{dx} = \left(\dfrac{\sqrt{x + 1}}{x^2}\right) \left(\dfrac{1}{2x + 2} - \dfrac{2}{x}\right)$

47. $\dfrac{dy}{dx} = \left(\dfrac{3x^2(2x + 3)^{1/3}}{(4x^3 - 3x + 1)^2}\right) \left(\dfrac{2}{x} + \dfrac{2}{6x + 9} - \dfrac{24x^2 - 6}{4x^3 - 3x + 1}\right)$

49. $y'' = \dfrac{2y' - 6x - 6y(y')^2}{3y^2 - x}$

51. $y = \pm\sqrt{\dfrac{5x^2 - 4}{4}}$; $y' = \dfrac{5x}{4y}$, $y'(2, 2) = \dfrac{5}{4}$, $y'(-2, 2) = -\dfrac{5}{4}$, $y'(2, -2) = -\dfrac{5}{4}$, $y'(-2, -2) = \dfrac{5}{4}$; yes, there is agreement.

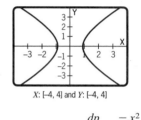

X: [-4, 4] and Y: [-4, 4]

53. (c)

55. $\dfrac{dp}{dx} = \dfrac{-x^2}{p^2}$

57. $p' = -\dfrac{1200}{1331}$

59. $\dfrac{dy}{dx} = \dfrac{4x}{x^2 + 3} + \dfrac{5}{10x + 2}$

Exercise Set 4.6, page 255

1. $\dfrac{dy}{dt} = -3$

3. $\dfrac{dy}{dt} = \dfrac{9}{2}$

5. $\dfrac{dy}{dt} = \dfrac{3}{\sqrt{2}}$

7. $\dfrac{dy}{dt} = \dfrac{9}{2\sqrt{2}}$

9. $\dfrac{dy}{dt} = 12$

11. $\dfrac{dy}{dt} = -3$

13. $\dfrac{dy}{dt} = -\dfrac{9}{2}$

15. $\dfrac{dy}{dt} = -\dfrac{5}{3}$

17. $\dfrac{dy}{dt} = 15$

19. $\dfrac{dy}{dt} = -\dfrac{14}{5}$

21. (a) $y'(2) = \sqrt{2}$ cm/sec
    (b) $y'(9) = 2/3$ cm/sec

23. (a) $A'(1) = 16$ cm²/sec
    (b) $A'(2) = 32$ cm²/sec

25. $\dfrac{dA}{dt} = 72$ cm²/sec

27. $\dfrac{dp}{dt} = -\dfrac{9}{5}$

29. $\dfrac{dR}{dt} = -\dfrac{5}{98}$

31. $\dfrac{dy}{dx} = -3$

33. $\dfrac{dy}{dx} = -\dfrac{2}{3}$

35. $\dfrac{dy}{dx} = \dfrac{-2y^2}{4xy + 1}$

37. $\dfrac{dy}{dx} = \dfrac{ye^x + 1}{1 - e^x}$

39. $\dfrac{dy}{dx} = \dfrac{2x}{x^2 + 2}$

13. Relative maximum, $(-2, 0.54)$; relative minimum, $(0, 0)$

14. Increasing for $x > 0$ and $x < -2$; decreasing for $-2 < x < 0$

15. Concave up for $x < -2 - \sqrt{2}$ and $x > -2 + \sqrt{2}$; concave down for $-2 - \sqrt{2} < x < -2 + \sqrt{2}$

Chapter 4 Test, page 256

1. $\dfrac{dy}{dx} = \dfrac{8x - 3y}{3x}$

2. $\dfrac{dy}{dt} = -\dfrac{16}{3}$

3. $\dfrac{dw}{dz} = 8ze^{z^2 + 5}$

4. $\dfrac{dr}{ds} = \dfrac{12s + 24}{s^2 + 4s}$

5. $\dfrac{dy}{dx} = 2 + \ln x^2$

6. (a) Maximum: approximately $\dfrac{1}{3}$

   (b) Maximum: $\dfrac{1}{e}$

7. $y' = \dfrac{e^x}{1 + e^x}$

8. (a) (i) Increasing for $x < 1$; decreasing for $x > 1$
   (ii) Concave up for $x > 2$; concave down for $x < 2$
   (iii)

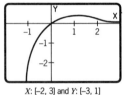

X: [-2, 3] and Y: [-3, 1]

   (b) (i) Increasing for $x < 1$; decreasing for $x > 1$
   (ii) Concave up for $x > 2$; concave down for $x < 2$
   (iii) Same as in part (a)

9. $y' = \dfrac{x + 2x \ln x + e^y}{1 + xe^y}$

10. 1193.46

11. $y' = \dfrac{[1/(x + 1)] - 2xe^y}{x^2 e^y + (1/y)}$

12.

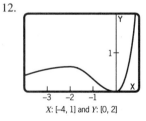

X: [-4, 1] and Y: [0, 2]

Exercise set 5.1, page 269

1. $5x + C$

3. $\sqrt{3}x + C$

5. $-x^2 + C$

7. $\dfrac{2\sqrt{2}}{3} x^{3/2} + C$

9. $2x^3 + C$

11. $\dfrac{3}{4} x^4 + C$

13. $-\dfrac{6}{x} + C$

15. $8u^2 + C$

17. $\dfrac{8}{3} t^{3/2} + C$

19. $\dfrac{2}{3} x^{3/2} + C$

21. $-\dfrac{1}{2x^2} + C$

23. $-\dfrac{3}{2x^2} + C$

25. $-\dfrac{1}{x} + C$

27. $-\dfrac{1}{x} - x^2 + C$

29. $8\sqrt{u} - \dfrac{2}{3} u^{3/2} + C$

31. $\dfrac{-4x^{-3}}{3} + x + \dfrac{x^2}{2} + C$

33. $f(x) = 2x + \dfrac{3}{2} x^2 + 10$

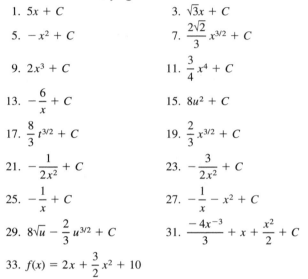

X: [-5, 5] and Y: [0, 30]

35. $y(u) = \dfrac{-10}{u} + 4$

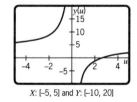

X: [-5, 5] and Y: [-10, 20]

37. $C(x) = 0.1x^2 + \dfrac{4}{3}x^3 + 10$

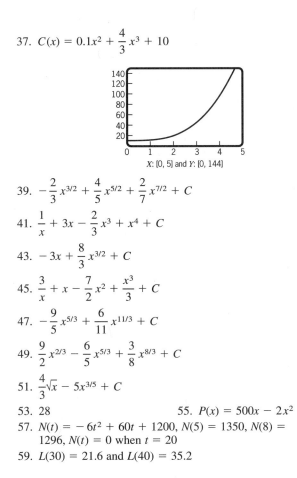

X: [0, 5] and Y: [0, 144]

39. $-\dfrac{2}{3}x^{3/2} + \dfrac{4}{5}x^{5/2} + \dfrac{2}{7}x^{7/2} + C$

41. $\dfrac{1}{x} + 3x - \dfrac{2}{3}x^3 + x^4 + C$

43. $-3x + \dfrac{8}{3}x^{3/2} + C$

45. $\dfrac{3}{x} + x - \dfrac{7}{2}x^2 + \dfrac{x^3}{3} + C$

47. $-\dfrac{9}{5}x^{5/3} + \dfrac{6}{11}x^{11/3} + C$

49. $\dfrac{9}{2}x^{2/3} - \dfrac{6}{5}x^{5/3} + \dfrac{3}{8}x^{8/3} + C$

51. $\dfrac{4}{3}\sqrt{x} - 5x^{3/5} + C$

53. 28

55. $P(x) = 500x - 2x^2$

57. $N(t) = -6t^2 + 60t + 1200$, $N(5) = 1350$, $N(8) = 1296$, $N(t) = 0$ when $t = 20$

59. $L(30) = 21.6$ and $L(40) = 35.2$

Exercise Set 5.2, page 279

1. $6 \ln|x| + C$

3. $4e^x + C$

5. $6e^x + x^2 + C$

7. $e^x + 2 \ln|x| + C$

9. $3 \ln|x| + C$

11. $\dfrac{1}{2}e^{2t} + \dfrac{1}{t} + C$

13. $\dfrac{1}{10e^{10x}} + 2 \ln|x| + C$

15. $\dfrac{-150}{e^{0.02t}} + C$

17. $\dfrac{-1}{8e^{2x}} - \dfrac{\ln|x|}{2} + C$

19. $-2e^x + \dfrac{e^{2x}}{2} + C$

21. $\dfrac{x^2}{2} + \ln|x| + C$

23. $\dfrac{-10}{3e^{0.3x}} - \dfrac{10}{e^{0.1x}} + 2 \ln|x| + C$

25. $\dfrac{1}{2}e^{2x} + x + \dfrac{11}{2}$

27. $x + 3 \ln|x| + 3$

29. 600,000 at $x = 1000$; $R(2000) = 800,000$ is maximum revenue

31. $D(x) = 6000e^{-0.05x} + 6360.82$

33. $R(x) = 4000x - \dfrac{5}{2}x^2$

35. (a) $P'(x) = 1000 - x$

(b) $P(x) = 1000x - \dfrac{x^2}{2} - 5000$, $P(100) = 90,000$

(c) Yes   (d) $P(10) = 4950$

37. $\dfrac{10}{3}x^{3/2} + C$

39. $2\sqrt{x} - \dfrac{2}{5}x^{5/2} + C$

Exercise Set 5.3, page 286

1. $dy = 0.2$, $\Delta y = 0.2$

3. $dy = 0.4$, $\Delta y = 0.41$

5. $dy = 0.075$, $\Delta y = 0.0761905$

7. $dy = 0.02$, $\Delta y = 0.02$

9. $dy = 0.04$, $\Delta y = 0.0401$

11. $dy = 0.0075$, $\Delta y = 0.00751244$

13. $du = (2t + 4)\,dt$

15. $du = 0.3e^{0.3x}\,dx$

17. $dv = -\dfrac{1}{(x-1)^2}\,dx$

19. $dp = \dfrac{2t}{(1+t)^2}\,dt$

21. $du = 0.6xe^{0.3x^2 + 6}\,dx$

23. $dv = -\dfrac{x}{(x^2 - 1)^{3/2}}\,dx$

25. $dp = \dfrac{t}{1 + t^2}\,dt$

27. $9x - 3x^2 + \dfrac{1}{3}x^3 + C$ or $\dfrac{(x-3)^3}{3} + C$

29. $9x - 6x^2 + \dfrac{4}{3}x^3 + C$ or $\dfrac{(2x-3)^3}{6} + C$

31. $4x + 6x^2 + 3x^3 + C$ or $\dfrac{(3x+2)^3}{9} + C$

33. $\dfrac{1}{5}e^{5x} + C$

35. $\dfrac{1}{3}\ln|3x + 2| + C$

37. $-7x + \dfrac{5}{2}x^2 + x^3 + C$

Exercise Set 5.4, page 294

1. $4(x + 3)^{3/2} + C$

3. $-\dfrac{4}{3}(4 - 3x)^{3/2} + C$

5. $\dfrac{1}{3}e^{3x^2} + C$

7. $\dfrac{-3}{10e^{5x^2}} + C$

9. $\dfrac{3}{28}(3 + 7x)^{4/3} + C$

11. $\dfrac{(2x^2 - 7)^{7/4}}{7} + C$

13. $\dfrac{(x^2 + 4)^{3/2}}{3} + C$

15. $\dfrac{2(1 + x^3)^{3/2}}{9} + C$

17. $x + 2 \ln|x - 1| + C$

19. $2x + 6 \ln|3 - x| + C$

21. $\dfrac{(x + 1)^6}{6} - \dfrac{(x + 1)^5}{5} + C$

23. $\dfrac{4}{5}(x - 5)^{5/2} + \dfrac{20}{3}(x - 5)^{3/2} + C$

25. $\dfrac{1}{2}\ln(1 + e^{2x}) + C$ 27. $-\dfrac{x}{3} - \dfrac{\ln|1 - 3x|}{9} + C$

29. $3(x + 1)^{4/3} - \dfrac{3}{7}(x + 1)^{7/3} + C$

31. $2e^{\sqrt{x-1}} + C$ 33. $\dfrac{1}{3}(\sqrt{x} + 3)^6 + C$

35. Because $1/(3x + 4)^2$ can be written as $(3x + 4)^{-2}$, the power rule can be used if $n \neq -1$; however, for $1/(3x + 4)$, $n = -1$ and the power rule does not apply.

37. $C(10) = 1900$

39. $R(x) = (x^2 + 16)^{1/2} - 4$; $R(3) = 1$

41. $N(t) = -80e^{0.02t} + 160$ 43. $-\dfrac{1}{4}e^{-4x} + C$

45. $\ln|x + 1| + C$

Exercise Set 5.5, page 305

1. 2   3. 7

5. 12   7. 30

9. $\dfrac{e^8 - 1}{2}$ 11. 31.5

13. $\dfrac{8}{3}$ 15. 3.159

17. 3.810   19. 1.39

21. 0.25   23. 0.5728

25. 0.6095   27. 4.20

29. 218.8   31. 10

33. 36   35. -18

37. 193 hours   39. 5500

41. \$2000; \$4000; \$2000   43. (a) -0.9 (b) -3

45. 60   47. $\dfrac{3}{8}(2x + 3)^{4/3} + C$

49. $\dfrac{2(x - 1)^{5/2}}{5} + \dfrac{2(x - 1)^{3/2}}{3} + C$

Exercise Set 5.6, page 318

1. 15   3. 12

5. 4   7. $\dfrac{14}{3}$

9. $\dfrac{76}{3}$ 11. 26.7991

13. 0.202733   15. $\dfrac{31}{2}$

17. $\dfrac{4}{3}$ 19. 18

21. $166\frac{2}{3}$ 23. 4.25

25. $9\frac{1}{3}$ 27. $\dfrac{4}{3}$

29. 4.5   31. 37/12

33. $21\frac{1}{3}$ 35. $49\frac{1}{3}$

37. $22\frac{2}{3}$ 39. 241.54

41. $\dfrac{-1}{3e^{x^3}} + C$ 43. $\dfrac{1}{10}$

Exercise Set 5.7, page 328

1. 780   3. 11.5

5. $\Delta x = \dfrac{1 - 0}{10} = .1$; $c_1 = .05$, $c_2 = .15$, $c_3 = .25 \ldots$
    $c_{10} = .95$; 0.3325

7. 3   9. 16

11. 10   13. 5.1875

15. 53.044

17. Answers will vary. Select $x = 1, 2, 3, 4, 5$; $F(1) = .667$, $F(2) = 1.886$; $F(3) = 3.464$; $F(4) = 5.333$; $F(5) = 7.453$. Graph is same as the graph of $f(x) = \dfrac{2}{3}x^{3/2}$

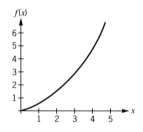

19. Answers will vary. Select $x = 1, 2, 3, 4, 5$; $F(1) = -.75$, $F(2) = 0$, $F(3) = 11.25$, $F(4) = 48$, $F(5) = 131.25$. The graph is the same as the graph of $f(x) = \dfrac{x^4}{4} - x^2$ for $x \geq 0$.

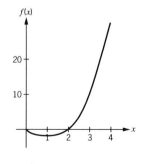

21. (c)

23. 75,875.80

25. $-6\sqrt{x} + \dfrac{2}{3} x^{3/2} + \dfrac{4}{5} x^{5/2} + C$

27. $\ln(1 + e^x) + C$

29. $\dfrac{10}{3} (x + 2)^{3/2} - 20(x + 2)^{1/2} + C$

31. -3.93147

33. 93

Exercise Set 5.8, page 340

1. (a) $\dfrac{1}{3}$ (b) $\dfrac{1}{2}$

3. 15,401.86

5. 16.229

7. 4.51188

9. (a)–(c)

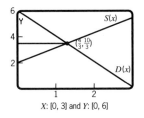

X: [0, 3] and Y: [0, 6]

(d) $CS = \dfrac{16}{9}, PS = \dfrac{8}{9}$

11. (a)–(c)

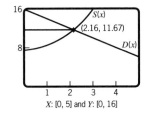

X: [0, 5] and Y: [0, 16]

(d) $CS = 4.70, PS = 6.73$

13. (a)–(c)

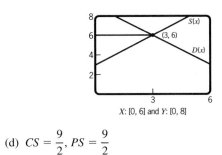

X: [0, 6] and Y: [0, 8]

(d) $CS = \dfrac{9}{2}, PS = \dfrac{9}{2}$

15. (a)–(c)

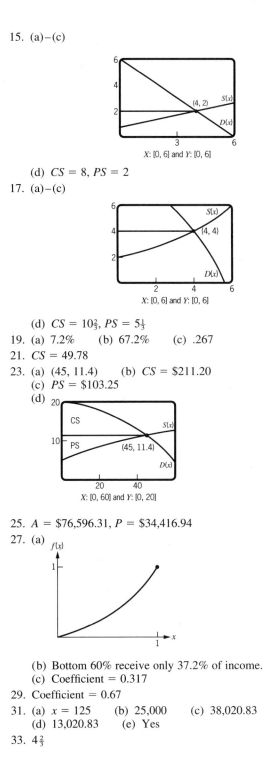

X: [0, 6] and Y: [0, 6]

(d) $CS = 8, PS = 2$

17. (a)–(c)

X: [0, 6] and Y: [0, 6]

(d) $CS = 10\frac{2}{3}, PS = 5\frac{1}{3}$

19. (a) 7.2%   (b) 67.2%   (c) .267

21. $CS = 49.78$

23. (a) (45, 11.4)   (b) $CS = \$211.20$
    (c) $PS = \$103.25$
    (d)

X: [0, 60] and Y: [0, 20]

25. $A = \$76,596.31, P = \$34,416.94$

27. (a)

(b) Bottom 60% receive only 37.2% of income.
(c) Coefficient = 0.317

29. Coefficient = 0.67

31. (a) $x = 125$ (b) 25,000   (c) 38,020.83
    (d) 13,020.83   (e) Yes

33. $4\frac{2}{3}$

Chapter 5 Test, page 342

1. $\dfrac{1}{2}x + \dfrac{4}{3}x^3 + \dfrac{1}{4}x^4 + C$

2. $\dfrac{3}{2}(1 + x)^{2/3} + C$

3. $\dfrac{4}{x} + \dfrac{7}{2x^2} + \dfrac{-2}{x^3} + C$

4. $-\dfrac{1}{3}(1 - x^2)^{3/2} + C$

5. $\dfrac{\ln|4x + 1|}{2} + C$

6. $5e^{4x} - \dfrac{2}{x^2} + C$

7. $2e^{3x^2} + C$

8. $2x + 3\ln|x - 1| + C$

9. $\dfrac{1}{3}(x^2 - 3)^{3/2} + C$

10. $\dfrac{(x + 3)^6}{6} - \dfrac{3(x + 3)^5}{5} + C$

11. 105

12. 6.22

13. $du = 3x(3x^2 - 1)^{-1/2}\,dx$

14. $\dfrac{4}{3}$

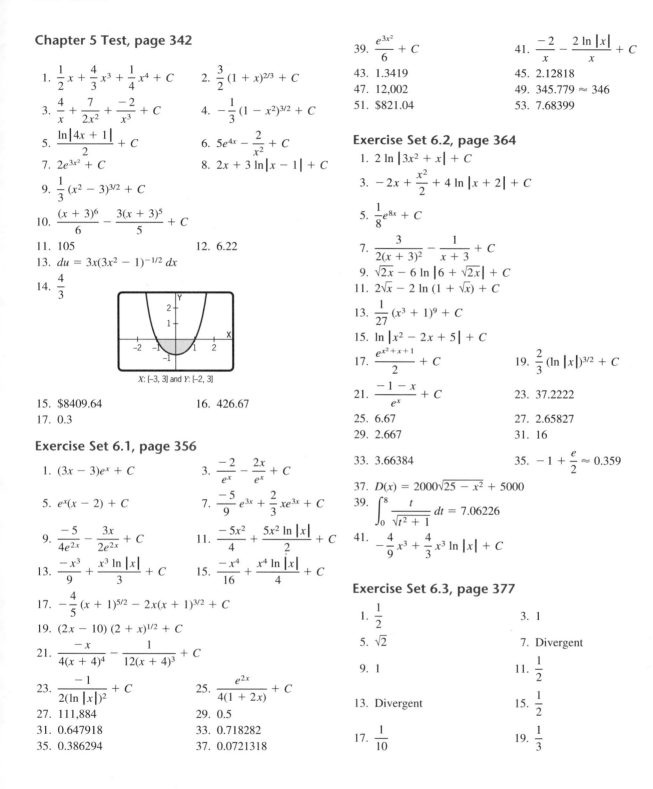

X: [-3, 3] and Y: [-2, 3]

15. $8409.64

16. 426.67

17. 0.3

Exercise Set 6.1, page 356

1. $(3x - 3)e^x + C$

3. $\dfrac{-2}{e^x} - \dfrac{2x}{e^x} + C$

5. $e^x(x - 2) + C$

7. $\dfrac{-5}{9}e^{3x} + \dfrac{2}{3}xe^{3x} + C$

9. $\dfrac{-5}{4e^{2x}} - \dfrac{3x}{2e^{2x}} + C$

11. $\dfrac{-5x^2}{4} + \dfrac{5x^2\ln|x|}{2} + C$

13. $\dfrac{-x^3}{9} + \dfrac{x^3\ln|x|}{3} + C$

15. $\dfrac{-x^4}{16} + \dfrac{x^4\ln|x|}{4} + C$

17. $-\dfrac{4}{5}(x + 1)^{5/2} - 2x(x + 1)^{3/2} + C$

19. $(2x - 10)(2 + x)^{1/2} + C$

21. $\dfrac{-x}{4(x + 4)^4} - \dfrac{1}{12(x + 4)^3} + C$

23. $\dfrac{-1}{2(\ln|x|)^2} + C$

25. $\dfrac{e^{2x}}{4(1 + 2x)} + C$

27. 111,884

29. 0.5

31. 0.647918

33. 0.718282

35. 0.386294

37. 0.0721318

39. $\dfrac{e^{3x^2}}{6} + C$

41. $\dfrac{-2}{x} - \dfrac{2\ln|x|}{x} + C$

43. 1.3419

45. 2.12818

47. 12,002

49. $345.779 \approx 346$

51. $821.04

53. 7.68399

Exercise Set 6.2, page 364

1. $2\ln|3x^2 + x| + C$

3. $-2x + \dfrac{x^2}{2} + 4\ln|x + 2| + C$

5. $\dfrac{1}{8}e^{8x} + C$

7. $\dfrac{3}{2(x + 3)^2} - \dfrac{1}{x + 3} + C$

9. $\sqrt{2x} - 6\ln|6 + \sqrt{2x}| + C$

11. $2\sqrt{x} - 2\ln(1 + \sqrt{x}) + C$

13. $\dfrac{1}{27}(x^3 + 1)^9 + C$

15. $\ln|x^2 - 2x + 5| + C$

17. $\dfrac{e^{x^2 + x + 1}}{2} + C$

19. $\dfrac{2}{3}(\ln|x|)^{3/2} + C$

21. $\dfrac{-1 - x}{e^x} + C$

23. 37.2222

25. 6.67

27. 2.65827

29. 2.667

31. 16

33. 3.66384

35. $-1 + \dfrac{e}{2} \approx 0.359$

37. $D(x) = 2000\sqrt{25 - x^2} + 5000$

39. $\displaystyle\int_0^8 \dfrac{t}{\sqrt{t^2 + 1}}\,dt = 7.06226$

41. $-\dfrac{4}{9}x^3 + \dfrac{4}{3}x^3\ln|x| + C$

Exercise Set 6.3, page 377

1. $\dfrac{1}{2}$

3. 1

5. $\sqrt{2}$

7. Divergent

9. 1

11. $\dfrac{1}{2}$

13. Divergent

15. $\dfrac{1}{2}$

17. $\dfrac{1}{10}$

19. $\dfrac{1}{3}$

21. 0

23. Divergent

25. 1

27. $-\dfrac{1}{2}$

29. Divergent

31. $33,333.33

33. $187,500

35. $166,666.67

37. $\displaystyle\int_0^\infty Re^{-rt}\,dt = \lim_{b\to\infty}\int_0^b Re^{-rt}\,dt = \lim_{b\to\infty} \left. -\dfrac{R}{r}e^{-rt}\right|_0^b$

$= 0 + \dfrac{R}{r} = \dfrac{R}{r}$

39. $\displaystyle\int_{-\infty}^0 0\,dx - \int_0^\infty \dfrac{1}{a}e^{-x/a}\,dx = \lim_{b\to\infty}\int_0^b \dfrac{1}{a}e^{-x/a}\,dx =$

$\lim_{b\to\infty} \left. -e^{-x/a}\right|_0^b = 0 + 1 = 1$

41. $\sigma^2 = \displaystyle\int_a^b x^2\left(\dfrac{1}{b-a}\right)dx - \left(\dfrac{a+b}{2}\right)^2 = \dfrac{(b-a)^2}{12}$

and $\sigma = \dfrac{b-a}{\sqrt{12}}$

43. $\sigma^2 = \displaystyle\int_a^b x^2\left(\dfrac{1}{a}c^{-x/a}\right)dx - a^2 = a^2$ and $\sigma = a$

45. 0.135

47. 0.63212

49. $\dfrac{1}{2}$

51. (a) $\dfrac{1}{2}$ (b) 25%   (c) 220

53. (a) 0.393   (b) 0.123

55. (a) $x = 1$ (b) $x = 1.628$

57. $\dfrac{15}{2}(e^x + 2)^{2/3} + C$

59. $(x - 2)e^{2x} + C$

61. $\dfrac{-x^5}{25} + \dfrac{x^5}{5}\ln|x| + C$

Exercise Set 6.4, page 386

1. $\dfrac{1}{10}\ln\left|\dfrac{x-5}{x+5}\right| + C$

3. $\dfrac{1}{12}\ln\left|\dfrac{6+x}{6-x}\right| + C$

5. $\ln\left|x + \sqrt{x^2 + 36}\right| + C$

7. $-\dfrac{1}{5}\ln\left|\dfrac{5 + \sqrt{25 + x^2}}{x}\right| + C$

9. $-\dfrac{\sqrt{49 + x^2}}{49x} + C$ 11. $\dfrac{\sqrt{x^2 - 49}}{49x} + C$

13. $\dfrac{1}{8}\cdot[x(2x^2 + 25)\sqrt{x^2 + 25} - 625\ln|x + \sqrt{x^2 + 25}|] + C$

15. $\sqrt{25 + x^2} - 5\ln\left|\dfrac{5 + \sqrt{25 + x^2}}{x}\right| + C$

17. $\dfrac{1}{2}[x\sqrt{x^2 + 49} - 49\ln|x + \sqrt{x^2 + 49}|] + C$

19. $\dfrac{1}{5}\ln\left|\dfrac{x-2}{x+3}\right| + C$

21. $2x^2 e^{x/2} - 8\ xe^{x/2} + 16e^{x/2} + C$

23. $\dfrac{1}{3}(x - 3)\sqrt{2x + 3} + C$

25. 0.52159

27. $\dfrac{1}{5}\ln|5x + \sqrt{25x^2 + 16}| + C$

29. $-\dfrac{8}{5}\ln\left|\dfrac{5 + \sqrt{25 - 4x^2}}{2x}\right| + C$

31. $\dfrac{1}{72}[3x(18x^2 + 16)\sqrt{9x^2 + 16} -$

$256\ln|3x + \sqrt{9x^2 + 16}|] + C$

33. $-\dfrac{1}{3}\dfrac{\sqrt{25x^2 + 16}}{x} + \dfrac{5}{3}\ln|5x + \sqrt{25x^2 + 16}| + C$

35. $\dfrac{1}{13}\ln\left|\dfrac{2x-3}{3x+2}\right| + C$

37. $PS = 135.21$

39. (a) $2051.63   (b) $2385.09   (c) $10,239.50

41. 150.85

43. 4.2426

45. 1.5

Exercise Set 6.5, page 392

1. Trapezoidal rule: 3.34375; numeric: 3.333

3. Trapezoidal rule: 1.21819; numeric: 1.21895

5. Simpson's rule: 3.33333; numeric: 3.333

7. Simpson's rule: 1.21895; numeric: 1.21895

9. 0.001302   11. 8700

13. 933.08

15. (a) Divergent   (b) Divergent

Chapter 6 Test, page 393

1. Divergent

2. $\dfrac{2}{15}(3x - 4)(x + 2)^{3/2} + C$

3. $\dfrac{1}{60}\ln\left|\dfrac{5x - 6}{5x + 6}\right| + C$

4. 1304.23

5. Simpson's rule: 4; numeric: 4

6. 21.5037

7. Trapezoidal rule: 22; numeric: 21.333

8. $e^x(2 - 2x + x^2) + C$

9. 12.66667

10. pdf $= \dfrac{1}{7}e^{-x/7}$, 0.681

Exercise Set 7.1, page 402

1. 1   3. 3

5. 9   7. 7

9. 7   11. 6

13.

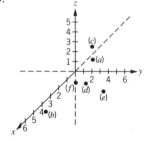

15. $3x + 7y = 14$, $(0, 2, 0)$, $\left(\dfrac{14}{3}, 0, 0\right)$; $3x + 2z = 14$,

$(0, 0, 7)$, $\left(\dfrac{14}{3}, 0, 0\right)$; $7y + 2z = 14$, $(0, 2, 0)$, $(0, 0, 7)$

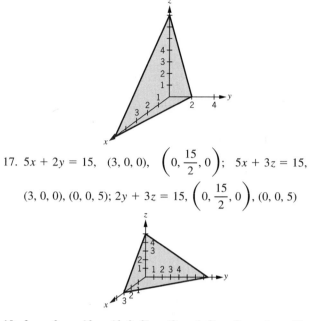

17. $5x + 2y = 15$, $(3, 0, 0)$, $\left(0, \dfrac{15}{2}, 0\right)$; $5x + 3z = 15$,

$(3, 0, 0)$, $(0, 0, 5)$; $2y + 3z = 15$, $\left(0, \dfrac{15}{2}, 0\right)$, $(0, 0, 5)$

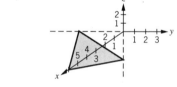

19. $2x - 3y = 12$, $(6, 0, 0)$, $(0, -4, 0)$; $2x - 4z = 12$,
$(6, 0, 0)$, $(0, 0, -3)$; $-3y - 4z = 12$, $(0, -4, 0)$,
$(0, 0, -3)$

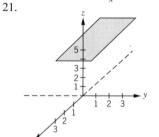

21.

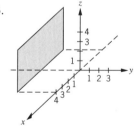

23.

25.

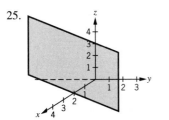

27.

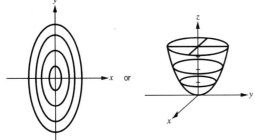

29.

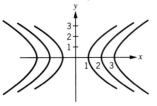

31. (a) $P(3.2) = 44$ (b) $P(4, 1) = 53$
(c) $P(3, 4) = 78$

33. $R(110, 10) = 11$. Ratio is 11 to 1.

35. 1600

37. (a) $\dfrac{6755}{13} = 519.615$ (b) $\dfrac{525}{16} = 32.8125$

Exercise Set 7.2, page 408

1. $f_x = 2$, $f_y = 0$, $f_x(3, 2) = 2$, $f_y(3, 2) = 0$

3. $f_x = 3$, $f_y = -2$, $f_x(3, 2) = 3$, $f_y(3, 2) = -2$

5. $f_x = -6x - 2y$, $f_y = -2x + 3y^2$,
$f_x(3, 2) = -22$, $f_y(3, 2) = 6$

7. $f_x = 3e^x$, $f_y = -2e^y$, $f_x(3, 2) = 60.257$,
$f_y(3, 2) = -14.778$

9. $f_x = \dfrac{3y + 6x}{xy + x^2}$, $f_y = \dfrac{3x}{xy + x^2}$, $f_x(3, 2) = \dfrac{8}{5}$,

$f_y(3, 2) = \dfrac{3}{5}$

11. $z_x = ywe^{xyw}$, $z_y = xwe^{xyw}$, $z_w = xye^{xyw}$

13. $z_x = \dfrac{2x + y}{x^2 + xy + w^2}$,

$z_y = \dfrac{x}{x^2 + xy + w^2}$, $z_w = \dfrac{2w}{x^2 + xy + w^2}$

15. $f_x(1, 1, 2) = 3$, $f_y(1, 1, 2) = 2$

17. $f_x\left(2, \dfrac{1}{2}, 1\right) = \dfrac{1}{2}$, $f_y\left(2, \dfrac{1}{2}, 1\right) = 2$

19. $A_i = nP(1 + i)^{n-1}$

21. (a) 28   (b) 22   (c) 6   (d) 26

23. $P_I(400, 10{,}000) = 2$, $P_C(400, 10{,}000) = 0.08$

25. (a) 130   (b) 76.923   (c) 10
(d) The rate of change of I with respect to M is increasing.

27.

Exercise Set 7.3, page 415

1. (a) $C_x = 30$. Each additional item A costs \$30 to produce.
(b) $C_x = 20$. Each addition item B costs \$20 to produce.
(c) $R_x = 50$. Each additional item A yields \$50 revenue.
(d) $R = 30$. Each additional item B yields \$30 revenue.
(e) $P_x = 20$. Profit for each item A produces a \$20 profit.
(f) $P_y = 10$. Profit for each item B produces a \$10 profit.

3. $R_x (8, 10) = 99.8$ is additional revenue by increasing production of Product A at (8, 10) $R_y (8, 10) = 59.84$; additional revenue by increasing production of Product B at (8, 10).

5. $P_x = 4x - 2y$; additional profit from selling one additional unit of item 1 for given value of x. $P_y = 4y - 2x$; additional profit from selling one additional unit of item 2 for given value of y.

7. Neither   **9.** Complementary

11. Complementary

13. (a) $f_l(l, c) = 45 \left(\dfrac{c^{1/4}}{l^{1/4}}\right)$ (b) $f_c(l, c) = 15 \left(\dfrac{l^{3/4}}{c^{3/4}}\right)$

15. $f_x = 8xy - 3y^3$, $f_y = 4x^2 - 9xy^2$, $f_x(1, 2) = -8$,
$f_y(1, 2) = -32$

17. $f_x = 4y$, $f_y = 4x - 5e^{-y}$, $f_x(1, 2) = 8$,
$f_y(1, 2) = 3.3233$

Exercise Set 7.4, page 419

1. $f_{xx} = 0$, $f_{yy} = 0$, $f_{xy}(1, 3) = 0$

3. $f_{xx} = 20y^3$, $f_{yy} = 60x^2y$, $f_{xy}(1, 3) = 540$

5. $f_{xx} = 10y$, $f_{yy} = -18xy$, $f_{xy}(1, 3) = -71$

7. $f_{xx} = 3e^x$, $f_{yy} = -2e^y$, $f_{xy}(1, 3) = 0$

9. $f_{xx} = 3y^2e^{xy}$, $f_{yy} = 3x^2e^{xy}$, $f_{xy}(1, 3) = 241.0264431$

11. $f_{xx} = \dfrac{18x(x^2 + 2y^2)}{(3x^2 + 2y^2)^2}$, $f_{yy} = \dfrac{4x(3x^2 - 2y^2)}{(3x^2 + 2y^2)^2}$,

$f_{xy}(1, 3) = 0.4081632655$

13. $f_{xx} = e^x \ln(3x + 2y) = \dfrac{6e^x(3x + 2y) - 9e^x}{(3x + 2y)^2}$,

$f_{yy} = \dfrac{-4e^x}{(3x + 2y)^2}$, $f_{xy}(1, 3) = 0.4027084189$

15. $f_{xx} = \dfrac{-1}{(2x + 3y)^{3/2}}$, $f_{yy} = \dfrac{-9}{4(2x + 3y)^{3/2}}$,

$f_{xy}(1, 3) = -0.04111518336$

17. $f_{xx} = 0$, $f_{yy} = \dfrac{2x}{y^3}$, $f_{xy}(1, 3) = -\dfrac{1}{9}$

19. $f_{xx} = e^{x-y}y^2$, $f_{yy} = (e^{x-y})(y^2 - 4y + 2)$,

$f_{xy}(1, 3) = -0.4060058498$

21. $f_{xxx} = 6y$, $f_{yyy} = 6x$, $f_{xyx} = 6x$, $f_{yxy} = 6y$

23. $f_{xxx} = e^{x+y}$, $f_{yyy} = e^{x+y}$, $f_{xyx} = e^{x+y}$, $f_{yxy} = e^{x+y}$

25. $f_{xxx} = \dfrac{2}{(x + y)^3}$, $f_{yyy} = \dfrac{2}{(x + y)^3}$, $f_{xyx} = \dfrac{2}{(x + y)^3}$,

$f_{yxy} = \dfrac{2}{(x + y)^3}$

27. $f_{ll} = \dfrac{-800C^{2/3}}{9l^{5/3}}$, $f_{CC} = \dfrac{-800l^{1/3}}{9C^{4/3}}$, $f_{lC} = \dfrac{800}{9l^{2/3}C^{1/3}}$

29. $f_x = 3y + \dfrac{2x}{x^2 + 2y^2}$, $f_x(1, 2) = \dfrac{56}{9}$,

$f_y = 3x + \dfrac{4y}{x^2 + 2y^2}$, $f_y(1, 2) = \dfrac{35}{9}$

Exercise Set 7.5, page 426

1. $(0, 1)$

3. $(1, 0)$, $(1, 1)$, $(1, -1)$

5. $(-3, -\frac{3}{2})$

7. $(\frac{1}{2}, -1)$, relative minimum

9. $(-1, 0)$, relative minimum

11. $(0, 0)$, relative minimum

13. $(0, 0)$, relative maximum

15. $(0, 0)$, relative maximum

17. $(0, 0)$, saddle point

19. $(2, 0)$, relative maximum

21. $(-2, 3)$, saddle point

23. $(0, 0)$, saddle point

25. $(2, -1)$, relative minimum

27. $(3, 2)$, relative maximum

29. $(0, 0)$, saddle point; $(0, 3)$, saddle point; $(3, 0)$, saddle point; $(1, 1)$ relative maximum

31. $(0, -1)$, relative maximum; $(0, 3)$, saddle point; $(4, -1)$, saddle point; $(4, 3)$, relative minimum

33. If the line segment is divided into three equal parts, the sum of the squares of their length is a minimum.

35. $C(2, 3) = 138$; \$138,000

37. Critical point is $\left(\dfrac{91}{38}, \dfrac{103}{19}\right)$. \$211.68

39. $f_{xx} = \dfrac{-2x^2 + 6y^2}{(x^2 + 3y^2)^2}$, $f_{xy} = \dfrac{-12xy}{(x^2 + 3y^2)^2}$

Exercise Set 7.6, page 435

1. $f(5, 5) = 75$

3. $f\left(3, \dfrac{3}{2}\right) = 18$

5. $f\left(\dfrac{2}{3}, 2\right) = 6$

7. $f\left(\dfrac{68}{3}, \dfrac{34}{3}\right) = \dfrac{157,216}{27}$

9. $f\left(\dfrac{5}{6}, \dfrac{10}{9}\right) = \dfrac{500}{243}$

11. $f(47, 47) = 2209$

13. $f\left(\dfrac{35}{17}, \dfrac{15}{17}\right) = \dfrac{50}{17}$

15. $f\left(\dfrac{146}{3}, \dfrac{146}{3}, \dfrac{146}{3}\right) = \dfrac{3,112,136}{27}$

17. $x = 33,670$, $y = 67,330$, $p = 2,137,700$

19. $x = \dfrac{412}{41}$, $y = \dfrac{510}{41}$, $U(x, y) = \dfrac{5484}{41}$

21. 132.64 by 132.64 by 56.84

23. 75 by 75

25. 14.14 by 14.14 by 7.07

27. Relative minimum: $f(0, 2) = -4$

29. Relative minimum: $f\left(\dfrac{10}{3}, -\dfrac{2}{3}\right) = -\dfrac{28}{3}$

Chapter 7 Test, page 436

1. 10

2.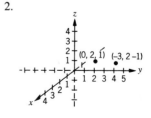

3.

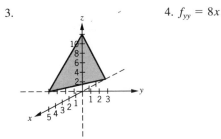

4. $f_{yy} = 8x$

5. $f_{xx} = \dfrac{6}{x^2 + 3y^2} - \dfrac{12x^2}{(x^2 + 3y^2)^2}$

6. $f_y = xe^{xy}$

7. $f_{xy} = 2x + 2y$

8.

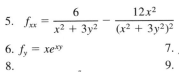

9. The critical point is $(2, -3)$

10. Relative minimum

11.

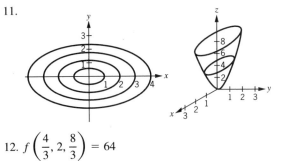

12. $f\left(\dfrac{4}{3}, 2, \dfrac{8}{3}\right) = 64$

Exercise Set 8.1, page 447

1. III

3. I

5. III

7. (a) Negative
   (b) Positive
   (c) Negative

9. (a) Negative
   (b) Negative
   (c) Positive

11. $\dfrac{2\pi}{3}$

13. $\dfrac{9\pi}{20}$

15. $\dfrac{19\pi}{10}$

17. $150°$

19. $34.38°$

21. $-92.819°$

23. $\dfrac{\sqrt{3}}{2}, \dfrac{1}{2}, \sqrt{3}$

25. $0, -1, 0$

27. $-1, 0$, undefined

29. $0.43837, -0.89879, -0.48773$

31. $-0.87158, -0.49026, 1.77778$

33. $0.78183, -0.62349, -1.25396$

35. $\sin \theta = \dfrac{12}{13}$, $\cos \theta = -\dfrac{5}{13}$, $\tan \theta = -\dfrac{12}{5}$, $\csc \theta = \dfrac{13}{12}$, $\sec \theta = -\dfrac{13}{5}$, $\cot \theta = -\dfrac{5}{12}$

37. $\sin \theta = 0$, $\cos \theta = -1$, $\tan \theta = 0$, $\csc \theta$ is undefined, $\sec \theta = -1$, $\cot \theta$ is undefined

39. $\sin \theta = -\dfrac{3}{\sqrt{13}}$, $\cos \theta = -\dfrac{2}{\sqrt{13}}$, $\tan \theta = \dfrac{3}{2}$, $\csc \theta = -\dfrac{\sqrt{13}}{3}$, $\sec \theta = -\dfrac{\sqrt{13}}{2}$, $\cot \theta = \dfrac{2}{3}$

41. $\dfrac{\pi}{3}$

43. $-\dfrac{\pi}{4}$

45. $\dfrac{2\pi}{3}$

47. 1.24507

49. 1.99325

51. -0.775397

53. (a) 300 (b) 293 (c) 241 (d) 152

Exercise Set 8.2, page 451

1. $A = 0.5, P = \pi$

3. $A = 1, P = 4$

5. $P = \pi$

7. $P = 2\pi$

9. $P = 2$

11.

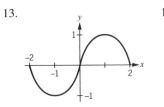

13.

15.

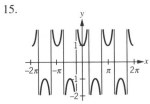

17.

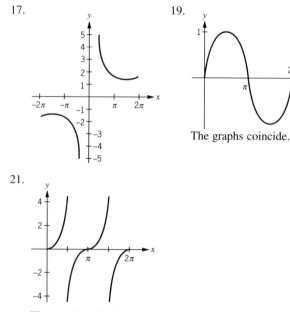

19.

The graphs coincide.

21.

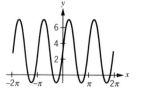

The graphs coincide.

23. $P = \pi$, A $= 4$, maximum $= 7$, minimum $= -1$

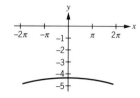

25. $P = \pi$, A $= 0.6$, maximum $= 0.1$, minimum $= -1.1$

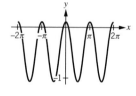

27. $P = 10\pi$, A $= .6$, maximum $= -4.4$,
minimum $= -4.81459$

29. $x = 0.2058$

31. $x = 0.6591$

33. $x = 1.6961$

35. $x = \dfrac{\pi}{4}, \dfrac{3\pi}{4}$

37. $x = \dfrac{\pi}{4}, \dfrac{5\pi}{4}$

39. $x = 0$ or $\dfrac{\pi}{2}$

41.

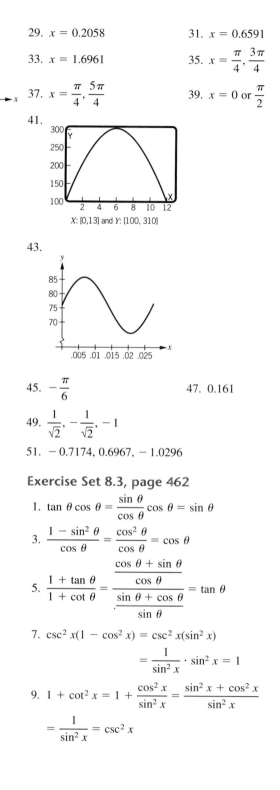

X: [0,13] and Y: [100, 310]

43.

45. $-\dfrac{\pi}{6}$

47. 0.161

49. $\dfrac{1}{\sqrt{2}}, -\dfrac{1}{\sqrt{2}}, -1$

51. $-0.7174, 0.6967, -1.0296$

Exercise Set 8.3, page 462

1. $\tan \theta \cos \theta = \dfrac{\sin \theta}{\cos \theta} \cos \theta = \sin \theta$

3. $\dfrac{1 - \sin^2 \theta}{\cos \theta} = \dfrac{\cos^2 \theta}{\cos \theta} = \cos \theta$

5. $\dfrac{1 + \tan \theta}{1 + \cot \theta} = \dfrac{\dfrac{\cos \theta + \sin \theta}{\cos \theta}}{\dfrac{\sin \theta + \cos \theta}{\sin \theta}} = \tan \theta$

7. $\csc^2 x(1 - \cos^2 x) = \csc^2 x(\sin^2 x)$

$= \dfrac{1}{\sin^2 x} \cdot \sin^2 x = 1$

9. $1 + \cot^2 x = 1 + \dfrac{\cos^2 x}{\sin^2 x} = \dfrac{\sin^2 x + \cos^2 x}{\sin^2 x}$

$= \dfrac{1}{\sin^2 x} = \csc^2 x$

11. $1 - 2 \sin^2 x = 1 - 2 (1 - \cos^2 x)$
$$= 1 - 2 + 2 \cos^2 x$$
$$= 2 \cos^2 x - 1$$

13. $\dfrac{1 - \cos x}{\sin x} = \dfrac{1 - \cos x}{\sin x} \cdot \dfrac{1 + \cos x}{1 + \cos x}$
$$= \dfrac{1 - \cos^2 x}{\sin x (1 + \cos x)} = \dfrac{\sin^2 x}{\sin x(1 + \cos x)} = \dfrac{\sin x}{1 + \cos x}$$

15. $\dfrac{\sec x + 1}{\sec x - 1} = \dfrac{\left(\dfrac{1}{\cos x} + 1\right)}{\left(\dfrac{1}{\cos x} - 1\right)} \cdot \dfrac{\cos x}{\cos x}$
$$= \dfrac{1 + \cos x}{1 - \cos x}$$

17. $\tan(2\pi - \theta) = \dfrac{\tan 2\pi - \tan \theta}{1 + \tan 2\pi \tan \theta}$
$$= \dfrac{0 - \tan \theta}{1 + 0 (\tan \theta)} = - \tan \theta$$

19. $\sin(\alpha - \beta) \sin(\alpha + \beta) = (\sin \alpha \cos \beta - \cos \alpha \sin \beta) \cdot$
$(\sin \alpha \cos \beta + \cos \alpha \sin \beta)$
$= \sin^2 \alpha \cos^2 \beta - \cos^2 \alpha \sin^2 \beta$
$= \sin^2 \alpha (1 - \sin^2 \beta) - (1 - \sin^2 \alpha) \sin^2 \beta$
$= \sin^2 \alpha - \sin^2 \alpha \sin^2 \beta - \sin^2 \beta + \sin^2 \alpha \sin^2 \beta$
$= \sin^2 \alpha - \sin^2 \beta$

21. $\tan(\theta/2) = \dfrac{1 - \cos \theta}{\sin \theta}$
$$= \dfrac{1}{\sin \theta} - \dfrac{\cos \theta}{\sin \theta}$$
$$= \csc \theta - \cot \theta$$

23. $1 - \sqrt{2}$

25. $\dfrac{\sqrt{2 - \sqrt{3}}}{2}$

27. $\sin \left(\dfrac{\pi}{2} - x\right) = \sin \dfrac{\pi}{2} \cdot \cos x - \cos \dfrac{\pi}{2} \sin x$
$$= \cos x$$

29. $-\dfrac{\pi}{6}$

31. Period $= 1$ amplitude $= 2$ maximum $= 2$
minimum $= -2$

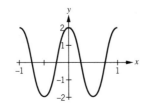

Exercise Set 8.4, page 472

1. $y' = 7 \cos 7x$

3. $y' = 2 \cos \pi x - 2 \pi x \sin \pi x$

5. $y' = 15 \sin 3x - 6 \cos 2x$

7. $y' = -[(1 + \cos(2x)] \cot x \csc x) - 2 \csc x \sin 2x$

9. $y' = 2xe^{\tan x^2} \sec^2 x^2$

11. $y' = \dfrac{3 - 2 \sin 2x}{3x + \cos 2x}$

13. $y' = e^{2x} \sec (1 + 3x)[2 + 3 \tan(1 + 3x)]$

15. $y' = \dfrac{-\csc^2 x}{2\sqrt{\cot x}}$

17. $y' = \dfrac{3x \cos 3x - \sin 3x}{2x^2}$

19. $y' = -5 \sec x \tan x$

21. $y' = -6(\cos x^2 - 2x)^2(1 + x \sin x^2)$

23. $y = 0.5403x + 0.30117$

25. $y = 0.4227x - 0.5556$

27. Horizontal tangent at $(7\pi/6, 5.40)$, $(11\pi/6, 4.03)$

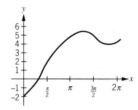

29. Horizontal tangent at $(2\pi/3, 1.913)$, $(4\pi/3, 1.228)$

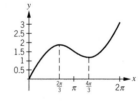

31. $(0, 0)$, $(1.318, 1.452)$, $(3.141, 0.002)$, $(4.965, -1.453)$

33. $(0.132, -1.781)$, $(0.921, -1.087)$, $(1.951, 0.666)$,
$(3.133, 1.797)$, $(3.905, 1.298)$, $(4.967, -0.452)$,
$(6.131, -1.777)$

35. $78,739.82

37. (a) $P'(x) = \dfrac{3\pi \sin (\pi x/26)}{13}$ (b) $P'(10) = 0.67787$,
$P'(32) = -0.48075$. Profit is increasing at 10 and decreasing at 32. (c) Relative maximum at $(26, 12)$

39. 2.617999

41. $T'(6) = -0.833$. Temperature is dropping at a rate of 0.8° per day on day 6.

43. $A = 3$, $P = \pi$, maximum $= 3$, minimum $= -3$

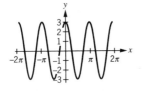

45. Is an identity

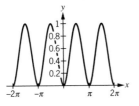

Exercise Set 8.5, page 478

1. $\dfrac{\sin 5x}{5} + C$

3. $-4 \cos \dfrac{x}{4} + C$

5. $\dfrac{-\cos(1 + 3x)}{3} + C$

7. $\dfrac{\tan 3x}{3} + C$

9. $\dfrac{(\sin x)^6}{6} + C$

11. $2\sqrt{\sin x} + C$

13. $-2 \cot \dfrac{x}{2} + C$

15. $x - 2 \ln |\csc x - \cot x| - \cot x + C$

17. $\dfrac{\cos^4 x}{4} + C$

19. $-\cos e^x + C$

21. $\ln |x + \sin x| + C$

23. 0

25. 1.41615

27. 3

29. 1

31. 2

33. 6.076

35. 8

37. -0.475447

39. 0

41. (a) 36,000   (b) 3000   43. $79.75

45. 432; 368

47. $f'(x) = 2e^{-x} \sec^2 2x - e^{-x} \tan 2x$

49. $f'(x) = -9[1 - \sec(3x)]^2 \sec(3x) \tan(3x)$

Chapter Test 8, Page 480

1. $\dfrac{-55\pi}{36}$

2. 114.55°

3. $\sin \theta = \dfrac{4}{5}$, $\cos \theta = -\dfrac{3}{5}$, $\tan \theta = -\dfrac{4}{3}$, $\csc \theta = \dfrac{5}{4}$,

$\sec \theta = -\dfrac{5}{3}$, $\cot \theta = -\dfrac{3}{4}$

4. (a) $P = \dfrac{20\pi}{3}$; $A = 0.8$; Max $= 2.8$; Min $= 1.2$

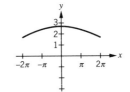

(b) $P = \dfrac{\pi}{2}$; $A = 3$; Max $= 1$; Min $= -5$

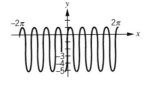

5. (a) $0, \pi$ (b) $x = 0.262, 1.309, 3.403, 4.451$

6. (a) $\dfrac{1 - x \sin x}{\cos x} \cdot \dfrac{1 + \sin x}{1 + \sin x} = \dfrac{1 - \sin^2 x}{\cos x(1 + \sin x)}$

$= \dfrac{\cos^2 x}{\cos x(1 + \sin x)} = \dfrac{\cos x}{1 + \sin x}$

(b) $(\sin x + \cos x)^2 + (\sin x - \cos x)^2 = \sin^2 x + 2 \sin x$
$\cos x + \cos^2 x + \sin^2 x - 2 \sin x \cos x + \cos^2 x =$
$2 \sin^2 x + 2 \cos^2 x = 2(\sin^2 x + \cos^2 x) = 2$

7. (a) $y' = 4x \cos 2x^2 + 3 \sin 3x$

(b) $y' = \dfrac{6x^2 + 6x \sin x^2}{2x^3 - 3 \cos x^2}$

(c) $y' = x^3 \sec^2 x + 3x^2 \tan x$

(d) $y' = 2e^{2x-3} \csc x^2 - 2xe^{2x-3} \cot x^2 \csc x^2$

8. $(0.9828, 0)$, $(4.1244, 0)$

9. At $x = 40$, $R(40) = 2.9$, so 2900.

10. (a) $\dfrac{-\cos(4x - 3)}{4} + C$

(b) $\dfrac{-(\cos x)^5}{5} + C$

(c) $-e^{\cos x} + C$ (d) $\dfrac{-\cos^2 2x}{4} + C$

11. (a) 0   (b) 0   12. 45.76

13. 200

14. $\dfrac{\sin\theta}{1-\cos\theta} \cdot \dfrac{1+\cos\theta}{1+\cos\theta} = \dfrac{\sin\theta\,(1+\cos\theta)}{1-\cos^2\theta}$

$= \dfrac{\sin\theta\,(1+\cos\theta)}{\sin^2\theta}$

$= \dfrac{1+\cos\theta}{\sin\theta}$

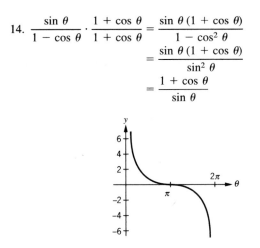

Index